DIFFERENTIAL GEOMETRY

Symposium in Pure Mathematics, ...

PROCEEDINGS OF SYMPOSIA
IN PURE MATHEMATICS

VOLUME XXVII, PART 1

DIFFERENTIAL GEOMETRY

AMERICAN MATHEMATICAL SOCIETY
PROVIDENCE, RHODE ISLAND
1975

PROCEEDINGS OF THE SYMPOSIUM IN PURE MATHEMATICS
OF THE AMERICAN MATHEMATICAL SOCIETY

HELD AT STANFORD UNIVERSITY
STANFORD, CALIFORNIA
JULY 30–AUGUST 17, 1973

EDITED BY

S. S. CHERN and R. OSSERMAN

Prepared by the American Mathematical Society
with the partial support of National Science Foundation Grant GP–37243

Library of Congress Cataloging in Publication Data CIP

Symposium in Pure Mathematics, Stanford University,
1973.
Differential geometry.

(Proceedings of symposia in pure mathematics;
v. 27, pt. 1–2)
"Final versions of talks given at the AMS Summer
Research Institute on Differential Geometry."
Includes bibliographies and indexes.
1. Geometry, Differential–Congresses.
I. Chern, Shiing-Shen, 1911– II. Osserman,
Robert. III. American Mathematical Society.
IV. Series.
QA641.S88 1973 516'.36 75-6593
ISBN 0-8218-0247-X (v. 1)

Printed in the United States of America

CONTENTS

Submanifolds

*General lecture given at the Institute.

Foliations

Algebraic and Piecewise Linear Topology

*General lecture given at the Institute.

Miscellaneous

Indexes

Preface

The papers in these PROCEEDINGS represent the final versions of talks given at the AMS Summer Research Institute on Differential Geometry, which took place at Stanford University, Stanford, California, from July 30 to August 17, 1973. This Institute was made possible by a grant from the National Science Foundation. The organizing committee consisted of Raoul H. Bott, Eugenio Calabi, S. S. Chern, Leon W. Green, Shoshichi Kobayashi, Tilla K. Milnor, Barrett O'Neill, Robert Osserman, James Simons, I. M. Singer, with the coeditors serving as cochairmen.

The activities were divided between general lectures and seminar talks. In these PROCEEDINGS the general lectures have been distributed among the various seminars, according to their subject matter. Each part of the PROCEEDINGS consists of a group of seminars, whose titles and chairmen are as follows:

Part 1:
- Riemannian geometry (J. Cheeger)
- Submanifolds (K. Nomizu)
- Foliations (B. L. Reinhart)
- Algebraic and piecewise-linear topology (T. F. Banchoff and H. R. Gluck)
- Miscellaneous (B. O'Neill and J. Simons)

Part 2:
- Complex differential geometry (S. Kobayashi)
- Partial differential equations (J. L. Kazdan and F. W. Warner)
- Homogeneous spaces (J. Wolf)
- Relativity (T. Frankel)

Generally papers are included in the seminars in which they were presented, although in certain cases the contents would make them more appropriate in another section. In cases where a complete version of the talk appears elsewhere, only an abstract is included here, together with a reference to the full paper.

A list of open problems submitted by participants was compiled by Leon Green. These are included at the end of Volume I in the Miscellaneous Section.

We should like to thank the seminar chairmen, and also the secretarial staff: Dorothy Smith and Muriel Toupin of AMS as well as Catherine Lowe and Elizabeth Plowman of the Stanford Mathematics Department, all of whose tireless efforts were a large factor in the success of the Institute.

S. S. CHERN
ROBERT OSSERMAN

JANUARY 1975

RIEMANNIAN GEOMETRY

Proceedings of Symposia in Pure Mathematics
Volume 27, 1975

DÉFORMATIONS LOCALEMENT TRIVIALES DES VARIÉTÉS RIEMANNIENNES

L. BERNARD BERGERY, J. P. BOURGUIGNON, AND J. LAFONTAINE

Introduction. L'exposé qui suit est une introduction à la théorie des *déformations localement triviales des variétés riemanniennes.* C'est le résultat d'un travail en commun des trois auteurs sur l'article d'E. Calabi: *On compact Riemannian manifolds with constant curvature.* I (Proc. Sympos. Pure Math., vol. III, 155–180 (1961)). Ce travail a fait l'objet d'une série d'exposés au Séminaire de Géométrie Riemannienne de M. Berger en 1971/72 à l'Université de Paris VII, sauf le paragraphe IV de la 2ième partie et l'appendice, qui exposent des résultats démontrés ultérieurement (Février 1973) par J. Lafontaine.

Une métrique riemannienne g sur une variété differentielle M admet en général "beaucoup" de déformations. En effet, si M est compacte et si h est un tenseur covariant symétrique d'ordre 2 sur M, la famille de métriques riemanniennes $g + th$ (pour t réel assez petit) est une déformation de g. Pour la théorie des déformations générales, on pourra se reporter aux travaux de D. Ebin (référence **[8]** de la bibliographie de la seconde partie).

Nous nous intéressons ici aux déformations qui préservent localement la structure riemannienne. La situation est alors complètement différente, et les principaux résultats sont des théorèmes de rigidité, ou d'existence et de finitude. Le premier travail dans cette voie est un article de E. Calabi sur les déformations des variétés compactes à courbure constante. Par une adaptation de la théorie des déformations de pseudo-groupes (rappelons qu'une métrique riemannienne à courbure constante peut se définir à l'aide d'un pseudo-groupe), il se ramène au calcul de la cohomologie à valeurs dans le faisceau Θ des champs de vecteurs de Killing, et donne une

AMS (MOS) subject classifications (1970). Primary 58H05, 53C10.

résolution de Θ qui présente l'avantage d'être très géométrique. A. Weil a démontré des théorèmes de rigidité dans le cadre des espaces homogènes (référence **[6]** de la première partie). Par la suite, P. A. Griffiths a montré qu'on pouvait généraliser la célèbre théorie des déformations des structures complexes en une théorie des déformations localement triviales des G-structures.

Les deux parties sont très différentes. La première suit l'article de Griffiths, en se restreignant au cas des variétés riemanniennes ($O(n)$-structures). Elle introduit les définitions générales et décrit les méthodes par lesquelles on ramène le problème de l'existence de déformations localement triviales au calcul de la cohomologie à valeurs dans le faisceau des champs de Killing.

La deuxième partie reprend et complète l'article de Calabi en utilisant les travaux de Singer et Sternberg sur les G-structures et les travaux d'Ebin cités plus haut, qui permettent de mieux dégager les idées.

Aux deux premiers paragraphes, nous exposons en détail la résolution de Θ donnée par Calabi. Le théorème de base est un lemme de Poincaré pour la dérivée de Lie de la métrique et la variation de la courbure, dont la démonstration par Calabi est incomplète, bien qu'on trouve ailleurs dans son travail toutes les idées nécessaires. Au paragraphe III, nous donnons un théorème du type Hodge-de Rham pour les $H^p(N, \Theta)$, que nous utilisons au paragraphe IV pour calculer H^1 dans différents cas. Le dictionnaire "variété compacte orientée à courbure constante négative—surface de Riemann compacte de genre supérieur ou égal à 2" permet ainsi de retrouver la dimension de l'espace des modules d'une surface de Riemann par des méthodes de géométrie différentielle.

Suivant les mêmes idées, nous obtenons dans l'appendice une version infinitésimale du théorème de Torelli, mais de démonstration "élémentaire" c-à-d. sans théorie des intersections.

1. THEORIE GENERALE DES DEFORMATIONS

L. BERNARD BERGERY

I. Notion générale de déformation.

DÉFINITION 1. Une déformation d'une structure riemannienne g sur une variété différentiable M est la donnée d'une fibration localement triviale $p : V \to D$ et d'une structure riemannienne g_t sur chaque fibre $M_t = p^{-1}(t)$, telles que $M = M_0$, $g = g_0$, où 0 est un point fixé de D, et g_t dépende différentiablement du paramètre t dans D.

Nous nous intéresserons surtout aux germes de déformations à un paramètre réel. Pour ne pas avoir à répéter des raisonnements de germes, nous poserons une fois pour toutes que D est un intervalle de $\boldsymbol{R}$, contenant 0, que nous restreindrons à volonté. Et V sera le produit $M \times D$, p étant la projection naturelle sur le deuxième facteur.

DÉFINITION 2. Deux déformations $(M \times D, g_t)$ et $(M \times D', g'_t)$ de (M, g) seront dites *équivalentes* s'il existe un intervalle $D'' \subset D \cap D'$ et un difféomorphisme F

de $M \times D''$ sur lui-même, respectant les fibres et induisant sur chaque fibre M_t ($t \in D''$) une isométrie de (M_t, g_t) sur (M_t, g'_t).

DÉFINITION 3. On appelera *déformation produit* la déformation telle que $g_t = g \, \forall \, t \in D$; et *déformation triviale* toute déformation équivalente à la déformation produit.

Nous supposerons toujours M paracompacte et connexe.

II. **Déformations localement triviales.** Comme il y a "beaucoup" de déformations non équivalentes d'une structure riemannienne, nous allons nous restreindre ici à certaines déformations particulières, auxquelles on peut appliquer les méthodes employées dans la théorie des déformations des structures complexes [**4**].

DÉFINITION 4. Une déformation sera dite *localement triviale* si tout point de M_0 possède un voisinage W dans V tel que $p|_W$, $W \to p(W) = D$, muni des structures induites, soit une déformation triviale. Un tel ouvert W sera appelé ouvert trivialisant.

EXEMPLE 1. Toute déformation de S^1 ou de $\boldsymbol{R}$ est localement triviale pour toute structure riemannienne.

EXEMPLE 2. Soit (M, g) une variété riemannienne compacte plate. Alors la déformation définie par $g_t = (1 + t)g$ est localement triviale mais n'est pas triviale puisque, si d est le diamètre de (M, g), le diamètre de (M_t, g_t) est $(1 + t)^{1/2} \times d$.

PROPRIÉTÉS. Une déformation localement triviale conserve les propriétés *locales* de la structure riemannienne; par exemple: analyticité, locale homogénéité, locale symétrie, ou courbure constante.

Mais elle ne conserve pas en général les propriétés globales (Exemple 2).

EXEMPLE 3. Soit $V = \{(t, x) \in R^2 \text{ tels que } -1 < xt < +1\}$, D l'axe des t et p la projection naturelle de V sur D. Alors $p: V \to D$ et les structures riemanniennes induites par la structure canonique de $\boldsymbol{R}^2$ définissent une déformation localement triviale de la structure canonique de $\boldsymbol{R}$, qui est complète, par des structures riemanniennes non complètes (Figure 1).

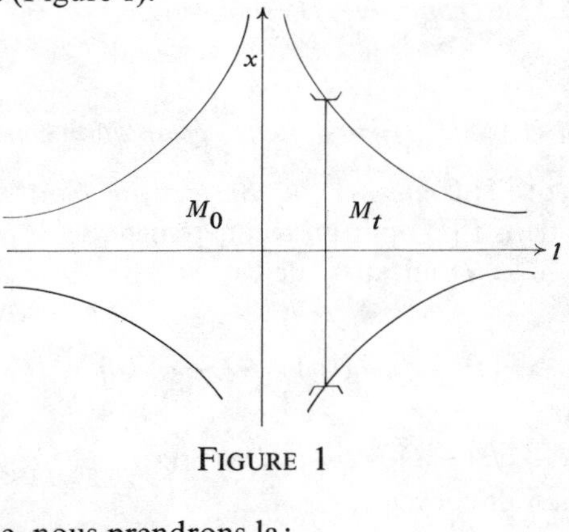

FIGURE 1

A cause de cet exemple, nous prendrons la:

DÉFINITION 5. Une déformation sera dite *complète* si la structure riemannienne g_t est complète quel que soit t dans D.

Nous allons maintenant décrire la situation en coordonnées locales.

III. Coordonnées de première espèce (ou eulériennes).

DÉFINITION 6. Nous appelerons système de coordonnées de première espèce d'une déformation localement triviale la donnée d'un recouvrement de $V = M \times D$ par des ouverts $U_i \times D$, les U_i étant des ouverts de V tels que $U_i \times D$ soit contenu dans un ouvert trivialisant W_i.

Nous noterons $W_i(t) = W_i \cap M_t$, h_i le difféomorphisme trivialisant : $W_i \to W_i(0) \times D$ et $h_i(t)$ sa restriction à $W_i(t)$. On peut considérer les $h_i(t)$ comme des familles à un paramètre de difféomorphismes d'un voisinage variable $W_i(t)$ de U_i sur un voisinage fixe $W_i(0)$ de U_i qui sont des isométries de g_t sur g. On est donc conduit à introduire le faisceau $\Gamma[t]$ sur M associé au préfaisceau qui à tout ouvert U de M associe l'ensemble $\Gamma[t](U)$ des germes en t de familles à un paramètre de g-isométries $f(t)$ d'un voisinage (variable) de U sur un voisinage (variable) de U, dépendant différentiablement du paramètre t et telles que $f(0)$ soit l'identité. C'est un faisceau en groupes non abéliens pour la composition des isométries.

Soit $E(M, g)$ l'ensemble des classes d'équivalence de germes de déformations localement triviales de (M, g) et $E^c(M, g)$ le sous-ensemble des déformations complètes. On définit l'application naturelle $H: E(M, g) \to H^1(M, \Gamma[t])$ de la façon suivante: soit (U_i, W_i, h_i) un système de coordonnées de première espèce pour la déformation $(M \times D, g_t)$ de (M, g). On forme la 1-cochaine (a_{ij}) à valeurs dans $\Gamma[t]$ pour le recouvrement U_i par: $a_{ij}(t) = h_i(t) \cdot h_j^{-1}(t)$ sur $U_i \cap U_j$. On vérifie facilement que a_{ij} est un 1-cocycle et que l'élément correspondant de $H^1(M, \Gamma[t])$ ne dépend que du germe de la déformation. On a alors le:

THÉORÈME 1. *H est surjective de $E(M, g)$ sur $H^1(M, \Gamma[t])$ et sa restriction à $E^c(M, g)$ est injective.*

Comme toute déformation est complète si M est compacte, on a le:

COROLLAIRE 1. *Si M est compacte, H est bijective.*

On a aussi:

COROLLAIRE 2. *Si $H^1(M, \Gamma[t]) = 0$, toute déformation complète est triviale.*

DÉMONSTRATION DU THÉORÈME 1. (a) *Surjectivité.* Soit $c_{ij}(t)$ les éléments d'un 1-cocycle à valeurs dans $\Gamma[t]$ pour un recouvrement de M par des cartes $\varphi_i: U_i \to \boldsymbol{R}^n$. On définit sur $M \times D$ un atlas, de cartes $U_i \times D$ avec les changements de cartes

$$(f_{ij}, \mathrm{id}): \varphi_j(U_j) \times D \to \varphi_i(U_i) \times D$$

où f_{ij} est définie par:

$$f_{ij}(x_j, t) = \varphi_i \cdot c_{ij}(t) \cdot \varphi_j^{-1}(x_j).$$

On a bien la condition de cocycle:

$$f_{ij} \cdot f_{jk} = \varphi_i \cdot c_{ij} \cdot \varphi_j^{-1} \cdot \varphi_j \cdot c_{jk} \cdot \varphi_k^{-1} = \varphi_i \cdot c_{ij} \cdot c_{jk} \cdot \varphi_k^{-1} = \varphi_i \cdot c_{ik} \cdot \varphi_k^{-1} = f_{ik}.$$

Donc les U_i, recollées par les $f_{ij}(x_j, t)$ pour t fixé suffisamment petit, forment une variété M_t difféomorphe à M sur laquelle on peut définir une structure riemannienne g_t en prenant $g_t|_{U_i} = g|_{U_i}$. Les (M_t, g_t) constituent une déformation de (M, g) dont l'image par H est exactement la classe de $(c_{ij}(t))$.

(b) *Injectivité*. Soit (U_i, W_i, h_i) un système de coordonnées de première espèce de la déformation $(M \times D, g_t)$; si $a_{ij}(t) = h_i(t) \cdot h_j^{-1}(t)$ est un cobord, on a $a_{ij}(t) = b_i^{-1}(t) \cdot b_j(t)$ où $b_i(t)$ appartient à $\Gamma[t](U_i)$. Donc $b_i(t) \cdot h_i(t)$ est défini sur un voisinage $V_i(t)$ de U_i et vérifie sur $U_i \cap U_j$ la relation $b_i(t) \cdot h_i(t) = b_j(t) \cdot h_j(t)$. On peut recoller ces difféomorphismes, pour t fixé, en une application $f_t \colon M_t \to M$ telle que $f_t|_{V_i(t)} = b_i(t) \cdot h_i(t)$. Et $f_t^*(g) = g_t$, donc f_t est une isométrie locale. Comme on a supposé de plus que g_t est complète pour tout t, on en déduit que f_t est surjective (voir [7]). Donc f_t est une isométrie globale de M_t sur M, dépendant différentiablement de t et la déformation est triviale.

Comme $\Gamma[t]$ n'est pas un faisceau très maniable, on introduit le faisceau Θ des germes de champs de vecteurs de Killing pour g sur M.

L'application $f(t) \to df/dt|_{t=0}$ induit un morphisme de faisceaux: $\Gamma[t] \to \Theta$ et donc une application $r : H^1(M, \Gamma[t]) \to H^1(M, \Theta)$.

DÉFINITION 7. Si γ est un germe de déformations, $r \cdot H(\gamma)$ s'appelle sa déformation infinitésimale.

IV. **Coordonnées de deuxième espèce (ou lagrangiennes).**

DÉFINITION 8. Nous appelerons système de coordonnées de deuxième espèce d'une déformation localement triviale $(M \times D, g_t)$ la donnée d'un recouvrement (V_i) de $M \times D$ formant atlas tel que:

(a) Il existe des difféomorphismes $f_i \colon V_i \to V_i(0) \times D$ préservant les fibres (où $V_i(t) = V_i \cap M_t$).

(b) Les changements de cartes respectent les fibres et sont donc de la forme: $(x_i, t) = (f_{ij}(x_j, t), t)$.

(c) $g_t|_{V_i(t)} = f_i^*(g|_{V_i(0)})$.

COMMENTAIRES. On peut décrire la différence entre les systèmes de coordonnées de première et de deuxième espèce de la façon suivante: en coordonnées locales, on a pris dans le premier cas un atlas fixe, le même pour toutes les M_t et on a fait varier les structures riemanniennes tandis que dans le deuxième cas, on prend comme atlas sur les M_t des cartes sur lesquelles la structure riemannienne reste fixe, avec des changements de cartes variables. Dans ce deuxième point de vue, la déformation est vue comme le glissement de plaques rigides les unes sur les autres. Ces deux systèmes de coordonnées ont été introduits par Calabi, qui les avaient appelés coordonnées eulériennes et lagrangiennes en analogie avec les deux formalismes de la mécanique classique.

On peut visualiser la situation avec les deux figures ci-contre:

EXEMPLE 4. On a déjà vu un exemple de système de coordonnées de deuxième espèce dans la démonstration de la surjectivité dans le Théoréme I: les $U_i \times D$ avec les changements de cartes $(f_{ij}(x_j, t), t)$.

FAISCEAUX ASSOCIÉS. On introduit alors les faisceaux suivants:

le faisceau $\bar{\Pi}$ sur $M \times D$ associé aux germes de champs de vecteurs sur $M \times D$

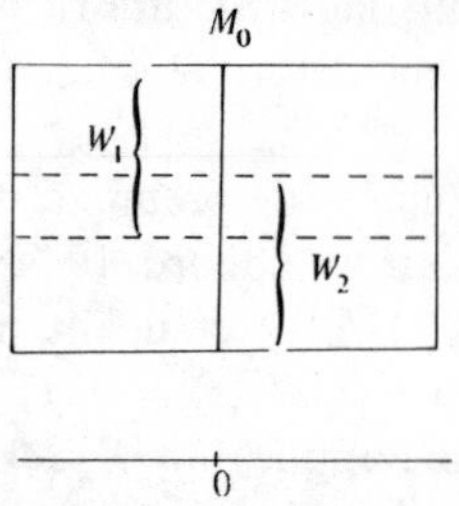

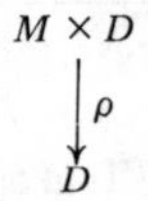

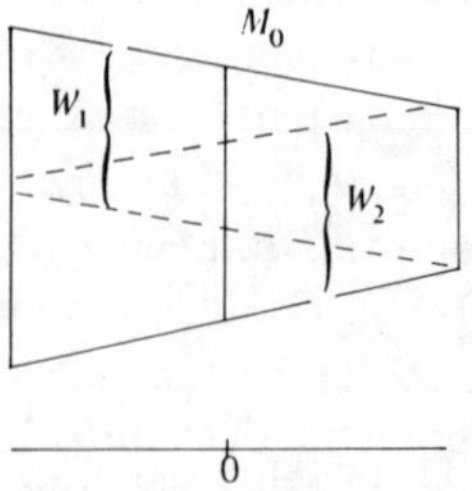

FIGURE 2: PREMIÈRE ESPÈCE FIGURE 3: DEUXIÈME ESPÈCE

tangents aux groupes à un paramètre de difféomorphismes qui envoient fibre en fibre isométriquement.

Le sous-faisceau $\bar{\Theta}$ de $\bar{\Pi}$ correspondant aux champs de vecteurs tangents aux difféomorphismes conservant chacune des fibres M_t et y induisant une isométrie.

Le faisceau T image réciproque par la projection $p\colon M \times D \to D$ du faisceau constant $\mathscr{C}_D$ sur D engendré par le champ de vecteurs d/dt.

PROPOSITION 1. *Le faisceau* T *s'injecte dans le faisceau quotient* Q *du faisceau* $\bar{\Pi}$ *par son sous-faisceau* $\bar{\Theta}$.

DÉMONSTRATION. Tout point x de $M \times D$ a un voisinage trivialisant W, difféomorphe à $W(0) \times D$ par un difféomorphisme α envoyant fibre en fibre. Maintenant sur D il existe un groupe à un paramètre de difféomorphismes φ_u tels que le champ de vecteurs tangents soit d/dt au voisinage de 0. Alors le groupe à un paramètre de difféomorphisme $\psi_u\colon (y, t) \to (y, \varphi_u(t))$ sur $W(0) \times D$, transporté sur W par α, induit une section locale de $\bar{\Pi}$ au voisinage de x dont la projection sur D est bien d/dt.

On introduit enfin:

Le faisceau Π image réciproque de T dans $\bar{\Pi}$ pour la projection des faisceaux $\bar{\Pi} \to Q$. Π est le faisceau associé aux germes de champs de vecteurs tangents aux groupes à un paramètre de difféomorphismes de $M \times D$ qui envoient fibre en fibre isométriquement et qui induisent sur D un groupe à un paramètre de difféomorphismes dont le champ de vecteurs tangents est proportionnel à d/dt.

Les faisceaux Θ_t, Π_t, T_t sur M_t restrictions de Θ, Π, T. On obtient le diagramme commutatif suivant, où les lignes sont exactes

$$\begin{array}{ccccccccc} 0 & \longrightarrow & \bar{\Theta} & \longrightarrow & \Pi & \longrightarrow & \mathrm{T} & \longrightarrow & 0 \\ & & \downarrow & & \downarrow & & \downarrow & & \\ 0 & \longrightarrow & \Theta_t & \longrightarrow & \Pi_t & \longrightarrow & \mathrm{T}_t & \longrightarrow & 0. \end{array}$$

On en déduit le diagramme commutatif (où $V = M \times D$):

$$\begin{array}{ccccccccc} H^0(V, \bar{\Theta}) & \to & H^0(V, \Pi) & \to & H^0(V, \mathrm{T}) & \xrightarrow{\delta} & H^1(V, \Theta) & \to \\ \downarrow & & \downarrow & & \downarrow & & \downarrow & \\ H^0(M_t, \Theta_t) & \to & H^0(M_t, \Pi_t) & \to & H^0(M_t, \mathrm{T}_t) & \xrightarrow{\delta^t} & H^1(M_t, \Theta_t) & \to \end{array}$$

où nous avons noté δ et δ_t les opérateurs de recollement.

THÉORÈME 2. *La déformation localement triviale et complète* $(M \times D, g_t)$ *est triviale si et seulement si* $\delta = 0$.

DÉMONSTRATION. Si la déformation est triviale, la suite exacte de faisceaux sur $M \times D$ se scinde et donc δ est nulle.

Inversement, le champ de vecteurs d/dt de D, qui est une section de $H^0(D, \tau_D)$, se relève en une section σ de $H^0(V, \tau)$ par définition de τ. Si $\delta = 0$, σ provient d'un élément γ de $H^0(V, \Pi)$. En intégrant γ, grâce à l'hypothèse de complétion, on trouve un groupe à un paramètre d'isométries conservant les fibres de V et couvrant le champ de vecteurs d/dt et donc la déformation est triviale.

REMARQUE. Si γ est une déformation, on a $\delta_0(d/dt) = r \cdot H(\gamma)$.

V. **Etude du faisceau** Θ. Rappelons quelques propriétés connues du faisceau Θ associé aux germes de champ de vecteurs de Killing d'une variété riemannienne (M, g):

(a) Θ est un faisceau en algèbres de Lie, de dimension finie inférieure à $n(n+1)/2$ (où n est la dimension de M) et cette borne supérieure n'est atteinte que dans le cas de la courbure constante.

(b) Si V et U sont des ouverts de M et $V \subset U$, l'application de restriction P_{UV}: $\Gamma(U, \Theta) \to \Gamma(V, \Theta)$ est injective.

(c) Si (M, g) est complète, $H^0(M, \Theta)$ est l'algèbre de Lie $\mathfrak{G}$ du groupe G des isométries de (M, g).

(d) Si (M, g) est analytique, tout germe dans Θ est analytique.

(e) Soit $x \in M$ et U_n une suite décroissante de voisinages de x, alors p_{ij}: $\Theta(U_i) \to \Theta(U_j)$ définie par la restriction $(j > i)$ est injective. Comme la dimension de $\Theta(U_i)$ est bornée, p_{ij} est bijective à partir d'un certain rang. En particulier, pour toute (M, g), il existe un recouvrement de M par des cartes $(U_x, x \in M)$ telles que $x \in U_x$ et $\Theta(U_x) \to \Theta_x$ soit un isomorphisme. Un tel recouvrement sera dit normal.

Nous allons introduire maintenant une condition de régularité pour g, qui permettra d'énoncer des théorèmes plus précis. Donnons d'abord un exemple montrant la nécessité d'une telle condition.

EXEMPLE 5. On considère la sphère de dimension 2 plongée dans R^3 en forme d'haltère telle que la partie médiane soit un cylindre de révolution (voir Figure 4). On peut définir une déformation de la structure riemannienne induite par le plongement en agrandissant la partie médiane dans sa longueur. C'est une déformation localement triviale nontriviale puisque le diamètre croit.

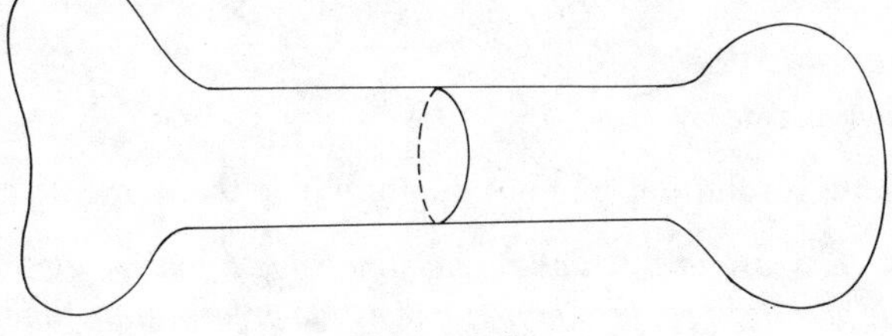

FIGURE 4

EXEMPLE 6. Une construction équivalente à celle de l'Exemple 5 montrerait que sur toute variété différentiable il existe des structures riemanniennes qui admettent des déformations localement triviales nontriviales.

VI. **La condition de régularité.**

DÉFINITION 9. La structure riemannienne g sur M sera dite *régulière* si la dimension de la fibre du faisceau Θ est constante.

PROPOSITION 2. *Si (M, g) est analytique ou localement homogène, elle est régulière.*

THÉORÈME 3. *Si (M, g) est régulière, Θ est un faisceau localement constant.*

DÉMONSTRATION. Soit $(U_x, x \in M)$ un recouvrement normal (voir § V(e)).

Quel que soit y dans U_x, la restriction $p_{xy}: \Theta(U_x) \to \Theta_y$ est injective et $\dim \Theta(U_x) = \dim \Theta_x = \dim \Theta_y$ donc p_{xy} est bijective. Cela entraine que pour tout ouvert V dans U_x, $\Theta(U_x) \to \Theta(V)$ est bijective, et donc Θ est localement constant.

Nous allons maintenant construire une résolution fine de Θ. Soit (M, g) une variété riemannienne régulière, Θ le faisceau associé, $(\tilde{M}, \tilde{g})$ le revêtement universel riemannien de (M, g) et $\tilde{\Theta}$ le faisceau associé à $(\tilde{M}, \tilde{g})$. $\tilde{g}$ est aussi une structure régulière, donc $\tilde{\Theta}$ est un faisceau localement constant, et même constant puisque $\tilde{M}$ est simplement connexe. On considère alors les faisceaux $\tilde{\Lambda}^q$ associés aux germes de q-formes différentielles extérieurs sur $\tilde{M}$. On forme $\tilde{G}^q = \Theta \otimes \tilde{\Lambda}^q$ et on définit les opérateurs $D^q: \tilde{G}^q \to \tilde{G}^{q+1}$ par $D^q(a \otimes u) = a \otimes du$ et

$$i: \tilde{\Theta} \to \tilde{G}^0 \text{ par } i(a) = a \otimes 1.$$

PROPOSITION 3.

$$0 \to \tilde{\Theta} \xrightarrow{i} \tilde{G}^0 \xrightarrow{D^0} \cdots \to D^q \xrightarrow{D^q} \cdots$$

est une résolution fine de $\tilde{\Theta}$.

DÉMONSTRATION. $\tilde{\Lambda}^q$ est un faisceau fin, donc $\tilde{G}^q$ aussi. Puis $D \cdot D(a \otimes u) = a \otimes ddu = 0$ et $D \cdot i(a) = a \otimes d1 = 0$. Enfin c'est une suite exacte de faisceaux d'après le lemme de Poincaré.

Si Γ est le groupe du revêtement $\pi: \tilde{M} \to M$, Γ agit sur $\tilde{M}$ par des isométries, donc Γ agit sur les faisceaux $\tilde{\Theta}$, $\tilde{\Lambda}^q$ et $\tilde{G}^q$. On définit les faisceaux G^q sur M ainsi: sur un ouvert U de M, $G^q(U)$ est l'ensemble des sections Γ-invariantes de $\tilde{G}^q$ sur $\pi^{-1}(U)$. Enfin Γ commute à D^q et i. D'où le:

THÉORÈME 4.

$$0 \to \Theta \xrightarrow{i} G^0 \to \cdots \to G^q \xrightarrow{D^q} \cdots$$

est une résolution fine de Θ.

A l'aide de cette résolution, on peut démontrer le théorème de rigidité suivant:

THÉORÈME 5. *Si g est une structure riemannienne régulière et complète sur une variété M dont le premier groupe d'homotopie est de présentation finie et si $H^1(M, \Theta) = 0$, toute déformation localement triviale complète de (M, g) est triviale.*

DÉMONSTRATION. Nous n'allons donner ici que l'esquisse de démonstration présentée par Griffiths dans le cas où M est compacte. Pour le cas plus général du théorème, voir [5].

Les sections des faisceaux G^q sont les sections C^∞ de fibrés vectoriels E^q sur M, sur lesquels les opérateurs D^q sont des opérateurs différentiels. Si on munit ces fibrés E^q de produits scalaires C^∞, on peut donc former les adjoints D^* des opérateurs D et les opérateurs $D \cdot D^* + D^* \cdot D$ sont alors elliptiques. En appliquant la théorie des équations différentielles elliptiques comme dans [3], on trouve donc que:

(1) dim $H^q(M_t, \Theta_t) \leqq$ dim $H^q(M, \Theta)$ pour tout t dans un voisinage de 0 dans D.

(2) si dim $H^q(M_t, \Theta_t)$ est indépendant de t, alors $\bigcup_{t \in D} H^q(M_t, \Theta_t)$ peut être muni d'une structure de fibré vectoriel sur $V = M \times D$ et $H^q(V, \bar{\Theta})$ est isomorphe à l'espace des sections C^∞ de ce fibré. En particulier, si $H^1(M, \Theta) = 0$, alors $H^1(V, \bar{\Theta}) = 0$ et $\delta = 0$. Il ne reste plus qu'à appliquer le Théorème 2.

VII. **Utilisation de la cohomologie des groupes**. Soit (M, g) une variété riemannienne régulière et complète. Son revêtement universel riemannien $(\tilde{M}, \tilde{g})$ est encore une variété riemannienne régulière et complète. De plus, le faisceau $\tilde{\Theta}$ des champs de vecteurs de Killing sur $\tilde{M}$ est localement constant par régularité et donc constant puisque $\tilde{M}$ est simplement connexe. Sa fibre est $\tilde{\mathfrak{G}}$, l'algèbre de Lie de $\tilde{G}$ le groupe des isométries de $\tilde{M}$. Le groupe Γ du revêtement $\tilde{M} \to M$ est un sous-groupe discret de $\tilde{G}$; il agit donc sur $\tilde{G}$ par la représentation adjointe de G sur $\tilde{\mathfrak{G}}$. Le résultat suivant est connu:

PROPOSITION 4. *$H^0(M, \Theta) = H^0(\Gamma, \tilde{\mathfrak{G}})$; $H^1(M, \Theta) = H^1(\Gamma, \tilde{\mathfrak{G}})$ et il y a une injection de $H^2(\Gamma, \tilde{\mathfrak{G}})$ dans $H^2(M, \Theta)$.*

En particulier, on en déduit le:

THÉORÈME 6. *Si (M, g) est une variété riemannienne régulière et complète et si $\Gamma = \Pi_1(M)$ est fini, toute déformation localement triviale complète de (M, g) est triviale.*

DÉMONSTRATION. Si Γ est fini, $H^1(\Gamma, \tilde{G}) = 0$ et on applique le Théorème 5.

REMARQUE 1. On voit en particulier que le revêtement universel riemannien $(\tilde{M}, \tilde{g})$ de (M, g) variété riemannienne régulière et complète n'a pas de déformations localement triviales complètes nontriviales. On peut donc obtenir toutes les déformations localement triviales complètes de (M, g) en prenant les quotients de $(\tilde{M}, \tilde{g})$ par les images de Γ pour les familles à un paramètre d'homomorphismes injectifs de Γ dans $\tilde{G}$, l'équivalence de deux déformations correspondant à la conjugaison dans $\tilde{G}$ par une famille à un paramètre d'éléments de $\tilde{G}$. On peut donc regarder les déformations des sous-groupes discrets des groupes de Lie (voir par exemple [6]).

REMARQUE 2. Le groupe d'isométries G de (M, g) est le centralisateur de Γ dans $\tilde{G}$ et son algèbre de Lie $\mathfrak{G}$ s'identifie à l'ensemble des éléments de $\tilde{\mathfrak{G}}$ stables pour l'action adjointe de Γ. On a donc une injection de Γ-modules du Γ-module trivial

$\mathfrak{G}$ dans $\tilde{\mathfrak{G}}$ et une injection de $H^1(\Gamma, \mathfrak{G}) = \mathfrak{G} \otimes H^1(M, R)$ dans $H^1(\Gamma, \tilde{\mathfrak{G}})$. Griffiths expose d'ailleurs une construction explicite des déformations de la forme $\gamma \otimes \omega$ à l'aide de revêtements.

VIII. **Existence de déformations.** L'un des principaux résultats de l'article de Griffiths est un théorème d'existence de déformation analogue à celui de [**4**]:

THÉORÈME 7. *Si (M, g) est une variété riemannienne régulière et compacte, et si $H^2(M, \Theta) = 0$, alors quel que soit α dans un voisinage de 0 dans $H^1(M, \Theta)$, il existe une déformation localement triviales de (M, g) dont α soit la déformation infinitésimale.*

Nous n'en donnerons pas ici la démonstration, fort longue, qui se fait en deux étapes:

(I) Montrer que toutes les obstructions à l'existence de déformations formelles sont dans $H^2(M, \Theta)$.

(II) Montrer la convergence.

BIBLIOGRAPHIE

1. E. Calabi, *On compact Riemannian manifolds with constant curvature.* I, Proc. Sympos. Pure Math., vol. 3, Amer. Math. Soc., Providence, R. I., 1971, pp. 155–180. MR **24** #A3612.

2. P. A. Griffiths, *Deformations of G-structures.* Part A. *General theory of deformations*, Math. Ann. **155** (1964), 292–315. MR **30** #3480.

———, *Deformations of G-structures.* Part B. *Deformations of geometric G-structures*, Math. Ann. **158** (1965), 326–351. MR **31** #6252.

3. K. Kodaira and D. C. Spencer, *Deformations of complex analytic structures.* I, II, Ann. of Math. (2) **67** (1958), 328–466. MR **22** #3009.

4. K. Kodaira, L. Nirenberg and D. C. Spencer, *On the existence of deformations of complex analytic structures*, Ann. of Math. (2) **68** (1958), 450–459. MR **22** #3012.

5. M. S. Raghunathan, *Deformations of linear connections and Riemannian metrics*, Differential Analysis, Bombay Colloq., Oxford Univ. Press, London, 1964, pp. 243–247; addendum, J. Math. Mech. **13** (1964), 1043–1045. MR **29** #5201; **32** #413.

6. A. Weil, *On discrete subgroups of Lie groups.* I, II, Ann. of Math. (2) **72** (1960), 369–384; ibid. (2) **75** (1962), 578–602. MR **25** #1214; #1242.

7. J. A. Wolf and P. A. Griffiths, *Complete maps and differentiable coverings*, Michigan Math. J. **10** (1963), 253–255. MR **27** #4179.

2. DEFORMATIONS DES VARIETES COMPACTES A COURBURE CONSTANTE

J. LAFONTAINE

0. Introduction et notations. On désignera par (M, g) une variété riemannienne compacte, par (X, g) une variété compacte à courbure constante; ∇ désigne la connexion riemannienne associée à la métrique g, R le tenseur de courbure quatre fois covariant. Enfin, si E est un fibré vectoriel sur M, $C^\infty(E)$ désigne l'espace vectoriel des sections C^∞ globales de E, et $\bar{E}$ le faisceau des germes de sections C^∞ de E, qui est un faisceau fin [**13**]. Au paragraphe III, $u \to u^\#$ désigne l'isomorphisme musical de T^*_xM sur T_xM associé à la structure riemannienne, $\xi \to \xi^\flat$ l'isomorphisme réciproque [**3**, p.21]. Rappelons [**8**, p. 129] que l'espace des sections C^∞ du fibré O^2T^*M des formes bilinéaires symétriques admet la décomposition

$$C^\infty(O^2T^*M) = \operatorname{Ker} \delta' \oplus \operatorname{Im} \delta^*,$$

orthogonale pour le produit scalaire global, où l'on a noté δ^* l'opérateur différentiel de $C^\infty(TM)$ dans $C^\infty(O^2T^*M)$ qui au champ de vecteurs ξ associe $\delta^*\xi = \frac{1}{2}\mathscr{L}_\xi g$, et δ' son adjoint formel. Cette décomposition sera d'ailleurs raffinée au paragraphe III, pour les variétés à courbure constante.

I. L'opérateur D_1.

I.1. On désigne par $\mathscr{R}$ l'opérateur de courbure de (M, g), c'est-à-dire le représentant du tenseur de courbure contravariant par rapport aux deux premiers indices et covariant par rapport aux deux derniers. C'est une section du fibré $\mathscr{L}(\Lambda^2T^*M, \Lambda^2T^*M)$, et dans le cas d'une variété (X, g) à courbure constante K, $\mathscr{R} = K\,\mathrm{Id}_{\Lambda^2T^*X}$ (pour notre convention de signe, K est positif pour la sphère canonique). On définit un opérateur différentiel $\mathscr{R}'$ de degré 2 de $C^\infty(O^2T^*M)$ dans $C^\infty(\mathscr{L}(\Lambda^2T^*M, \Lambda^2T^*M))$ de la façon suivante : Si $t \to g(t)$ est une variation de métrique telle que $g(0) = g$ et $(d/dt)g(t)|_{t=0} = h$, on pose

$$\mathscr{R}'(h) = \frac{d}{dt}\mathscr{R}(t)\Big|_{t=0}.$$

Au cas où $h = \delta^*\xi$, on peut prendre $g(t) = \varphi^*_{t/2}\,g$, où φ_t est le flot du champ de vecteurs ξ, et on a $\mathscr{R}'(\delta^*\xi) = \frac{1}{2}\mathscr{L}_\xi\,\mathscr{R}$ dans le cas général, et $\mathscr{R}'\circ\delta^* = 0$ si et seulement si (M, g) est à courbure constante.

Toutes ces propriétés sont locales, et sont encore vraies si les sections considérées sont des sections locales.

Il est commode de remplacer $\mathscr{R}'\cdot h$ par le tenseur quatre fois covariant associé, noté $D_1\cdot h$. D'après [**1**, § 3], D_1h a pour expression dans le cas général

$$\begin{aligned} D_1h(X, Y, Z, T) = \tfrac{1}{2}[&\nabla_Y\nabla_Z h(X, T) - \nabla_Y\nabla_T h(X, Z) - \nabla_X\nabla_Z h(Y, T) \\ &+ \nabla_X\nabla_Z h(Y, Z) - R(h^\#(X), Y, Z, T) - R(X, h^\#(Y), Z, T)]. \end{aligned}$$

On définit également un opérateur différentiel R' de $C^\infty(O^2T^*M)$ dans $C^\infty(\Lambda^2T^*M \otimes \Lambda^2T^*M)$ en posant $R'h = (d/dt)R(t)\big|_{t=0}$, où R est le représentant quatre fois covariant du tenseur de courbure.

Dans le cas d'une variété (X, g) à courbure constante K on a

$$\begin{aligned} D_1h(X, Y, Z, T) = R'h(X, Y, Z, T) \\ + K[g(X, T)h(Y, Z) + g(Y, Z)h(X, T) \\ - g(X, Z)h(Y, T) - g(Y, T)h(Y, Z)]. \end{aligned}$$

On a alors la propriété d'intégrabilité locale suivante qui constitue un lemme de Poincaré pour les opérateurs δ^* et D_1 et dont la démonstration occupera le reste du paragraphe.

THÉORÈME I. *Soit U un ouvert de X. Soit h une section du fibré O^2T^*U. Pour qu'il* *existe un ouvert $V \subset U$ et un champ de vecteurs ξ sur V tels que $h = \delta^*\xi$, il faut et il suffit que $D_1 h = 0$.*

I.2. RAPPELS SUR LE REPÈRE MOBILE (VOIR [**12**, CHAPTER 7], ET [**6**, CHAPITRE 20, § 6]). Soit M une variété C^∞ de dimension n, et $B(M)$ son fibré des repères. Une structure riemannienne sur M peut être définie par la donnée d'un recouvrement ouvert $(U_\alpha)_{\alpha\in A}$ et pour chaque indice α d'une section s_α de $B(M)$ au dessus de U_α, de façon que, pour tout x de $U_\alpha \cap U_\beta$, $s_\beta(x) = s_\alpha(x)\, a_{\alpha\beta}(x)$, avec $a_{\alpha\beta} \in C^\infty(U_\alpha \cap U_\beta, O(n))$.

Si $(\theta_k^\alpha(x))_{1\leqq k\leqq n}$ désigne le corepère dual du "repère mobile" $s_\alpha(x) = (X_k^\alpha(x))_{1\leqq k\leqq n}$, la restriction à U_α de la métrique riemannienne s'exprime par

$$\sum_{i=1}^{p} \theta_i^\alpha \otimes \theta_i^\alpha.$$

Un repère mobile sur un ouvert U étant donné, rappelons l'expression sur U de la connexion riemannienne ∇ et de la courbure en fonction des θ_k. Le résultat fondamental suivant est dû à Elie Cartan.

LEMME 1. *Il existe sur U une matrice antisymétrique (ω_{ij}) de 1-formes et une seule telle que*

$$d\theta_i = -\sum_{j=1}^{n} \omega_{ij} \wedge \theta_j \qquad (1 \leqq i \leqq n). \tag{1}$$

Cette relation résulte évidemment de la formule $d\bar\theta = -\bar\omega \wedge \bar\theta$, où $\bar\theta$ est la forme fondamentale et $\bar\omega$ la forme de connexion du fibré des repères orthonormés sur M, si on prend l'image réciproque par s des deux membres; de plus, il est facile de voir que $\bar\omega$ et les ω_{ij} se déterminent mutuellement [**14**, Chapter I]. Le Lemme 1 est aussi la traduction du lemme algébrique suivant, [**11**, p.22] essentiel pour ce qui suit.

LEMME 2. *Soit V un espace euclidien réel de dimension finie, $\mathfrak{g}$ l'espace vectoriel des endomorphismes antisymétriques, et ∂ l'application linéaire de* Hom$(V, \mathfrak{g})$ *dans* Hom$(V \wedge V, V)$ *définie par*

$$(\partial T)(u, v) = T_u(v) - T_v(u)$$

pour un élément $u \to T_u$ de Hom$(V, \mathfrak{g})$. *Alors ∂ est un isomorphisme.*

DÉMONSTRATION. Il suffit de montrer que ∂ est injective. Si $\partial T = 0$, alors $\forall u$, $v \in V$, $T_u v = T_v u$. En notant (,) le produit scalaire de V, on a, $\forall u, v, w \in V$,

$$(T_w v, u) = (T_v w, u) = -(T_v u, w) = -(T_u v, w) = (T_u w, v) = (T_w u, v) = -(T_w v, u).$$

Ainsi, $(T_w v, u) \equiv 0$, et par conséquent $T_w = 0$.

Donnons-nous maintenant sur V une base orthonormée de base duale $(\alpha_i)_{1\leqq i\leqq n}$. Par rapport à cette base, un élément $u \to T_u$ de Hom$(V, \mathfrak{g})$ est donné par une matrice antisymétrique (β_{ij}), $1 \leqq i, j \leqq n$, de formes linéaires. De plus $\forall u, v \in V$,

$$\left(\sum_{j=1}^{n} \beta_{ij} \wedge \alpha_j\right)\cdot(u \wedge v) = \sum_{j=1}^{n} [\beta_{ij}(u)\, \alpha_j(v) - \beta_{ij}(v)\, \alpha_j(u)]$$

est la i-ième composante du vecteur $T_u v - T_v u$, d'où le Lemme 1.

La connection ∇ s'exprime simplement au moyen des ω_{ij}. On voit facilement que $\nabla_{X_i} X_j = \sum_{k=1}^{n} \omega_{kj}(X_i) X_k$. Si de plus on pose

$$\Omega_{ij} = d\omega_{ij} + \sum_{k=1}^{n} \omega_{ik} \wedge \omega_{kj} \tag{2}$$

(on reconnait là encore l'image réciproque par la section s de l'égalité $\bar{\Omega} = d\bar{\omega} + \bar{\omega} \wedge \bar{\omega}$, où $\bar{\Omega}$ est la forme de courbure dans le fibré des repères orthonormés), on vérifie que $R(X_i, X_j, X_k, X_1) = \Omega_{k1}(X_i, X_j)$ pour tout quadruplet de vecteurs du repère mobile, et on en déduit que

$$R(X_i \wedge X_j) = \frac{1}{2} \sum_{k,1} \Omega_{k1}(X_i, X_j)\,(X_k \wedge X_1)$$

et

$$R = \frac{1}{2} \sum_{i,j} \Omega_{ij} \otimes \theta_i \wedge \theta_j.$$

En particulier, $\Omega_{ij} = K\theta_i \wedge \theta_j$ si M est à courbure constante K. Pour plus de détails, voir [**12**, Chapter 7].

I.3. Calcul de D_1 par la méthode du Repère Mobile. Soit U un ouvert de (X, g) pour lequel la structure riemannienne est donnée par un corepère mobile $(\theta_i)_{1 \leq i \leq n}$. Toute forme bilinéaire symétrique h sur U s'écrit d'une façon unique sous la forme $h = 2 \sum_{i=1}^{n} \theta_i \otimes \eta_i$, les η_i étant des 1-formes telles que $\sum_{i=1}^{n} \theta_i \wedge \eta_i = 0$, ce qui traduit la symétrie. De plus, si t est un réel assez petit, les 1-formes $\theta_i(t) = \theta_i + t\eta_i$ constituent un corepère mobile donnant une variation $g(t)$ de la métrique g telle que $(d/dt)g(t)\big|_{t=0} = h$. En dérivant en t les deux termes de l'égalité (1) écrite pour la métrique $g(t)$, il vient

$$d\eta_i = -\sum_{j=1}^{n}\left[\psi_{ij} \wedge \theta_j + \sum \omega_{ij} \wedge \eta_j\right],$$

avec $\psi_{ij} = (d/dt)\omega_{ij}(t)\big|_{t=0}$ et d'après le Lemme 2, l'application $(\eta_i)_{1 \leq i \leq n} \to (\psi_{ij})_{1 \leq i \leq j \leq n}$ est un opérateur différentiel C^∞. En dérivant (2) et en posant $\Omega'_{ij} = (d/dt)\Omega_{ij}(t)\big|_{t=0}$ il vient

$$\Omega'_{ij} = d\psi_{ij} + \sum_{k=1}^{n} \omega_{ik} \wedge \psi_{kj} + \psi_{ik} \wedge \omega_{kj}$$

et on en déduit que $D_1 h = \frac{1}{2} \sum_{i,j} \varphi_{ij} \otimes \theta_i \wedge \theta_j$, avec $\varphi_{ij} = \Omega'_{ij} - K[\eta_i \wedge \theta_j + \theta_i \wedge \eta_j]$.

Rapprochée des formules de I.2, cette relation montre bien que les déformations infinitésimales h telles que $D_1 h = 0$ sont celles qui preservent la structure de variété à courbure constante.

I.4. Démonstration du Théorème I dans le Cas d'une Variété Plate. Tout

x de U est contenu dans un ouvert $V \subset U$ contractile, et tel que sur V la métrique riemannienne puisse être donnée par un corepère orthonormé $\theta_i = dx_i$, où $(x_i)_{1 \leqq i \leqq n}$ est un système de coordonnées sur V.

Soit $h = 2 \sum_{i=1}^n dx_i \otimes \eta_i$. Comme $\omega_{ij} = 0$, la condition $D_1 h = 0$ se réduit à $d\psi_{ij} = 0$. D'après le lemme de Poincaré, on peut écrire successivement

$$\psi_{ij} = d\chi_{ij} \qquad (\text{avec } \chi_{ij} + \chi_{ji} = 0),$$

$$d\eta_i = -\sum_j d\chi_{ij} \wedge dx_j = -d\Big(\sum_j \chi_{ij}\, dx_j\Big),$$

d'où $\eta_i = -\sum_j \chi_{ij}\, dx_j + df_i$.

En utilisant alors la condition $\sum_i dx_i \wedge \eta_i = 0$, on en déduit que

$$h = \sum_{i,j} \left(\frac{\partial f_i}{\partial x_j} + \frac{\partial f_j}{\partial x_i}\right) dx_i \otimes dx_j = \mathscr{L}_\xi\, g$$

où ξ est le champ de vecteurs dont l'expression en coordonnées locales est $(f_1, \cdots, f_n)$.

Remarquons que le théorème analogue est encore vrai pour une variété munie d'une structure Lorentzienne plate. Nous utiliserons ce fait au paragraphe suivant.

I.5. Démonstration du Théorème I dans le Cas où $K \neq 0$. La propriété à démontrer étant locale, il suffit d'étudier le cas du revêtement riemannien simplement connexe $(\tilde{X}, \tilde{g})$ de (X, g). Ce dernier peut être identifié à l'hypersurface de R^{n+1} définie par l'équation

$$\sum_{i=0}^n x_i^2 = 1/K \qquad (\text{et } x_0 > 0, \text{ si } K < 0)$$

munie de la métrique riemannienne induite par la forme bilinéaire $(x, y) = (\operatorname{sgn} K) x_0 y_0 + \sum_{i=1}^n x_i y_i$. Nous supposerons désormais que $K = 1$ c'est-à-dire que $(\tilde{X}, \tilde{g})$ est la sphére canonique (S^n, g_0) ; le cas où la courbure est négative se traite d'une façon analogue si on utilise le modèle ci-dessus de $(\tilde{X}, \tilde{g})$, et la remarque finale de I.4.

Soit j la rétraction $x \to x/r$, avec $r = (x, x)^{1/2}$, de $R^{n+1} - \{0\} = E$ sur S^n. Nous allons montrer que le diagramme

$$\begin{array}{ccccccccc}
0 \longrightarrow & \Theta(E) & \xrightarrow{i} & \overline{TE} & \xrightarrow{\delta^*} & \overline{\bigcirc^2 T^*E} & \xrightarrow{D_1} & \overline{\Lambda^2 T^*E \otimes \Lambda^2 T^*E} \\
& \uparrow J_0 & & \uparrow J_0 & & \uparrow J_1 & & \uparrow J_2 \\
0 \longrightarrow & \Theta(S^n) & \xrightarrow{i} & \overline{TS^n} & \xrightarrow{\delta^*} & \overline{\bigcirc^2 T^*S^n} & \xrightarrow{D_1} & \overline{\Lambda^2 T^*S^n \otimes \Lambda^2 T^*S^n}
\end{array}$$

est commutatif, les J_k étant des injections naturelles définies par homogénéité que nous allons préciser.

Si ξ est un champ de vecteurs sur un ouvert U de S^n de flot φ_t, le champ $\tilde{\xi}(x) = r\xi(x/r)$ défini sur $j^{-1}(U)$ a pour flot $\tilde{\varphi}_t(x) = r\,\varphi_t(x/r)$; nous poserons $\tilde{\xi} = J_0(\xi)$. Soit d'autre part ρ le champ de vecteurs sur E défini par $\rho(x) = (x_1, \cdots, x_n)$. Pour qu'un champ de vecteurs $\theta = (f_1, \cdots, f_n)$ sur $j^{-1}(U)$ soit de la forme $J_0(\xi)$, il faut et il suffit que:

(a_0) $[\rho, \theta] = \theta$ (identité d'Euler pour les f_i, qui doivent être homogènes et de degré 1).

(b_0) $\rho \cdot \theta = 0$ (i.e. $\theta(x)$ est tangent à la sphère de centre 0 et de rayon r).

De plus, J_0 est un isomorphisme entre l'espace vectoriel des champs de vecteurs sur U et celui des champs de vecteurs sur $j^{-1}(U)$ satisfaisant aux conditions (a_0) et (b_0); tout champ de Killing sur $j^{-1}(U)$ satisfait à ces conditions et est l'image d'un champ de Killing sur U. Le premier carré du diagramme est donc bien commutatif. Si maintenant h est une forme bilinéaire, on pose $J_1(h) = r^2 j^* h$.

LEMME 3. $J_1 \circ \delta^*_{S^n} = \delta^*_E \circ J_0$.

DÉMONSTRATION. Il revient au même de montrer que pour tout champ de vecteurs ξ sur un ouvert de S^n, $J_1(\mathscr{L}_\xi\, g_0) = \mathscr{L}_{J_0\xi}\, \tilde{g}_0$, g_0 et $\tilde{g}_0$ étant les métriques canoniques de S^n et E respectivement. Comme $g_0 = r^2 j^* g_0 + dr \otimes dr$, on a, en revenant aux flots φ_t de ξ et $\tilde{\varphi}_t$ de $\tilde{\xi}$,

$$\begin{aligned}\varphi_t^* \tilde{g}_0 &= r^2 \tilde{\varphi}_t^*(j^* g_0) + dr \otimes dr \quad \text{(puisque } \tilde{\varphi}_t \text{ respecte les sphères de centre 0)}\\ &= r^2 j^*(\varphi_t^* g_0) + dr \otimes dr\end{aligned}$$

d'où

$$\frac{d}{dt}\, \tilde{\varphi}_t^* g_0 \Big|_{t=0} = r^2 j^* \Big(\frac{d}{dt}\, \varphi_t^* g_0 \Big|_{t=0} \Big).$$

Plus généralement, si g est une métrique riemannienne sur un ouvert U de S^n, $J_1(g) + dr \otimes dr = \tilde{g}$ est une métrique riemannienne sur $j^{-1}(U)$ induisant g sur U; et si $x \to (X_1(x), \cdots, X_n(x))$ est un repère mobile orthonormé associé à g ayant pour corepère dual $(\theta_1, \ , \theta_n)$, $(X_1(x/r), \ , X_n(x/r), \rho(x))$ est un repère orthonormé associé à $\tilde{g}$ ayant pour corepère dual $\tilde{\theta}_i = rj^* \theta_i$ $(i \leqq n)$, $\tilde{\theta}_{n+1} = dr$. Soit ω_{ij} la (n, n) matrice antisymétrique du Lemme 1 associée aux θ_i. On a

$$d\tilde{\theta}_i = rd(j^*\theta_i) + dr \wedge j^*\theta_i = - \sum_{k=1}^{n} j^*\omega_{ik} \wedge rj^*\theta_k - j^*\theta_i \wedge dr$$

et

$$d(dr) = 0 = \sum_{k=1}^{n} j^*(\theta_k) \wedge rj^*\theta_k.$$

La $(n + 1, n + 1)$ matrice associée aux $(\tilde{\theta}_i)_{1 \leqq i \leqq n+1}$ est donc d'après le Lemme I (unicité)

$$\begin{aligned}\tilde{\omega}_{ik} &= j^*\omega_{ik} && (i \leqq n \text{ et } k \leqq n);\\ \tilde{\omega}_{i,n+1} &= - \tilde{\omega}_{n+1,i} = j^*\theta_i && (i \leqq n).\end{aligned}$$

LEMME 4. *Pour une section locale h ou un germe de section de* $\Lambda^2 T^* S^n \otimes \Lambda^2 T^* S^n$, *posons* $J_2(h) = r^2 j^* h$. *Alors* $F_2 \circ D_1 = D_1 \circ F_1$.

DÉMONSTRATION. Si $(\theta_i)_{1 \leqq i \leqq n}$ est un corepère local de S^n donnant la métrique riemannienne canonique, h peut s'écrire comme en I.3 sous la forme $2\sum_{i=1}^{n}\theta_i\otimes\eta_i$ avec $\sum_{i=1}^{n}\theta_i\wedge\eta_i=0$, et dans ces conditions $J_1(h)=2r^2\sum_{i=1}^{n}j^*\theta_i\otimes j^*\eta_i$. D'après ce qui précède, la matrice antisymétrique associée au système de formes $(rj^*\eta_1,\ ,\ rj^*\eta_n,\ 0)$ est

$$\tilde{\psi}_{ij}=j^*\psi_{ij}\quad(i\leqq n \text{ et } j\leqq n);$$
$$\tilde{\psi}_{i,n+1}=-\tilde{\psi}_{n+1,i}=-j^*\eta_i$$

où ψ_{ij} est la matrice de variation de la connexion associée à la variation de métrique donnée par les $(\eta_i)_{1\leqq i\leqq n}$. Un calcul un peu long, mais sans difficulté, fondé sur les formules de I.3 donne alors le résultat.

Le Théorème I sera donc complètement démontré si on prouve le

LEMME 5. *Soit $\tilde{h}$ un germe de section de O^2T^*E tel que $\tilde{h}=J_1(h)$, où h est un germe de section de $O^2T^*S^n$. Si $D_1\tilde{h}=0$, il existe un germe de champ de vecteurs $\tilde{\xi}$ sur E tel que $\tilde{h}=\delta_E^*\tilde{\xi}$ et qui soit de la forme $J_0\xi$, où ξ est un germe de champ de vecteurs sur S^n.*

REMARQUE. Ce lemme est une forme équivariante (par rapport aux homothéties) du résultat prouvé en I.4, ce qui donne l'idée de sa démonstration.

DÉMONSTRATION. Remarquons d'abord que pour tout ouvert U de S^n, J_1 est un isomorphisme de l'espace vectoriel $C^\infty(U,\ O^2T^*S^n)$ sur le sous-espace des h de $C^\infty(j^{-1}(U),\ O^2T^*E)$ vérifiant les conditions

(a_1) $\mathscr{L}_\rho(h)=2h$ (homogénéité de degré 2).

(b_1) $h(x)(\rho(x),v)=0$ pour tout x de $j^{-1}(U)$ et v de $T_xj^{-1}(U)$. Si on écrit h sous la forme $\sum_{i=1}^{n+1}dx_i\otimes\eta_i$, la condition ($a_1$) devient $\mathscr{L}_\rho\eta_i=\eta_i$ pour tout i, et du fait que h est symétrique, la condition (b_1) peut s'exprimer sous une des deux formes équivalentes

(b_1') $\eta_i(\rho)=0$ pour tout i; (b_1'') $\sum_{i=1}^{n+1}x_i\eta_i=0$.

Reprenons maintenant le raisonnement de I.4.

Les ψ_{ij} sont maintenant des formes homogènes de degré 0, et par suite on peut prendre les χ_{ij} homogènes de degré 0, puis les f_i homogènes de degré 1, c'est à dire vérifiant la condition (a_0) ∘ (b_1'') s'écrit

$$\sum_{i=1}^{n}\left(x_i\frac{\partial f_i}{\partial x_j}+x_i\chi_{ij}\right)=0\quad\text{pour tout } j,$$

d'où

$$\sum_{j=1}^{n+1}\sum_{i=1}^{n+1}x_i\,x_j\frac{\partial f_i}{\partial x_j}=0\ =\sum_{i=1}^{n+1}\sum_{j=1}^{n+1}x_i\,x_j\frac{\partial f_i}{\partial x_j}=\sum_{i=1}^{n+1}x_if_i,$$

d'après l'identité d'Euler pour les f_i.

Le champ $\tilde{\xi}=(f_1,\cdots,f_n)$ satisfaisant aux conditions (a_0) et (b_0) est bien de la forme $J_0(\xi)$.

II. Description de la résolution fine Θ donnée par Calabi. Aucune démonstration ne sera donnée dans ce paragraphe, dont le but est de servir de guide à la lecture du paragraphe 4 de [5].

II.1. L'OPÉRATEUR T_R. On définit un endomorphisme T_R du fibré bigradué $\sum_{p,q\geqq 0} \Lambda^p T^*M \otimes \Lambda^q T^*M$ par la formule

$$T_R F(X_1,\cdots, X_{p-1}, Y_0, Y_1,\cdots, Y_q) = \sum_{i=0}^{q} (-1)^i F(X_1,\cdots, X_{p-1}, Y_i, Y_0,\cdots, \hat{Y}_i,\cdots Y_q)$$

si F est dans la composante $\Lambda^p T^*M \otimes \Lambda^q T^*M$. T_R est bihomogène et de bidegré $(-1, +1)$, et indique que l'on fait un transfert à droite.

EXEMPLES. Si $F \in T^*M \otimes T^*M$,

$$T_R F(Y_0, Y_1) = F(Y_0, Y_1) - F(Y_1, Y_0).$$

Si $F \in \Lambda^2 T^*M \otimes \Lambda^2 T^*M$,

$$T_R F(X, Y, Z, T) = F(X, Y, Z, T) - F(X, Z, Y, T) + F(X, T, Y, Z).$$

Pour un élément décomposable $u \wedge v \otimes w$ de $\Lambda^2 T^*M \otimes \Lambda^p T^*M$, $T_R(u \wedge v \otimes w) = u \otimes v \wedge w - v \otimes u \wedge w$.

Un élément φ de $\Lambda^p T^*M \otimes \Lambda^q T^*M$ est dit *primitif à droite* si $T_R \varphi = 0$.

Ainsi, les éléments primitifs à droite de $T^*M \otimes T^*M$ sont les formes bilinéaires symétriques, et ceux de $\Lambda^2 T^*M \otimes \Lambda^2 T^*M$ les tenseurs de courbure. On a le

LEMME 6 (THÉORÈME 2 DE [5]). *T_R est une application surjective de $\Lambda^{p+1} T^*M \otimes \Lambda^{q-1} T^*M$ dans $\Lambda^p T^*M \otimes \Lambda^q T^*M$ dès que $p < q$.*

II.2. LES FAISCEAUX Φ_q ET LES OPÉRATEURS DIFFÉRENTIELS D_q $(q \geqq 2)$. La restriction à $\Lambda^p T^*M \otimes \Lambda^q T^*M$ de la dérivation covariante est un opérateur différentiel $D^{p,q}$ de $\Lambda^p T^*M \otimes \Lambda^q T^*M$ dans $\Lambda^p T^*M \otimes \Lambda^q T^*M \otimes T^*M$. On notera D_q l'opérateur différentiel de $\Lambda^2 T^*M \otimes \Lambda^q T^*M$ dans $\Lambda^2 T^*M \otimes \Lambda^{q+1} T^*M$ obtenu en antisymétrisant $D^{2,q}$ par rapport aux $q + 1$ derniers indices. Si une section locale φ de $\Lambda^2 T^*M \otimes \Lambda^q T^*M$ s'écrit

$$\varphi = \frac{1}{2} \sum_{i,j} \theta_i \wedge \theta_j \otimes \varphi_{ij},$$

$(\theta_i)_{1\leqq i\leqq n}$ étant un corepère local orthonormé, on a par un calcul direct

$$D_q \varphi = \frac{1}{2} \sum_{i,j} \theta_i \wedge \theta_j \otimes \varphi'_{ij}$$

avec

$$\varphi'_{ij} = d\varphi_{ij} + \sum_{k=1}^{n} \omega_{ik} \wedge \varphi_{kj} + \sum_{k=1}^{n} (-1)^{q+1} \varphi_{ik} \wedge \omega_{kj}.$$

On définit de manière analogue un opérateur D'_q de $T^*M \otimes \Lambda^q T^*M$ dans $T^*M \otimes \Lambda^{q+1} T^*M$; si $\psi = \sum_{i=1}^{n} \theta_i \otimes \psi_i$ est une section locale de $T^*M \otimes \Lambda^q T^*M$,

$$D'_q \psi = \sum_{i=1}^{n} \theta_i \otimes \left(d\psi_i + \sum_{j=1}^{n} \omega_{jk} \wedge \psi_k \right).$$

On vérifie que le diagramme

$$\begin{array}{ccc} \Lambda^2T^*M \otimes \Lambda^qT^*M & \xrightarrow{D_q} & \Lambda^2T^*M \otimes \Lambda^{q+1}T^*M \\ T_R \downarrow & & \downarrow T_R \\ T^*M \otimes \Lambda^{q+1}T^*M & \xrightarrow{D'_{q+1}} & T^*M \otimes \Lambda^{q+2}T^*M \end{array}$$

est anticommutatif, et donc que D_q transforme tout élément primitif à droite en un élément primitif à droite ; et il est clair d'après le paragraphe I que D_1 transforme toute forme bilinéaire symétrique en un tenseur de courbure. Si on appelle Φ_q le sous-fibré de $\Lambda^2T^*M \otimes \Lambda^qT^*M$ formé par les éléments primitifs à droite, on prouve la généralisation suivante du Théorème I:

THÉORÈME II. *Si (X, g) est à courbure constante, de dimension n,*

$$0 \longrightarrow \Theta \xrightarrow{i} \overline{TX} \xrightarrow{D_0} \bar{\Phi}_1 \xrightarrow{D_1} \bar{\Phi}_2 \xrightarrow{D_{n-1}} \bar{\Phi}_n \longrightarrow 0$$

est une résolution fine du faisceau Θ (on a posé $D_0 = \delta^*$, et $\Phi_1 = O^2T^*X$, notation cohérente d'après II.1).

D'après le théorème de Rham abstrait [**13**, Chapter 5] les groupes de cohomologie $H^q(X, \Theta)$ sont isomorphes aux groupes de cohomologie du complexe

$$C^\infty(TX) \xrightarrow{D_0} C^\infty(\Phi_1(X)) \xrightarrow{D_1} C^\infty(\Phi_2(X)) \longrightarrow \cdots$$
$$\xrightarrow{D_{n-1}} C^\infty(\Phi_n(X)) \longrightarrow 0$$

associé à la résolution.

Remarquons que tout notre calcul a reposé sur des théorèmes algébriques et sur le lemme de Poincaré. Mais le cas à courbure constante semble être le seul où une telle réduction de la géométrie soit possible.

III. Un théorème du type Hodge-de Rham pour les $H^q(X, \Theta)$. On pourra consulter [**13**, Chapter 6] pour tout ce qui concerne les opérateurs différentiels sur les variétés et leurs symboles.

III.1. SYMBOLES DES OPÉRATEURS D_q ET DE LEURS ADJOINTS FORMELS. L'opérateur δ^* et son adjoint δ' ont pour symboles respectifs

$$\sigma_u\delta^*(\xi) = u\circ\xi^b \quad \text{et} \quad \sigma_u\delta'(h) = h(u^\#, \cdot)$$

où $u \in T_x^*M$, $\xi \in T_xM$, et $h \in O^2T_x^*M$ (voir [**4**] ; il suffit de remarquer que δ^* est la dérivée covariante symétrisée).

De l'expression de D_1 donnée en I.1, on déduit les symboles de D_1 et de D_1^*,

$$\begin{aligned} \sigma_u(D_1)(h)(X, Y, Z, T) &= \tfrac{1}{2}[u(X)u(Z)h(Y, T) + u(Y)u(T)h(X, Z) \\ &\qquad - u(Y)u(Z)h(X, T) - u(X)u(T)h(Y, Z)], \\ \sigma_u(D_1^*)(S)(X, Y) &= 2S(u^\#, X, u^\#, Y), \end{aligned}$$

où $S \in \Lambda^2T_x^*M \otimes \Lambda^2T_x^*M$ et a les symétries d'un tenseur de courbure.

Les formules de II.2 donnent immédiatement l'expression de symboles de D_q et D_q^* ($q \geqq 2$) pour les tenseurs décomposables:

$$\sigma_u(D_q)(\alpha \otimes \beta) = \alpha \otimes (u \wedge \beta), \qquad \sigma_u(D_q^*) = \alpha \otimes \operatorname{int}_{u*}\beta.$$

Les propriétés suivantes en résultent.

LEMME 7. *Le noyau de $\sigma_t(D_1)$ est l'ensemble des h de $O^2T_x^*M$ qui peuvent s'écrire $h = u \circ t$, où $u \in T_x^*M$.*

DÉMONSTRATION. Il est immédiat qu'un tel h appartient au noyau. Réciproquement, pour que le tenseur de courbure $\sigma_t(D_1)h$ soit nul, il faut et il suffit que

$$\sigma_t(D_1)(h)(X, Y, X, Y) = (t(X))^2 h(Y, Y) + (t(Y))^2 h(X, X) - 2t(X)t(Y)h(X, Y)$$

le soit, quels que soient X et Y dans T_xM, puisqu'un tenseur de courbure est déterminé par la courbure sectionnelle **[12]**.

Si on prend X dans $\operatorname{Ker}(u)$ et $Y = t^\#$, on voit que h s'annule sur $\operatorname{Ker}(t)$; h est bien de la forme cherchée.

LEMME 8. *Pour tout $t \neq 0$ de T_x^*M, on a*

$$\operatorname{Ker} \sigma_t(D_q) \cap \operatorname{Ker} \sigma_t(D_{q-1}^*) = 0 \quad (1 \leqq q \leqq n).$$

DÉMONSTRATION. (a) Si $q = 1$, il suffit de vérifier, d'après le lemme précédent, que si $\sigma_t(\delta')(\xi^b \circ t) = 0$, $\xi = 0$. Mais si $(\xi^b \circ t)(t^\#, X) = \frac{1}{2}(t(\xi)t(X) + \xi^b(X)|t|^2)$ est nul pour tout X de T_xM, ξ^b s'annule sur $\operatorname{Ker}(t)$ et sur $t^\#$, donc sur T_xM tout entier.

(b) Cas où $q = 2$. Pour tout tenseur de courbure S appartenant à $\operatorname{Ker} \sigma_t(D_1^*)$, la 2-forme $S(t^\#, X, t^\#, Y)$ est nulle. Si le tenseur

$$(\sigma_t(D_2)S)(X,Y,Z,T,U) = S(X,Y,Z,T)t(U) + S(X,Y,U,Z)t(T) + S(X,Y,T,U)t(Z)$$

est nul, il en est alors de même de la 3-forme $S(t^\#, Y, T, U)$ (faire $X = Z = t^\#$), puis de $S(X, Y, Z, T)$ (faire $U = t^\#$).

(c) Cas où $q > 2$. La propriété à démontrer est la conséquence de l'égalité $\sigma_t(D_q^*D_q + D_{q-1}D_{q-1}^*) = |t|^2 \operatorname{Id}$; on vérifie facilement que les deux membres prennent la même valeur sur les tenseurs décomposables.

III.2. COHOMOLOGIE DE Θ ET OPÉRATEURS ELLIPTIQUES. Rappelons le théorème fondamental de la théorie des opérateurs elliptiques [**13**, Chapter 6].

THÉORÈME. *Soit Δ un opérateur elliptique de $C^\infty(A)$ dans $C^\infty(B)$, A et B étant deux fibrés vectoriels riemanniens de même base compacte. Alors*

$$C^\infty(A) = \operatorname{Ker}\Delta \oplus^\perp \operatorname{Im} \Delta^*$$

(*somme directe topologique pour la topologie de Fréchet de $C^\infty(A)$, la décomposition étant orthogonale pour le produit scalaire global*), *et* Ker Δ *est de dimension finie.*

Nous nous servirons aussi du lemme suivant:

LEMME 9. *Soit P un opérateur différentiel de $C^\infty(A)$ dans $C^\infty(B)$, A et B satisfaisant* *aux mêmes hypothèses que plus haut. Les opérateurs P, et $(P^*P)^n$ ont même noyau, pour tout entier $n \neq 0$.*

DÉMONSTRATION. Si $P^*Ph = 0$, $\langle P^*Ph, h \rangle = \|Ph\|^2 = 0$, donc $Ph = 0$. Si $(P^*P)^n h = 0$ pour $n \geqq 2$, $\langle (P^*P)^n h, (P^*P)^{n-2} h \rangle = \|(P^*P)^{n-1} h\|^2 = 0$ donc $(P^*P)^{n-1}h = 0$. Algébriquement, la même propriété est vraie au niveau des symboles, et se démontre de la même façon.

D'après les Lemmes 7 et 8, chacun des opérateurs différentiels $(\delta^*\delta', D_1)$, $(D_1^*, D_2^* D_2)$ et (D_{q-1}^*, D_q) a un symbole injectif, et par suite les opérateurs $T_1 = D_1^*D_1 + (\delta^*\delta')^2$, $T_2 = (D_2^*D_2)^2 + D_1D_1^*$, $T_q = D_q^*D_q + D_{q-1}D_{q-1}^*$ $(q > 2)$ sont elliptiques et autoadjoints, et on peut leur appliquer le théorème cité plus haut. De plus, $\operatorname{Ker}(T_q) = \operatorname{Ker}(D_{q-1}^*) \cap \operatorname{Ker}(D_q)$ (d'après le Lemme 8 pour $q \leqq 2$) est un espace vectoriel de dimension finie. On a prouvé le

THÉORÈME III. *Pour toute variété riemannienne compacte (M, g), les espaces $C^\infty(\Phi_q(M))$ admettent les décompositions suivantes, orthogonales pour le produit scalaire global*

$$C^\infty(O^2T^*M) = \operatorname{Ker} T_1 \oplus \operatorname{Im} T_1,$$
$$C^\infty(\Phi_q(M)) = \operatorname{Ker} T_q \oplus \operatorname{Im} T_q \quad (q \geqq 2).$$

Dans le cas d'une variété (X, g) à courbure constante, d'après le Théorème II et la remarque qui le suit, on a le

THÉORÈME III *bis*. *Les espaces $C^\infty(\Phi_q(X))$ admettent les décompositions orthogonales*

$$C^\infty(O^2T^*X) = \operatorname{Ker} T_1 \oplus \operatorname{Im} D_1^* \oplus \operatorname{Im} \delta^* \qquad (q = 1),$$
$$C^\infty(\Phi_q(X)) = \operatorname{Ker} T_q \oplus \operatorname{Im} D_q^* \oplus \operatorname{Im} D_{q-1} \quad (q \geqq 2),$$

et les espaces Ker T_q *sont isomorphes aux* $H^q(X, \Theta)$.

IV. Exemples.

IV.1. LES TORES PLATS. Nous allons calculer $H^1(T^n, \Theta)$ en reprenant globalement les calculs de I.4. Soit $(\theta_i)_{1 \leqq i \leqq n}$ un corepère orthonormé global constitué de formes parallèles et par suite harmoniques, et $h = 2 \sum_{i=1}^n \theta_i \otimes \eta_i$ une 2-forme symétrique sur T^n (on a donc $\sum_{i=1}^n \theta_i \wedge \eta_i = 0$). Si $D_1 h = 0$, on peut écrire successivement avec les notations de I.4: $\psi_{ij} = d\chi_{ij} + \alpha_{ij}$, où les α_{ij} sont harmoniques.

$$d\eta_i = \sum_{j=1}^n (d\chi_{ij} \wedge \theta_j + \alpha_{ij} \wedge \theta_j) = -d\Big(\sum_{j=1}^n \chi_{ij}\, \theta_j\Big) - \sum_{j=1}^n \alpha_{ij} \wedge \theta_j.$$

Les 2-formes $\sum_{j=1}^n \alpha_{ij} \wedge \theta_j$ sont harmoniques (sur une variété riemannienne plate, les formes harmoniques constituent une algèbre), et comme ce sont des bords elles sont nulles. Par suite, $\eta_i = \sum_{j=1}^n \chi_{ij}\, \theta_j + df_i + \alpha_i$, où les α_i sont harmoniques, $h = \mathscr{L}_\xi g + h'$ où h' est dans le sous-espace de $C^\infty(O^2T^*T^n)$ engendré par les produits symétriques de 1-formes harmoniques.

Comme d'autre part δ' s'annule sur ce sous-espace, il est égal à $\mathrm{Ker}(T_1)$. On a donc comme cas particulier du Théorème III la:

PROPOSITION 1. *$H^1(T^n, \Theta)$ est isomorphe au sous-espace de $C^\infty(O^2T^*T^n)$ engendré par les produits symétriques de 1-formes harmoniques; en particulier, $H^1(T^n, \Theta)$ est une bijection avec l'espace vectoriel des 1-jets de déformations plates de T^n, et a pour dimension $n(n+1)/2$.*

IV.2. LES VARIÉTÉS A COURBURE CONSTANTE STRICTEMENT POSITIVE. Il est classique que tout germe de champ de Killing sur (S^n, g_0) (et plus généralement sur une variété $\tilde{X}$ simplement connexe complète à courbure constante) se prolonge d'une façon unique en un champ de Killing sur toute la variété. Le faisceau Θ est alors isomorphe au faisceau constant associé à l'espace vectoriel $R^{n(n+1)/2}$ et les groupes de cohomologie $H^q(\tilde{X}, \Theta)$ sont isomorphes aux produits tensoriels $H^q(\tilde{X}, \boldsymbol{R}) \otimes R^{n(n+1)/2}$. En particulier, $H^1(S^n, \Theta) = 0$ si $n > 1$.

Toute variété à courbure constante positive est un quotient S^n/Γ, où Γ est un sous-groupe fini de $O(n+1)$ opérant sans point fixe sur S^n [**14**, Chapter II, p. 69]. On peut reprendre le raisonnement du paragraphe I.5. en remplaçant $E = R^{n+1} - \{0\}$ et S^n par les variétés quotients E/Γ et S^n/Γ respectivement. Sachant que $H^1(E, R) = H^1(E/\Gamma, R) = 0$, on peut refaire globalement les calculs du Lemme 5, et on obtient la

PROPOSITION 2 [**9**, B p. 337]. *Pour toute variété X à courbure constante positive, $H^1(X, \Theta) = 0$; en particulier, X est indéformable.*

REMARQUE. On peut démontrer les Propositions 1 et 2 en faisant appel à la cohomologie des groupes opérant sur un espace vectoriel ([7, **B** pp. 335–336] où l'exposé précédent). Notre méthode, plus simple, reste dans l'esprit du travail de Calabi: se ramener à des calculs sur les formes différentielles et la cohomologie de de Rham.

IV.3. LES VARIÉTÉS DE DIMENSION 2. Dans ce cas, si τ désigne la courbure scalaire, l'opérateur de courbure est $\frac{1}{2}\tau\ \mathrm{Id}_{\Lambda^2TX}$. On définit, de façon analogue à $\boldsymbol{R}'$ et R', un opérateur différentiel τ' de $C^\infty(O^2T^*X)$ dans $C^\infty(X)$ (variation première de la courbure scalaire). On a alors

$$D_1h(X, Y, Z, T) = \tfrac{1}{2}\tau'(h)\,[g(X, Z)\,g(Y, T) - g(X, T)\,g(Y, Z)]$$

et d'après [**1**, § 3.4],

$$\tau'(h) = \Delta(\mathrm{tr}\ h) + \delta\delta'h - K\ \mathrm{tr}\ h$$

où on a désigné par δ l'adjoint formel de l'opérateur de différentiation extérieure, et par $\Delta = \delta d$ le laplacien sur les fonctions. Rappelons [**3**, Chapter II, §C] d'autre part que la première valeur propre non nulle du laplacien est $2K$ pour (S^2, g_0), et $6K$ pour $(P^2(R), g_0)$. On voit alors, en distingant les cas $K > 0$, $K < 0$, $K = 0$, que sur $C^\infty(O^2T^*X)$ le système d'équations $D_1h = 0$, $\delta'h = 0$ est équivalent au système $\delta'h = 0$, $\mathrm{tr}(h) = 0$ si K est non nul, et au système $\delta'h = 0$, $\mathrm{tr}(h) = cte$ si K est nul.

Appelons N le sous-espace des h de $C^\infty(O^2T^*X)$ telles que $\delta' h = 0$ et $\operatorname{tr}(h) = 0$. Autrement dit, N est l'espace vectoriel des sections globales de O^2T^*X qui sont orthogonales aux variations infinitésimales conformes, puisque ces dernières sont de la forme $fg + \delta^*\xi$, où $f \in C^\infty(X)$, $\xi \in C^\infty(TX)$ (cf. [**2**, § 12]). En appliquant le Théorème III bis, on obtient la

PROPOSITION 3. *Soit (X, g) une variété riemannienne compacte de dimension* 2 *à courbure constante K. Le groupe $H^1(X, \Theta)$ est isomorphe à N si $K \neq 0$, et à $N \oplus Rg$ si $K = 0$.*

REMARQUE. 1° On sait déjà que $H^1(S^2, \Theta) = 0$: on obtient dans ce cas une version infinitésimale du théorème de la représentation conforme.

2° Sur tout ouvert U de X, l'espace des solutions du système $\delta' h = 0$, $\operatorname{tr}(h) = 0$ ne dépend que de la classe conforme de la métrique. En effet, si $e^a g$ est une métrique riemannienne conforme à g, on voit par un calcul facile que

$$\delta'_{e^a g} = e^{-a}\,[\delta'_g + (da)\operatorname{tr}_g].$$

Nous utiliserons ce fait au cours de la démonstration du Théorème IV.

Pour le calcul de $H^1(X, \Theta)$, supposons d'abord X orientable. Rappelons qu'à toute métrique riemannienne sur une variété orientée de dimension 2 est associée une structure complexe canonique, qui ne dépend que de la classe conforme de la métrique: si J_x désigne la rotation de $+\pi/2$ dans le plan euclidien orienté T_xX, le champ d'endomorphismes $x \to J_x$ définit une structure presque complexe intégrable. Soit $T_x^*X^c$ le fibré cotangent de X pour cette structure complexe. Une section holomorphe du fibré $O^2T^*X^c$ s'appelle une *différentielle quadratique holomorphe*; pour toute carte complexe (U, z) de X, $h|_U = f(z)dz \odot dz$, où f est une fonction holomorphe. On a évidemment une injection canonique de $O^2T^*X^c$ dans $O^2T^*X \otimes \mathbf{C}$.

THÉORÈME IV. *Si (X, g) est orientée, si $H^0(X, O^2T^*X^c)$ est l'espace vectoriel des différentielles quadratiques holomorphes sur X,*

$$N \otimes \boldsymbol{C} = H^0(X, O^2T^*X^c) \oplus \overline{H^0(X, O^2T^*X^c)}.$$

REMARQUE. Un changement d'orientation permute les deux espaces facteurs de cette décomposition, qui ne dépend donc pas de l'orientation.

DÉMONSTRATION. Considérons d'abord le cas où X est à courbure strictement négative. Soit P le demi-plan complexe supérieur, muni de sa structure complexe usuelle, et de la métrique riemannienne $y^{-2}(dx \odot dx + dy \odot dy)$, dont la courbure est -4. Le groupe G des automorphismes holomorphes de P est formé par les transformations

$$z \to (az + b)(cz + d)^{-1}, \qquad a, b, c, d \in R,\ ad - bc = 1.$$

C'est aussi le groupe des isométries de P qui conservent l'orientation. Il existe alors un sous-groupe Γ de G, proprement discontinu et sans points fixes, tel que

$X = P/\Gamma$, comme variété riemannienne et comme variété complexe (cf. [7, §§ 4 et 5]). Soit h une 2-forme bilinéaire symétrique sur X à valeurs complexes, vérifiant les équations $\operatorname{tr}(h) = 0$, $\delta' h = 0$. La projection canonique p de P sur X étant un revêtement riemannien, la 2-forme p^*h vérifie les mêmes équations sur p. D'après la remarque 2 suivant la proposition, p^*h vérifie encore ces équations si on remplace la métrique $y^{-2}(dx \odot dx + dy \odot dy)$ par la métrique plate $dx \odot dx + dy \odot dy$. Dans ces conditions, p^*h, qui est de trace nulle, peut s'écrire

$$p^*h = f dz \odot dz + f' \overline{dz} \odot \overline{dz}, \quad \text{où } f \text{ et } f' \text{ sont des fonctions } C^\infty.$$

La nullité de $\delta'(p^*h)$ équivaut alors à

$$\partial_1(f+f') + i\partial_2(f-f') = 0, \qquad i\partial_1(f-f') - \partial_2(f+f') = 0$$

où on à désigné par ∂_1 et ∂_2 les opérateurs de dérivation par rapport à x et à y. On a donc

$$\partial_1 f + i\,\partial_2 f = 0 \quad \text{et} \quad \partial_1 f' - i\,\partial_2 f' = 0.$$

Autrement dit, f est holomorphe et f' antiholomorphe. Les formes $f\,dz \odot dz$ et $f' \overline{dz} \odot \overline{dz}$ sont alors Γ-invariantes, puisque p^*h l'est, et se redescendent sur X. Les cas $K = 0$, $K > 0$ se traitent d'une façon analogue.

COROLLAIRE 1. *Si X est orientable et de genre $\gamma > 1$, $H^1(X, \Theta) \cong R^{6\gamma-6}$.*

DÉMONSTRATION. Il suffit de remarquer que l'espace vectoriel des différentielles quadratiques holomorphes sur X est alors de dimension complexe $3\gamma - 3$, d'après le théorème de Riemann-Roch [**10**, § 7].

COROLLAIRE 2. *Si X est non orientable et de genre $\gamma > 1$, $H^1(X, \Theta) \cong R^{3\gamma-3}$.*

DÉMONSTRATION. Il existe un sous-groupe Γ, proprement discontinu et sans points fixes, du groupe engendré par G et la transformation $z \to -\bar{z}$, tel que $X = P/\Gamma$, et un sous-groupe distingué $\tilde{\Gamma}$ de Γ, d'indice 2, tel que le revêtement orientable $\tilde{X}$ de X soit $P/\tilde{\Gamma}$ [**6**, tome 3, Chapitre 16, §28, Exercice 2]. Il en résulte que toute forme bilinéaire symétrique complexe h sur qui est Γ-invariante et vérifient $\delta' h = \operatorname{Tr} h = 0$ peut s'écrire $h = h' + \gamma^* h'$, où h' est une différentielle quadratique holomorphe $\tilde{\Gamma}$-invariante et γ une transformation de Γ renversant l'orientation; h' ne dépend pas du choix de γ dans $\Gamma - \tilde{\Gamma}$ et l'application $h \to h'$ est une bijection.

Passons maintenant au calcul de $H^2(X, \Theta)$, qui, d'après le Lemme 8 et le Théorème III bis, est isomorphe à $\operatorname{Ker}(D_1^*)$. En dimension 2, toute section globale de $\Phi_2(X)$ s'écrit d'une manière unique sous la forme fS, où $f \in C^\infty(X)$ et $S(X, Y, Z, T) = g(X, Z)g(Y, T) - g(X, T)g(Y, Z)$, $D_1^*(fS) = 2\tau'^*(f)$, avec, d'après [**4**, § 5],

$$\tau'^*(f) = (\Delta f)g + \delta^* df - Kfg.$$

En prenant la trace des deux membres, on voit que pour que $\tau'^*(f) = 0$, il est nécessaire que $\Delta f - 2Kf = 0$. Distinguons les différents cas:

(a) *La sphère S^2.* La première valeur propre du laplacien est $2K$, elle est de multiplicité 3, et les fonctions propres correspondantes satisfont à la relation $\delta^* df =$

$-Kfg$ [3, p. 181]: ce sont donc les solutions de l'équation $\tau'^*(f) = 0$. On retrouve bien le fait que $H^2(S^2, \Theta)$ est de dimension 3.

(b) Pour le projectif $P^2(R)$ et pour les variétés à courbure négative, $\Delta f - 2Kf = 0$ implique $f = 0$.

(c) Pour une variété plate, les solutions de l'équation $\Delta f - 2Kf = 0$ sont les fonctions constantes, qui inversement vérifient bien l'équation $\tau'^*(f) = 0$. En résumé

PROPOSITION 4. $H^2(P^2(R), \Theta) = 0$; *$H^2(X, \Theta)$ est isomorphe à R si X est plate, et est nul si X est à courbure négative.*

APPLICATION. Si X est à courbure négative, d'aprés [9] où le Théorème 7 de l'exposé précédent, $H^1(X, \odot)$ paramétrise les 1-jets de déformations globales conservant la courbure; en particulier, si X est orientable et de genre γ, on retrouve un résultat de la théorie de Teichmüller (cf. par exemple [7, § 7]).

Si X est plate, il est encore vrai d'après IV.I (le cas des bouteilles de Klein plates se traite de la même façon, et H^1 est alors isomorphes à R^2) que $H^1(X, \Theta)$ paramétrise les 1-jets de déformations globales conservant la courbure, bien que $H^2(X, \Theta)$ soit non nul. Le Théorème 7 de l'exposé précédent n'est donc pas le meilleur possible.

BIBLIOGRAPHIE

1. M. Berger, *Quelques formules de variation pour une structure riemannienne*, Ann. Sci. École Norm. Sup. (4) **3** (1970), 285–294. MR **43** #3969.

2. ———, *Du côté de chez Pu*, Ann. Sci. École Norm. Sup. (4) **5** (1972), 1–44. MR **46** #8119.

3. M. Berger, P. Gauduchon and E. Mazet, *Le spectre d'une variété riemannienne*, Lecture Notes in Math., vol. 194, Springer-Verlag, Berlin and New York, 1971. MR **43** #8025.

4. M. Berger and D. Ebin, *Some decompositions of the space of symmetric tensors on a Riemannian manifold*, J. Differential Geometry **3** (1969), 379–392. MR **42** #993.

5. E. Calabi, *On compact, Riemannian manifolds with constant curvature.* I, Proc. Sympos. Pure Math., vol. 3, Amer. Math. Soc., Providence, R.I., 1961, pp. 155–180. MR **24** #A3612.

6. J. Dieudonné, *Fondements de l'analyse moderne*, Cahiers Scientifiques, fasc. 28, Gauthier-Villars, Paris, 1963. MR **28** #5149.

7. C. J. Earle and J. Eells, *A fibre bundle description of Teichmüller theory*, J. Differential Geometry **3** (1969), 19–43. MR **43** #2737a.

8. D. Ebin, *Espace des métriques riemanniennes et mouvement des fluides via les variétés d'applications*, Cours professé à l'École Polytechnique et à l'Université Paris VII en 71–72, Edité par le Centre de Mathématiques de l'École Polytechnique.

9. P. A. Griffiths, *Deformations of G-structures*, Part A: *General theory of deformations*, Math. Ann. **155** (1964), 292–315. MR **30** #3480.

———, *Deformations of G-structures.* Part B: *Deformations of geometric G-structures*, Math. Ann. **158** (1965), 326–351. MR **31** #6252.

10. R. C. Gunning, *Lectures on Riemann surfaces*, Princeton Math. Notes, Princeton Univ. Press, Princeton, N.J., 1966. MR **34** #7789.

11. I.M. Singer and S. Sternberg, *The infinite groups of Lie and Cartan.* I. *The transitive groups*, J. Analyse Math. **15** (1965), 1–144. MR **36** #911.

12. M. Spivak, *A comprehensive introduction to differential geometry.* Vol. II, M. Spivak, Brandeis Univ., Waltham, Mass., 1970, Chaps. 7 and 8. MR **42** #6726.

13. F. W. Warner, *Foundations of differentiable manifolds and Lie groups*, Scott, Foresman, Glenview, Ill., 1971. MR **45** #4312.
14. J. A. Wolf, *Spaces of constant curvature*, McGraw-Hill, New York, 1967. MR **36** #829.

3. APPENDICE. MODULES INFINITESIMAUX DES SURFACES DE RIEMANN[1]

J. LAFONTAINE

I. Introduction (cf. [10, § 5 et 8], [15], [17, § 1]). Soit X une variété compacte orientée de dimension 2 et de genre $\gamma > 1$, donnée une fois pour toutes. Il a été rappelé en IV.3 que sur X les structures complexes et les structures riemanniennes à courbure constante (négative, d'après la formule de Gauss-Bonnet puisque $\gamma > 1$), se correspondent bijectivement. Nous parlerons par abus de langage de le surface de Riemann (X, g). La résolution de Calabi permet ainsi de retrouver des résultats de la théorie des déformations des surfaces de Riemann ([7], [21], [23]) notamment une version infinitésimale—mais de démonstration simple du théorème de Torelli pour les surfaces de Riemann non hyperelliptiques.

Rappelons d'abord quelques précisions sur le dictionnaire complexe-riemannien en dimension 2, en gardant les notations de IV.3. Sur (X, g), l'opérateur $*$ de Rham pour les 1-formes est donné par $*\alpha = \alpha(J)$; les 1-formes α et β ont pour produit scalaire global

$$(1) \qquad \langle \alpha \cdot \beta \rangle = \int_X (\alpha, \bar{\beta}) vg = \int_X \alpha \wedge *\bar{\beta}.$$

Une 1-forme α est une différentielle abélienne, c'est-à-dire une section holomorphe de T^*X, si et seulement si elle est fermée et de type (1, 0); c'est une forme harmonique si et seulement si elle est fermée ainsi que $*\alpha$. Les parties réelles et imaginaires d'une différentielle abélienne sont donc harmoniques; réciproquement, toute forme harmonique réelle α est la partie réelle d'une différentielle abélienne et d'une seule, à savoir la 1-forme $\alpha - i * \alpha$. Notons que toutes ces notions ne dépendent que de la classe conforme de g. L'espace vectoriel des différentielles abéliennes est de dimension complexe γ, celui des 1-formes harmoniques réelles de dimension réelle 2γ.

Donnons-nous maintenant 2γ cycles singuliers $a_1, b_1, \cdots, a_\gamma, b_\gamma$ représentant une base du groupe d'homologie $H_1(X, Z)$, et ayant pour intersections $a_i \times a_j = b_i \times b_j = 0$, $a_i \times b_j = \delta_{ij}$, c'est-à-dire une base d'homologie canonique. Le cup-produit de deux classes de cohomologie représentées par les 1-formes fermées α et β est donné par

$$(2) \qquad \int_X \alpha \wedge \beta = \sum_{k=1}^{\gamma} \int_{a_k} \alpha \int_{b_k} \beta - \int_{a_k} \beta \int_{b_k} \alpha.$$

[1]La numérotation des propositions, des théorèmes et des réfrences fait suite à celle du 2e exposé; les notations sont les mêmes.

DÉFINITION. *On appelle surface marquée une variété compacte de dimension* 2 *munie d'une orientation et d'une base d'homologie en degré* 1.

Nous supposerons désormais X marquée au moyen de la base canonique a_1, $b_1, \cdots, a_\gamma, b_\gamma$. Si de plus X est une surface de Riemann, et si $(\omega_i)_{1 \leqq i \leqq \gamma}$ est une base de l'espace des différentielles abéliennes, posons

$$\omega_{ij} = \int_{a_j} \omega_i, \qquad \omega_{i,j+\gamma} = \int_{bj} \omega_i \qquad (I \leqq i, j \leqq \gamma).$$

On obtient ainsi une matrice $\Omega = (\Omega' \ \Omega'')$ à γ lignes et 2γ colonnes, qu'on appelle la *matrice des périodes* de la base $(\omega_i)_{1 \leqq i \leqq \gamma}$. A l'aide des relations (1) et (2) on démontre facilement qu'il existe une base et une seule de l'espace des différentielles abéliennes dont la matrice des périodes soit de la forme $(I \ Z)$, et que la matrice Z est alors symétrique et de partie imaginaire définie négative [**17**, § 1]. On dit que Z est la matrice des périodes canonique de la surface de Riemann marquée X, à laquelle on a ainsi associé $\gamma(\gamma + 1)/2$ nombres complexes. Nous pouvons maintenant énoncer le

THÉORÈME DE TORELLI (CF. [**17**], [**20**], [**22**] POUR LA DÉMONSTRATION). *Soit X une surface marquée au moyen d'une base d'homologie canonique. Si deux structures complexes sur X ont même matrice des périodes canonique, elles sont isomorphes.*

REMARQUE. A toute surface de Riemann X de genre $\gamma > 1$, on associe un tore complexe appellé la variété de Jacobi de X, soit $J(X)$ ([**10**, § 7], [**17**]). C'est le quotient de $\boldsymbol{C}^\gamma$ par le réseau qu'engendrent les 2γ vecteurs de la matrice Ω, qui sont indépendants puisque $\operatorname{Im} Z < 0$. Mais il existe des surfaces de Riemann non isomorphes ayant même variété de Jacobi (cf. [**20** bis]). Pour connaître la structure complexe, on a besoin non seulement du réseau donnant la variété de Jacobi, mais aussi d'une base distinguée de ce réseau.

II. Cas des surfaces de Riemann non hyperelliptiques. Le théorème de Torelli infinitésimal est alors une conséquence du théorème suivant, dû à M. Noether (1880?), dont on trouvera dans [**18**] une démonstration plus récente.

THÉORÈME. *Sur une surface de Riemann non hyperelliptique, l'espace des différentielles quadratiques holomorphes est engendré par les produits symétriques de différentielles abéliennes.*

REMARQUE. On vérifie facilement que pour des raisons combinatoires cette propriété est encore vraie en genre 2, bien que toute surface de Riemann soit alors hyperelliptique (cf. [**10**, pp. 228–229]).

Traduit en langage riemannien, le théorème de Noether donne la

PROPOSITION 5. *Si (X, g) est orientée et non hyperelliptique, $H^1(X, \Theta)$ est isomorphe au sous-espace de $C^\infty(O^2T^*X)$ engendré par les formes bilinéaires symétriques $\alpha \circ \beta - *\alpha \circ *\beta$, où les 1-formes α et β sont harmoniques. Cette propriété reste vraie si X est orientée et de genre* 2.

DÉMONSTRATION. Il suffit d'appliquer le Théorème IV et le théorème de Noether, en remarquant que les parties réelle et imaginaire d'un produit symétrique de différentielles abéliennes sont bien de la forme $\alpha\circ\beta - *\alpha\circ*\beta$, où α et β sont harmoniques.

THÉORÈME V. *Soit X une surface marquée au moyen d'une base d'homologie canonique. Soit $t \to g(t)$ une déformation localement triviale d'une métrique à courbure constante $(X, g(0))$ non hyperelliptique telle que la matrice des périodes canonique de $(X, g(t))$ soit égale pour tout t à la matrice des périodes canonique de $(X, g(0))$. Il existe alors un champ de vecteurs ξ sur X tel que*

$$\frac{d}{dt} g(t)\Big|_{t=0} = \delta^*\xi.$$

DÉMONSTRATION. Posons $h = (d/dt)g(t)|_{t=0}$. Pour une déformation localement triviale, $D_1h = 0$, et d'après le Théorème III et la Proposition 3 on peut écrire $h = \delta^*\xi + h_1$, où $\xi \in C^\infty(TX)$ et h_1 appartient au sous-espace $N = \delta'^{-1}(0) \cap \mathrm{Tr}^{-1}(0)$ de $C^\infty(O^2T^*X)$. Si φ_t est le flot du champ de vecteurs ξ, la variation $g_1(t) = \varphi^*_{-t/2}(g(t))$ a les mêmes propriétés que $g(t)$ et de plus $(d/dt)g_1(t)|_{t=0} \in N$. Il suffit de montrer que si $h \in N$, $h = 0$.

Soit $(\omega_k(t) = \omega_k^1(t) + i\,\omega_k^2(t))$ $(1 \leqq k \leqq \gamma)$ la base canonique de l'espace des différentielles abéliennes pour la structure complexe associée à la métrique $g(t)$. Les formes $\omega_k(t)$ et $\omega_k(0)$ ayant les mêmes périodes par rapport à une base d'homologie sont cohomologues; il en est de même de $\omega_k^r(t)$ et $\omega_k^r(0)$ pour $r = 1$ ou 2. Rappelons d'autre part que $\omega_k^2(t) = - *^t\omega_k^1(t)$.

En appliquant les formules (1) et (2) écrites pour la métrique $g(t)$ aux forme $\omega_k^r(t)$ $(1 \leqq k \leqq \gamma, r = 1$ ou $2)$, on voit que la déformation $g(t)$ conserve les produits scalaires globaux de ces formes deux à deux. Plus généralement, pour toute 1-forme harmonique α (pour la métrique $g(0)$) est cohomologue à une et une seule forme harmonique pour la métrique $g(t)$, soit $\alpha(t)$, et on a

$$\langle\alpha(t), \beta(t)\rangle_t = \langle\alpha, \beta\rangle_0.$$

En prenant les dérivées des 2 membres par rapport à t pour $t = 0$, on obtient

$$\langle\alpha', \beta\rangle_0 + \langle\alpha, \beta'\rangle_0 + \int_X (\alpha, \beta)\mathrm{Tr}\, h\, v_{g(0)} - \langle h, \alpha\circ\beta\rangle = 0$$

où l'on a posé

$$\alpha' = \frac{d}{dt}\alpha(t)\Big|_{t=0}, \qquad \beta' = \frac{d}{dt}\beta(t)\Big|_{t=0}.$$

Mais $\alpha(t)$ et $\beta(t)$ étant pour tout t cohomologues à α et β respectivement, α' et β' sont des bords; de plus $\mathrm{tr}(h) = 0$. On a donc $\langle h, \alpha\circ\beta\rangle = 0$ pour tout couple (α, β) de formes harmoniques, et par suite $h = 0$ d'après la Proposition 5.

III. Cas des surfaces de Riemann hyperelliptiques. Enonçons d'abord la formule des points fixes d'Atiyah et Bott [**16**] dont nous nous servirons au cours de ce paragraphe.

Soit

$$E_0 \xrightarrow{D_0} E_1 \xrightarrow{D_1} \cdots \xrightarrow{D_{n-1}} E_n \longrightarrow 0$$

un complexe elliptique sur une variété compacte M, les fibrés E_i étant formés de tenseurs covariants. *On ne suppose pas que les opérateurs différentiels D_i ont le même degré.* Soit f un endomorphisme du complexe, c'est-à-dire une application C^∞ de M dans M telle que $f^* \circ D_i = D_i \circ f^*$ pour tout i. On désigne par $H_i(f)$ l'endomorphisme qu'induit f sur le i-ième espace vectoriel de cohomologie du complexe. Supposons enfin que f n'a qu'un nombre fini de points fixes, et que ces points fixes sont simples; autrement dit, si $f(A) = A$, le graphe de f est transverse à la diagonale en (A, A), ou encore $\det(1 - df_A) \neq 0$. Soit $f_{i,A}$ l'endomorphisme de la fibre $(E_i)_A$ induit par f en tout point fixe. On a alors

$$\sum_{i=0}^{n} (-1)^i \operatorname{Tr} H^i(f) = \sum_{\{A \mid f(A) = A\}} \sum_{i=0}^{n} (-1)^i \frac{\operatorname{Tr} f_{i,A}}{|\det(1 - df_A)|}.$$

Appliquée au complexe de Rham, cette formule redonne la formule de Lefschetz. Pour pouvoir 1 appliquer au complexe de Calabi, il suffit de prendre pour f une isométrie.

Une surface de Riemann de genre $\gamma > 1$ est dite *hyperelliptique* si on peut la représenter comme un revêtement ramifié à 2 feuillets de $P^1(C)$; cela équivaut à l'existence d'un automorphisme involutif—en langage riemannien une isométrie involutive conservant l'orientation – ayant $2\gamma + 2$ points fixes, et cet automorphisme est alors unique (cf. [**10**, §§7 et 10], [**17**, §2]). La caractérisation suivante des surfaces hyperelliptiques est certainement folklorique.

PROPOSITION 6. *Pour qu'une surface de Riemann de genre $\gamma > 1$ soit hyperelliptique, il faut et suffit qu'il existe un automorphisme involutif f tel que $f^*_\alpha = -\alpha$ pour toute 1-forme harmonique α, ou, ce qui revient au même, pour tout différentielle abélienne.*

DÉMONSTRATION. Rappelons d'abord que les points fixes d'un automorphisme d'une surface de Riemann sont en nombre fini (cf. [**16**, p. 37]) et que si $f^2 = \mathrm{Id}$, alors $df_A = -\mathrm{Id}$ en tout point fixe A. Dans l'isomorphisme entre $H^k(X, R)$ et l'espace vectoriel des formes harmoniques de degré k donné par le théorème de Hodge-de Rham [**13**], $H^k(f)$ devient la restriction de f^* aux formes harmoniques de degré k, puisque f est une isométrie.

Appliquons maintenant la formule des points fixes au complexe de de Rham,

$$2 - \operatorname{Tr} H^1(f) = \operatorname{Card} \{A, f(A) = A\}.$$

Mais $H^1(f)$ est un endomorphisme involutif d'un espace vectoriel de dimension 2γ: $H^1(f) = -\mathrm{Id}$ si et seulement si f a $2\gamma + 2$ points fixes.

En particulier, sur une surface hyperelliptique, les formes bilinéaires $\alpha \circ \beta - *\alpha \circ *\beta$, où α et β sont harmoniques, sont invariantes par l'automorphisme hyperelliptique. La proposition suivante, classique en géométrie complexe (cf. [**21**, p. 14]), montre qu'elles ne peuvent engendrer N si $\gamma > 2$.

PROPOSITION 7. *Soit (X, g) une surface de Riemann hyperelliptique. L'automorphisme hyperelliptique f est un endomorphisme du complexe de Calabi; $H^1(X, \Theta)$ est la somme directe des deux sous-espaces propres de $H^1(f)$ pour les valeurs propres* $+1$ *et* -1, *qui sont de dimension* $2(2\gamma - 1)$ *et* $2(\gamma - 2)$ *respectivement.*

DÉMONSTRATION. Rappelons qu'en dimension 2 le complexe de Calabi s'écrit

$$T^*X \to O^2T^*X \to O^2(\Lambda^2T^*X) \to 0$$

(nous avons utilisé l'isomorphisme musical $TX \to^{b} T^*X$ pour n'avoir que des tenseurs covariants), et que

$$H^0(X, \Theta) = H^2(X, \Theta) = 0, \qquad H^1(X, \Theta) \cong R^{6\gamma-6};$$

ici encore, par l'isomorphisme entre $H^1(X, \Theta)$ et N, $H^1(f)$ devient f^*. La formule d'Atiyah-Bott donne

$$-\operatorname{Tr} H^1(f) = \tfrac{1}{4}(2\gamma+2)(2-3+1) = -(2\gamma+2)$$

d'où le résultat cherché, sachant que $(H^1(f))^2 = \mathrm{Id}$.

REMARQUES. 1° On a encore un résultat analogue au théorème de Noether: le sous-espace propre relatif à la valeur propre $+1$ est engendré par les 2-formes $\alpha\circ\beta - *\alpha\circ*\beta$ [21].

2° Soit (X, g) une variété à courbure constante de dimension 2 non orientable, et $(\tilde{X}, \tilde{g})$ son revêtement riemannien orientable. En appliquant la formule d'Atiyah-Bott à $(\tilde{X}, \tilde{g})$ et au changement d'orientation θ, on voit que $\operatorname{Tr} H^1(\theta) = 0$; les sous-espaces propres de $H^1(\tilde{X}, \tilde{\Theta})$ pour les valeurs propres $+1$ et -1 de $H^1(\theta)$ sont donc tous deux de dimension $3\gamma - 3$, ce qui donne une autre démonstration du fait que $\dim H^1(X, \Theta) = 3\gamma - 3$.

BIBLIOGRAPHIE SUPPLEMENTAIRE

15. L. V. Ahlfors and L. Sario, *Riemann surfaces*, Princeton Math. Ser., no. 26, Princeton Univ. Press, Princeton, N. J., 1960. Chaps. I and V. MR **22** #5729.

16. M. F. Atiyah and R. Bott, *A Lefschetz fixed point formula for elliptic complexes*. I, Ann. of Math. (2) **86** (1967), 374–407. MR **35** #3701.

17. R. C. Gunning, *Lectures on Riemann surfaces, Jacobi varieties*, Math. Notes., Princeton Univ. Press, Princeton, N. J., 1972.

18. R. S. Hamilton, *Non-hyperelliptic Riemann surfaces*, J. Differential Geometry **3** (1969), 95–101. MR **40** #2847.

19. T. Kato, *Perturbation theory for linear operators*, Die Grundlehren der math. Wissenschaften, Band 132, Springer-Verlag, New York, 1966. MR **34** #3324.

20. H. H. Martens, *Torelli's theorem and a generalization for hyper-elliptic surfaces*, Comm. Pure Appl. Math. **16** (1963), 97–110. MR **27** #2623.

20. bis.———, *From the classical theory of Jacobian varieties*, Proc. Fifteenth Scandinavian Congress (Oslo, 1968), Lecture Notes in Math., Vol. 118, Springer-Verlag, Berlin, 1970, pp. 74–98. MR **41** #1738.

21. H. E. Rauch, *A transcendental view of the space of algebraic Riemann surfaces*, Bull. Amer. Math. Soc. **71** (1965), 1–39. MR **35** #4403.

22. A. Weil, *Séminaire Bourbaki.* 9ième année: 1956/57, Exposé 151, *Sur le théorème de Torelli*; 10e année: 1957/58, Exposé 168, *Modules des surfaces de Riemann*, Secrétariat mathématique, Paris, 1959. MR **28** #1090.

École Normale Supérieure

Centre de Mathématiques de l'École Polytechnique

Université Paris VII

Proceedings of Symposia in Pure Mathematics
Volume 27, 1975

SOME CONSTRUCTIONS RELATED TO H. HOPF'S CONJECTURE ON PRODUCT MANIFOLDS

JEAN PIERRE BOURGUIGNON

1. Introduction. M will denote a C^∞ *compact* riemannian manifold. We will be interested in the existence of riemannian metrics with positive sectional curvature K on product manifolds.

H. Hopf conjectured (cf. [**6**]): "On $S^2 \times S^2$, there does not exist any metric with $K > 0$."

$S^2 \times S^2$ is the first case to be considered since for any compact manifold N, $M = S^1 \times N$ cannot admit a positively curved metric by Myers' theorem (cf. [**9**, p. 365]).

"If M admits a metric with $K > 0$, $\pi_1(M)$ is finite."

In fact very few topological restrictions are known for compact positively curved manifolds. We recall the two versions of Synge's lemma (cf. [**9**, p. 365]):

"If dim M is even, then $\pi_1(M)$ is either $\boldsymbol{Z}_2$ or $\{1\}$,"

and (cf. [**8**, p. 63]).

"If dim M is odd, M is orientable."

As we want to look at a generalized version of H. Hopf's conjecture on product manifolds, we will have to take into account these restrictions.

For example for any $m, n \in \boldsymbol{N}$, $P_m(\boldsymbol{R}) \times P_n(\boldsymbol{R})$, the product of real projective spaces, does not admit a metric with $K > 0$ (by Synge's lemma if $m + n$ is even and since $P_{\text{even}}(\boldsymbol{R})$ is not orientable if $m + n$ is odd): compare [**10**].

REMARK. The analogous question for negatively curved manifolds is solved since $\pi_1(M_1 \times M_2)$ contains some abelian subgroups isomorphic to $\boldsymbol{Z} \times \boldsymbol{Z}$, a situation

AMS (MOS) subject classifications (1970). Primary 58H05; Secondary 53C35.

which cannot occur on a negatively curved manifold by Preissmann's theorem [11].

At this point I should mention the following result due to A. Weinstein [12]: "A metric on $S^2 \times S^2$ induced by an immersion in $\boldsymbol{R}^6$ cannot be positively curved."

Now I would like to report some results related to the conjecture due to Mrs. A. Deschamps, Mrs. P. Sentenac and myself ([3] and [4]) using deformation of product metrics and their connection with a construction of J. Cheeger [5].

2. Deforming the metric. Let $M = M_1 \times M_2$. The lower index i for $i = 1$ (resp. 2) relates the object to M_1 (resp. M_2).

Let $g_0 = g_1 \times g_2$ where g_i is a riemannian metric on M_i. Suppose (M_1, g_1) and (M_2, g_2) are positively curved.

Let $\{X, Y\}$ be a g_0-orthonormal basis of a plane π. If $X = X_1 + X_2$, $Y = Y_1 + Y_2$, then

$$K(\pi) = g_1(R_1(X_1, Y_1)X_1, Y_1) + g_2(R_2(X_2, Y_2)X_2, Y_2).$$

Therefore all the planes have positive curvature except those for which one can find a basis of the form $\{X_1, Y_2\}$; they are called *mixed planes.*

We will try to increase the curvature of these planes by considering a variation $g(t)$ of the metric g_0 (i.e. a curve of metrics such that $g(0) = g_0$, always supposed sufficiently many times differentiable).

DEFINITION. A variation $g(t)$ of g_0 is *nonnegative* (resp. *stationary*) at first order if for all mixed planes π_m,

$$\left.\frac{dK_{g(t)}(\pi_m)}{dt}\right|_{t=0} \geqq 0 \qquad (\text{resp.} = 0).$$

It is easy to see that these conditions involve only the first jet of the variation (generally denoted by h^1).

LEMMA (M. BERGER [1]). *Every nonnegative variation at first order is stationary at first order.*

The idea of the proof is to use the special form of $\dot{K}(h^1)$ on the mixed planes, then to integrate on the unit sphere bundle of M_1, then of M_2 and to apply the maximum principle.

The idea of our method is to go on by induction, but we have to make the following *basic remark* relative to the significance for a variation $g(t)$ of g_0 of being stationary at first order and nonnegative at the second order (with the obvious meaning for this last expression).

Suppose that there exists a vector field V transverse to TM_1 for example; let $\mathscr{V}_t$ denote its flow. Consider the variation of g_0, $g(t) = \mathscr{V}_t^*(g_0)$. As by $\mathscr{V}_t$ every mixed plane π_m is moved into a nonmixed plane and as

$$K_{\mathscr{V}_t^*(g_0)}(\pi_m) = K_{g_0}(\mathscr{V}_t(\pi_m)),$$

$g(t)$ is stationary at first order and positive at second order. But from a geometric point of view, we have not done anything except turning around M on itself. To go on, we must take into account the action of the diffeomorphism group $\mathscr{D}(M)$ on the space $\mathscr{M}$ of metrics on M.

D. G. Ebin proved (cf. [7]) that this action admits a slice: the one he uses is given infinitesimally at a metric g_0 by Ker div_{g_0} (where $\mathrm{div}_{g_0}\colon O^2T^*M = T_{g_0}\mathscr{M} \to T^*M$ is the divergence on symmetric tensors).

The natural *geometric* question which arises is: "What are the symmetric tensors h such that for all mixed planes π_m, $\dot{K}_{g_0}(h)(\pi_m) = 0$ and $\mathrm{div}_{g_0} h = 0$?"

It turns out that div_{g_0} does not take sufficiently into account the product structure of (M, g_0) to get a nice answer.

We define $\mathrm{div}^p_{g_0} h = \mathrm{div}_{g_1} h_1 + \mathrm{div}_{g_2} h_2$ where h_i denotes the restriction of h to TM_i and we try to substitute Ker $\mathrm{div}^p_{g_0}$ to Ker div_{g_0} for an infinitesimal slice at g_0. To go further, we have to distinguish two completely different cases: (M, g_0) has no infinitesimal isometry and (M, g_0) has some (this distinction is made clearer by a construction of Cheeger: see § 5 and [5]).

3. (M, g_0) has no infinitesimal isometry. We suppose in this paragraph that (M, g_0) has no infinitesimal isometry; therefore (M_1, g_1) and (M_2, g_2) do not have any either.

In [3] we prove

THEOREM 1. Ker $\mathrm{div}^p_{g_0}$ *is an infinitesimal slice at* g_0.

THEOREM 2. *If for all mixed planes* π_m, $\dot{K}_{g_0}(h)(\pi_m) = 0$ *and* $\mathrm{div}^p_{g_0}(h) = 0$, *then* $h = h_1 \times h_2$ *where* h_i *is a tensor field on* M_i.

These two theorems allow us to go on by induction and to prove

THEOREM 3. *Let* $g(t)$ *be a nonnegative at order* k *variation of* g_0. *Suppose that the* k*-jet of* $g(t)$ *is in* $\bigoplus_{i=1}^k$ Ker $\mathrm{div}^p_{g_0}$, *then* $g(t) = g_1(t) \times g_2(t)$ *where* $g_i(t)$ *is a variation of* g_i *on* M_i.

COROLLARY 4. *There is no deformation* $g(t)$ *of* g_0 *analytic in* t *such that for* $\varepsilon > 0$ *and* $t \in]0, \varepsilon[$, $K_{g(t)} > 0$.

In fact the theorem we would like to prove is:

"The product metrics without infinitesimal isometries are not in the closure of the positively curved metrics" which is a local version near the product metrics of the generalized Hopf conjecture.

For that we should solve the following

QUESTION. Can each point of $\partial\{g \in \mathscr{M} \mid K_g > 0\}$ be attained by an analytic path inside $\{g \in \mathscr{M} \mid K_g > 0\}$?

4. M has infinitesimal isometries. Ker $\mathrm{div}^p_{g_0}$ is not any more an infinitesimal slice at g_0 and also Theorem 2 is no more true as we will see.

Let $\mathscr{I}(M)$ be the Lie algebra of infinitesimal isometries on M. As $\mathscr{I}(M) \neq 0$, we can suppose $\mathscr{I}(M_1) \neq 0$, for example.

We denote by ξ_1 a Killing form on M_1 (dual of an infinitesimal isometry) and η_2 a 1-form on M_2.

PROPOSITION 5. *$g(t) = g_0 + t\xi_1 \circ \eta_2$ is a variation of g_0 stationary at first order, which is nonnegative at second order.*

We again have to consider transversality to the orbit of g_0 under $\mathscr{D}(M)$: if η_2 is the gradient of a function, it is easy to see that $g(t)$ is tangent to the orbit, but if η_2 is coclosed, $g(t)$ is tangent to Ebin's slice (such a variation will be called geometric). In particular if $\mathscr{I}(M_2) \neq 0$, we can take η_2 to be also a Killing form on M_2.

One can show (cf. [4]):

THEOREM 6. *Let M_1 be riemannian homogeneous and* dim $M_2 \geqq 2$. *Then there exists a geometric variation $g(t)$ of g_0 such that for all mixed planes π_m, $K_{g(t)}(\pi_m)$ is an increasing function of t.*

(In fact we take $g(t) = g_0 + \sum_{i=1}^{k} (t^i/i!)\, \xi_1^i \circ \eta_2^i$ where the ξ_1^i are in $\mathscr{I}(M_1)$ and the η_2^i are coclosed.)

This does not imply that for $t \neq 0$ and for all planes π, $K_{g(t)}(\pi) > 0$, since we do not know that we have uniformity in the mixed planes. In fact in the case of $(S^m \times S^n$, *can* $\times\, g_2)$, where can means the canonical metric of the sphere S^m and g_2 is invariant by the antipodal map, our construction can give rise to a variation of a metric of $P^m(\mathbf{R}) \times P^n(\mathbf{R})$ space which does not admit any positively curved metric (cf. § 1).

By a more careful examination of K, one can even see that "very often" in a neighborhood of the mixed planes some planes become negatively curved.

One should also emphasize that the even or odd dimensionality of the factors affects very much the computation, because of the completely different behavior of the infinitesimal isometries in these two cases. ("If dim M is even, any infinitesimal isometry has a zero": M. Berger [1].)

We relate now these results with a construction of J. Cheeger [5].

5. Cheeger's construction. Let G be a closed subgroup of the isometry group of (M, g_0), therefore compact (M has no longer to be a product). The action of G on $G \times M$ given by $(x, m) \to (x \cdot y^{-1}, y \cdot m)$ is free and totally discontinuous. Moreover the orbit space $G \backslash G \times M$ is diffeomorphic to M and inherits a natural metric $\bar{g}$ if G is endowed with a bi-invariant metric b, since G acts by isometries: $(G \times M, b \times g_0) \to (M, \bar{g})$ is a riemannian submersion. Therefore $(M, \bar{g})$ has nonnegative sectional curvature and in general fewer planes with zero curvature than g_0.

It is possible to introduce a parameter t in the construction by considering $(G, t^{-1} b)$.

Let TO (resp. NO) be the subbundle of TM tangent (resp. normal) to the orbits of G. We denote by A_t the element of $\mathrm{Hom}(TM, TM)$ given by $A_t|_{NO} = \mathrm{Id}|_{NO}$ and $A_t|_{TO}$ is the image under the action of the isomorphism from TG to T^*G defined by $t^{-1}b$.

Then $g(t) = g_0 \cdot (\mathrm{Id}_{TM} + A_t)^{-1}$.

REMARK. This construction (which involves global properties of M) has the obvious advantage of providing nonnegatively curved metrics, though, up to now, no positively curved metric on a product has been obtained in this way. On the other hand to be of any interest when M is a product manifold, it requires $\mathscr{I}(M_1)$ and $\mathscr{I}(M_2)$ to be at the same time nontrivial.

One can think of applying again the construction to $(M, \bar{g})$, but $(M, \bar{g})$ has fewer isometries than (M, g_0) leaving less freedom for the construction.

One can also mention the following connection between Cheeger's construction and ours: "The first jet of every geometric strongly nonnegative variation of $(S^2 \times S^2$, can $\times$ can) is the first jet of a Cheeger's variation." (Strongly nonnegative means that at second order the variation does not make appear some planes with negative curvature.) This is definitively false for other cases. The proof uses essentially bounds for the first eigenvalue of the laplacian (cf. [**2**, p. 179]).

Finally I would like to mention the special case of $S^3 \times S^3$ (Spin 4) which seems to be particularly well adapted to Cheeger's construction. Moreover if a positively curved metric does exist on $S^3 \times S^3$, this will give a negative answer to the question: "Does $K > 0$ imply that the Euler characteristic is positive?" (cf. [**6**] and [**13**]) which was another conjecture of H. Hopf.

REFERENCES

1. M. Berger, *Trois remarques sur les variétés riemanniennes à courbure positive*, C. R. Acad. Sci. Paris Sér. A–B **263** (1966), A76–A78. MR **33** #7966.
2. M. Berger, P. Gauduchon and E. Mazet, *Le spectre d'une variété riemannienne*, Lecture Notes in Math., no. 194, Springer-Verlag, Berlin and New York, 1971. MR **43** #8025.
3. J. P. Bourguignon, A. Deschamps and P. Sentenac, *Conjecture de H. Hopf sur les produits de variétés*, Ann. Sci. École Norm. Sup. (4) **5** (1972), fasc. 2.
4. ———, *Quelques variations particulières d'un produit de métriques*, Ann. Sci. École Norm. Sup. (4) **6** (1973), fasc. 1.
5. J. Cheeger, *Some examples of manifolds of nonnegative curvature* (preprint).
6. S. S. Chern, *The geometry of G-structures*, Bull. Amer. Math. Soc. **72** (1966), 167–219. MR **33** #661.
7. D. G. Ebin, *The manifold of Riemannian metrics*, Proc. Sympos. Pure Math., vol. 15, Amer. Math. Soc., Providence, R.I., 1970, pp. 11–40. MR **42** #2506.
8. S. Kobayashi, *Transformation groups in differential geometry*, Springer-Verlag, Berlin and New York, 1972.
9. S. Kobayashi and N. Nomizu, *Foundations of differential geometry*. Vols. I, II, Interscience, New York, 1963, 1969. MR **27** #2945; **38** #6501.
10. J. W. Milnor, *Morse theory*, Ann. of Math. Studies, no. 51, Princeton Univ. Press, Princeton, N. J., 1963. MR **29** #634.
11. A. Preissmann, *Quelques propriétés globales des espaces de Riemann*, Comment. Math. Helv. **15** (1943), 175–216. MR **6**, 20.
12. A. Weinstein, *Positively curved n-manifolds in R^{n+2}*, J. Differential Geometry **4** (1970), 1–4. MR **41** #9154.
13. ———, *Remarks on curvature and the Euler integrand*, J. Differential Geometry **6** (1971/72), 259–262. MR **45** #7641.

CENTRE DE MATHÉMATIQUES DE L'ÉCOLE POLYTECHNIQUE, PARIS

STATE UNIVERSITY OF NEW YORK AT STONY BROOK

Proceedings of Symposia in Pure Mathematics
Volume 27, 1975

CONNECTIONS, HOLONOMY AND PATH SPACE HOMOLOGY

KUO-TSAI CHEN*

The purpose of this work is two-fold—first, to present a generalization of the notion of a linear connection and its associated holonomy map, and, next, to describe the method of formal power series connections as an efficient device to compute loop space homology through differential forms.

Such a generalized connection will be locally defined through connection forms of arbitrary positive degrees. The associated holonomy map sends not only loops but also, for example, singular simplices of the smooth loop space into the full linear algebra or some other kind of algebra. Furthermore, the vector bundle in consideration need not be of finite dimension, and the base space can be a differentiable space other than a manifold.

Instead of presenting in generality, we shall introduce our notion of a linear connection through a finite dimensional vector bundle and then treat, roughly speaking, the case of a product bundle, whose fiber is a graded algebra. There the holonomy map will be from a cubical chain complex of the loop space to the graded algebra. We equip the graded algebra with a differential and ask for a sufficient condition to make the holonomy map a chain map. The condition turns out to be that of a twisting cochain. (See Brown [3].)

We use k to denote the field of real (or complex) numbers. If, for the graded algebra, we have a formal power series algebra $k[[X_1,\cdots, X_m]]$, where $X_1,\cdots,X_m$

AMS (MOS) subject classifications (1970). Primary 53B15, 55D35; Secondary 53C65, 55H20, 58A99.

Key words and phrases. Linear connection, holonomy map, curvature, twisting cochain, loop space homology, cobar construction, iterated integral, formal power series connection.

*Supported in part by National Science Foundation grant GP 34257.

are noncommutative indeterminates, then we have formal power series connections, whose connection forms are formal power series in $X_1, \cdots, X_m$ with coefficients being differential forms of positive degrees. The associated holonomy map of a formal power series connection is a formal power series whose coefficients are iterated integrals. We shall establish a sufficient condition for the associated holonomy map to induce a faithful representation of the loop space homology algebra.

In order to illustrate the usefulness of the method of formal power series connections, we apply it to simply connected compact 4-manifolds, Lie groups and spheres. In particular, Hopf invariants are obtained in a manner, which allows generalization. We shall also obtain sufficient conditions to make the loop space homology algebra free.

The material in §§ 1—4 covers connections for finite dimensional vector bundles. Theorem 4.1 relates the logarithmic differential of the holonomy map with the curvature of the connection. In the locally flat case, the holonomy gives rise to cohomology classes of the loop space.

In §§ 5—7, we treat product bundles whose fiber is a graded algebra. Theorem 7.1 makes the notion of a twisting cochain differentiably meaningful. The rest of this article deals with formal power series connections.

Homotopy classes of transport of usual connections have been known to be of importance in classification problems of fiber bundles. (See Stasheff **[9]**.) Brown **[3]** used a twisting cochain to construct what amounts to a topological version of a holonomy map on the homology level. The method is valid only for simply connected spaces and is somewhat indirect because of the use of the cobar construction. A more direct topological version of holonomy map through a twisting cochain was given in **[5]**. In a way, the notion of a twisting cochain generalizes that of a locally flat connection. The topological significance of locally flat connections has been extensively studied. (See, for example, Kamber and Tondeur **[8]**.)

It should be mentioned that Theorem 9.1 is the result of an effort to compare the method of formal power series connections with Sullivan's method of minimal models **[10]**.

Some of the basic notions used in this article have been introduced earlier in **[4]**, where they are described in greater detail.

The author wishes to take this opportunity to express his gratitude to Professor André Weil for his kind interest and encouragement ever since the early stage of the author's work on iterated integrals.

Throughout this paper, manifolds and vector bundles are differentiable. The word "differentiable" stands for C^∞. Most of our arguments are also valid for the real or complex analytic case. The field k of real (or complex) numbers will be the ground field for all vector spaces.

1. Let X be a manifold. (As a matter of fact, X can be a differentiable space as defined in **[4]**.) We shall call a map $\phi: U \to X$ a plot if U is a closed convex set of arbitrary finite dimension and if ϕ is a differentiable map. The differentiable structure of X is determined by the family of all plots of X.

Let E be a vector bundle over X, whose fiber is isomorphic with an n-dimensional vector space V, n being finite. Consider $U \times V$ as a product bundle. Any differentiable vector bundle map

$$\tilde{\phi}, U \times V \to E$$

over a plot $\phi : U \to X$ will be called a trivializing plot of E.

Any graded vector space will be understood to be graded by natural integers. For any element w of degree p, write

$$Jw = (-1)^p w.$$

Furthermore, we extend J to an involutary endomorphism of the graded vector space.

Recall that the graded vector space $\Lambda(X; E)$ of E-valued forms on X is a left module over the exterior algebra $\Lambda(X)$. According to Bott and Chern [2], a (linear) connection of E can be defined to be a graded derivation

$$D : \Lambda(X; E) \to \Lambda(X; E)$$

of degree 1, i.e. a graded map such that

$$D(w \wedge u) = dw \wedge u + Jw \wedge Du, \qquad \forall w \in \Lambda(X), u \in \Lambda(X; E).$$

Thus, given any trivializing plot $\tilde{\phi} : U \times V \to E$, there is an associated graded derivation

$$D_{\tilde{\phi}} : \Lambda(U) \otimes V \to \Lambda(U) \otimes V.$$

If $\tilde{\theta} : U' \times V \to U \times V$ is a differentiable vector bundle map over a differentiable map $\theta : U' \to U$, then there is an induced map

$$\tilde{\theta}^* : \Lambda(U) \otimes V \to \Lambda(U') \otimes V.$$

For the trivializing plots $\tilde{\phi}$ and $\tilde{\phi}\tilde{\theta}$, the compatibility condition

(1.1) $$D_{\tilde{\phi}\tilde{\theta}}\, \tilde{\theta}^* = \tilde{\theta}^*\, D_{\tilde{\phi}}$$

must be satisfied.

Conversely, the associated derivations of the family of all trivializing plots of E together with the compatibility condition define a connection of E. We now define a generalized notion of connections.

For any differential graded algebra $A = \{A^p\}_{p \geqq 0}$ with differential d of degree 1, regard $A \otimes V$ as a left A-module. By a derivation of $A \otimes V$, we shall mean a (not necessarily graded) map $f \colon A \otimes V \to A \otimes V$ such that

$$f(1 \otimes v) \in A^+ \otimes V \quad \text{and} \quad f(w \otimes v) = dw \otimes v + Jw \wedge f(1 \otimes v),$$

where $A^+ = \sum_{p>0} A^p$.

It should be mentioned that, in a different framework from ours, A. Frölicher and A. Nijenhuis have considered graded derivations of higher degree [7].

DEFINITION. A connection D of a vector bundle E over X assigns to each trivializing plot $\tilde{\phi} : U \times V \to E$ a derivation (not necessarily graded)

$$D_{\tilde{\phi}} : \Lambda(U) \otimes V \to \Lambda^+(U) \otimes V \subset \Lambda(U) \otimes V$$

such that the compatibility condition (1.1) holds.

If $\{e^1, \cdots, e^n\}$ is a basis for V, then

$$D_{\tilde{\phi}}(1 \otimes e^\lambda) = -\sum_\mu \omega_\mu^\lambda \otimes e^\mu,$$

where each ω_μ^λ is not necessarily a 1-form but can be an arbitrary element (either homogeneous or heterogeneous) of $\Lambda^+(X)$.

The associated curvature D^2 is then given by

$$D_{\tilde{\phi}}^2 (1 \otimes e^\lambda) = -\sum_\mu \kappa_\mu^\lambda \otimes e^\mu$$

where

(1.2)
$$\kappa_\mu^\lambda = d\omega_\mu^\lambda - \sum_\nu J\omega_\nu^\lambda \wedge \omega_\mu^\nu$$

Observe that, if D, D_0, D_1 are connections of a vector bundle E and if f is a polynomial, then $(f(D^2) - 1)D_0 + f(D^2)D_1$ is also a connection of E.

2. Every path $I \to X$ is understood to be continuous and piecewise differentiable. Denote by $P(X)$ the set of paths in X. For any map $\alpha : U \to P(X)$, let $\phi_\alpha : U \times I \to X$ be given by $(\xi, t) \mapsto \alpha(\xi)(t)$.

By a plot of $P(X)$, we mean a map $\alpha : U \to P(X)$, U being a closed convex set of finite dimension, such that, for some partition $0 = t_0 < \cdots < t_r = 1$ of I, each restricted map $\phi | U \times [t_{i-1}, t_i]$ is a plot of X. We use the family of all plots to define the differentiable structure of $P(X)$. (See [4].) For the sake of convenience, let $P(X)$ have the compact open topology.

Let α, $\beta : U \to P(X)$, $\alpha' : U' \to P(X)$ be plots. If $\alpha(\xi)(1) = \beta(\xi)(0)$, $\forall \xi \in U$, define the plot

(2.1)
$$\alpha\beta : U \to P(X)$$

such that $(\alpha\beta)(\xi)$ is the product path $\alpha(\xi) \cdot \beta(\xi)$, $\forall \xi \in U$. Define

(2.2)
$$\alpha^{-1} : U \to P(X)$$

such that $\alpha^{-1}(\xi)(t) = \alpha(\xi)(1 - t)$, $\forall \xi \in U$, $t \in I$.

If there exists $x \in X$ such that $\alpha(\xi)(1) = \alpha'(\xi')(0) = x$, $\forall \xi \in U$, $\xi' \in U'$, define the plot

(2.3)
$$\alpha \times \alpha' : U \times U' \to P(X)$$

such that $(\alpha \times \alpha')(\xi, \xi')$ is the product path $\alpha(\xi) \cdot \alpha'(\xi')$.

Denote by $P(X; x_0, x_1)$ the subspace of the path space $P(X)$ consisting of paths from x_0 to x_1. We say that α is a plot of $P(X; x_0, x_1)$ if the composition

$$U \xrightarrow{\alpha} P(X\,;x_0, x_1) \xrightarrow{\subset} P(X)$$

is a plot of $P(X)$.

3. Let $\mathrm{Hom}(E, E)$ be the vector bundle over $X \times X$ such that the fiber over (x_0, x_1) is the n^2-dimensional vector space $\mathrm{Hom}(E_{x_0}, E_{x_1})$ of all linear maps from E_{x_0} to E_{x_1}. Denote by $P(E)$ the vector bundle over $P(X)$ which is the pullback of $\mathrm{Hom}(E, E)$ via the map $P(X) \to X \times X$ given by $\gamma \mapsto (\gamma(0), \gamma(1))$.

Choose a base point $x_0 \in x$. Then $\Omega(X) = P(X\,;x_0, x_0)$ is the space of loops at x_0. A p-form w on $\Omega(X)$ is defined by associating to each plot $\alpha : U \to \Omega(X)$ a p-form $w_\alpha \in \Lambda(U)$ satisfying the compatibility condition $w_{\alpha\theta} = \theta^* w_\alpha$ for all differentiable maps $\theta : U' \to U$. Denote by $\Lambda^p(\Omega(X))$ the vector space of p-forms on $\Omega(X)$. Denote by $\Lambda(\Omega(X))$ the vector space of all infinite sums

$$w_0 + \cdots + w_p + \cdots, \qquad w_p \in \Lambda^p(\Omega(X)),$$

so that $\Lambda(\Omega(X))$ is the direct product of all $\Lambda^p(\Omega(X))$, $p \geqq 0$. The exterior multiplication and the exterior differentiation are defined in the obvious manner.

Let D be a given connection (in the generalized sense). We may construct the associated transport of D as an element of $\Lambda(P(X); P(E))$. For simplicity, we shall consider only the holonomy map T, which is the restriction of the transport to $\Omega(X)$. Now we proceed to construct T as an element of $\Lambda(\Omega(X)) \otimes \mathrm{End}\, E_{x_0}$.

Let $\alpha : U \to P(X)$ be a plot so that, for some partition $0 = t_0 < \cdots < t_r = 1$ of I, each restriction

$$\phi_i = \phi_\alpha \,|\, U \times [t_{i-1}, t_i]$$

is a plot of X. Let

$$\tilde{\phi} : U \times I \times V \to E \tag{3.1}$$

be a vector bundle map over ϕ_α such that each restriction

$$\tilde{\phi}_i = \tilde{\phi} \,|\, U \times [t_{i-1}, t_i] \times V$$

is a trivialization plot of E over the plot ϕ_i. Then $\tilde{\phi}$ gives rise to a trivializing plot

$$\tilde{\alpha} : U \times \mathrm{Hom}(V, V) \to P(E)$$

such that $\tilde{\alpha}(\xi, h) \in \mathrm{Hom}(E_{\alpha(\xi)(0)}, E_{\alpha(\xi)(1)})$ is the homomorphism given by

$$\tilde{\phi}(\xi, 0, v) \mapsto \tilde{\phi}(\xi, 1, hv) \qquad \forall\, v \in V.$$

Moreover, for every trivializing plot $\tilde{\alpha}$ of $P(E)$, we can construct a vector bundle map $\tilde{\phi}$ as given in (3.1).

Recall that $\{e^1, \cdots, e^n\}$ is a basis of V. Write

$$D_{\tilde{\phi}_i}(1 \otimes e^\lambda) = -\sum \omega_\mu^\lambda \otimes e^\mu.$$

Then each ω_μ^λ can be regarded as being piecewise defined on $U \times I$. We may write

$$\omega_\mu^\lambda = dt \wedge \omega'^\lambda_\mu + \omega''^\lambda_\mu$$

where both $\omega'^{\lambda}_{\mu} = \omega'^{\lambda}_{\mu}(t)$ and $\omega''^{\lambda}_{\mu} = \omega''^{\lambda}_{\mu}(t)$ are $\Lambda(U)$-valued functions of t. (By a $\Lambda(U)$-valued function of t, we mean a sum of terms of the type $a(\xi, t)d\xi^{i_1} \wedge \cdots \wedge d\xi^{i_r}$.)

Solve the system of linear differential equations

$$d\tau^{\lambda}_{\mu}/dt = \sum \tau^{\lambda}_{l} \wedge \omega'^{l}_{\mu}(t), \qquad \tau^{\lambda}_{\mu}(0) = \delta^{\lambda}_{\mu}, \lambda, \mu = 1, \cdots, n,$$

where the unknown functions τ^{λ}_{μ} are $\Lambda(U)$-valued. A unique solution $\tau = (\tau^{\lambda}_{\mu}(t))$ exists and gives rise to an element $\tau(1)$ of $\Lambda(U) \otimes \mathrm{Hom}(V, V)$. Verify that $\tau(1)$ depends only on the trivializing plot $\tilde{\alpha}$ and not on the particular choice of $\tilde{\phi}$.

4. The pullback of the vector bundle $P(E)$ via the inclusion $\Omega(X) \subset P(X)$ is the product bundle $\Omega(X) \times \mathrm{End}\, E_{x_0}$. For simplicity, identify V with E_{x_0} and $\mathrm{Hom}(V, V)$ with $\mathrm{End}\, E_{x_0}$.

For every plot $\alpha : U \to \Omega(X)$, there is a trivializing plot

$$\tilde{\alpha} = \alpha \times 1 : U \times \mathrm{Hom}(V, V) \to \Omega(X) \times \mathrm{End}\, E_{x_0}$$

and a vector bundle map $\tilde{\phi}$ of the type (3.1). Write $T_{\alpha} = \tau(1)$ so that T becomes an element of $\Lambda(\Omega(X)) \otimes \mathrm{End}\, E_{x_0}$.

Recall that $\kappa^{\lambda}_{\mu} = d\omega^{\lambda}_{\mu} - \sum J\omega^{\lambda}_{\nu} \wedge \omega^{\nu}_{\mu}$. Write

$$\kappa^{\lambda}_{\mu} = dt \wedge \kappa'^{\lambda}_{\mu} + \kappa''^{\lambda}_{\mu}$$

where both κ'^{λ}_{μ} and κ''^{λ}_{μ} are $\Lambda(U)$-valued functions of t. Thus $\kappa'(t) = (\kappa'^{\lambda}_{\mu})$ is a $\Lambda(U) \otimes \mathrm{Hom}(V, V)$-valued function of t. The proof of Theorem 2.3.2 of **[4]** is also valid for the next assertion.

THEOREM 4.1. *Let T be the transport of a connection of a vector bundle E over X. If $\alpha : U \to \Omega(X)$ is a plot, then*

$$(dT \wedge T^{-1})_{\alpha} = -\int_0^1 J\tau \wedge \kappa' \wedge \tau^{-1}\, dt.$$

COROLLARY 1. *If the curvature D^2 is zero, then $dT = 0$.*

In the next two corollaries, D will be a usual linear connection so that $T \in \Lambda^0(\Omega(X)) \otimes \mathrm{End}\, V$. Therefore, $\det T$ is an element of $\Lambda^0(\Omega(X))$ and $\mathrm{Tr}\, \kappa$ is a closed 2-form on X.

COROLLARY 2. *If $\alpha : I \to \Omega(X)$ is a plot, then*

$$\det T_{\alpha(1)}/\det T_{\alpha(0)} = \exp - \int_{\phi_{\alpha}} \mathrm{Tr}\, \kappa,$$

where $\phi_{\alpha} : I \times I \to X$ is taken as a singular 2-cube.

The last corollary is implicitly known, though it does not seem to be given in the literature.

5. Let $A = \{A_s\}_{s \geq 0}$ be a graded algebra. By an element of A, we mean a formal infinite sum $a = \sum_{s \geq 0} a_s$. In other words, we do not make distinction between the graded algebra A and the product ΠA_s. If a and b are elements of A, then the product $c = ab$ is given by $c_s = \sum_{i+j=s} a_i b_j$.

An example of such a graded algebra is the formal power series algebra $k[[X]] = k[[X_1,\cdots, X_m]]$, where $X_1,\cdots, X_m$ are noncommutative indeterminates. An element of $k[[X]]$ is of the type

$$a_0 + \sum a_i X_i + \sum a_{ij} X_i X_j + \cdots, \qquad a\text{'s} \in k.$$

For simplicity, assume that each vector space A_s is of finite dimension. Define $\Lambda(X; A)$ to be the vector space of all formal sums $u = \sum_{s\geqq 0} u_s$, where $u_s \in \Lambda(X) \otimes A_s$. Then $\Lambda(X;A)$ is an algebra, whose multiplication is again denoted by $\wedge$. Moreover, $\Lambda(X;A)$ is a left $\Lambda(X)$-module. The left module action will be also denoted by $\wedge$.

Define $Ju = \sum Ju_s$, where Ju_s is the image of u_s under the map

$$\Lambda(X) \otimes A_s \xrightarrow{J \otimes 1} \Lambda(X) \otimes A_s.$$

Similarly define $du = \sum du_s$.

DEFINITION. An A-valued connection $D : \Lambda(X; A) \to \Lambda(X; A)$ is a map over k satisfying the following conditions:

(a) There exists an element $\omega = \sum \omega_s \in \Lambda(X; A)$, $\omega_s \in \Lambda^+(X) \otimes A_s$, such that the composite map

$$\Lambda(X) \otimes A_s \subset \Lambda(X; A) \xrightarrow{D} \Lambda(X; A)$$

is given by $w \otimes a \mapsto dw \otimes a - Jw \wedge (\omega \wedge (1 \otimes \mathrm{a}))$.

(b) For any $u = \sum u_s \in \Lambda(X; A)$,

$$Du = Du_0 + Du_1 + \cdots + Du_s + \cdots.$$

REMARK. The right-hand side of the above equation makes sense because each Du_s is of the type $\sum_{r\geqq s} v_r$.

We call ω the connection form of D and $\kappa = d\omega - J\omega \wedge \omega$ the curvature form of D.

6. The associated transport of the A-valued connection D can be constructed by using a graded modification of the procedure described in § 3 as follows:

Let $\alpha : U \to P(X)$ be a plot, and ω_{ϕ_α} be obtained by pasting together $\phi_i^* \omega \in \Lambda(U \times [t_{i-1}, t_i]; A)$, $i = 1,\cdots, r$. Write

$$\omega_{\phi_\alpha} = dt \wedge \omega' + \omega''.$$

Solve the linear differential equation

$$d\tau/dt = \tau \wedge \omega', \qquad \tau(1) = 1,$$

where $\tau = \sum_{s\geqq 0} \tau_s(t)$, each $\tau_s(t)$ being a $\Lambda(U) \otimes A_s$-valued function of t. This differential equation is equivalent to the system of differential equations

$$d\tau_0/dt = \tau_0 \wedge \omega_0', \qquad \tau_0(0) = 1,$$

and, for $s > 0$,

$$d\tau_s/dt = \sum \tau_i \wedge \omega'_{s-i}, \qquad \tau_s(0) = 0.$$

Therefore there exists a unique solution. We define $T_\alpha = \tau(1) \in \Lambda(U; A)$ so that the transport T is an element of $\Lambda(P(X); A)$. Theorem 4.1 is also valid for the A-valued connection D.

Let $\alpha\beta$, α^{-1}, $\alpha \times \alpha'$ be plots given by (2.1), (2.2) and (2.3) respectively. Then there are the formulas

(6.1) $$T_{\alpha\beta} = T_\alpha \wedge T_\beta;$$

(6.2) $$T_{\alpha^{-1}} = T_\alpha^{-1};$$

(6.3) $$T_{\alpha\times\beta} = T_\alpha \times T_\beta.$$

(See similar formulas in § 2, **[4]**.)

7. We shall now assume that $A = \{A_s\}$ is double graded so that $A_s = \{A_{sq}\}_{q\geqq 0}$ is itself a graded vector space. Moreover $A_{sq}\cdot A_{s'q'} \subset A_{s+s'\ q+q'}$.

REMARK. At this point we may remove the condition that A_s is of finite dimension but demand that each A_{sq} is of finite dimension.

Each element of A can be written as $a = \sum a_s$, where each a_s is itself a formal infinite sum $a_s = \sum a_{sq}$.

For example, if, for the formal power series algebra $k[[X]]$, we assign to each noncommutative indeterminate X_i an integer $\deg X_i \geqq 0$, then $A = k[[X]]$ becomes a double graded algebra such that A_{sq} has as a basis the set of all monomials $X_{i_1}\cdots X_{i_s}$, $\deg X_{i_1} + \cdots + \deg X_{i_s} = q$.

We shall further assume that A is equipped with a differential ∂ satisfying the following conditions:

(a) If $b \in A_{sq}$, then $\partial b = \sum_{i\geqq s} b_i$, where $b_i \in A_{i\ q-1}$.
(b) For any $a = \sum_s \sum_q a_{sq}$, $\partial a = \sum\sum \partial a_{sq}$.
(c) The product rule holds, i.e. $\partial(a_{sq}a_{s'q'}) = \partial a_{sq}a_{s'q'} + (-1)^q a_{sq}\,\partial a_{s'q'}$.

The algebra $\Lambda(X; A)$ consists of elements of the type

$$w = \sum\sum w_{sq}, \qquad w_{sq} \in \Lambda(X) \otimes A_{sq}.$$

The differential ∂ can be extended to a differential $\partial : \Lambda(X; A) \to \Lambda(X; A)$.

The connection form of an A-valued connection D now takes the appearance of $\omega = \sum\sum\omega_{sq}$. We shall make an additional requirement that

(7.1) $$\omega_{sq} \in \Lambda^{q+1}(X) \otimes A_{sq}.$$

Then the curvature form $\kappa = \sum\sum\kappa_{sq}$ is such that $\kappa_{sq} \in \Lambda^{q+2}(X) \otimes A_{sq}$.

Verify that the transport $T = \sum\sum T_{sq} \in \Lambda(P(X); A)$ is such that $T_{sq} \in \Lambda^q(P(X)) \otimes A_{sq}$.

If $\alpha : U \to P(X)$ is a plot, let $\langle T, \alpha\rangle_{sq} \in A_{sq}$ be obtained by integrating the A_{sq}-valued q-form T_{sq} over α. Here we assume that U is compact. If $\dim U \neq q$, it is understood that $\langle T, \alpha \rangle = 0$. Thus $\langle T, \alpha \rangle = \sum\sum \langle T, \alpha \rangle_{sq}$ is an element of A.

Let $C_*(\Omega(X))$ be the normalized differentiable cubical chain complex of the loop space $\Omega(X)$. Then there is a map

$$T : C_*(\Omega(X)) \to A \tag{7.2}$$

given by $\alpha \to \langle T, \alpha \rangle$ for every differentiable cube $\alpha : I^q \to \Omega(X)$. Moreover

$$\langle T, \alpha \times \alpha' \rangle = \langle T, \alpha \rangle \langle T, \alpha' \rangle.$$

The map (7.2) will be called the holonomy map of the A-valued connection D.

THEOREM 7.1. *Let A be a double graded algebra equipped with a differential ∂ as described above. Let D be an A-valued connection on X such that*

$$\partial\omega + \kappa = 0 \tag{7.3}$$

where ω and κ are respectively the connection and curvature forms of D. Then the holonomy map as given in (7.2) *is a chain map.*

REMARK. The condition (7.3) is equivalent to

$$\partial\omega + d\omega = J\omega \wedge \omega.$$

Such a connection form ω can be called a twisting cochain. (See [3].)

PROOF. Set $u(t) = -\partial\tau(t)$, where $\tau(t)$ is given in § 6. Then

$$du/dt = u \wedge \omega' + J\tau \wedge \kappa', \qquad u(0) = 0.$$

The above nonhomogeneous linear differential equation has a unique solution so that

$$u(s) = \left(\int_0^s J\tau \wedge \kappa' \wedge \tau^{-1}\, dt\right) \wedge \tau(s).$$

Since Theorem 4.1 is also valid for A-valued connections,

$$\partial T_\alpha = \partial\tau(1) = -\int_0^1 J\tau \wedge \kappa' \wedge \tau^{-1} dt \wedge T_\alpha = (dT)_\alpha.$$

If $\alpha : I^q \to \Omega(X)$ is a differentiable cube, then

$$\partial\langle T, \alpha \rangle = \langle \partial T, \alpha \rangle = \langle dT, \alpha \rangle = \langle T, \partial\alpha \rangle.$$

Hence the theorem is proved.

REMARK. Throughout the proof, we have not used the condition that $\partial\partial = 0$. If the image of the holonomy map T given in (7.2) is "dense" in A, then the condition that $\partial\partial = 0$ is a consequence of the theorem.

8. Let us make the formal power series algebra $k[[X]]$ double graded by assigning to each indeterminate X_i a nonnegative degree $\deg X_i$. We may remove the condition that the set of indeterminates $\{X_1, \cdots, X_m\}$ is finite but require that, for each $q \geqq 0$, there are only a finite number of indeterminates X_i with $\deg X_i = q$.

A $k[[X]]$-valued connection D is called a formal power series connection. Its connection form appears as

$$\omega = \sum w_i X_i + \cdots + \sum w_{i_1 \cdots i_r} X_{i_1} \cdots X_{i_r} + \cdots$$

such that, if $q = \deg X_{i_1} + \cdots + \deg X_{i_r}$, then $w_{i_1 \cdots i_r} \in \Lambda^{q+1}(X)$.

The associated transport can be written as a formal power series of the type

$$T = 1 + \sum \int w_i X_i + \sum \int (w_i w_j + w_{ij}) X_i X_j$$
$$+ \sum \int (w_i w_j w_k + w_{ij} w_k + w_i w_{jk} + w_{ijk}) X_i X_j X_k + \cdots$$

where the coefficients are iterated integrals. (See § 2, **[4]**.) The holonomy map $T\colon C_*(\Omega(X)) \to k[[X]]$ is given by

$$\alpha \mapsto \langle 1, \alpha \rangle + \sum \left\langle \int w_i, \alpha \right\rangle X_i + \cdots$$

where $\langle 1, \alpha \rangle = \varepsilon\alpha$, and $\varepsilon : C_*(\Omega(X)) \to Z \subset k$ is the augmentation of $C_*(\Omega(X))$.

REMARK. If $\deg X_i > 0$ for every indeterminate X_i, then $\langle T, \alpha \rangle$ is a finite sum and belongs to the free algebra $k[X]$, which is a subalgebra of $k[[X]]$. Therefore the holonomy map can be taken to be of the type

$$T : C_*(\Omega(X)) \to k[X].$$

Denote by $\mathfrak{J}$ the ideal of $k[[X]]$ consisting of all elements of the type $a = a_0 + \sum a_i X_i + \sum a_{ij} X_i X_j + \cdots$ with $a_0 = 0$. We say that a map T has a dense image in $k[[X]]$ if, for any $s \geqq 0$, the composite map

$$C_*(\Omega(X)) \xrightarrow{T} k[[X]] \to k[[X]]/\mathfrak{J}^{s+1}$$

is surjective.

PROPOSITION 8.1. *If* $w_1, \cdots, w_m$ *are linearly independent modulo* $d\Lambda^0(x)$ *(i.e. no nontrivial linear combination of* $w_1, \cdots, w_m$ *is an exact* 1*-form), then the holonomy map of the above-mentioned formal power series connection has a dense image in* $k[[X]]$.

PROOF. Let $q_i = \deg X_i$. Construct plots $\alpha_i : \mathfrak{J}^{q_i} \to \Omega(X)$ such that the $m \times m$ matrix $(\langle \int w_i, \alpha_j \rangle) = (\int_{\phi_{\alpha_j}} w_i)$ is nonsingular. Let $\chi : k[[X]] \to k[[X]]$ be the algebra endomorphism determined by $X_j \to \langle T, \alpha_j \rangle - \langle 1, \alpha_j \rangle$. Since the matrix $(\langle \int w_i, \alpha_j \rangle)$ is nonsingular, it can be shown that χ is an automorphism. Moreover $\chi\mathfrak{J} = \mathfrak{J}$.

The composite homomorphism

$$C_*(\Omega(X)) \xrightarrow{T} k[[X]] \xrightarrow{\chi^{-1}} k[[X]]$$

is such that $\alpha_j \to \langle 1, \alpha_j \rangle + X_j$ and has a dense image in $k[[X]]$. Hence, so does T.

9. THEOREM 9.1. *Let the connection form*

$$\omega = \sum w_i X_i + \sum w_{ij} X_i X_j + \cdots$$

of a formal power series connection D satisfies the following conditions:

(a) $1, w_1, \cdots, w_m$ *form a basis of a cochain subcomplex A of* $\Lambda(X)$, *whose cohomology is isomorphic to the de Rham cohomology of X via the inclusion.*

(b) *Each* ∂X_i *is a finite sum.*

(c) *X is simply connected, and* $\deg w_i > 1$, $i = 1, \cdots, m$.

If $k[[X]]$ *has a differential* ∂ *as described in* § 7 *such that* $\partial\omega + \kappa = 0$, *then the holonomy map induces an isomorphism*

$$H_*(\Omega(X)) \otimes k \approx H(k[X]).$$

PROOF. Write $dw_i = \sum c_{ij}w_j$, $c_{ij} \in k$. The condition $\partial\omega + \kappa = 0$ implies that $\partial x_j + \sum c_{ij}X_i \in \mathfrak{J}^2$. Let C' be the dual chain complex of the cochain complex A. Let $\{1, x_1, \cdots, x_m\}$ be the dual basis. Then the boundary operator of C' is given by $1 \mapsto 0$ and $x_j \mapsto -\sum c_{ij}x_i$. Denote by $H_{*-1}(C')$ the graded vector space such that

$$(H_{*-1}(C'))_q = H_{q+1}(C') \text{ or } 0$$

according as $q \geqq 0$ or $q < 0$.

Let I be the ideal of the free algebra $k[X]$ generated by the elements $X_1, \cdots, X_m$. Then the descending filtration $k[X] \supset I \supset I^2 \supset \cdots$ of the chain complex $k[X]$ gives rise to a spectral sequence whose E^1 terms are

$$E^1_s \approx \bigotimes^s H_{*-1}(C') \approx \bigotimes^s H_{*-1}(X\,; k).$$

As in § 4.4, [**4**], let $\hat{C}_*(X)$ be the chain complex of those differentiable simplices of X that map the 1-skeleton of the standard simplex to the base point x_0. Moreover the group of degenerate 1-chains is factored out so that $\hat{C}_1(X) = 0$. For every n-simplex σ in $C_*(X)$ there is an $(n-1)$-cube $\bar{\sigma}$ in $C_*(\Omega(X))$ as constructed in 4.6, [**4**].

Let F be the cobar construction on $\hat{C}_*(X)$. Let the map $\Theta : F \to C_*(\Omega(X))$ be given by

$$[\sigma_1|\cdots|\sigma_s] \mapsto (\cdots((\bar{\sigma}_1 \times \bar{\sigma}_2) \times \bar{\sigma}_3)\cdots) \times \bar{\sigma}_s.$$

Though Θ is not a chain map, the composite map

$$F \xrightarrow{\;\Theta\;} C_*(\Omega(X)) \xrightarrow{\;T\;} k[[X]]$$

is a chain map whose image lies in $k[X]$. (See § 4.6, [**4**].) Let $T' : F \otimes k \to k[X]$ be induced by $T \circ \Theta$. By comparing the E^1 terms of the spectral sequences of $F \otimes k$ and $k[X]$, we conclude that T' induces an isomorphism $H(F) \otimes k \approx H(k[X])$. The proof is completed by observing that, since X is simply connected, $H(\Omega(X))$ and $H(F)$ are naturally isomorphic.

REMARK. We just mention that, if conditions (b) and (c) of the theorem are removed, then there are isomorphisms

$$H_0(k[[X]]/\mathfrak{J}^{s+1}) \approx k\pi_1(X)/\mathfrak{J}'^{s+1}, \qquad s \geqq 0,$$

where $k\pi_1(X)$ is the group algebra of the fundamental group, and $\mathfrak{J}$ is the augmentation ideal, which is generated by all $g - 1$, $g \in \pi_1(X)$. Such isomorphisms can be obtained through methods in the work of [**6**].

The next theorem is a corollary of Theorem 9.1.

THEOREM 9.2. *Let a formal power series connection on a simply connected manifold X be given by*

$$\omega = \sum w_i X_i + \sum w_{ij} X_i X_j + \cdots$$

such that $1, w_1, \cdots, w_m$ *are closed forms representing a basis for the de Rham cohomology. If* $\kappa = 0$, *then the loop space homology algebra* $H_*(\Omega X) \otimes k$ *is a free algebra isomorphic with* $k[X]$.

PROOF. Let $k[X]$ be equipped with the trivial differential. Since $\kappa = 0$, the equation $\partial\omega + \kappa = 0$ holds. Hence we may apply Theorem 9.1.

COROLLARY. *Let X be a simply connected manifold. Let*

$$\mathfrak{Q} = \{q > 0 : H_q(X; k) \neq 0\}.$$

If the inequality $(q_1 - 1) + \cdots + (q_r - 1) \neq q - 2$ *holds for any* $q_1, \cdots, q_r, q \in \mathfrak{Q}$, $r \geqq 1$, *then* $H_*(\Omega X) \otimes k$ *is a free algebra.*

10. In order to apply Theorem 9.1, we start with forms $w_1, \cdots, w_m$ satisfying condition (a) of the theorem. We need to construct a formal power series connection ω and a differential ∂ of $k[[X]]$ such that $\omega = \sum w_i X_i + \cdots$ and $\partial\omega + \kappa = 0$.

EXAMPLE 1. For the n-sphere S^n, $n > 1$, it is known that $H_*(\Omega S^n)$ is a free algebra with a single generator of degree $n - 1$. Let $\gamma_n : I^{n-1} \to \Omega(S^n)$ be a differentiable $(n - 1)$-cube representing the generator. Then its suspension $S\gamma_n : I^n \to S^n$ represents a generator of $H_n(S^n)$.

Let w_n denote a closed n-form on S^n such that $\int_{S^n} w_n = 1$. Let X_n be an indeterminate of degree $n - 1$. Set $\partial X_n = 0$. Then, for the formal power series connection $\omega = w_n X_n$, we have $\kappa = 0$ and the condition $\partial\omega + \kappa = 0$ holds. Therefore the holonomy map

$$T: C_*(\Omega(S^n)) \to k[X_n]$$

is a chain map given by

$$\alpha \mapsto \langle 1, \alpha\rangle + \left\langle \int w_n, \alpha \right\rangle X_n + \left\langle \int w_n w_n, \alpha \right\rangle X_n^2 + \cdots.$$

In particular $T\gamma_n = X_n$. Thus T induces

$$H_*(\Omega S^n) \xrightarrow{\approx} Z[X_n] \subset k[X_n].$$

Let $f: S^{2n-1} \to S^n$ be a differentiable map, where n is even. There is a commutative diagram

$$\begin{array}{ccc} H_*(\Omega S^{2n-1}) & \xrightarrow{(\Omega f)_*} & H_*(\Omega S^n) \\ \downarrow \approx & & \downarrow \approx \\ Z[X_{2n-1}] & \xrightarrow{\bar{f}} & Z[X_n], \end{array}$$

wherc the homomorphism $\bar{f}$ is given by

$$\begin{aligned} \bar{f}(X_{2n-1}) &= T((\Omega f)_*\gamma_{2n-1}) \\ &= \langle 1, (\Omega f)_*\gamma_{2n-1}\rangle + \left\langle \int w_n, (\Omega f)_*\gamma_{2n-1}\right\rangle X_n \\ &\quad + \left\langle \int w_n w_n, (\Omega f)_*\gamma_{2n-1}\right\rangle X_n^2 + \cdots \\ &= \left\langle \int w_n w_n, (\Omega f)_*\gamma_{2n-1}\right\rangle X_n^2. \end{aligned}$$

Thus $\langle \int f^*w_n f^*w_n, \gamma_{2n-1}\rangle$ must be an integer, which turns out to be the Hopf invariant. The twice iterated integral $\int f^*w_n f^*w_n$ is essentially the one constructed by J. H. C. Whitehead **[11]**.

EXAMPLE 2. Let X be a simply connected manifold. Let b_q denote the qth Betti number. Suppose that $b_r = m > 0$, $b_{2r} = 1$ for some $r > 1$, and that all other Betti numbers in positive dimensions vanish. For simplicity, assume that $\dim X < 3r - 1$.

Let $w_1, \cdots, w_m$ be closed r-forms, whose cohomology classes $[w_1], \cdots, [w_m]$ form a basis for $H^r(X; k)$. Let w be a closed $2r$-form whose cohomology class $[w]$ does not vanish. Let the cup product be given by

$$[Jw_i] \wedge [w_j] = -c_{ij}[w], \qquad c_{ij} \in k.$$

Choose $(2r-1)$-forms w_{ij} such that $dw_{ij} - Jw_i \wedge w_j = c_{ij}w$. Let $X_1, \cdots, X_m, X'$ be noncommutative indeterminates with $\deg X_i = r - 1$ and $\deg X' = 2r - 1$. Define the differential ∂ for $k[[X_1, \cdots, X_m, X']]$ such that $\partial X_i = 0$ and $\partial X' = \sum c_{ij}X_iX_j$. Then, for the formal power series connection

$$\omega = \sum w_iX_i + wX' + \sum w_{ij}X_iX_j,$$

we have $\partial\omega + \kappa = 0$. Therefore, according to Theorem 9.1, the homology of the chain complex $k[X]$ is isomorphic with $H_*(\Omega X) \otimes k$.

Observe that the chain complex $k[X]$, up to an isomorphism, is completely determined by the cohomology algebra $H^*(X; k)$. In the particular case of X being a simply connected compact 4-manifold, $H_*(X; k)$ is consequently isomorphic with the homology of a differential graded algebra $k[X_1, \cdots, X_m, X']$ whose differential ∂ is given by $\partial X_i = 0$ and $\partial X' = X_1^2 + \cdots + X_l^2 - X_{l+1}^2 - \cdots - X_m^2$.

EXAMPLE 3. Let the de Rham complex of a manifold X have a finite dimensional subcomplex A, which is closed under the exterior multiplication. Assume that $A^0 = k$. (If X is a compact Lie group, then the exterior algebra of left invariant forms is such a subcomplex of the de Rham complex. More generally, the Hodge complex of a compact symmetric Riemannian manifold is closed under exterior multiplication.)

Choose for the subcomplex A a basis $\{1, w_1, \cdots, w_m\}$. For any formal power series connection of the type

$$\omega = \sum w_i X_i + \sum w_{ij} X_i X_j + \sum w_{ijk} X_i X_j X_k + \cdots$$

such that $w_{ij}, w_{ijk}, \cdots$ are elements of A, we are going to construct a differential ∂ for $k[[X]]$ such that $\partial\omega + \kappa = 0$.

Write $\omega = \sum w_\lambda \, \omega(\lambda)$, where $\omega(\lambda)$, $\lambda = 1, \cdots, m$, are elements of $k[[X]]$. Observe that

$$\omega(\lambda) \equiv X_\lambda \bmod \mathfrak{J}^2$$

and

$$\kappa = \sum w_\lambda \left(\sum c_{i\lambda}\omega(i) + \sum c_{jk\lambda}\omega(j)\omega(k)\right), \qquad c\text{'s} \in k.$$

Thus the condition $\partial\omega + \kappa = 0$ demands that

$$\partial\omega(\lambda) + \sum c_{i\lambda}\omega(i) + \sum c_{jk\lambda}\omega(j)\omega(k) = 0.$$

From these equations, we may recursively determine the coefficients of the formal power series ∂X_λ, $\lambda = 1, \cdots, m$. If we take the special case $\omega = \sum w_i X_i$, then ∂X_λ is a finite sum.

Finally we observe that the formal power series connection $\omega = \sum w_i X_i$ provides a canonical way of computing the loop space homology of a simply connected compact Lie group.

BIBLIOGRAPHY

1. A. Asada, *Connection of flat vector bundles*, J. Fac. Sci. Shinshu Univ. **2** (1967), 109–116. MR **38** #2799.

2. R. Bott and S. S. Chern, *Hermitian vector bundles and the equidistribution of the zeroes of their holomorphic sections*, Acta Math. **114** (1965), 71–112. MR **32** #3070.

3. E. H. Brown, Jr., *Twisted tensor products*. I, Ann. of Math. (2) **69** (1959), 223–246. MR **21** #4423.

4. K.-T. Chen, *Iterated integrals of differential forms and loop space homology*, Ann. of Math. (2) **97** (1973), 217–246.

5. ———, *Free subalgebras of loop space homology and Massey products*, Topology **11** (1972), 237–243. MR **45** #4411.

6. ———, *Iterated integrals, fundamental groups and covering spaces*, Trans. Amer. Math. Soc. **205** (1975) (to appear).

7. A. Frölicher and A. Nijenhuis, *Theory of vector-valued differential forms*. I. *Derivations in the graded ring of differential forms*, Nederl. Akad. Wetensch. Proc. Ser. A **59** = Indag. Math. **18** (1956), 338–359. MR **18**, 569.

8. F. W. Kamber and Ph. Tondeur, *On flat bundles*, Bull. Amer. Math. Soc. **72** (1966), 846–849. MR **33** #6631.

9. J. D. Stasheff, "*Parallel*" *transport in fibre spaces*, Bol. Soc. Mat. Mexicana (2) **11** (1966), 68–84. MR **38** #5219.

10. D. Sullivan, *Differential forms and topology of manifolds* (forthcoming).

11. J. H. C. Whitehead, *An expression of Hopf's invariant as an integral*, Proc. Nat. Acad. Sci. U.S.A. **33** (1947), 117–123. MR **8**, 525.

UNIVERSITY OF ILLINOIS, URBANA

Proceedings of Symposia in Pure Mathematics
Volume 27, 1975

SPIN FIBRATIONS OVER MANIFOLDS AND GENERALISED TWISTORS

A. CRUMEYROLLE

Abstract. In §1 my purpose is to find existence conditions for a spin structure on a manifold in the purely differential geometry scope, without algebraic topology machinery and avoiding systematically any matricial formalism.

A spin structure can be considered as a G-structure (using a classical terminology). Groups called by myself "groups of spinoriality" play an essential part in this problem. We study these in §2 with a few details.

§3 is devoted to introduce the notion of twistors, generalising studies already made in particular cases. Twistorial fibrations naturally go with spin fibrations.

Broadly, twistors of order 2 are for the conformal group, in signature (p,q) which are spinors for the isometric group in signature $(p + 1, q + 1)$.

I. Pin Q-Spin Structures on Riemannian or Pseudo-Riemannian Manifolds

1. Recall of classical definitions. R^{2r} is endowed with a quadratic form of arbitrary signature. Classically the group Pin Q constitutes a 2-fold covering of the group $O(Q)$. If $\gamma \in$ Pin Q, we define $p(\gamma)(x) = \gamma x \gamma^{-1}$, $x \in R^{2r}$, $p(\gamma) \in O(Q)$. We invite the uninformed reader to consult **[2]** and **[3a]** about basic topics in Clifford algebras and spinors.

V is a real paracompact n-dimensional smooth pseudo-riemannian (in particular riemannian) manifold, its fundamental tensor field is called, abusively, Q. $\xi(E, V, O(Q), \pi)$, or simply ξ, is the bundle of orthonormal frames.

Definition 1. *V admits a strict, meaning* Pin *Q-spin, structure if there exists a principal fibre bundle $\eta(P, V,$ Pin $Q, q)$ and a principal morphism h from η onto ξ*

AMS (MOS) subject classifications (1970). Primary 53C50.

such that the following diagram, where R_γ and $R_{p(\gamma)}$ denote right translations, is commutative.

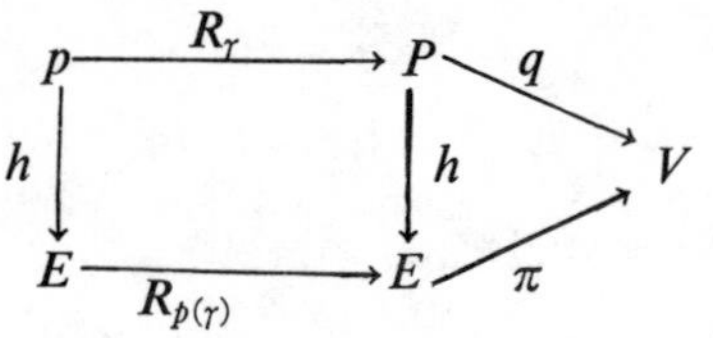

It is always possible to define both fibrations η and ξ by means of the same trivialising neighbourhoods $(U_\alpha)_{\alpha\in A}$, and local cross-sections z_α, Ω_α, with transition functions $\gamma_{\alpha\beta}$ and $p(\gamma_{\alpha\beta})$ respectively

$$z_\beta(x) = z_\alpha(x)\cdot\gamma_{\alpha\beta}(x), \qquad \gamma_{\alpha\beta}(x) \in \text{Pin } Q.$$

$$h(z_\beta(x)) = \Omega_\beta(x) = h(z_\alpha(x))\cdot p(\gamma_{\alpha\beta}(x)) = \Omega_\alpha(x)\cdot p(\gamma_{\alpha\beta}(x)).$$

Consider the standard Clifford algebra $C(Q)$ of R^n, and the complexified algebra $C(Q')$, Q' being the complexification of Q. Let $(x_1,\cdots, x_r, y_1,\cdots, y_r)$ ($\{x_i, y_j\}$ for brevity) be a "real" Witt's frame of C^n, i.e. naturally associated with an orthonormal basis of R^n, according to the construction given in [**3a**, p. 34].

The sequence

$$\{x_{i_1}\, x_{i_2} \cdots x_{i_h}\, y_{j_1}\, y_{j_2} \cdots y_{j_k}\}, \qquad \begin{matrix} 1 \leqq i_1 < i_2 < \cdots \leqq r, \\ 1 \leqq j_1 < j_2 < \cdots \leqq r, \end{matrix}$$

($\{x_{i_1} \cdots y_{j_1} \cdots\}$ for brevity) is a basis of $C(Q')$.

The choice of the above basis realises a linear isomorphism φ between $C(Q')$ and C^{2^n}.

We associate with the principal bundle η a spinor bundle $\zeta(S, V, \text{Pin } Q, \hat{q})$, of which the typical fibre is C^{2^r}, the group Pin Q operates effectively in C^{2^r} (C^{2^r} is an irreducible $C(Q')$ representation space). It is permissible to choose any irreducible representation of $C(Q')$ in C^{2^r}, and convenient to choose the representation corresponding to the left action of $C(Q')$ in the minimal ideal $C(Q')f$, $f = y_1y_2\cdots y_r$. $C(Q')f$, of which the $x_{i_1}\, x_{i_2} \cdots x_{i_k}\, f$ (briefly, $x_{(i)}\, f$) constitute a standard basis, will be called typical spinor space.

By restriction of φ to $C(Q')\, f$, we obtain a linear identification of this space with C^{2^r}.

Over an open set U of V, endowed with the cross-section $z : x \to z(x)$ of η, a spinor field ψ will be defined by a smooth application ψ from P into C^{2^r}, $z \to \psi(z)$, such that [**6**]: $\psi(z) = \varphi(u)$,

$$\psi(z\cdot\gamma^{-1}) = \gamma\cdot\psi(z) = \varphi(\gamma u), \qquad \forall\gamma \in \text{Pin } Q. \tag{1}$$

We can denote by ψ_x the restriction of ψ to $S_x = \hat{q}^{-1}(x)$ and observe that

$$(\gamma\cdot\psi_x(z))^{i_1\cdots i_k}\, x_{i_1} \cdots x_{i_k}\, f = (\psi_x(z))^{i_1\cdots i_k}(\gamma x_{i_1} \cdots x_{i_k}\, f). \tag{2}$$

2. Necessary existence conditions for a strict meaning Pin Q**-spin structure.** Let $x \to z_x$ be a local cross-section over U, a trivialising open set in the bundle

$\eta \cdot z_x = \nu(x, g(x)) = \nu^x(g(x))$, $g(x) \in \text{Pin } Q$; according to the associated bundles construction $(z_x, x_{(i)} f)$, identified by $(z_x\gamma^{-1}, \gamma^{-1}, \gamma x_{(i)} f)$, is a cross-section over U in the bundle ζ, denoted $[z_x, x_{(i)} f]$ or $\mu^x(x_{(i)} f)$. Take also $\Omega_x = h(z_x)$.

If (U_α) is a trivialising atlas in the pseudo-riemannian basis bundle, it is always possible to suppose that there exists over U_α a cross-section z_α in η ; we take again $\Omega_\alpha(x) = h(z_\alpha(x))$, z_α, Ω_α, admitting respectively $\gamma_{\alpha\beta}$ and $p(\gamma_{\alpha\beta})$ for transition functions, according to § 1 above.

It is therefore possible to write ($W'_\alpha(x)$ being the "real" Witt's frame naturally associated with $\Omega_\alpha(x)$ and with a little abuse) $h(z_\alpha(x)) = W'_\alpha(x) = \theta^x_\alpha\{(x_i, y_j)\}$, where the θ^x_α admit $p(\gamma_{\alpha\beta})$ ($p(\gamma_{\alpha\beta}(x)) \in O(Q)$) for transition functions.

The Ω_α are also local cross-sections-frames in ξ_C, and allow us to build local basis cross-sections in the Clifford bundles $\text{Cli}(V \cdot Q)$, $\text{Cli}(V \cdot Q')$ with fibres over x called $C(Q)_x$ and $C(Q')_x$ respectively.

The $\gamma_{\alpha\beta}$ are transition functions for cross-sections in ζ. According to the definition for Clifford bundles, θ^x_α defines an isomorphism from $C(Q)$ onto $C(Q)_x$.

Recall it is possible in pure algebra to define a Pin Q-spin trivial structure, by an equivalence class of pairs $(\mathscr{R}, g)$, where $\mathscr{R}$ is a real orthonormal frame, g an element of Pin Q, being specified that $(\mathscr{R}, g) \sim (\mathscr{R}', g')$ means $\mathscr{R}' = \sigma(\mathscr{R})$, $p(\gamma) = \sigma$, with $g' = \gamma g$, $g, g', \gamma \in \text{Pin } Q$ (cf. [**4**, pp. 129, 132]). We already used this point of sight in [**3a**]. Select a pair $(\mathscr{R}, g)$ in an equivalence class which fixes a Pin Q-spinorial frame. To various possible choices there correspond isomorphic spin spaces [**4**]. If we are working with Witt's frames, we can define a trivial Pin Q-spin structure by means of an equivalence class of pairs, $(W, g) \sim (W', g')$ meaning $W' = \sigma(W)$, with $p(\gamma) = \sigma$ and $g' = \gamma g$, $\gamma \in \text{Pin } Q$, but with g, g', belonging to Pin Q'. In the above class, there will always be "real" Witt's frames, because O (Q') operates transitively in the "real" or complex Witt's set of frames.

We can therefore define a trivial Pin Q-spin structure from a not "real origin" Witt's frame as well.

If there exists over V a Pin Q-spin structure, this structure induces in the tangent space at x a Pin Q-spin structure (in the purely algebraic meaning) defined by (W_x, g_x), W_x Witt's frame, $g_x \in \text{Pin } Q'$, depending differentially on x.

With anterior notations, at $x \in U_\alpha \cap U_\beta$ we must obtain two origin frames

$$(\theta^x_\alpha \{\lambda_\alpha(x)(x_i, y_i)\lambda_\alpha^{-1}(x)\} = W^x_\alpha, g_\alpha(x)),$$
$$(\theta^x_\beta \{\lambda_\beta(x)(x_i, y_j)\lambda_\beta^{-1}(x)\} = W^x_\beta, g_\beta(x)),$$

$\lambda_\alpha(x), \lambda_\beta(x) \in \text{Pin } Q'$, $g_\alpha(x), g_\beta(x) \in \text{Pin } Q'$, λ_α, λ_β defined respectively over U_α and U_β and g_α, g_β over a neighbourhood of x, included in $U_\alpha \cap U_\beta$, which determine necessarily the same trivial Pin Q-spin structure in the tangent space at x.

Let $p(g_{\alpha\beta})$ be transition functions of W^x_α, W^x_β, with values in Pin Q, $p(\lambda_\alpha g_{\alpha\beta} \lambda_\beta^{-1}) = p(\gamma_{\alpha\beta})$; take also

$$\theta^x_\alpha(\lambda_\alpha(x)(x_i, y_j)\lambda_\alpha^{-1}(x)) = \varphi^x_\alpha(x_i, y_j),$$

where φ^x_α defines an isomorphism from $C(Q')$ onto $C(Q')_x$.

According to our previous developments

$$(3) \qquad p(g_\beta(x)) = p(g_{\alpha\beta}(x)\, g_\alpha(x)).$$

At x the origin frame of ζ may be identified with $\varphi^x_\alpha\{x_{i_1} \cdots x_{i_h} f\}$ if we consider that x belongs to U_α.

In effect if $\alpha^{i_1 \cdots i_h}(x)$ are components for any spinor in the origin frame $(W^x_\alpha, g_\alpha(x))$ and if we identify it with $\varphi^x_\alpha(\alpha^{i_1 \cdots i_h}(x) x_{i_1} \cdots x_{i_h} f)$ in the frame deduced from $\gamma \in \text{Pin } Q'$, its expression becomes

$$\varphi^x_\alpha(\gamma^{-1}\, \alpha^{i_1 \cdots i_h}(x)\, \gamma x_{i_1} \cdots x_{i_h} f),$$

equal according to (2) with the above value.

Consider the spinor identified in the first frame with $\varphi^x_\alpha(f)$, in the second frame it must be identified with $\varphi^x_\beta(\varepsilon g^{-1}_{\alpha\beta} f)$ $(\varepsilon = \pm 1)$, which gives

$$(4) \qquad f = \varepsilon f g^{-1}_{\alpha\beta}(x)$$

and implies, using the principal anti-automorphism Clifford algebra,

$$(5) \qquad g_{\alpha\beta}(x) f = \pm f.$$

That means $g_{\alpha\beta}(x)$ belongs to a subgroup H of Pin Q, which p sends in $O(Q)$, onto the subgroup called in **[3b]** "group of spinoriality" $\mathfrak{S}$. We observe $\mathfrak{S} \subset SO(Q)$.

It is possible to present the proof differently: With the notations used in § 1, if $\psi(W^x_\alpha, g_\alpha(x)) = \varphi(f)$, then $\psi(W^x_\beta, g_\beta(x)) = \varphi(\varepsilon\, g^{-1}_{\alpha\beta}(x) f)$, since the spinor thus defined at x is an element of $C(Q')_x$ determined, without any ambiguity, by $\varphi^x_\alpha(f) = \varphi^x_\beta(\varepsilon\, g^{-1}_{\alpha\beta}(x) f)$, etc.

Therefore we have

$$\varphi^x_\beta(f) = \varphi^x_\alpha(g_{\alpha\beta}(x))\, \varphi^x_\alpha(f)\, \psi^x_\alpha(g^{-1}_{\alpha\beta}(x))$$

which we shall write $f_\beta(x) = \hat{g}_{\alpha\beta}(x) f_\alpha(x)\, \hat{g}^{-1}_{\alpha\beta}(x)$, obtaining, according to (5):

$$(6) \qquad f_\beta(x) = N(g_{\alpha\beta}(x))\, f_\alpha(x).$$

We observe that $p(g_{\alpha\beta})$ are transition functions for cross-sections in the bundle complexified ξ_C of ξ. The cocycle $p(\gamma_{\alpha\beta})$ which defines ξ and the cocycle $p(g_{\alpha\beta})$ are cohomologous in $O(Q')$. It is impossible to affirm more here, in general. Thus we have obtained

PROPOSITION 1. (1°) *If there exists on V a strict meaning* Pin *Q-spin structure, then there exists on V, modulo a factor ± 1, an isotropic r-vector field, pseudo-cross-section in the bundle* Cli (V, Q'), *therefore a vector subfibering in* Cli(V, Q').

(2°) *If the pseudo-riemannian complexified bundle ξ_C admits local cross-sections, over a trivialising open set U_α, with transition functions $p(g_{\alpha\beta})$, $g_{\alpha\beta}(x) \in$* Pin *Q, such that, if $x \in U_\alpha \cap U_\beta \to f_\alpha(x)$ defines locally the isotropic r-vector field above, then*

$$f_\beta(x) = N(g_{\alpha\beta}(x)) f_\alpha(x), \qquad f_\beta(x) = \hat{g}_{\alpha\beta}(x) f_\alpha(x)\, \hat{g}^{-1}_{\alpha\beta}(x).$$

(3°) *The structural group of the bundle* ξ *is reducible, in* $O(Q')$, *to a "group of spinoriality".*

3. Identification of ζ with a subvector bundle of Cli(V, Q'). The local frames $W_\alpha^x = \varphi_\alpha^x\{x_i, y_j\}$ can be considered as projections of local frames $\psi_\alpha^x\{x_{(i)} f\}$ in ζ. We have identified the frame $\psi_\alpha^x\{x_{(i)} f\}$ with $\varphi_\alpha^x\{x_{(i)} f\}$ and

$$\varphi_\beta^x\{x_{(i)} f\} = \varphi_\alpha^x\{g_{\alpha\beta}(x)\, x_{(i)}\, f\, g_{\alpha\beta}^{-1}(x)\}$$
$$= \varphi_\alpha^x\{g_{\alpha\beta}(x)\, x_{(i)}\, f\},$$

so that this identification permits us to tell that for reduced fibrations, with transition functions $p(g_{\alpha\beta}(x))$, the morphism from ζ into ξ_C is principal linear.

At some spinor with components $u^{i_1 \cdots i_h}(x)$ in the origin frame, it is possible to associate, over U_α, the cliffordian cross-section

$$u^{i_1 \cdots i_h}(x)\, \varphi_\alpha^x(x_{i_1} \cdots x_{i_h} f),$$

still denoted by $(u^{i_1 \cdots i_h} x_{i_1} \cdots x_{i_h} f)_\alpha^x$, this having only intrinsic meaning for reduced trivialisations.

Observe, at last, that conversely, if $\psi_\alpha^x(x_{(i)} f)$ can be identified with $\psi_\alpha^x(x_{(i)} f)$, necessarily $f_\beta(x) = N(g_{\alpha\beta}(x)) f_\alpha(x)$.

Over $\theta_\alpha^x\{x_i, y_j\}$ we have $\varphi_\alpha^x(\lambda_\alpha^{-1} x_{(i)} f)$, we can write

$$\theta_\alpha^x\{x_{(i)}\, \lambda_\alpha(x)\}\, f_\alpha(x) = \{\xi_{i_1} \cdots \xi_{i_h}\, \hat{\lambda}_\alpha(x)\, f_\alpha(x)\},$$

in agreement with [**3a**], if we take $\theta_\alpha^x(x_i, y_j) = (\xi_i, \eta_j)_x$.

4. Sufficient conditions for existence of a Pin **Q-spin structure.**

PROPOSITION 2. *Let* $(U_\alpha, \varphi_\alpha)_{\alpha \in A}$ *be a trivialising atlas for the pseudo-riemannian bundle* ξ_C *on* V, *with transition functions* $p(g_{\alpha\beta}(x)) \in O(Q)$. *If there exists over* V *an isotropic r-vector pseudo-field (pseudo-cross-section in* Cli(V, Q')), *locally defined by means of* $x \in U_\alpha \to f_\alpha(x)$, *such that, if for* $x \in U_\alpha \cap U_\beta \neq \emptyset$ *we have*

$$f_\beta(x) = \hat{g}_{\alpha\beta}(x)\, f_\alpha(x)\, \hat{g}_{\alpha\beta}^{-1}(x), \qquad \varphi_\alpha^x(g_{\alpha\beta}(x)) = \hat{g}_{\alpha\beta}(x),$$
$$f_\beta(x) = N(g_{\alpha\beta}(x))\, f_\alpha(x)$$

then the manifold V *admits a strict meaning* Pin *Q-spin structure.*

From $f_\beta(x)\, \hat{g}_{\beta\alpha}(x) = \hat{g}_{\alpha\beta}(x)\, f_\alpha(x)$, and since the intersection of any right minimal ideal with any left minimal ideal is 1-dimensional [**2**, p. 71], we deduce

$$N(\hat{g}_{\alpha\beta}(x))\, f_\alpha(x)\, \hat{g}_{\alpha\beta}(x) = \hat{g}_{\alpha\beta}(x)\, f_\alpha(x) = \mu(x) \in \boldsymbol{C}^*;$$

then $\hat{g}_{\alpha\beta}(x)\, f_\alpha(x) = \pm f_\alpha(x)$ and $f_\beta(x)\, \hat{g}_{\alpha\beta}^{-1}(x) = \pm f_\alpha(x)$.

Therefore $g_{\alpha\beta}(x) f = \pm f$ and $g_{\alpha\beta}(x) \in H$.

If, at $x \in U_\alpha \cap U_\beta$, $\varphi_\alpha^x(y_1' y_2' \cdots y_r') = f_\alpha(x)$, we can complete the vector set $y_1', y_2', \cdots, y_r'$, with $x_1', x_2', \cdots, x_r'$, so that $\varphi_\alpha^x\{x_i', y_j'\}$ and $\varphi_\beta^x\{x_i', y_j'\}$ constitute Witt frames in the complexified bundle ξ_C, with transition functions $p(g_{\alpha\beta})$. (This is a trivial consequence of the Witt theorem.) Therefore we shall omit the prime and suppose $\varphi_x^\alpha(y_1 \cdots y_r) = f_\alpha(x)$.

Make up over U_α, the local cross-section in $\mathrm{Cli}(V, Q') : x \to (x_{(i)} f)^x_\alpha = \varphi^x_\alpha(x_{(i)} f)$, for any $\alpha \in A$.

If $x \in U_\alpha \cap U_\beta$, $f_\beta(x)\, \hat{g}^{-1}_{\alpha\beta}(x) = \pm f_\alpha(x)$,

$$(x_{(i)} f)^x_\beta = \pm\, \hat{g}_{\alpha\beta}(x)\,(x_{(i)} f)^x_\alpha,$$

where the sign + or − is known without ambiguity.

At $x \in V$, we can associate differentially a 2^r subspace, in the tangent space at x, by means of the isomorphisms: $\varphi^x_\alpha : \varphi^x_\alpha(x_{(i)} f) = (x_{(i)} f)^x_\alpha$ and the transition functions of φ^x_α are $p(g_{\alpha\beta})$. Therefore we have constructed a spinorial bundle, over V, with fibre $\boldsymbol{C}^{2^r}$. With the frame $\{x_{(i)} f\}^x_\alpha$, we associate the frame $\varphi^x_\alpha(\{x_i, y_j\}) = p\{x_{(i)} f\}^x_\alpha$.

$$\begin{aligned}\varphi^x_\beta\,\{x_i, y_j\} &= \hat{g}_{\alpha\beta}(x)\varphi^x_\alpha\{x_i, y_j\}\hat{g}^{-1}_{\alpha\beta}(x)\\ &= p\{g_{\alpha\beta}(x)\, x_{(i)} f\}^x_\alpha.\end{aligned}$$

It is possible to determine λ_α, $\lambda_\alpha(x) \in \mathrm{Pin}\, Q'$, such that

$$p\{\lambda_\alpha\, x_{(i)} f\}^x_\alpha = \varphi^x_\alpha(\lambda_\alpha\{x_i, y_j\}\,\lambda^{-1}_\alpha)_x,$$

with $\varphi^x_\alpha(\lambda_\alpha\{x_i, y_j\}\lambda^{-1}_\alpha)_x$, "real" Witt frames in ξ_C. We have a Pin Q-spin structure in the strict meaning.

PROPOSITION 3. *Assume the structural group of an orthonormal bundle of frames reduces in $O(Q')$ to a group of spinoriality, then the manifold admits a* Pin *Q-spin structure in the strict meaning.*

If we have transition functions $p(g_{\alpha\beta})$, $g_{\alpha\beta}(x) \in H$, according to $g_{\alpha\beta}(x) f = \pm f$, φ^x_α sending $C(Q')$ onto $C(Q')_x$:

$$\begin{aligned}\varphi^x_\alpha(\theta_{\alpha\beta}(x) f) &= \pm f_\alpha(x),\\ \hat{g}_{\alpha\beta}(x) f_\alpha(x) &= \pm f_\alpha(x) \Rightarrow f_\alpha(x)\,\hat{g}^{-1}_{\alpha\beta}(x) = \pm f_\alpha(x),\end{aligned}$$

and, from $f_\beta(x) = \hat{g}_{\alpha\beta}(x) f_\alpha(x)\hat{g}^{-1}_{\alpha\beta}(x)$, we deduce

$$\begin{aligned}f_\beta(x) &= \pm\, \hat{g}_{\alpha\beta}(x) f_\alpha(x) = \pm f_\alpha(x),\\ f_\beta(x)\,\hat{g}^{-1}_{\alpha\beta}(x) &= \pm f_\alpha(x)\,\hat{g}^{-1}_{\alpha\beta}(x) = \pm f_\alpha(x),\end{aligned}$$

and starting with this result it is possible to take up again the proof of Proposition 2.

5. Other definitions for Pin **Q-spin structures.** We suppose the manifold V, n-dimensional, basis of a principal bundle, with structural group Pin Q. At any pair of local cross-sections $x \in U_\alpha \to z_\alpha(x)$, $x \in U_\beta \to z_\beta(x)$, with $z_\beta(x) = z_\alpha(x)\cdot\gamma_{\alpha\beta}(x)$, $x \in U_\alpha \cap U_\beta \neq \emptyset$, we can associate a differentiable application $x \in U_\alpha \cap U_\beta \to p(\gamma_{\alpha\beta}(x)) \in O(Q)$ and thus define a principal bundle ξ', with group $O(Q)$, and transition functions $p(\gamma_{\alpha\beta}(x))$. We can consider it as the bundle of frames for an associate pseudo-riemannian bundle, with fibre $\boldsymbol{R}^{2r}$, $O(Q)$ operating in this fibre accordingly : $a \to \gamma a \gamma^{-1}$ ($\gamma \in \mathrm{Pin}\, Q$, $a \in \boldsymbol{R}^{2r}$). η determines this bundle, modulo an isomorphism, in $O(Q)$.

ξ having the same meaning as above, one gives also the definition [1]:

DEFINITION 2. *V admits a* Pin *Q-spin structure (in the strict meaning), if the bundle ξ is isomorphic with the bundle ξ', this isomorphism being principal and according to the group $O(Q)$.*

Otherwise, some cocycle, equivalent in $O(Q)$ to $p(\gamma_{\alpha\beta})$, defines the fibration ξ.

Compare definitions (1) and (2).

If there exists a Pin Q-spin structure in the meaning (1), we can associate to z_α the local cross-section in ξ', over $U_\alpha: x \to \{z_\alpha(x), (x_i, y_j)\}$, it is understood that

$$\begin{aligned}\{z_\beta(x), (x_i, y_j)\} &= \{z_\alpha(x)\cdot\gamma_{\alpha\beta}(x), (x_i, y_j)\}\\ &= \{z_\alpha(x), \gamma_{\alpha\beta}(x)\,(x_i, y_j)\,\gamma_{\alpha\beta}^{-1}(x)\},\end{aligned}$$

and thus we define a principal morphism h' from η onto ξ'. ξ and ξ', admitting the same transition functions relative to the open covering U_α, are isomorphic.

Conversely, if there exists a Pin Q-spin structure in Definition 2, since a morphism from η onto ξ' always exists, there also exists a morphism from η onto ξ and we again find (1). We also use the following definition found in the literature ([**5**], [**6a**], [**7**]).

DEFINITION 3. *Let ξ be the principal pseudo-riemannian bundle with structural group $O(Q)$. It is said that V admits a* Pin *Q-spin structure, if it is possible to build a principal bundle η with group* Pin *Q, basis V, such that if $\gamma_{\alpha\beta}$, $\gamma_{\alpha\beta}(x) \in$* Pin *Q, are its transition functions relative to trivialising open sets (U_α), the $p(\gamma_{\alpha\beta})$ (or a cocycle equivalent in $O(Q)$) define the fibration ξ (with the same trivialising "opens" U_α).*

It is immediate that (2) implies (3).

Conversely, suppose that $p(\gamma'_{\alpha\beta})$ defines the fibration ξ, $p(\gamma'_{\alpha\beta})$ cocycle equivalent in $O(Q)$ to $p(\gamma_{\alpha\beta})$, $\gamma_{\alpha\beta}$ defining η. It is possible to construct a cross-section system z_α in η with transition functions $\gamma_{\alpha\beta}$, and associate with this cross-section z_α, cross-sections in ξ' so that a morphism is defined from η onto ξ', ξ' and ξ are isomorphic and one can find again (2).

6. $H^1(V, C^*)$ is the set of isomorphism classes of principal bundles, with groups C^*, dim $V = 2r$. Assume Q definite in §§ 6 and 7.

PROPOSITION 4. *If the manifold V admits an almost-complex structure, it admits also a* Spin *Q-spin structure when*

(a) *$H^1(V, C^*) = 1$, or*

(b) *the structural group $U(r, C)$ reduces to $Su(r, C)$ in $O(Q')$ (cf. § 2, $Su(r, C) \cong \mathfrak{S}$).*

Let $(U_\alpha, \theta^\alpha)$ be a trivialising atlas of ξ. It is allowable to suppose V admits an almost-hermitian structure, then V carries an isotropic r-vector field (modulo a complex factor with module 1) locally:

$$\begin{aligned}f_\alpha(x) &= \theta_\alpha^x(f),\\ f_\beta(x) &= \hat{g}_{\alpha\beta}(x)\,f_\alpha(x)\,\hat{g}_{\alpha\beta}^{-1}(x), &\quad g_{\alpha\beta}(x) \in \text{Spin } Q,\\ &= \mu_{\alpha\beta}(x)\,f_\alpha(x), &\quad |\mu_{\alpha\beta}(x)| = 1.\end{aligned}$$

Then $\mu_{\alpha\beta}N(g_{\alpha\beta})$ defines cocycle valued in C^* (here $N(g_{\alpha\beta}) = 1$), therefore there exists a principal fibration with trivialising open sets U_α and transition functions $\mu_{\alpha\beta}N(g_{\alpha\beta})$. In both cases this fibration is trivial, $\mu_{\alpha\beta}(x)$ represents the determinant of $p(g_{\alpha\beta}(x))$ contracted to $f(b)$.

Then there exist applications $x \in U_\alpha \to \sigma_\alpha^2(x) \in C^*$, such that if $x \in U_\beta \cap U_\beta$,

$$\sigma_\beta^2(x) = \sigma_\alpha^2(x) \cdot \mu_{\alpha\beta}(x)\, N(g_{\alpha\beta}(x)).$$

According to a result pointed out in [**3b**, p. 313], there exists $\delta_\alpha(x) \in \text{Spin } Q$, such that

$$\frac{1}{\sigma_\alpha(x)} f = \delta_\alpha(x)\, f,$$

and over U_α,

$$\hat{\delta}_\alpha(x)\, f_\alpha(x) = \frac{1}{\sigma_\alpha(x)} f_\alpha(x), \quad \text{if } \hat{\delta}_\alpha(x) = \theta_\alpha^x(\delta_\alpha(x)).$$

This result implies [**3b**]:

$$f'_\alpha(x) = \hat{\delta}_\alpha(x)\, f_\alpha(x)\, \hat{\delta}_\alpha^{-1}(x) = \frac{1}{\sigma_\alpha^2(x)} N(\delta_\alpha(x))\, f_\alpha(x),$$

$$f'_\beta = N(\hat{\gamma}_{\alpha\beta}(x))\, f'_\alpha(x) = \hat{\gamma}_{\alpha\beta}(x)\, f'_\alpha(x)\, \hat{\gamma}_{\alpha\beta}^{-1}(x) \quad \text{with } \gamma_{\alpha\beta} = \delta_\alpha^{-1}(x)\, g_{\alpha\beta}(x)\, \delta_\alpha(x)$$

(the $\gamma_{\alpha\beta}(x)$ correspond to local isomorphisms not explicit here, associated with the $\gamma_{\alpha\beta}(x)$).

We conclude according to Proposition 2 that V admits a Spin Q-spin structure (built starting from f'_α).

REMARKS. In Proposition 4, (a) is a strong condition. S^2 does not satisfy (a) but admits nevertheless both structures.

$H^1(V, C^*)$ and $H^2(V, Z)$ are isomorphic, therefore (a) is equivalent with $H^2(V, Z) = 0$.

(a) implies that the first Chern class is null, consequently the second Stiefel-Whitney class is null.

In § 2, we prove $\mathfrak{S} \cong Su(r, C)$ so that (b) is also necessary.

In a lecture delivered in Rome (January 1973) B. Kostant developed the notion of symplectic spinors. His Proposition 3.3.1, p. 12, can be obtained by the sharpening of Proposition 4. If the bundle of rank 1, defined by cocycle $\mu_{\alpha\beta}$, is the tensorial square product of a complex bundle of rank 1, then an almost complex manifold V admits a Spin Q-spin structure (and conversely). We think that our approach is very convenient to obtain essential Kostant results.

COROLLARY. *Assume V is a kählerian Ricci flat manifold. Then V admits a* Spin *Q-spin structure.*

It is an immediate consequence of the reduction holonomy group theorem in

$O(Q)$. If V is Ricci flat this last group is included in $Su(r, \mathbf{C})$ [**6b**, p. 261].

It is easy to prove the following result:

Any n-manifold ($n = 2r$) admitting a Pin Q-spin structure (Q definite) and a proper "spin-euclidean" riemannian connection is Ricci flat (cf. [**3c**]). The same result is true with Q normal, hyperbolic.

7. PROPOSITION 5. *Any riemannian manifold, even dimensional, admits a* Pin *Q-spin structure if and only if the following two properties are satisfied*:

(1) *There exists on V an r-vector field : $x \to f_x$, cross-section in the exterior algebra of the complexified tangent bundle $T^{\mathbf{C}}(V)$.*

(2) *f_x and its conjugate $\bar{f}_x$ determine a direct decomposition in $T^{\mathbf{C}}_x(V)$, $\forall\, x \in V$.*

NECESSARY CONDITION. Part (1) is already known. Concerning part (2), we remark, if Q is definite in the standard Clifford algebra, f being defined by means of a "real" Witt frame, then $f\bar{f} \neq 0$, equivalent to $F' \cap \bar{F}' = 0$, if F' is the maximal totally isotropic subspace of f [**2**] (s.t.i.m. of f). With the notations of § 2,

$$\theta^x_\alpha(\lambda_\alpha(x)\, f \lambda_\alpha^{-1}(x)) = \varphi^x_\alpha(f) = f_\alpha(x),$$

$\lambda_\alpha(x)\, f\, \lambda_\alpha^{-1}(x)$ defines a s.t.i.m. of $C(Q')$. If this s.t.i.m. exists $f'(x) = \mu(x)\, \lambda_\alpha(x) \cdot f\, \lambda_\alpha^{-1}(x)$, $\mu(x) \in \mathbf{C}^*$, built by means of a "real" Witt frame, then $f'(x)\, \bar{f}(x) \neq 0$ and $G'(x) \cap \bar{G}'(x) = 0$ if $G'(x)$ is the s.t.i.m. of $f'(x)$, which implies that $f'_\alpha(x) = \theta^x_\alpha(f'(x))$ is such that $f'_\alpha(x)\, \bar{f}_\alpha(x) \neq 0$ and $f_\alpha(x) \bar{f}_\alpha(x) \neq 0$.

SUFFICIENT CONDITION. Assume f_x, $\bar{f}_x$ determine a direct decomposition of $T^{\mathbf{C}}_x(v)$ according to $T^{\mathbf{C}}_x = F'_x \oplus \bar{F}'_x$, where the F'_x, $\bar{F}'_x$ depend differentially on x; if $u_x \in T_x$ and

$$u_x = v_x + w_x, \qquad v_x \in F'_x,\ w'_x \in \bar{F}'_x,$$

$$v_x = v^i_x(\xi_i)_x, \qquad (\xi_i)_x \text{ a frame of } F'_x,$$

introduce the following bilinear form $B(x)$ on $T^{\mathbf{C}}_x$:

$$B(x)\,(u(x), u'(x)) = \textstyle\sum_1^r v^i_x\, w'^i_x + w^i_x\, v'^i_x$$

with evident notations. B is a real bilinear form, definite, $x \to f_x$ is an isotropic r-vector field in the associated Clifford bundle. Then we use Proposition 2.

Proposition 5 gives a condition independent of the riemannian metric.

8. **Manifolds with** Pin **Q'-spin structure.** Their definition is obtained, replacing Pin Q, $O(Q)$ respectively by Pin Q', $O(Q')$. Pin Q', we recall, is a 4-fold covering of $O(Q')$, ε used in (4) (§§ 1 and 4) equals now ± 1 or $\pm i$, so that condition (6) becomes $f_\beta(x) = \pm f_\alpha(x)$, and in $O(Q')$ a group of spinoriality is defined by projection, starting from $g_{\alpha\beta}(x)\, f = \varepsilon f$.

Propositions 1 and 2 (§§ 1 and 2) state easily replacing (6) by $f_\beta(x) = \pm f_\alpha(x)$ and observing that $g_{\alpha\beta}(x) \in$ Pin Q':

PROPOSITION 6. *V admits a* Pin *Q'-spin structure if and only if there exists on V a s.t.i.m. field.*

PROOF. Easy.

Observe the reduction of a structural group to a subgroup in Spin Q'. We have a Spin Q'-spin structure.

9. Manifolds with Pin **Q-spin structure in the broad meaning.** In $SO(Q)$ we call an enlarged group of spinoriality $\mathfrak{S}_e$, any subgroup, projection by p of the set $H_e \subset$ Pin Q such that:

$$gf = \mu(g)\,f, \qquad \mu(g) \in \boldsymbol{C}^*,\ f \text{ being an isotropic } r\text{-vector.}$$

$gf = \pm\,\mu f$ is equivalent to $gfg^{-1} = N(g)\,\mu^2 f$, therefore $\mathfrak{S}_e$ is the stabilizer in the space F' determined by f. It is possible to suppose f constructed with a "real" Witt frame [3c].

In the case where Q is definite, one sees easily $|\mu(g)| = 1$, $\mathfrak{S}_e$ is identified with the real representation of $U(r, \boldsymbol{C})$ (cf. part II, §2 below).

If we assume that the structural group of ξ reduces in $O(Q')$ to an enlarged group of spinoriality, there exist on V a s.t.i.m. field and a fortiori a Pin *Q'-spin structure.*

IMMEDIATE PROOF. A manifold with Pin Q-spin structure in the broad meaning is such that the structural group reduces in $O(Q')$ to a group $\mathfrak{S}_e$.

If we take in Proposition 2, $p(g_{\alpha\beta}(x)) \in O(Q)$,

$$f_\beta(x) = \hat{g}_{\alpha\beta}(x)\,f_\alpha(x)\,\hat{g}_{\alpha\beta}^{-1}(x),$$
$$f_\beta(x) = \mu_{\alpha\beta}(x)\,f_\alpha(x), \qquad \mu_{\alpha\beta}(x) \in \boldsymbol{C}^*,$$

then V admits a Pin Q-spin structure in the broad meaning.

In particular, if V admits an almost-complex structure, V admits also a Pin *Q-spin structure in the broad meaning.*

If $H^1(V, \boldsymbol{C}^*) = 1$ and if V admits a Pin Q-spin structure in the broad meaning also V admits one in the strict meaning.

10. Spin **Q-spin structure in odd dimensions.** V is $(2r + 1)$-dimensional, orientable. Definition 1 remains the same, Spin Q and $SO(Q)$ replacing respectively Pin Q and $O(Q)$. $C^+(Q)$ is central, simple [2]. $C^+(Q')$ $(n = 2r + 1)$ is isomorphic with $C(Q')$ $(n = 2r)$. We introduce a Witt frame $(x_1, \cdots, x_r, y_1, \cdots, y_r, z)$ with z nonisotropic and the representation of $C^+(Q')$ in the space $\{x_{i_1} \cdots x_{i_h} f, f = y_1 \cdots y_r\}$; ζ is defined in the same way.

In the study of necessary and sufficient existence conditions, only a few details are modified, one arrives at identical statements, the $g_{\alpha\beta}(x)$ belonging to Spin Q.

Spin structures in the broad meaning generalize in odd dimensions to almost-cocomplex structures (Bouzon, Sasaki).

II. GROUPS OF SPINORIALITY

The standard space is $2r$-dimensional. The following developments are purely algebraic. Some details are omitted in proofs.

1. H is the subgroup constituted by elements $\gamma \in$ Spin Q, with $\gamma f = \pm f$, f is defined in Part I and determines the s.t.i.m. F'. $p(H) = \mathfrak{S}$, $\mathfrak{S}$ is the group of spinoriality associated with f.

If we replace f by any r-vector in F', H is not modified. If another s.t.i.m. is put in the place of F', H is replaced by a conjugate subgroup.

LEMMA 1. *Two subgroups H and H_1, associated respectively with f and f_1, are conjugate in* Pin Q.

We can therefore consider groups of spinoriality constructed by means of r isotropic vectors which determine a s.t.i.m. and belong to a "real" Witt frame.

PROPOSITION 1. *In elliptic signature $\mathfrak{S}$ becomes identified with $Su(r, \boldsymbol{C})$. $\mathfrak{S}$ is the set of elements with determinant* 1 *in the stabilizer of some s.t.i.m.* dim $\mathfrak{S} = r^2 - 1$. *$\mathfrak{S}$ is connected and simply connected.*

PROPOSITION 2. *In signature $(k, n-k)$ $(k < n-k$, k positive terms and $r \geqq 2)$ $\mathfrak{S}$ is isomorphic with the subgroup of elements in $SL(n, \boldsymbol{R})$ with matrix*

$$\begin{Vmatrix} \alpha & -\bar{\mu} & \lambda & \mu \\ 0 & \beta & \nu & 0 \\ 0 & 0 & \rho & 0 \\ 0 & 0 & -\bar{\nu} & \bar{\beta} \end{Vmatrix}$$

such that

$\alpha \in M_k(\boldsymbol{R})$, $\det \alpha = \pm 1$,
$\beta \in M_{r-k}(\boldsymbol{C})$, $\beta {}^t\bar{\beta} = \mathrm{Id}$, $\det \beta = \det \alpha$,
$\alpha {}^t\rho = \mathrm{Id}$, $\lambda \in M_k(\boldsymbol{R})$,
$\mu \in \boldsymbol{C}^{k(r-k)}$, $\nu = \boldsymbol{C}^{(r-k)k}$,
$\nu = -\beta {}^t\mu\rho$, ${}^t\rho\lambda + {}^t\lambda\rho = {}^t\nu\bar{\nu} + {}^t\bar{\nu}\nu$.
$\mathfrak{S}$ has two connected components and becomes identified with the set of elements with determinant 1 *in the stabilizer of some s.t.i.m.* dim $\mathfrak{S} = (r^2 - 2) + k(k-1)/2$.

PROPOSITION 3. *If Q is a neutral form $(k = r)$ $\mathfrak{S}$ is isomorphic with the subgroup of elements in $SL(n, \boldsymbol{R})$ with matrix*

$$\begin{Vmatrix} \alpha & \lambda \\ 0 & \rho \end{Vmatrix},$$

such that

$\alpha \in M_r(\boldsymbol{R})$, $\det \alpha = 1$,
$\alpha {}^t\rho = \mathrm{Id}$,
${}^t\rho\lambda + {}^t\lambda\rho = 0$.
$\mathfrak{S}$ is connected and becomes identified with the set of elements with determinant 1 *in the stabilizer of some s.t.i.m.* dim $\mathfrak{S} = r^2 - 1 + r(r-1)/2$.

2. Enlarged group of spinoriality. It is the set of elements in the stabilizer of some s.t.i.m. for action of $SO(Q)$.

$$H_e = \{\gamma \in \mathrm{Spin}\, Q, \gamma f = \chi e^{i\theta} f\}, \qquad p(H_e) = \mathfrak{S}_e.$$

If Q is definite $\chi = 1$, $p(H_e) \cong U(r, \boldsymbol{C})$.

$$\dim \mathfrak{S}_e = r^2 + k(k-1)/2, \qquad 0 \leqq k \leqq r.$$

If $k \neq 0$ an enlarged group is not a generalised unitary group. $\mathfrak{S}_e$ is connected if $k = 0$ or r, has 2 connected components otherwise.

III. Generalised Twistors

1. Preliminaries. Lemmas 1, 2 and 3 are taken from Chevalley [2]. It is possible to give more simple proofs omitted here.

Let $\{x_i, y_j\}$ be a Witt frame of $E' = E_c = F \oplus F'$ and σ an isometry of E' which conserves any element of F. $F = (x_1, x_2, \cdots, x_r)$, $F' = (y_1, \cdots, y_r)$. G_0 (resp. G_0') is the subgroup of Pin Q (resp. Pin Q') with elements γ, such that $N(\gamma) = 1$.

LEMMA 1. *$\sigma \in O(Q')$ reduces in F to identity if and only if*

$$\sigma(x_i) = x_i,$$
$$\sigma(y_i) = a_i^j x_j + y_i, \qquad \|a_j^i\| \text{ antisymmetric}$$

for all $i = 1, 2, \cdots, r$.

Note that the data of F does not determine uniquely F' such that $F \oplus F'$ is Witt's decomposition of $E_c = E'$.

LEMMA 2. *If $u \in \Lambda^2 F$, then* $\exp u \in G_0'^{+}$ *and $p(\exp(u))$ conserves any element of F. Conversely any element of $O(Q')$ which conserves any elements of F belongs to $p(G_0'^{+})$ and can be written: $p(\exp(u))$, $u \in \Lambda^2 F$.*

LEMMA 3. *If V, V', V'' are s.t.i.m. such that $V \cap V' = V \cap V''$, then there exists $\sigma \in p(G_0'^{+})$ conserving any element of V and sending V' into V''.*

LEMMA 4. *Assume Q definite. Let F_1 be any s.t.i.m. such that $F_1 \oplus F' = E'$. We can transform F' in F_1 by means of an element $g \in$* Pin Q', *with $g^{-1} Hg \subset H$.*

LEMMA 5. *Assume Q indefinite. Let F_1 be any s.t.i.m. such that $F' \cap F_1 = F' \cap \bar{F}'$ ($\bar{F}'$ complex conjugate of F'). Then there exists $g \in$* Pin Q' *with $g F' g^{-1} = F_1$ and $g^{-1} Hg \subset H$.*

LEMMA 6. *Let g, γ be elements in* Pin Q'. *If $g\gamma$ sends f onto f' and belongs to the normaliser N of H in* Pin Q', *if γ sends also f onto f', then γ belongs to N.*

2. Twistorial fibrations canonically associated with a Pin **Q-spin fibration on a manifold.** Generalising the terminology used by R. Penrose [8], we shall call a q-twistor (or twistor of order q) any element in some vector space isomorphic with the direct sum of 9 spinor spaces. With the choice of the spinorial representation above, a twistor is an element of a left ideal in $C(Q')$.

If there exists on V a Pin Q-spin structure, then there exists in the tangent complexified exterior bundle, modulo a scalar factor, an isotropic r-vector field locally defined by: $x \in U_\alpha \to f_\alpha(x) = \varphi_\alpha^x(f)$, φ_α^x being an isomorphism from $C(Q')$ onto $C(Q')_x$ (cf. Part I), the U_α constituting the atlas for fibration.

Look for, if they exist, *without consideration concerning the topology of V (V*

manifold with a Pin Q-spin structure) other isotropic r-vectors $f' \in C(Q')$ such that also $\varphi_\alpha^x(f')$ is the restriction at U_α of an isotropic r-vector field (modulo a scalar factor). If there exists such an f', then $f' = gfg^{-1}$, $g \in \text{Pin } Q'$, and $N(\hat{g}_{\alpha\beta}(x)) \cdot \varphi_\alpha^x(gfg^{-1}) = \varphi_\beta^x(gfg^{-1})$, where $g_{\alpha\beta}$ are transition-functions of φ_α^x (§ 1).

Necessarily: $(g^{-1}\, g_{\alpha\beta}(x)\, g) f = \pm f$, $\forall g_{\alpha\beta}(x) \in \text{Pin } Q$, $g_{\alpha\beta}(x) f' = \pm f'$; therefore g is in the normaliser of H in Pin Q'. Then there will exist other "pseudo-isotropic" r-vector fields, if $\mathfrak{S}$ is also a group of spinoriality for f', with $f' \neq \lambda f$, $\lambda \in \boldsymbol{C}^*$.

Now, if $\gamma \in \mathfrak{S}$, $\gamma f = \pm f$ is equivalent to $\gamma \bar{f} = \pm \bar{f}$, then there always exist two isotropic r-vectors $f, \bar{f}$, which define the same group of spinoriality $\mathfrak{S}$, $f \neq \lambda \bar{f}$, if $0 \leqq k < r$. Thus it follows:

PROPOSITION 1. *If there exists on V a* Pin *Q-spin fibration there exists also a 2-twistorial fibration, Whitney sum of two* Pin *Q-spin fibrations if* $0 \leqq k < r$.

We observe that the second fibration, thus obtained is not necessarily complex conjugate for the given fibration, because φ_α^x sends $C(Q')$ onto $C(Q')_x$.

PROPOSITION 2. *If there exists a* Pin *Q-spinorial structure, the complexified Clifford bundle* Cli(V, Q') *contains a subbundle in twistors of order 2^{r-k}, if the quadratic form Q admits signature $(k, n-k)$ $(0 \leqq k < n-k)$. This twistorial bundle is the direct sum of 2^{r-k}* Pin *Q-spin subbundle, isomorphic in* Pin Q'. *In the elliptic case,* Cli(V, Q') *is therefore the direct sum of* 2^r Pin *Q-spin bundles.*

The proof uses the preceding lemmas.

DEFINITION. *The spin fibrations deduced of some spin fibration by modifying only the choice of f in $C(Q')$ are called "canonically associated" with the given fibration.*

Proposition 3 completes our results:

PROPOSITION 3. *There do not exist other* Pin *Q-spin fibrations canonically associated with some given* Pin *Q-spin fibration.*

3. Isotropic twistors. Our purpose is to find again essential results given by Penrose [8] and generalise these for any dimension and any signature. In [8] Penrose considers only Minkowski space, twistors of order 2, in a purely algebraic context.

We have pointed out in [3c] the possibility of constructing a fundamental hermitian form $\mathscr{H}$ in the spinor space, taking

$$\beta(\overline{uf})\, vf = a\mathscr{H}(uf, vf)\gamma f$$

where a is a scalar constant, β the principal anti-automorphism, $uf, vf \in C(Q')f$, $\overline{uf}$ is the complex conjugate of uf, γf a pure spinor suitably select determined by this conjugation [3c].

On a manifold owing a G_0-spin structure, there exists a hermitian-form field [3c] so that algebraic considerations below extend to like manifolds.

We come again to E_n, n-dimensional on R, with quadratic form Q.

DEFINITION. *A twistor, with valence* $\left[\begin{smallmatrix}1\\0\end{smallmatrix}\right]$, *is a cliffordian element* $uf + u'f'$, *where f and f' are two isotropic r-vectors determining minimal distinct left ideals.*

We introduce the hermitian structure given by $\mathscr{H}$ and write [3c]:

$$\beta\overline{(uf + u'f')}\,(vf + v'f') = a\mathscr{H}(uf, vf)\gamma f + a'\mathscr{H}'(u'f', v'f')\gamma' f' + a_1\mathscr{H}_1(u'f', vf)\,\gamma_1 f + a_1'\mathscr{H}_1'(uf, v'f')\gamma_1' f'$$

where a, a', a_1, a_1' are scalar constants, γf, $\gamma' f'$, $\gamma_1 f$, $\gamma_1' f'$ are pure spinors determined respectively by $(f, \bar{f})$, $(f', \bar{f}')$, $(f, \bar{f}')$, $(f', \bar{f})$. It is possible to choose a, a', a_1, a_1' such that

$$\mathscr{H}(uf, vf) = \overline{\mathscr{H}(vf, uf)},$$
$$\mathscr{H}(u'f', v'f') = \overline{\mathscr{H}'(v'f', u'f')},$$
$$\mathscr{H}_1'(vf, u'f') = \overline{\mathscr{H}_1(u'f', vf)}.$$

In the decomposition above γ, γ', γ_1, γ_1' being fixed, only numeric coefficients are determined. We obtain:

PROPOSITION 4. *There exists on the twistor space of valence* $\left[\begin{smallmatrix}1\\0\end{smallmatrix}\right]$, *a hermitian form, natural extension for the hermitian spinorial form*:

$$\langle uf + u'f' \mid vf + v'f' \rangle = \mathscr{H}(u, v) + \mathscr{H}'(u', v') + \mathscr{H}_1(u', v) + \overline{\mathscr{H}_1(v', u)}.$$

(*written with a little abuse.*)

Penrose considers in **[8]** elements u, u', v, v', belonging to the real Minkowski space E, real and isotropic. Here, without restriction in the generality, we consider u, u', v, v' isotropic for the "metric"

$$\langle uf + u'f' \mid uf + u'f' \rangle = 2\mathscr{R}_e\mathscr{H}_1(u', v),$$

and call isotropic twistor with valence $\left[\begin{smallmatrix}1\\0\end{smallmatrix}\right]$ any twistor such that

$$\langle uf + u'f', uf + u'f' \rangle = 0.$$

This definition generalises the Penrose definition. In particular, we can prove:

PROPOSITION 5. *In the Minkowski space, the stabiliser for any isotropic twistor, and any group of spinoriality are isomorphic.*

4. Identification between twistors (of even order) and spinors. We are able to identify the twistor-space $C(Q')f \oplus C(Q')f'$, and the spinor space for a Clifford algebra, constructed on E_{n+2}, $(n + 2)$-dimensional (with a new representation equivalent to the representation used above).

It is particularly convenient to consider E_n with metric Q_n of signature (p, q) and space E_{n+2} with metric Q_{n+2} of signature $(p + 1, q + 1)$ by reason of the well-known result:

The Clifford algebra $C(Q_{n+2})$, *denoted* $C_{p+1,\,q+1}$, *is isomorphic with the tensorial product:* $C_{p,q} \otimes C_{1,1}$.

Any twistor, of order $2l$, built on E_{2r}, with "metric" Q_{2r}, signature (p, q) is a spinor on $E_{2(r+l)}$ with "metric" Q_{2r+2l}, signature $(p + l, q + l)$.

More precisely, if $l = 1$:

PROPOSITION 6. *One can find a representation of* $C_{p+1,q+1}$ *in* $C(Q')f \oplus C(Q')f'$

space of twistors of order 2, such that $C(Q')f$ and $C(Q')f'$ are respectively even and odd semi-spinors for $C_{p+1,q+1}$.

5. The part of the conformal groups.

PROPOSITION 7. *The conformal group of E_n endowed with a "metric" of signature (p, q) is locally isomorphic with the isometric group of E_{n+2}, endowed with a "metric" of signature $(p + 1, q + 1)$.*

This property is available for any parity of n $(n > 2)$.

The proof consists of verifying isomorphisms of Lie algebras. Broadly, one can say that twistors of order 2 are for the conformal group of E_n, $Q_n(p, q)$, what spinors are for the group of isometries of E_{n+2}, $Q_{n+2}(p + 1, q + 1)$, and one adapts this result, to the fibrations on manifolds.

N.B. In **[3c]** the author gives more technical results concerning derivations, and usual topics almost only developed previously in a space-time manifold for the most part by means of local matricial calculus.

BIBLIOGRAPHY

1. M. F. Atiyah, R. Bott and A. Shapiro, *Clifford modules*, Topology **3** (1964), suppl. 1, 3–38. MR **29** #5250.

1. bis. Ch. Barbance, Thèse, Paris, 1969.

2. C. Chevalley, *The algebraic theory of spinors*, Columbia Univ. Press, New York, 1954. MR **15**, 678.

3a. A. Crumeyrolle, *Structures spinorielles*, Ann. Inst. H. Poincaré Sect. A **11** (1969), 19–55. MR **42** #6737.

3b. ———, *Groupes de spinorialité*, Ann. Inst. H. Poincaré Sect. A **14** (1971), 309–323. MR **45** #9260.

3c. ———, *Dérivations, formes et opérateurs usuels sur les champs spinoriels de variété différentiables de dimension paire*, Ann. Inst. H. Poincaré Sect. A **16** (1972), 171–201.

4. Y. Choquet-Bruhat, *Géométrie différentielle et systèmes extérieurs*, Monographies Universitaires de Mathématiques, no. 28, Dunod, Paris, 1968. MR **38** #5118.

4 bis. R. Hermann, *Vector bundles in mathematical physics*. I, Benjamin, New York, 1970. MR **42** #1454.

5. Y. Kosmann, *Dérivée de Lie des spineurs*, Thèse, Paris, 1970; Ann. Mat. Pura Appl. (4) **91** (1972), 317–395.

6a. A. Lichnerowicz, *Champs spinoriels et propagateurs en relativité générale*, Bull. Soc. Math. France **92** (1964), 11–100. MR **29** #6913.

6b. ———, *Théorie globale des connexions et des groupes d'holonomie*, Edizioni Cremonese, Rome, 1957. MR **19**, 453.

7. R. Palais, *Seminar on the Atiyah-Singer index theorem*, Ann. of Math. Studies, no. 57, Princeton Univ. Press, Princeton, N.J., 1965, Chap. IV. MR **33** #6649.

8. R. Penrose, *Twistor algebra*, J. Mathematical Phys. **8** (1967), 345–366. MR **35** #7657.

9. N. E. Steenrod, *Topology of fibre bundles*, Princeton Math. Ser., vol. 14, Princeton Univ. Press, Princeton, N.J., 1951. MR **12**, 522.

UNIVERSITÉ PAUL SABATIER, TOULOUSE CEDEX, FRANCE

Proceedings of Symposia in Pure Mathematics
Volume 27, 1975

LOCAL CONVEX DEFORMATIONS OF RICCI AND SECTIONAL CURVATURE ON COMPACT MANIFOLDS

PAUL EWING EHRLICH*

In [1], Aubin stated a theorem which implied as a corollary that given any smooth Riemannian manifold (M, g_0) with nonnegative Ricci curvature and all Ricci curvatures positive at some point, M admits a metric of everywhere positive Ricci curvature. However, the proof given in [1] appears to be in doubt. Motivated by the corollary of [1], we decided to investigate the question of whether a given metric of nonnegative Ricci (or sectional) curvature can be perturbed near a point of everywhere positive Ricci (or sectional) curvature to increase the area of positive curvature without changing the metric except on a small neighborhood of the given point (and, of course, such that the new metric has nonnegative curvature everywhere). We give only an abstract of our results here. Details will appear in [2].

We will assume below that (M^n, g_0) is compact and $n \geqq 2$. We will say that a variation $g(t)$ of g_0 for t in $(-\varepsilon, \varepsilon)$ *has support* in $D \subset M$ iff $g(t) = g_0$ in $M - D$ for all t and, for all p in D, $\exists\, v \in M_p$ with $g(t)(v, v) \neq g(v, v)$ for all $t \neq 0$.

Consider a variation $g(t)$ of g_0 analytic in t of order at least 3 through C^4 metrics with support in a "nice" D such as a "small" g_0-metric ball. Let $\rho\colon D \to R$ be the distance to Bd(D). In computing $\mathrm{Ric}' := (d/dt)\, \mathrm{Ric}_{g(t)}|_{t=0}$ only the 1-jet of $g(t)$ occurs in the formula for Ric′. Thus, we may consider only variations of the form $g(t) = g_0 + th^1$, where h^1 is a symmetric 2-tensor, with support in D to determine

---

AMS (MOS) subject classifications (1970). Primary 53C20.
*Supported by an NSF Traineeship at SUNY at Stony Brook.

Ric′ for all possible variations with support in D. Then the condition "$g(t)$ has support in D" and is a variation through C^4 metrics implies that near Bd(D) we may write

$$g(t) = g_0 + t\,\rho^3 h$$

in order to compute Ric′ near Bd(D) for arbitrary C^4 variations with support in D. A computation of Ric′ for such a variation shows that the Hessian of ρ occurs in the lowest order term in ρ in the first derivative. This motivates the study of *local convex deformations*, that is, variations $g(t)$ of g_0 with support in g_0-convex metric disks, so that the Hessian of ρ will have a definite sign.

From our calculation of Ric′, we find a standard local convex deformation $g(t)$ of any metric g_0 for M with $\text{Ric}_{g_0} \geqq 0$ with the following property:

Given a g_0-convex disk D of g_0-radius R, $g(t)$ is defined with support in D and there exists a constant $\varepsilon = \varepsilon(n)$ (depending only on n = dimension of M) and $t_0 > 0$ such that $\text{Ric}_{g(t)} > 0$ for all t with $0 < t \leqq t_0$ in the "g_0-outer annulus" of all points in $D - \text{Bd}(D)$ a g_0-distance less than or equal to εR from Bd(D).

We next prove a local minorization of the convexity radius function on the space of Riemannian metrics.

THEOREM 1. *Given (M, g_0) with M compact, there exist constants $C(g_0) > 0$ and $\delta(g_0) > 0$ such that if g is any metric for M that is $\delta(g_0)$ close to g_0 in the C^4 topology on the space of metrics for M, then any g-disk of g-radius $\leqq C(g_0)$ is g-convex.*

With Theorem 1, it is possible to see that we can spread the positive Ricci curvature from the point of everywhere positive Ricci curvature to all of M in a finite number of standard deformations thus proving

THEOREM 2. *Given (M, g_0) with nonnegative Ricci curvature and all Ricci curvatures positive at some point, then for any $r \geqq 4$, by a finite number of local convex conformal variations, g_0 can be perturbed to a C^r metric for M with everywhere positive Ricci curvature. Hence, M admits a smooth metric of everywhere positive Ricci curvature.*

Similarly, using local convex conformal variations, we obtain

THEOREM 3. *Given (M, g_0) with nonpositive Ricci curvature and all Ricci curvatures negative at some point, then M admits a metric of everywhere negative Ricci curvature.*

THEOREM 4. *Suppose (M, g_0) is λ-Ricci pinched with $0 < \lambda < 1/(n-1)$ and at some point for all vectors the pinching is not attained. Then M admits a metric which is λ'-Ricci pinched with $\lambda' > \lambda$.*

By studying arbitrary local convex variations, it is possible to see that the upper bound of $1/(n-1)$ for λ from the conformal variation cannot be improved upon.

This obstruction to improving Ricci pinching from the variational formulas is interesting in light of results of Yau [3], which show the Euler class and signature of M are obstructions to certain compact manifolds admitting metrics which are $\geqq (5/6)^{1/2}$ Ricci pinched.

By the method of Weinstein [4] integrating over the isometry group of M in the case that this group is compact, "equivariant" versions of Theorems 2—4 can be obtained.

Consider (M, g_0) with nonnegative sectional curvature $K_{g_0} \geqq 0$. In the proof of Theorem 2, the variational formula shows the existence of the outer annulus with $\mathrm{Ric}_{g(t)} > 0$. This was indicated by the fact that $\mathrm{Ric}' > 0$ in this outer annulus. This motivates us to define a local convex deformation $g(t)$ of g_0 with support in D to be *positive at first order* in some neighborhood $\mathscr{U}$ of $\mathrm{Bd}(D)$ iff the first derivative

$$K' := (d/dt)\, K_{g(t)}\big|_{t=0}$$

of the sectional curvature function $K_{g(t)}$ is positive on all 2-planes P with $\pi(P)$ in $\mathscr{U}$, where $\pi : G_2(M) \to M$ is the Grassmann bundle of all 2-planes in TM.

Flatten S^n, $n \geqq 3$, in a closed convex ball N about the north pole to get a metric g_0 for S^n with $K_{g_0}(P) = 0$ for all 2-planes P with $\pi(P)$ in N and $K_{g_0}(P) > 0$ for all 2-planes P with $\pi(P)$ in $S^n - N$. Let D be a small convex disk centered at p in $S^n - N$ with $\mathrm{Int}(D) \cap \mathrm{Int}(N) \neq \emptyset$.

THEOREM 5. *There is no positive at first order variation $g(t)$ through C^4 metrics with support in D for any neighborhood of* $\mathrm{Int}(N) \cap \mathrm{Bd}(D)$.

Finally, for dimension 2, motivated by the results of Kazden and Warner like that discussed elsewhere in these PROCEEDINGS, we consider the following:

Problem. Given (M^2, g) compact with nonpositive Gauss curvature and negative Gauss curvature at some point, can we find a smooth function $u : M \to R$ so that if $g(t) = e^{2tu}g_0$ then $K'(q) < 0$ for all q with zero g_0-Gauss curvature at q?

This problem is trivially seen to have a solution, and thus

THEOREM 6. *Given a compact 2-manifold (M, g_0) with nonpositive Gauss curvature and some point of negative Gauss curvature, then, by a global conformal variation, we can produce a metric for M with everywhere negative Gauss curvature.*

REFERENCES

1. T. Aubin, *Métriques riemanniennes et courbure*, J. Differential Geometry **4** (1970), 383–424. MR **43** #5452.

2. P. E. Ehrlich, *Metric deformations of Ricci and sectional curvature on compact manifolds*, Thesis, State University of New York at Stony Brook, Spring 1974.

3. S.T. Yau, *Curvature restrictions on four manifolds* (to appear).

4. A. Weinstein, *Positively curved deformations of invariant Riemannian metrics*, Proc. Amer. Math. Soc. **26** (1970), 151–156. MR **41** #7582.

STATE UNIVERSITY OF NEW YORK AT STONY BROOK

Proceedings of Symposia in Pure Mathematics
Volume 27, 1975

TRANSGRESSIONS, CHERN-SIMONS INVARIANTS AND THE CLASSICAL GROUPS

JAMES L. HEITSCH AND H. BLAINE LAWSON, JR.*

Generalizing his work with S. S. Chern, J. Simons has defined a family of natural transformations from principal bundles with connection over a manifold M to characters on the integral cycles of M. Provided that certain Pontrjagin forms vanish identically, these characters determine IR/Z cohomology classes, which for the tangent frame bundle of a riemannian manifold constitute a set of obstructions to conformally immersing the manifold into Euclidean space.

We give a short proof that the Simons characters associated to any bi-invariant metric connection on a compact Lie group G are identically zero. At the same time we are able to compute the Chern-Simons invariants in the frame bundle of G, and by using these more delicate obstructions we obtain:

THEOREM. *Let* $G = SO(2k + 1)$ *or* $U(2k + 1)$ *and let* Γ *be a discrete subgroup of* G ($\Gamma = \{identity\}$ *is allowed*). *Then* G/Γ *with a metric induced from a bi-invariant metric on* G *does not conformally immerse in Euclidean space* (*or the sphere*) *in codimension* $2k-1$.

UNIVERSITY OF CALIFORNIA, LOS ANGELES

AMS (MOS) subject classifications (1970). Primary 53A30, 55F40; Secondary 22E10, 22E15.
*Submitted to J. Differential Geometry.

Proceedings of Symposia in Pure Mathematics
Volume 27, 1975

A CLASS OF COMPACT MANIFOLDS WITH POSITIVE RICCI CURVATURE

HORACIO HERNÁNDEZ-ANDRADE

I. Introduction. The problem we are dealing with is part of an important question in global Riemannian geometry: To study compact manifolds of strictly positive curvature. So far, one of the main difficulties has been the lack of enough examples. In fact, all known examples of *compact* manifolds with positive sectional curvature are diffeomorphic to locally homogeneous spaces. It is somewhat surprising that even in the much more general case of positive Ricci curvature, no examples other than locally homogeneous spaces are known (up to diffeomorphism). In particular, it is an interesting question whether or not exotic spheres (which can never be homogeneous) admit metrics of positive sectional curvature, or at least positive Ricci curvature. Only this year, Cheeger [2] has found the first nonhomogeneous manifolds of a positive Ricci curvature (including the Kervaire spheres), although by means of a very isolated construction which is essentially intrinsic.

In this paper we shall give a very large new class of (in general not locally homogeneous) compact manifolds with positive Ricci curvature, including a very rich class of exotic spheres; compare §§ III, IV. These examples are Brieskorn varieties endowed with a metric induced from euclidean space after a modification of the standard embedding. The methods involved seem to promise to produce even more new examples of positively curved manifolds among certain algebraic varieties.

To determine properties of the curvature tensor explicitly for a given Riemannian manifold is, in general, very difficult. Basically our approach is very classical, namely, to study Riemannian submanifolds of euclidean spaces which are globally defined by equations. Our key observation is that there are many interesting

AMS (MOS) subject classifications (1970). Primary 53C20; Secondary 53C40.

examples of such manifolds which are *orthogonal* intersections of level surfaces of simple functions.

I would like to take the opportunity to thank my advisor, Professor D. Gromoll, for his kindness and essential help in many ways, to Professors W. Meyer and J. Cheeger for help in the very early stage of the work.

II. The Ricci curvature of a variety. In this section we first review the fundamental curvature quantities of a Riemannian manifold M. Our main objective is to estimate the curvature of M in *computable* terms. This problem seems to be least difficult for Riemannian submanifolds which are globally defined by equations and which, for our purposes, we call *varieties*. Even though our approach is very classical and familiar in the case of hypersurfaces, hardly any results have been known for higher codimension. We develop some new ideas on how to obtain estimates for the curvature of certain varieties in terms of the defining ideal of functions. We finally prepare all formulas for application in our study of Brieskorn varieties in § III. For basic definitions and facts in Riemannian geometry, we refer to [5], [8].

We consider an n-dimensional Riemannian manifold M with metric $\langle\ ,\ \rangle$ and Levi-Civita connection ∇. All data will always be of sufficiently high differentiability class, say C^∞ for convenience.

For $p \in M$, let M_p denote the tangent space of M at p. The *curvature tensor* R of M assigns to vector fields X, Y, Z a new vector field,

$$R(X, Y)Z = \nabla_X(\nabla_Y Z) - \nabla_Y(\nabla_X Z) - \nabla_{[X,Y]} Z. \tag{1}$$

One may consider R as a skew symmetric 2-form which assigns to $u, v \in M_p$ a skew adjoint endomorphism $R(u, v)$ of M_p.

Let $\delta \subset M_p$ be a 2-dimensional linear subspace of M_p and $u, v \in \delta$ independent vectors. Then the *sectional curvature* K_δ of M with respect to the plane δ is the real number defined by:

$$K_\delta = K(u, v) = \frac{\langle R(u, v)v, u\rangle}{\|u\|^2\,\|v\|^2 - \langle u, v\rangle^2}. \tag{2}$$

K_δ depends only on the plane δ.

The *Ricci tensor* S is the 2-form on M defined by:

$$S(u, v) = \mathrm{trace}(w \to R(w, u)v), \tag{3}$$

where $u, v, w \in M_p$. So, if $e_1, \cdots, e_n$ is any orthonormal basis of M_p,

$$S(u, v) = \sum_{j=1}^{n} \langle R(e_j, u)\,v, e_j\rangle. \tag{3'}$$

In particular, S is symmetric. The *Ricci curvature* $\mathrm{Ric}(u)$ is a real number associated with the 1-dimensional linear subspace of M_p generated by a vector $u \neq 0$,

$$\mathrm{Ric}(u) = (1/\|u\|^2)\, S(u, u). \tag{4}$$

If $\|u\| = 1$, we obtain using (2) and (3′),

$$\mathrm{Ric}(u) = \sum_{j=1}^{n-1} K(e_j, u), \tag{4′}$$

for any orthonormal basis $e_1, \cdots, e_{n-1}, u$ of M_p.

REMARK. Some authors call $(n-1)^{-1}\mathrm{Ric}(u)$ the Ricci curvature, the reason being that this number is exactly the average of all sectional curvatures with respect to planes δ which contain u. In our context, such a normalization is irrelevant.

By definition, M has positive (negative) Ricci curvature everywhere if and only if the Ricci tensor S is positive (negative) definite at any point $p \in M$.

The *scalar curvature* of M is a real value function s on M defined by

$$s = \mathrm{trace}\ S, \tag{5}$$

where the trace is taken pointwise with respect to the inner product $\langle\ ,\ \rangle$. Hence,

$$s(p) = \sum_{j=1}^{n} \mathrm{Ric}(e_j) = 2 \sum_{1 \leqq i < j \leqq n} K(e_i, e_j), \tag{5′}$$

for any orthonormal basis $e_1, \cdots, e_n$ of M_p.

Now we turn to the situation where M is a Riemannian submanifold of some Riemannian manifold $\tilde{M}$ of dimension $n + k$. So $M \subset \tilde{M}$, and the inclusion map i is an isometric imbedding. We identify the tangent space M_p with $i_* M_p \subset \tilde{M}_p$. The *normal* space of M at p is the orthogonal complement $M_p^\perp$ of M_p in $\tilde{M}_p$. For $w \in \tilde{M}_p$, we write $w^T \in M_p$, $w^\perp \in M_p^\perp$ for the tangent and normal components of $w = w^T + w^\perp$.

Let $\tilde{\nabla}$ denote the Levi-Civita connection of $\tilde{M}$. One has the following relation

$$\nabla_X Y = (\tilde{\nabla}_X Y)^T, \tag{6}$$

for vector fields X, Y on M, which we may also consider as vector fields in $\tilde{M}$ along M, i.e., along the inclusion i.

By applying (6) to (1), one can express the curvature tensor R of M in terms of the curvature tensor $\tilde{R}$ of $\tilde{M}$ and in terms of derivates involving only $\tilde{\nabla}$,

$$R(X, Y)Z = [\tilde{R}(X, Y)Z]^T - (\tilde{\nabla}_X(\tilde{\nabla}_Y Z)^\perp)^T + (\tilde{\nabla}_Y(\tilde{\nabla}_X Z)^\perp)^T. \tag{7}$$

Now one introduces the *second fundamental tensor* of M at p as a 1-form A on M_p with values in linear transformations $M_p^\perp \to M_p$ by

$$A(u) N_p = (\tilde{\nabla}_u N)^T, \tag{8}$$

where $u \in M_p$, $N_p \in M_p^\perp$, and N is any normal vector field along M with $N(p) = N_p$. The right-hand side of (8) depends only on N_p. Let A^* denote the 1-form on M_p with values in linear transformations $M_p \to M_p^\perp$ defined by $A^*(u) = A(u)^*$, the adjoint of $A(u)$. One has

$$A^*(u)v = -(\tilde{\nabla}_u Y)^\perp, \tag{9}$$

where $v \in M_p$, and Y is any tangential field along M with $Y(p) = v$. Using (8) and (9) in (7) we obtain the *Gauss equations*,

$$R(u, v) = \tilde{R}(u, v)^T + A(u) \circ A^*(v) - A(v) \circ A^*(u). \tag{10}$$

Of course, the "product" $A(u) \circ A^*(v)$ is the composition of linear transformations. Let us consider $\Delta = R - \tilde{R}^T$ as the curvature difference of M and $\tilde{M}$ at p, and let $A \wedge A^*$ denote the skew 2-form on M_p, with values in endomorphisms $M_p \to M_p$ given by

$$A \wedge A^*(u, v) = A(u) \circ A^*(v) - A(v) \circ A^*(u).$$

Then the Gauss equations (10) take the more suggestive form

$$\Delta = R - \tilde{R}^T = A \wedge A^*. \tag{11}$$

To work with the curvature difference Δ explicitly, one often chooses a basis $N_1, \cdots, N_k$, for $M_p^\perp$ and considers the selfadjoint endomorphisms A_λ of M_p with $A_\lambda u = A(u) N_\lambda$ for $1 \leqq \lambda \leqq k$. In many cases, it need not necessarily be advantageous to assume that the fixed basis is orthonormal, for example, when M is a variety. For any $a \in M_p^\perp$, we have the identity

$$a = \sum_{\lambda, \mu=1}^{k} \Phi_{\lambda\mu} \langle a, N_\lambda \rangle N_\mu,$$

where $\Phi_{\lambda\mu}$ is the inverse matrix of $\langle N_\mu, N_\lambda \rangle$. Since $\langle A^*(u)w, N_\lambda \rangle = \langle w, A(u) N_\lambda \rangle = \langle w, A_\lambda u \rangle$,

$$A^*(u)\, w = \sum_{\lambda, \mu=1}^{k} \Phi_{\lambda\mu} \langle A_\lambda u, w \rangle N_\mu.$$

Therefore, (10) becomes

$$\Delta(uv)\, w = \sum_{\lambda, \mu=1}^{k} \Phi_{\lambda\mu}(\langle A_\lambda v, w \rangle A_\mu u - \langle A_\lambda u, w \rangle A_\mu v). \tag{12}$$

If $N_1, \cdots, N_k$ are *orthonormal*,

$$\Delta(u, v)\, w = \sum_{\lambda=1}^{k} (\langle A_\lambda v, w \rangle A_\lambda u - \langle A_\lambda u, w \rangle A_\lambda v). \tag{12'}$$

To deal with the curvature tensor R in many applications, the Gauss equations can only be used successfully if, at least locally, k independent normal fields of M can be found such that the transformations A_λ in (12) are computable, provided in addition, that $\tilde{R}$ is known. Moreover, the codimension k should be small compared to $n = \dim M$. The most favorable situation seems to be when M is defined by fairly simple "equations". This was studied to some extent by Dombrowski, whose paper [3] inspired our investigation.

Recall that the gradient of a function f defined in some open subset U of $\tilde{M}$ is the vector field ∇f on U such that $\langle \nabla f, X \rangle = Xf$ for all vector fields X on U. At $p \in \tilde{M}$ one has $\nabla f_{|p} = \sum_{i=0}^{n+k} (e_i f)\, e_i$ for any orthonormal basis $e_1, \cdots, e_{n+k}$ of $\tilde{M}_p$.

We say that a subset M of $\tilde{M}$ is a variety in $\tilde{M}$ if there is an open neighborhood U of M in $\tilde{M}$ and an ideal I_M of functions in the ring $C^\infty(U)$ of real valued C^∞-functions on U such that:

(a) $M = \{p \mid f(p) = 0 \text{ for all } f \in I_M\}$,

(b) I_M is generated by $f_1, \cdots, f_k$ and $\nabla f_{1|p}, \cdots, \nabla f_{k|p}$ are linearly independent for all $p \in M$.

Clearly, $k \leqq \dim \tilde{M}$, and necessarily, M is a closed submanifold of U, of codimension k, since 0 is a regular value of $F : U \to R^k$ by (b), $F = (f_1, \cdots, f_k)$, and $M = F^{-1}(0)$ by (a).

PROPOSITION II.1. *The submanifold M of $\tilde{M}$ is a variety if and only if the normal bundle ν_M of M is trivial.*

PROOF. We only sketch the straightforward argument. Clearly, if M is a variety, the vector fields $\nabla f_1, \cdots, \nabla f_k$ give a parallelization of ν_M. Conversely, assume that ν_M is trivial. Then we can find $k = \operatorname{codim} M$ orthonormal vector fields $N_1, \cdots, N_k$ along M.

The exponential map exp of $\tilde{M}$ maps some open neighborhood V of the zero section in ν_M diffeomorphically onto a neighborhood W of M in $\tilde{M}$. Let ν_i denote the subbundle of ν_M whose fiber $\nu_{i|p}$ over $p \in M$ is the linear hyperplane in $M_p^\perp$ perpendicular to $N_{i|p}$, $1 \leqq i \leqq k$. Let $V_i = V \cap \nu_i$. Then exp maps V_i diffeomorphically onto a hypersurface W_i such that $M = W_1 \cap \cdots \cap W_k$. Moreover, M is an "orthogonal intersection of hypersurfaces", i.e., the normal vectors $N_{i|p}$ of W_i at $p \in M$ are orthogonal. Let $\hat{V}_i$ denote the trivial normal line bundle of W_i in $\tilde{M}$ and consider the section $\hat{N}_i$ over W_i such that $\|\hat{N}_i\| = 1$ and $\hat{N}_i \mid M = N_i$. We have the function $\varphi_i : \hat{V}_i \to R$ with $\varphi_i(w) = \langle w, \hat{N}_i \rangle$. Again the exponential map exp of $\tilde{M}$ maps an open neighborhood $\hat{U}_i$ of the zero section in $\hat{V}_i$ diffeomorphically onto a neighborhood U_i of W_i in $\tilde{M}$. Let f_i be the restriction of $\varphi_i \circ (\exp \mid U_i)^{-1}$ to $U = U_1 \cap \cdots \cap U_k$. Then $\nabla f_{i|p} = N_{i|p}$ for $p \in M$ and $f_i^{-1}(0) = W_i$. Hence M is a variety with defining ideal I_M generated by $f_1, \cdots, f_k$ in $C^\infty(U)$.

We have two immediate consequences.

COROLLARY II.2. (1) *Any submanifold M of $\tilde{M}$ is locally a variety, i.e., every point $p \in M$ has an open neighborhood in M which is a variety in $\tilde{M}$.*

(2) *The defining ideal I_M of any variety M in $\tilde{M}$ has an "orthonormal generator system", i.e., I_M can be generated by $f_1, \cdots, f_k$ such that $\nabla f_1, \cdots, \nabla f_k$ are orthonormal along M.*

We want to emphasize that the whole point for introducing varieties is that they are *globally defined by functions*. Global curvature estimates can be obtained just by studying first and second derivates of k generating functions, as discussed next. This is usually a less difficult problem than working in the general case, in particular, when the defining functions and the ambient space are fairly "simple". We should mention that the second part of Corollary II.2 is mainly of theoretical interest. In examples, generating functions hardly ever form an orthonormal system, and explicit orthonormalization is in general too complicated to be of any practical use. However, one of our crucial observations is that sometimes the defining functions of a variety M in $\tilde{M}$ can be slightly modified to yield at least an orthogonal generator system for a new ideal $I_{M'}$ which defines a variety M' diffeomorphic to M.

For a real valued function f defined on some open subset U of $\tilde{M}$ and $p \in U$, the *hessian tensor* is the selfadjoint endomorphism $H_f : \tilde{M}_p \to \tilde{M}_p$ with

$$H_f v = \nabla_v \nabla f. \tag{13}$$

H_f is given by the matrix of second partial derivates of f at p in the case $\tilde{M} = R^k$.

Let M be a variety in $\tilde{M}$ defined by the functions $f_1, \cdots, f_k$ on $U \supset M$. Setting $H_\lambda = H_{f_\lambda}$ and $N_\lambda = \nabla f_\lambda \mid M$, we obtain from (12) and (13) for the curvature difference tensor Δ of M in $\tilde{M}$,

$$\Delta (u, v) w = \sum_{\lambda=1}^{k} \frac{1}{\|\nabla f_\lambda\|^2} (\langle H_\lambda v, w\rangle H_\lambda^T u - \langle H_\lambda u, w\rangle H_\lambda^T v). \tag{14}$$

Here, $H_\lambda^T : M_p \to M_p$ is the tangential projection of H_λ, i.e., $H_\lambda^T u = (H_\lambda u)^T$. From now on we restrict attention to the case where M is a variety in flat euclidean space $\tilde{M} = R^{n+k}$, so $\Delta = R$. Furthermore, we always assume that $\nabla f_1, \cdots, \nabla f_k$ are mutually orthogonal along M. Then for the sectional curvature of M, we get by using (14) in (2),

$$K_\delta = K (u, v) = \sum_{\lambda=1}^{k} \frac{1}{\|\nabla f_\lambda\|^2} (\langle H_\lambda u, u\rangle \langle H_\lambda v, v\rangle - \langle H_\lambda u, v\rangle^2), \tag{15}$$

where $u, v \in M_p$ orthonormal. Similarly, using (14′) in (3′) and (5′) yields for Ricci and scalar curvature:

$$s(u, u) = \sum_{\lambda=1}^{k} \frac{1}{\|\nabla f_\lambda\|^2} (\langle H_\lambda u, u\rangle \operatorname{tr} H_\lambda^T - \|H_\lambda^T u\|^2), \tag{16}$$

$$s(p) = \sum_{\lambda=1}^{k} \frac{1}{\|\nabla f_\lambda\|^2} ((\operatorname{tr} H_\lambda^T)^2 - \operatorname{tr} (H_\lambda^T)^2). \tag{17}$$

Now let $\tilde{M} = C^n = R^n \oplus iR^n \cong R^{2n}$ complex n-space, where we identify $z = (z_1, \cdots, z_n) \in C^n$ with $(x_1, \cdots, x_n; y_1, \cdots, y_n) \in R^{2n}$, $z_k = x_k + iy_k$ · Let $\langle u, v\rangle = \sum_{k=1}^n u_k \bar{v}_k$ denote the canonical hermitian inner product for C^n. Then the real part $\operatorname{Re}\langle u, v\rangle$ is the canonical inner product for $R^{2n} \cong C^n$. Suppose $f : U \to C$ is a holomorphic function defined on some open subset U of C^n. Setting $\varphi = \operatorname{Re} f$ and $\psi = \operatorname{Im} f$, $f = \varphi + i\psi$ satisfies the Cauchy-Riemann equations,

$$\nabla \varphi = - i \nabla \psi. \tag{18}$$

Hence,

$$\operatorname{Re} \langle \nabla \varphi, \nabla \psi\rangle = 0, \tag{19}$$

so $\nabla \varphi$, $\nabla \psi$ are always *orthogonal* in R^{2n}. The *complex gradient* of f is the vector field $\nabla f = (\overline{\partial f/\partial z_1}, \cdots, \overline{\partial f/\partial z_n})$ on U. We have

$$\nabla f = \tfrac{1}{2}(\nabla\varphi - i\nabla\psi) = \nabla\varphi = -i\nabla\psi. \tag{20}$$

The *complex hessian* H of f at $p \in U$ is a C-linear endomorphism of $C^n \cong \tilde{M}_p$ given by the matrix $((\partial^2 f/\partial z_j\, \partial z_k)\,|\,p)$ with respect to the canonical basis of C^n, $1 \leqq j, k \leqq n$. Now suppose, we are also given a real valued function g on U, such that the ideal generated by $\varphi = \operatorname{Re} f$, $\psi = \operatorname{Im} f$, g defines a variety M in C^n. So M is a real submanifold of codimension 3 in C^n which is the intersection of the complex hypersurface $f^{-1}(0)$ and the real hypersurface $g^{-1}(0)$. Now let us assume in addition that the complex gradient ∇f and the real gradient ∇g are complex orthogonal, $\langle \nabla f, \nabla g\rangle = 0$. Then we obtain the following result for the curvature of M. Using the Cauchy-Riemann equations, the complex hessian H of f and the real hessians H_φ, H_ψ are easily seen to be related as follows:

$$\langle Hu, \bar{v}\rangle = \operatorname{Re}\langle H_\varphi u, v\rangle + i \operatorname{Re}\langle H_\psi u, v\rangle. \tag{20'}$$

LEMMA II.3. *Let u be a tangent vector of M at p. Then*

$$\begin{aligned} S(u, u) = \frac{1}{\|\nabla f\|^2}(&- \operatorname{Re}\langle Hu, \bar{u}\rangle \overline{\langle H\ \nabla g/\|\nabla g\|\overline{\nabla g},\ /\|\nabla g\|\rangle} - 2\|Hu\|^2 \\ &+ 2|\langle Hu, \nabla f/\|\nabla f\|\rangle|^2 + |\langle Hu, \overline{\nabla g/}\|\nabla g\|\rangle|^2) \\ &+ \|\nabla g\|^{-2}(\langle H_g u, u\rangle \operatorname{tr} H_g^T - \|H_g^T u\|^2), \end{aligned} \tag{21}$$

$$\begin{aligned} s(p) = \frac{4}{\|\nabla f\|^2}\Bigg(&\left\|H\frac{\nabla g}{\|\nabla g\|}\right\|^2 + 2\left\|H\frac{\nabla f}{\|\nabla f\|}\right\|^2 - \left|\left\langle H\frac{\nabla f}{\|\nabla f\|}, \frac{\overline{\nabla f}}{\|\nabla f\|}\right\rangle\right|^2 \\ &-\left|\left\langle H\frac{\nabla f}{\|\nabla f\|}, \frac{\overline{\nabla g}}{\|\nabla g\|}\right\rangle\right|^2 - \operatorname{tr} H\bar{H}\Bigg) \\ &+ \frac{1}{\|\nabla g\|^2}((\operatorname{tr} H_g^T)^2 - \operatorname{tr}(H_g^T)^2). \end{aligned} \tag{22}$$

PROOF. The above formulas follow straightforwardly from (16), (17), and (20), using a complex orthonormal basis $e_1, \cdots, e_{n-2}$, $(\nabla g/\|\nabla g\|)\,|\,p$, $(\nabla f/\|\nabla f\|)\,|\,p$ for C^n. Note that $e_1, ie_1, \cdots, e_{n-2}, ie_{n-2}, i(\nabla g/\|\nabla g\|)\,|\,p$ is a real orthonormal basis of the tangent space of M at p.

In applications, formulas (21) and (22) can be used advantageously in particular when the function g is very simple, as we will study now in the next section.

III. Results for Brieskorn varieties. Let $n \geqq 3$ and $a_1 \geqq a_2 \geqq \cdots \geqq a_n \geqq 2$ integers. Consider the polynomial $f : C^n \to C$, $f(z) = \sum_{k=1}^n z_k^{a_k}$, and the function $g : C^n \to R$, $g(z) = \sum_{k=1}^n z_k \bar{z}_k - 1$. The intersection of the complex hypersurface $f^{-1}(0)$, which is singular only at $z = 0$, and the euclidean sphere $g^{-1}(0)$ is a compact submanifold of C^n with real codimension 3, the *Brieskorn variety* $V(a_1, \cdots, a_n)$. These manifolds have been studied extensively in recent years from various viewpoints and turned out to be extremely important and interesting. (For a general account, compare [9].) In differential geometry, of course, one is interested in their curvature behavior. Clearly $V(a_1, \cdots, a_n)$ is a variety in our sense defined by the ideal generated by $(\operatorname{Re} f, \operatorname{Im} f, g)$. It is very easy to see, and has been observed by many people, that $V(a_1, \cdots, a_n)$ with its induced metric always has sectional curvature of either sign. Our starting point was to study the Ricci tensor of V, which in some very special cases turned out to be positive, for example, when $a_1 = \cdots = a_n = 2$. However, already in the fairly simple case $a_1 = 3$, $a_2 = \cdots = a_n = 2$ (which contains exotic spheres), the formulas became too complicated to work with. In our attempt to make computations more accessible, we discovered that there is a simple modification of the imbedding of V in C^n which is an *orthogonal* intersection of a perturbed complex hypersurface and an ellipsoid. Then the fairly simple formula (21) is applicable, and that makes it possible to obtain estimates for the Ricci curvature.

Let $\alpha_1, \cdots, \alpha_n > 0$ and $\omega_1, \cdots, \omega_n > 0$ real numbers. Then we consider the generalized Brieskorn variety $V'(a_1, \cdots, a_n) = f^{-1}(0) \cap g^{-1}(0)$, where $f(z) = \sum_{k=1}^n \alpha_k z_k^{a_k}$, $g(z) = \sum_{k=1}^n \omega_k z_k \bar{z}_k - 1$. Note that on V', the complex gradient ∇f and the real gradient ∇g are independent over C: For $z \neq 0$,

$$\nabla f_{|z} = (\alpha_1 a_1 \bar{z}_1^{a_1-1} \cdots, \alpha_n a_n \bar{z}_n^{a_n-1}) \neq 0,$$
$$\nabla g_{|z} = 2(\omega_1 z_1, \cdots, \omega_n z_n) \neq 0.$$

Now if for some $0 \neq \mu \in C$ and $z \in V'$, $\nabla f_{|z} = \mu \nabla g_{|z}$, we would have

$$\alpha_k a_k \bar{z}_k^{a_k-1} = 2\mu\omega_k z_k, \quad \text{so } \alpha^k \bar{z}_k^{a_k} \mu = (2\mu\omega_k/a_k) z_k \bar{z}_k.$$

Summing over k yields $0 = 2\mu \sum_{k=1}^n (\omega_k/a_k) z_k \bar{z}_k$, which is impossible. Moreover, we have the following fact.

PROPOSITION III.1. *$V'(a_1, \cdots, a_n)$ and $V(a_1, \cdots, a_n)$ are diffeomorphic and isotopic in C^n.*

PROOF. Let $H : C^n \times [0, 1] \to C \times R$ defined by

$$H(z, t) = \left(\sum_{k=1}^n \alpha_k(t) z_k^{a_k}, \sum_{k=1}^n \omega_k(t) z_k \bar{z}_k \right),$$

where $\alpha_k(t) = 1 - t + t\alpha_k > 0$, $\omega_k(t) = 1 - t + t\omega_k > 0$. By the above argument, 0 is a regular value of H. So, $Q = H^{-1}(0)$ is a compact submanifold with boundary $V \times 0 \cup V' \times 1$. The restriction λ of the function $(z, t) \to t$ to Q has no critical points, $\lambda^{-1}(0) = V \times 0$, $\lambda^{-1}(1) = V' \times 1$. Hence, Q is diffeomorphic to $V \times [0, 1]$, and this completes the argument.

Now once and for all, given $a_1 \geqq a_2 \geqq \cdots \geqq a_n \geqq 2$ we choose $\alpha_k = 1/a_k(a_k - 1)$ and $\omega_k = 1/a_k$ and consider always the Brieskorn variety,

(23) $$V'(a_1, \cdots, a_n) = f^{-1}(0) \cap g^{-1}(0),$$

where $f(z) = \sum_{k=1}^n z_k^{a_k}/a_k(a_k - 1)$, $g(z) = \sum_{k=1}^n z_k \bar{z}_k/(a_k - 1)$, endowed with the induced Riemannian structure as a submanifold of C^n. We have

(24) $$\nabla f_{|z} = (\bar{z}_1^{a_1-1}/(a_1 - 1), \cdots, \bar{z}_n^{a_n-1}/(a_n - 1)),$$

(25) $$\nabla g_{|z} = 2(z_1/a_1, \cdots, z_n/a_n).$$

Hence, if $z \in V'$ it follows using (23) that

(26) $$\langle \nabla f, \nabla g \rangle_{|z} = 0.$$

We are now in a situation where the techniques developed at the end of §II can be applied. In particular, we want to find lower bounds for the Ricci curvature of V' using the formula (21). For this purpose, we need the following quantities:

$$\|\nabla f\|^2 = \sum_{k=1}^{n} \frac{|z_k|^{2(a_k-1)}}{(a_k-1)^2}, \tag{27}$$

$$\|\nabla g\|^2 = 4\sum_{k=1}^{n} \frac{|z_k|^2}{a_k^2}, \tag{28}$$

$$H = \begin{pmatrix} z_1^{a_1-2} & & \mathbf{0} \\ & \ddots & \\ \mathbf{0} & & z_n^{a_n-2} \end{pmatrix}, \tag{29}$$

$$\operatorname{Re}\langle H_g u, v\rangle = 2\operatorname{Re}\sum_{k=1}^{n} \frac{u_k \bar{v}_k}{a_k}. \tag{30}$$

The norm of H at z is the number

$$\|H\| = \sup_{C^n \ni u \neq 0} \frac{\|Hu\|}{\|u\|} = \max_k |z_k|^{a_k-2}. \tag{31}$$

LEMMA III.2. *A lower bound for the Ricci curvature of V' is given by*

$$\operatorname{Ric}(u) \geqq -3 \sup_{Z\in V'} \frac{\|H\|^2}{\|\nabla f\|^2} + \frac{a_n}{a_1^2}(2n-4). \tag{32}$$

In the case of equal exponents $a_1 = \cdots = a_n = p$ we have a better bound,

$$\operatorname{Ric}(u) \geqq -2 \sup_{Z\in V'} \frac{\|H\|^2}{\|\nabla f\|^2} + \frac{2n-4}{p}. \tag{32'}$$

PROOF. We may assume $\|u\| = 1$. Thus, from (21) we obtain

$$\begin{aligned} S(u,u) \geqq &-\frac{1}{\|\nabla f\|^2}\left(|\langle Hu, \bar{u}\rangle|\left|\left\langle H\frac{\nabla g}{\|\nabla g\|}, \frac{\overline{\nabla g}}{\|\nabla g\|}\right\rangle\right| + 2\|Hu\|\right) \\ &+ \frac{1}{\|\nabla g\|^2}(\langle H_g u, u\rangle \operatorname{tr} H_g^T - \|H_g^T u\|^2). \end{aligned} \tag{33}$$

Now

$$|\langle Hu, \bar{u}\rangle|\,|\langle H\nabla g/\|\nabla g\|, \overline{\nabla g}/\|\nabla g\|\rangle| \leqq \|H\|^2.$$

Note that $\langle H\nabla g/\|\nabla g\|, \overline{\nabla g}/\|\nabla g\|\rangle = 0$ in the case of equal exponents.

For the second term in (33) we get, using (15),

$$(1/\|\nabla g\|^2)(\langle H_g u, u\rangle \operatorname{tr} H_g^T - \|H_g^T u\|^2) \geqq (2n-4)K_{\min},$$

where $K_{\min}$ is the minimal sectional curvature of the ellipsoid $g^{-1}(0)$ in $R^{2n} \cong C^n$. Then the second term in the lower bound for $\operatorname{Ric}(u)$ in (32) follows from

$$K_{\min} = a_n/a_1^2. \tag{34}$$

To see that let u, v be orthonormal tangent vectors of the ellipsoid at $z \in g^{-1}(0)$. Then

$$K(u, v) = \frac{\langle H_g u, u\rangle \langle H_g v, v\rangle - \langle H_g u, v\rangle^2}{\|\nabla g\|^2}.$$

From (28) we obtain

$$\|\nabla g\|^2 = 4 \sum_{k=1}^{n} \frac{|z_k|^2}{a_k^2} \leqq \frac{4}{a_n} \sum_{k=1}^{n} \frac{|z_k|^2}{a_k} = \frac{4}{a_n}.$$

So

$$\|\nabla g\|^2 \leqq 4/a_n, \tag{35}$$

with equality holding for $z_0 = (0, \cdots, (a_n)^{1/2})$. On the other hand, setting $u_k' = \alpha_k u_k$, $v_k' = \alpha_k v_k$ for $1 \leqq k \leqq 2n$, $\alpha_k = (2/a_k)^{1/2}$ for $1 \leqq k \leqq n$, $\alpha_{k+n} = \alpha_k$,

$$\begin{aligned}\langle H_g u, u\rangle \langle H_g v, v\rangle - \langle H_g u, v\rangle^2 &= \langle u', u'\rangle \langle v', v'\rangle - \langle u', v'\rangle^2 \\ &= \sum_{1\leqq k<l\leqq 2n} (u_k' v_l' - u_l' v_k')^2 \\ &= \sum_{1\leqq k<l\leqq 2n} \alpha_k^2 \alpha_l^2 (u_k v_l - u_l v_k)^2 \\ &\geqq \min_{k<l} \alpha_k^2 \alpha_l^2 \sum_{1\leqq k<l\leqq 2n} (u_k v_l - u_l v_k)^2 \\ &= \min_{k<l} \alpha_k^2 \alpha_l^2.\end{aligned}$$

So, $\langle H_g u, u\rangle \langle H_g v, v\rangle - \langle H_g u, v\rangle^2 \geqq 4/a_1^2$, using the Lagrange identity twice and observing that u, v orthonormal. Equality holds at $z_0 = (0, \cdots, 0, (a_n)^{1/2})$ for the tangent vectors $u = (1, 0, \cdots, 0)$, $v = iu$, of the ellipsoid. Hence,

$$K(u, v) \geqq a_n/a_1^2 = K_{\min}.$$

REMARK. In the case $a_{n-1} = a_n$, the estimate for the second term in (32) can be substantially improved as follows:

$$\mathrm{Ric}(u) \geqq -3 \frac{\|H\|^2}{\|\nabla f\|^2} + \frac{a_n}{a_1}\left(\frac{1}{a_1} + 2\sum_{k=2}^{n-2} \frac{1}{a_k} + \frac{1}{a_n}\right), \tag{32''}$$

where the sum $\sum_{k=2}^{n-2} (1/a_k)$ is understood to be zero for $n = 3$.

PROOF. Choosing an orthonormal basis $e_1, \cdots, e_{2n}$ for $R^{2n} \cong C^n$ such that $e_1, \cdots, e_{2n-3}$ span the tangent space of V' at z, we have in (33),

$$\begin{aligned}\tfrac{1}{2} \operatorname{tr} H_g^T &= \tfrac{1}{2} \operatorname{tr} H_g - \tfrac{1}{2} \sum_{k=2n-2}^{2n} \langle H_g e_i, e_i\rangle \\ &\geqq \tfrac{1}{2} \operatorname{tr} H_g - 3/a_n = 2(1/a_1 + \cdots + 1/a_n) - 3/a_n \\ &= 2(1/a_1 + \cdots + 1/a_{n-1}) - 1/a_n = \rho > 0.\end{aligned}$$

Since $\|H_g^T u\|^2 \leqq \|H_g u\|^2$,

$$\begin{aligned}&\langle H_g u, u\rangle \operatorname{tr} H_g^T - \|H_g^T u\|^2 \\ &\qquad \geqq \langle (2\rho H_g - H_g^2) u, u\rangle \geqq 4 \min_k \frac{1}{a_k}\left(\rho - \frac{1}{a_k}\right) = 4\frac{1}{a_1}\left(\rho - \frac{1}{a_1}\right),\end{aligned}$$

using $a_{n-1} = a_n \cdot$ Combining this estimate with (35) yields (32″).

We now turn to a result on the Ricci curvature of Brieskorn varieties defined by homogeneous polynomials.

THEOREM III.3. *For any $b \geqq 2$, there exists an integer $N(b)$ such that in any dimension $n \geqq N(b)$, $V'(a_1, \cdots, a_n)$, $a_1 = \cdots = a_n = b$, has strictly positive Ricci curvature.*

PROOF. Fixing $b \geqq 2$ it suffices to show

$$\frac{2n-4}{b} > 2 \sup_{Z \in V'} \frac{\|H\|^2}{\|\nabla f\|^2} \tag{36}$$

for n sufficiently large, according to (32′). To prove this we shall construct a suitable upper bound for $\|H\|^2/\|\nabla f\|^2$. From (27) and (31) we obtain

$$\frac{\|H\|^2}{\|\nabla f\|^2} = \frac{(\max|z_k|^{b-2})^2}{(b-1)^2 \sum_{k=1}^n |z_k|^{2(b-1)}} = (b-1)^2 \frac{(\max_k|z_k|^2)^{b-2}}{\sum_{k=1}^n |z_k|^{2(b-1)}}.$$

Setting $\rho_k = |z_k|^2/b$ we have $\sum_{k=1}^n \rho_k = 1$ by (23). We may assume $\rho_1 \geqq \rho_2 \geqq \cdots \geqq \rho_n \geqq 0$. Then,

$$\frac{\|H\|^2}{\|\nabla f\|^2} = \frac{(b-1)^2}{b} \frac{{\rho_1}^{b-2}}{\sum_{k=1}^n {\rho_k}^{b-1}} = \frac{(b-1)^2}{b} A. \tag{37}$$

To maximize (37) we have to minimize the reciprocal

$$1/A = \rho_1(1 + (\rho_2/\rho_1)^{b-1} + \cdots + (\rho_n/\rho_1)^{b-1})$$

under the above constraints. Put $0 \leqq \delta_k = \rho_k/\rho_1 \leqq 1$ for $2 \leqq k \leqq n$. So we have to consider

$$1/A = \rho_1(1 + \delta_2^{b-1} + \cdots + \delta_n^{b-1}),$$

with the constraint $\rho_1(1 + \delta_2 + \cdots + \delta_n) = 1$, $1/n \leqq \rho_1 \leqq 1$. One easily verifies that $1/A$ assumes its minimum necessarily when $\delta_2 = \cdots = \delta_n$. Hence,

$$\frac{1}{A} \geqq \min \alpha(x), \qquad \alpha(x) = \frac{1 + (n-1)x^{b-1}}{1 + (n-1)x}, \qquad 0 \leqq x = \delta_2 \leqq 1.$$

Since $\alpha(0) = \alpha(1) = 1$ and $x^{b-1} \leqq x$, α assumes its minimum at some interior point $0 < x_0 < 1$. Observe $\alpha(x) \equiv 1$ for $b = 2$, so let $b \geqq 3$. Differentiation yields that x_0 is the unique positive root of the equation

$$(p-2)(n-1)x^{b-1} + (b-1)x^{b-2} - 1 = 0.$$

Therefore,

$$1/A \geqq \alpha(x_0) = (b-1)x_0^{b-2}. \tag{38}$$

As we cannot deal explicitly with x_0 for $b \geqq 5$, we shall find a lower bound for $\alpha(x_0)$, which suffices to prove the theorem. Clearly, $A \leqq (b-1)^{-1}y_0^{b-2}$, where

$y_0 = 1/x_0$ satisfies

$$Q(y) = y^{b-1} - (b-1)y - (b-2)(n-1) = 0.$$

It follows that $(b-1)^{-1} y_0{}^{b-2} = 1 + (b-1)^{-1}(b-2)(n-1)/y_0$, so

$$A \leqq 1 + \frac{b-2}{b-1}(n-1)\frac{1}{y_0}. \tag{39}$$

We have to find a lower bound for y_0. Since Q has only one positive root and $Q(0) < 0$, we have $y < y_0$ whenever $Q(y) < 0$. Choose $y_1 = [(b-2)(n-1)]^{1/(b-1)}$, clearly $Q(y_1) < 0$. Hence, (39) implies

$$\begin{aligned} A &\leqq 1 + \frac{b-2}{b-1}(n-1)\frac{1}{y_1} \\ &= 1 + \frac{1}{b-1}[(b-2)(n-1)]^{(b-2)/(b-1)}. \end{aligned} \tag{40}$$

Combining (37) and (40) yields (36), which completes the arugment.

We will now derive a general result for a large class of Brieskorn varieties.

THEOREM III.4. *Given arbitrary integers $a_1 \geqq a_2 \geqq \cdots \geqq a_m \geqq 2$, then there exists an integer $N(a_1, \cdots, a_m)$ such that the Brieskorn variety $V'(a_1, \cdots, a_m, a_{m+1}, \cdots, a_{m+n})$, $a_{m+1} = \cdots = a_{m+n} = 2$, has strictly positive Ricci curvature whenever $n \geqq N(a_1, \cdots, a_m)$.*

PROOF. As in the proof of the last theorem, it suffices to show

$$\frac{4}{a_1^2}(n+m-2) > 3 \sup_{Z \in V'} \frac{\|H\|^2}{\|\nabla f\|^2}, \tag{39}$$

for n sufficiently large, according to (32). We shall show that $\|H\|^2/\|\nabla f\|^2$ has an upper bound which only depends on $a_1, \cdots, a_m$ (and not on n). Then (39) follows. Clearly,

$$\|H\|^2 = (\max |z_k|^{a_k-2})^2 = \max_k |z_k|^{2(a_k-2)} < a_1^{a_1-2},$$

since $|z_k|^2 < a_k$ for all k by (23). On the other hand,

$$\|\nabla f\|^2 = \sum_{k=1}^{m} \frac{|z_k|^{2(a_k-1)}}{(a_k-1)^2} + \sum_{k=m+1}^{m+n} |z_k|^2.$$

Using (23) again, we conclude that if $\sum_{k=m+1}^{m+n} |z_k|^2 \leqq 1$, then $\sum_{k=1}^{m} |z_k|^2/a_k \geqq \frac{1}{2}$, so

$$\sum_{k=1}^{m} \frac{|z_k|^{2(a_k-1)}}{(a_k-1)^2}$$

has a positive minimum $B(a_1, \cdots, a_m)$ on the compact set in C^m defined by $\frac{1}{2} \leqq \sum_{k=1}^{m} |z_k|^2/a_k \leqq 1$. Hence,

$$\|\nabla f\|^2 \geqq \min(B(a_1, \cdots, a_m), 1).$$

REMARK. The original proof of this theorem is geometrical using the fact that the orthogonal group $O(n)$ acts transitively on the Grassmannian $G_{n,2}$ of 2-planes in R^n.

IV. Further remarks. We shall briefly discuss the scope of the results proved in the last section and sketch some applications. We mention first that there are Brieskorn varieties which do not admit *any* metric with positive Ricci curvature (in fact, not even with nonnegative Ricci curvature). This follows from Orlik's result in [7] saying that the fundamental group π_1 of $V(a_1, a_2, a_3)$ is infinite for $1/a_1 + 1/a_2 + 1/a_3 \leqq 1$ and from Myer's classical theorem according to which π_1 must be finite if there exists a metric of positive Ricci curvature. In higher dimensions, similar counterexamples are not known. However, it is easy to see that $V'(a_1, \cdots, a_n)$ with $a_1 = \cdots = a_n = b$ has negative Ricci curvature at $z = (b/n)^{1/2}(1, \cdots, 1, -1, \cdots, -1)$ for n even, b odd, and $\dim V' = 2n - 3 < 2(b-1)^2$. Thus, $\mathrm{Ric}(v) < 0$ for $0 \neq v \perp i\nabla g$ and $\mathrm{Ric}(i\nabla g) > 0$. The scalar curvature of V' at z is strictly negative.

On the other hand, working more explicitly with our estimate that appears in the proof of Theorem III.3, one obtains that the Ricci curvature of $V(b, \cdots, b)$ is positive for $n > [2(b-1)^2]^{b-1} + 2$.

Detailed computations give better bounds, for example $N(3) \leqq 11$. The fact that the Ricci tensor of $V'(b, \cdots, b)$ is only necessarily positive if the degree b of f is small compared to the dimension n looks somewhat related to the behavior of the first Chern class of the associated projective variety (over which V' is a circle bundle).

In this context we mention that the "affine quadric" $V'(2, \cdots, 2)$ is the Stiefel manifold of real 2-frames in R^n with some homogeneous metric. The computation of the Ricci curvature is a very special and simple case in our general framework. For a unit tangent vector u at $z \in V'$, $n \geqq 3$, we obtain from (21),

$$\mathrm{Ric}(u) = (n-3) + \langle u, iz/(2)^{1/2}\rangle^2.$$

Hence, $n - 3 \leqq \mathrm{Ric}(u) \leqq n - 2$, where the bounds are assumed at any point.

In the general case of Theorem III.4, a very rough explicit lower bound for the number of squares that have to be added to yield positive Ricci curvature is given by

$$N(a_1, \cdots, a_m) \leqq (2ma_1)^{a_1}.$$

If one is interested in substantially better bounds, then it seems to be more efficient to deal with each example in question separately, rather than to work in the general case which is difficult from a numerical point of view. For example, if $m = 1$, then $N(3) \leqq 5$. The general estimates given in § III cannot be improved by too much, and even fairly optimal estimates for special examples indicate that $N(a_1, \cdots, a_m)$ will be very large compared to a_1. Aside from this, it does not seem to be possible to generalize Theorem III.4 in any obvious way, say to add cubes instead of squares. Upper bounds for $\|H\|^2/\|\nabla f\|^2$ will depend on the dimension of V' such that (32) cannot be satisfied. We also remark that using $g(z) = \sum_{k=1}^{n} z_k \bar{z}_k / a_k - r^2$ with some other radius $r > 0$ in (23) does not make much difference in the basic estimates. In fact, $r = 1$ appears to be fairly optimal.

Of course, the essential content of Theorem III.4 is the *existence* of a very large infinite class of compact manifolds among Brieskorn varieties which carry a metric with strictly positive Ricci curvature in a fairly natural way. Even though we get only finitely many examples in a given dimension n, their number increases rapidly with n. Also, in a certain sense, we can interpret the result to the extent that almost all Brieskorn varieties admit positive Ricci curvature. This is because we may start out with any $V(a_1, \cdots, a_m)$, and "adding squares" makes $V(a_1, \cdots, a_m, 2, \cdots 2)$ very similar to $V(a_1, \cdots, a_m)$, except that the dimension rises and topological invariants are shifted as for suspensions.

As an important application, we consider exotic spheres. It is known that every odd dimensional homotopy sphere that bounds a parallelizable manifold is diffeomorphic to some Brieskorn variety $V(a_1, \cdots, a_m)$. For example, the varieties $V(6k - 1, 3, 2, 2, 2)$, $1 \leqq k \leqq 28$, represent all the 28 differentiable structures on the 7-sphere. There is a simple condition on $a_1, \cdots, a_m$ such that $V(a_1, \cdots, a_m)$ is a topological sphere, and differentiable structures can be distinguished by invariants computable from $a_1, \cdots, a_m$; compare **[1]** for details. It follows from that easily that if $V(a_1, \cdots, a_m)$ is an exotic sphere, m odd, then so is $V(a_1, \cdots, a_m, 2, 2)$. Hence, adding an even number of squares, if m odd, produces distinct exotic spheres from distinct exotic spheres to start with. Moreover, in fact adding an even number of squares in $V(6k - 1, 3, 2)$, $k \leqq 1$, gives already all the exotic spheres that bound parallelizable manifolds in the corresponding dimensions (congruent -1 modulo 4). In the remaining dimensions (congruent $+1$ modulo 4) there is at most one exotic sphere among Brieskorn varieties, the Kervaire sphere $V(3, 2, \cdots, 2)$. Applying our results, we obtain:

THEOREM IV.1. *Among the exotic spheres of odd dimension that bound a parallelizable manifold there are infinitely many that admit a metric with strictly positive Ricci curvature. These examples include in particular all the Kervaire spheres in dimension* $4d + 1$. *Moreover, in dimension* $4d - 1$ *there are at least* $\alpha_d \geqq 0$ *such exotic spheres, where* α_d *is weakly increasing and* $\alpha_d \to \infty$ *for* $d \to \infty$.

In contrast to that, Hitchen **[6]** pointed out that there are examples of exotic spheres in infinitely many dimensions which do not even admit any metric of positive scalar curvature. The very interesting problem whether or not there are exotic spheres with positive sectional curvature K is still unsolved. However, recently Gromoll and Meyer **[4]** constructed a metric on the Milnor sphere $\Sigma^7 \cong V(5, 3, 2, 2, 2)$ with $K \geqq 0$, where the set of points with $K > 0$ for all planes is open and dense. They make use of a completely different description of Σ^7 and intrinsic methods. Their example has strictly positive Ricci curvature by the way, which we cannot prove since our bounds are not strong enough in this case.

There are not many topological consequences one can derive for a compact manifold with positive Ricci curvature. One application is (using Morse theory

[8]) that the loop spaces of $V(a_1, \cdots, a_m, 2, \cdots, 2)$ and $V(p, p, \cdots, p)$ have the homotopy type of a finite CW-complex with finitely many cells attached in each dimension (n large).

We conclude by mentioning that we have also studied the scalar curvature s of the Brieskorn varieties V'. It is not true that always $s \leqq 0$; compare examples given above. It seems one can only quantitatively improve the results of § III in the case of scalar curvature, which we will discuss elsewhere.

Bibliography

1. E. Brieskorn, *Beispiele zur differential Topologie von Singularitäten*, Invent. Math. **2** (1966), 1–14. MR **34** #6788.

2. J. Cheeger, *Some examples of manifolds of non-negative curvature*, J. Differential Geometry **8** (1973), 623–628.

3. P. Dombrowski, *Krümmungsgrössen gleichungsdefinierter Untermannigfaltigkeiten Riemannscher Mannigfaltigkeiten*, Math. Nachr. **38** (1968), 133–180. MR **39** #7536.

4. D. Gromoll and W. Meyer, *An exotic sphere with positive curvature* (to appear).

5. D. Gromoll, W. Klingenberg and W. Meyer, *Riemannsche Geometrie im Grossen*, Lecture Notes in Math., no. 55, Springer-Verlag, Berlin and New York, 1968. MR **37** #4751.

6. N. Hitchen, Thesis, Oxford, 1972.

7. P. Orlik, *Weighted homogeneous polynomials and fundamental groups*, Topology **9** (1970), 267–273. MR **41** #6251.

8. J. Milnor, *Morse theory*, Ann. of Math. Studies, no. 51, Princeton Univ. Press, Princeton, N.J., 1963. MR **29** #634.

9. ———, *Singular points of complex hypersurfaces*, Ann. of Math. Studies, no. 61, Princeton Univ. Press, Princeton, N.J., 1968. MR **39** #969.

State University of New York at Stony Brook

Proceedings of Symposia in Pure Mathematics
Volume 27, 1975

TANGENT BUNDLES WITH SASAKI METRIC

MU-CHOU LIU

1. Introduction. Let M be a Riemannian manifold, and let $\Pi: TM \to M$ be the tangent bundle of M with Sasaki metric. The puprose of this note is to study the following: (i) the holonomy groups of TM, (ii) the identity component $A^0(TM)$ of affine transformation groups of TM. The main results are stated in Theorems 1 and 4. Theorem 1 is a consequence of elementary theory of first order vector differential equations, Theorem 7.1 of [**3**, p. 83], Proposition 10.1 and Theorem 10.2 of [**3**, p. 95]. In proving Theorem 4, we need the conceptions and arguments of Yano and Kobayashi [**9**, pp. 241–242], a theorem of Kobayashi [**3**, p. 242], a result of Tanno on infinitesimal isometry of TM [**8**] and Theorems 1 and 3. We remark that Theorem 4 is much stronger than Theorem 8 [**6**] concerning general affine transformations in $A^0(M)$. However, it is still an open question on the fiber-preserving property of general affine transformations of TM when M is irreducible, complete and nonflat. All manifolds considered here are connected and smooth. We will follow notations used in [**3**], [**5**], [**6**].

2. Holonomy group of TM. In this section, we only study the holonomy group of TM, when M is an irreducible, complete nonflat Riemannian manifold. Let K: $T^2M \to TM$ be the connection map of the Riemannian connection on M with local connector Γ. Let $K_T : T^3M \to T^2M$ be the connection map of the Sasaki connection with local connector Γ_T. It follows directly from formulae [7.4] of [**7**] that we have the following lemma.

AMS (MOS) subject classifications (1970). Primary 53B20, 53C05, 53C20.

Key words and phrases. Tangent bundles, Sasaki metric, holonomy groups, parallel translation, horizontal curve, complete irreducible Riemannian manifold, affine transformation, infinitesimal affine transformation, geodesics.

LEMMA 1. *In the coordinate neighborhood $\Pi^{-1}(U)$ of TM, the local connector Γ_T of K_T is given by the formula*

$$\begin{aligned}\Gamma_T(x, \xi)((y, \eta), (z, \zeta)) = (&\Gamma(y, z) + \tfrac{1}{2}R(\Gamma(\xi, y), \xi, z) + \tfrac{1}{2}R((\Gamma(\xi, z), \xi, y))\\ &+ \tfrac{1}{2}R(\eta, \xi, z) + \tfrac{1}{2}R(\zeta, \xi, y), D\Gamma(\xi, y, z) + \Gamma(\eta, z) + \Gamma(\zeta, y)\\ &+ \tfrac{1}{2}R(\xi, z, y) + \tfrac{1}{2}R(\xi, y, z) + \tfrac{1}{2}\Gamma(\xi, R(\xi, \Gamma(\xi, y), z))\\ &+ \tfrac{1}{2}\Gamma(\xi, R(\xi, \Gamma(\xi, z), y)) - \tfrac{1}{2}\Gamma(\xi, R(\eta, \xi, z))\\ &- \tfrac{1}{2}\Gamma(\xi, R(\zeta, \xi, y))).\end{aligned}$$

We first fix notations as follows:

$P = L(M)$: bundle of linear frames of M,
$\tilde{P} = L(TM)$: bundle of linear frames of TM with Sasaki metric,
$\Phi^*(x)$: local holonomy group of M at x,
$\Phi^*(u)$: local holonomy group of P at u,
$\tilde{\Phi}^*(\tilde{x})$: local holonomy group of TM at $\tilde{x} = (x, \xi)$,
$\tilde{\Phi}^*(\tilde{u})$: local holonomy group of $\tilde{P}$ at $\tilde{u}$,
$\Phi^0(u)$: restricted holonomy group of P at u,
$\tilde{\Phi}^0(\tilde{u})$: restricted holonomy group of $\tilde{P}$ at $\tilde{u}$.

It follows from [3, p. 94] that $\Phi^*(x) = \Phi^0(x, U)$ and $\tilde{\Phi}^*(\tilde{x}) = \Phi^0(\tilde{x}, \Pi^{-1}(U))$ for a sufficiently small neighborhood U of x with local coordinates $\{x^1, x^2, \cdots, x^n\}$. Let $\xi = (\xi^i)$ be a vector field on M and $U(t) = (U^i_j(t))$ a curve in P over a curve $\alpha(t)$ in M. Then ξ is a parallel vector if and only if

$$d\xi/dt + \Gamma(d\alpha/dt, \xi) = 0 \tag{1}$$

where $\alpha(t)$ is any curve in M. $U(t)$ is a horizontal curve in P if and only if

$$dU/dt + \Gamma(d\alpha/dt, U) = 0. \tag{2}$$

Since equations (1) and (2) are linear equations, their solutions are given as follows:

$$\xi(t) = \left\{\exp \int_0^t \Gamma\left(\frac{d\alpha}{ds}\right) ds\right\} \xi(0),$$
$$U(t) = \left\{\exp \int_0^t \Gamma\left(\frac{d\alpha}{ds}\right) ds\right\} U(0).$$

Let $C(x, U)$ denote the loop space at x contained in U. Then we have

$$\begin{aligned}\Phi^*(x) &= \left\{a = \exp \int_0^1 \Gamma\left(\frac{d\alpha}{ds}\right) ds \,\Big|\, \alpha(t) \in C(x, U)\right\},\\ \Phi^*(u) &= \{u^{-1}au \mid a \in \Phi^*(x)\} = u^{-1}\Phi^*(x)u.\end{aligned} \tag{3}$$

Let $\tilde{\alpha}(t) = (\alpha(t), \xi(t))$ be a curve in TM, $X(t) = (\alpha(t), \xi(t), z(t), \zeta(t))$ a vector field along $\tilde{\alpha}(t)$, and $\tilde{U}(t) = \begin{pmatrix} A(t) & B(t) \\ C(t) & D(t)\end{pmatrix}$ a curve in $\tilde{P}$. Then $X(t)$ is parallel along $\tilde{\alpha}(t)$ and $\tilde{U}(t)$ is a horizontal curve over $\tilde{\alpha}(t)$, if and only if the following equations hold.

$$dX/dt + \Gamma_T(d\tilde{\alpha}/dt, X(t)) = 0, \tag{4}$$

$$d\tilde{U}/dt + \Gamma_T(d\tilde{\alpha}/dt, \tilde{U}(t)) = 0. \tag{5}$$

If $\tilde{\alpha}(t) = (\alpha(t), 0)$ is a curve in $M \subset TM$, then equations (4) and (5) are simply reduced to

$$\begin{aligned} dz/dt + \Gamma(d\alpha/dt, z) &= 0, & d\zeta/dt + \Gamma(d\alpha/dt, \zeta) &= 0, \\ dA/dt + \Gamma(d\alpha/dt, A) &= 0, & dB/dt + \Gamma(d\alpha/dt, B) &= 0, \\ dC/dt + \Gamma(d\alpha/dt, C) &= 0, & dD/dt + \Gamma(d\alpha/dt, D) &= 0. \end{aligned}$$

Their solutions are given as follows:

$$X(t) = \begin{pmatrix} a_t & 0 \\ 0 & a_t \end{pmatrix} X(0), \tag{6}$$

$$\tilde{U}(t) = \begin{pmatrix} a_t & 0 \\ 0 & a_t \end{pmatrix} \tilde{U}(0), \tag{7}$$

where $a_t = \exp \int_0^t \Gamma(d\alpha/dS)\, ds$. Taking changes of local coordinate systems into consideration, equations (6) and (7) imply the following

LEMMA 2. *If $\tilde{U}(t)$ is a horizontal curve in $\tilde{P}$ over a curve $\tilde{\alpha}(t) = (\alpha(t), 0)$ in M with*

$$\tilde{U}(0) = \begin{pmatrix} I_m & 0 \\ 0 & I_m \end{pmatrix},$$

then

$$\tilde{U}(t) = \begin{pmatrix} A(t) & 0 \\ C(t) & A(t) \end{pmatrix}$$

where $A(t)$ is a horizontal curve in P with $A(0) = I_m$, $C(t)$ is a curve in $gl(m, R)$, and I_m is an identity matrix of order m.

Let $\tilde{\beta}(t) = (x, t\xi)$ be a curve in TM joining $(x, 0)$ and (x, ξ). Let $\tilde{V}(t)$ be a horizontal curve in $\tilde{P}$ over $\tilde{\beta}(t)$ with $V(0) = \tilde{e} = (x, I_{2m})$. Then equation (5) becomes

$$\begin{aligned} dA/dt &= 0, & dB/dt &= 0, \\ dC/dt + \Gamma(\xi, C) &= 0, & dD/dt + \Gamma(\xi, D) &= 0, \end{aligned}$$

which implies that

$$\tilde{V}(t) = \begin{pmatrix} I_m & 0 \\ 0 & \exp t\Gamma(\xi) \end{pmatrix}. \tag{8}$$

Let $f(t)$ be a piecewise function on $[0, 1]$ such that (i) $f(0) = f(1) = 0$, (ii) $\int_0^1 f'(t)dt = 0$ and (iii) $\int_0^1 f'(t)f(t)dt = k$, where the real number k may assume any value. Let $\tilde{\gamma}(t) = (x, \xi + f(t)\eta)$ be a closed curve at (x, ξ) in TM and let $X = (x, \xi + f(t)\eta, z(t), \zeta(t))$ be a parallel vector field along $\tilde{\gamma}(t)$. Then equation (4) becomes

$$\begin{aligned} dz/dt + \tfrac{1}{2}R(f'(t)\,\eta, \xi, z) &= 0, \\ d\zeta/dt + \Gamma(f'(t), \eta, z) - \tfrac{1}{2}\Gamma(\xi + f(t)\,\eta, R(f'(t)\eta, \xi, z)) &= 0. \end{aligned}$$

From properties (ii) and (iii) of f, we have

$$X(1) = \begin{pmatrix} I_m & 0 \\ -\frac{1}{2}k\,\Gamma(\eta, R(\eta, \xi)) & I_m \end{pmatrix} X(0) \equiv \begin{pmatrix} I_m & 0 \\ b & I_m \end{pmatrix} X(0)$$

which implies that the element $\begin{pmatrix} I_m & 0 \\ b & I_m \end{pmatrix}$ is in $\Phi^*((x, \xi))$. From (3), (8) and the theorem of [3, p. 95], it follows that the element

$$\tilde{V}(1)^{-1} \begin{pmatrix} I_m & 0 \\ b & I_m \end{pmatrix} \tilde{V}(1) = \begin{pmatrix} I_m & 0 \\ \{\exp - \Gamma(\xi)\} \cdot \{\frac{1}{2}k\Gamma(\eta, R(\eta, \xi))\} & I_m \end{pmatrix}$$

is in $\tilde{\Phi}^0(\tilde{e})$. From Lemma 2, for $a \in \Phi^*(\alpha(t))$ the element

$$\tilde{U}_t^{-1} \begin{pmatrix} a & 0 \\ 0 & a \end{pmatrix} \tilde{U}_t = \begin{pmatrix} h_t & 0 \\ d_t & h_t \end{pmatrix}$$

is in $\Phi^*(\tilde{U}(t))$ where $d_t = -A_t^{-1} C_t A_t^{-1} a A_t + A_t^{-1} a C_t$ and $h_t = A_t^{-1} a A_t$. By Lemma 2 and the theorem of [3, p. 95] the element $\begin{pmatrix} h_t & 0 \\ d_t & h_t \end{pmatrix}$ is in $\tilde{\Phi}^0(\tilde{e})$ and the element h_t is in $\Phi^0(e)$ where $e = (x, I_m)$. Summarizing the above results, we obtain

LEMMA 3. (i) *For any $h \in \Phi^0(e)$, there is an element $d \in gl(m, R)$ such that $\begin{pmatrix} h & 0 \\ d & h \end{pmatrix} \in \tilde{\Phi}^0(\tilde{e})$.* (ii) *If there exist vectors ξ, η in T_xM for some $x \in M$ such that $\Gamma(\eta, R(\eta, \xi))$ is not a zero matrix, then $\tilde{\Phi}^0(\tilde{e})$ contains a matrix of the form*

$$\begin{pmatrix} I_m & 0 \\ b & I_m \end{pmatrix} \quad \text{with } b \neq 0.$$

We note that if M is an irreducible nonflat Riemannian manifold, then the element b of Lemma 3 (ii) always exists. Let $\Phi^0(M)$ and $\Phi^0(TM)$ be restricted holonomy groups of M and TM respectively.

THEOREM 1. *Let M be an irreducible nonflat Riemannian manifold. For any $x \in M$, $\Phi^0(TM)$ acts irreducibly on $T_x(TM)$.*

PROOF. We note that $T_x(TM) \cong R^m \oplus R^m$ and the action of $\Phi^0(TM)$ on $T_x(TM)$ is equivalent to the action of $\tilde{\Phi}^0(\tilde{e})$ on $R^m \oplus R^m$. Let $W \neq \{0\}$ be a vector subspace of $R^m \oplus R^m$ invariant under $\tilde{\Phi}^0(\tilde{e})$. Since $\tilde{\Phi}^0(\tilde{e})$ contains an element of the form $\begin{pmatrix} h & 0 \\ d & h \end{pmatrix}$ with $h \in \Phi^0(e)$ and $d \neq 0$, we may assume that a vector (y, z) is in W with $y \neq 0$. Since $\Phi^0(e)$ is irreducible, by (i) of Lemma 4, we have $R^m \oplus V \subseteq W$ where V is a subspace of R^m. Since there is an element

$$\tilde{b} = \begin{pmatrix} I_m & 0 \\ b & I_m \end{pmatrix} \in \tilde{\Phi}^0(\tilde{e}) \quad \text{with } b \neq 0 \text{ and } R^m \oplus V \subseteq W,$$

there is an element $y \in R^m$ such that $by \neq 0$ and $(\tilde{b} - I_{2m})(y, z) = (0, by) \in W$. Again by the irreducibility of $\Phi^0(e)$ and (i) of Lemma 3, we obtain $0 \oplus R^m \subseteq W$. Thus $R^m \oplus R^m \subseteq W$ and $\tilde{\Phi}^0(e)$ acts irreducibly on $T_x(TM)$. Q.E.D.

By Theorems 1 and 2 [6], if M is a complete simply connected manifold, then TM is also a complete simply connected manifold. Using Theorem 1, we obtain the de Rham decomposition theorem on TM.

THEOREM 2. (i) *If M is a complete connected, simply connected irreducible nonflat Riemannian manifold, then TM is also a complete connected, simply connected irreducible nonflat Riemannian manifold.*

(ii) *If $M = M_0 \times \cdots \times M_r$ is a de Rham decomposition of a complete, connected, simply connected nonflat Riemannian manifold M, then $TM = TM_0 \times TM_1 \times \cdots \times TM_r$ is the de Rham decomposition of TM.*

3. Infinitesimal affine transformations of TM. Because of long complicated calculations involved, we state without proof the following theorem on general infinitesimal affine transformations of TM.

THEOREM 3. *Let M be a Riemannian manifold and let TM be the tangent bundles of M with Sasaki metric. If any infinitesimal affine transformation $X(x, \xi)$ is an analytic function of ξ, then a vector field $X(x, \xi) = (F(x, \xi), G(x, \xi))$ is an infinitesimal affine transformation on TM if and only if*

$$F(x, \xi) = f(x) + A(x)\cdot\xi,$$
$$G(x, \xi) = g(x) + P(x)\xi + Df(x)\cdot\xi - \Gamma(\xi, A\xi),$$

such that f, g, A, P satisfy the following conditions:

(i) $\nabla A(z, \eta) + \frac{1}{2}R(\eta, g, z) = 0.$
(ii) $P\cdot R(\eta, \xi, z) = R(\eta, \xi, Pz) = R(P\eta, \xi, z) = 0.$
(iii) $\nabla R(A\xi, \eta, \xi, z) + R(\eta, \xi, \nabla A(z, \xi)) - \nabla A(R(\eta, \xi, z), \xi) = 0.$
(iv) $\nabla P(z, \eta) = 0$ *and* $C(z, \eta) = 0.$
(v) $R(A\xi, z, \eta) = R(\xi, z, A\eta) = A\cdot R(\xi, z, \eta) = 0.$
(vi) $\nabla g\cdot R(\eta, \xi, z) = 0.$
(vii) $\frac{1}{2}R(\nabla g\cdot y, \xi, z) + \frac{1}{2}R(\nabla g\cdot z, \xi, y) + \nabla^2 A(z, y, \xi) = 0.$
(viii) $E(y, z) + \frac{1}{2}R(g, z, y) + \frac{1}{2}R(g, y, z) = 0.$
(ix) $\frac{1}{2}R(\xi, \nabla A(y, \xi), z) + \frac{1}{2}R(\xi, \nabla A(z, \xi), y) + R(\xi, y, \nabla A(z, \xi))$
$- R(\nabla A(\xi, \xi), y, z) - \frac{1}{2}\nabla A(R(y, \xi, z), \xi) - \frac{1}{2}\nabla A(R(z, \xi, y), \xi) = 0.$

Here

$$C(y, z) = D^2 f(y, z) + D\Gamma(f, y, z) + \Gamma(Df\cdot y, z) + \Gamma(y, Df\cdot z) - Df\cdot\Gamma(y, z),$$
$$E(y, z) = D^2 g(y, z) + D\Gamma(g, y, z) + \Gamma(G\cdot y, z) + \Gamma(y, Dg\cdot z) - Dg\cdot\Gamma(y, z).$$

We fix notation as follows:

$a(M)$: space of infinitesimal affine transformations of M,
$i(M)$: space of infinitesimal isometries of M,
$A^0(M)$: identity component of affine transformation group of M,
$I^0(M)$: identity component of isometry group of M.

THEOREM 4. *Let M be an irreducible complete Riemannian manifold. Then*
(i) $i(TM) = a(TM) = \{X | X(x, \xi) = (f(x), Df(x)\cdot\xi), f\in i(M)\},$
(ii) $I^0(TM) = A^0(TM) = \{F | F = h_*, h\in I^0(M)\}.$

PROOF OF (i). If $X(x, \xi)$ is an infinitesimal affine transformation on TM, then

$$X(x, \xi) = (f(x) + A(x)\cdot\xi,\ g(x) + P(x)\cdot\xi - \Gamma(\xi, A(x)\cdot\xi))$$

where f, g, A, P satisfy conditions (i)–(ix). By condition (iv) of Theorem 3, Theorem 1 and the lemma of [9, p. 241], we have $P = aI + bJ$ where J is an almost integrable complex structure on M. By condition (ii) of Theorem 3, we get $a = b = 0$, and hence $P = 0$. By condition (iv) of Theorem 3, $f \in a(M)$. It remains to show that $A = g = 0$. Since A is a (1, 1) tensor field and g is a vector field, locally we have $A = (A^i_j)$ and $g = (g^i)$. By Theorem 1 and the theorem of [3, p. 242], X is also an infinitesimal isometry of TM. It follows from Theorem 3 [8] that we have

$$A^i_j = -\nabla^i g_j, \tag{9}$$

$$\nabla A(z, y) + \nabla A(y, z) = 0. \tag{10}$$

Conditions (i), (viii) of Theorem 3 and equation (10) yield

$$D^2 g(z, y) + D\Gamma(g, z, y) + \Gamma(Dg\cdot y, z) + \Gamma(y, Dg\cdot z) - Dg\cdot\Gamma(z, y) = 0 \tag{11}$$

which implies that g is an infinitesimal affine transformation. By the theorem of [3, p. 242], g is also an infinitesimal isometry which gives $A^i_j = -\nabla^i g_j = \nabla_j g^i$ or

$$A = \nabla g. \tag{12}$$

Now equation (11) may be written as

$$\nabla^2 g(z, y) + R(g, y, z) = 0. \tag{13}$$

Equations (12), (13) and condition (i) of Theorem 3 together yield

$$\nabla A(z, y) = \nabla^2 g(z, y) = 0. \tag{14}$$

So A is a parallel (1, 1) tensor on M. By Theorem 1 and the lemma of [9, p. 241], we have $A = cI + dJ$. By condition (v) of Theorem 3, we obtain $c = d = 0$, and hence $A = 0$. Since $A = \nabla g$, it follows that g is a parallel vector field on M. Since M is irreducible and nonflat, there is no parallel vector field on M. Hence $g = 0$ and we have proved that any infinitesimal affine transformation X is of the form

$$X(x, \xi) = (f(x), Df(x)\cdot\xi) \quad \text{where } f \in i(M) \tag{15}$$

which is a complete lift of f.

PROOF OF (ii). Let f be an infinitesimal affine transformation on M, and let ϕ_t be a 1-parameter family of local affine transformations on M generated by f. Then ϕ_{t*} is a 1-parameter family of local affine transformations on TM generated by the complete lift $f^c = (f, Df)$ of f. It follows that $A^0(TM)$ is generated by the set $\{h_* | h \in U\}$ where U is an open neighborhood of identity in $A^0(M)$. So for any $F \in A^0(TM)$, F is a fiber-preserving affine transformation of TM. By Theorem 3 [6], $F = \pm h_*$. Q.E.D.

References

1. P. Dombrowski, *On the geometry of the tangent bundle*, J. Reine Angew. Math **210** (1962), 73–88. MR **25** #4463.

2. H. Eliasson, *Geometry of manifolds of maps*, J. Differential Geometry **1** (1967), 169–194. MR **37** #2268.

3. S. Kobayashi and K. Nomizu, *Foundations of differential geometry*. Vols. I, II, Wiley, New York, 1963, 1968. MR **27** #2945; **38** #6501.

4. O. Kowalski, *Curvature of the induced Riemannian metric on the tangent bundle of a Riemannian manifold*, J. Reine Angew. Math. **250** (1970), 124–129.

5. M. Liu, *On the affine map of tangent bundle*, Indiana Univ. Math. J. (to appear).

6. ———, *Affine maps of tangent bundle with Sasaki metric*, Tensor (to appear).

7. S. Sasaki, *On the differential geometry of tangent bundles of Riemannian manifolds*. I,II, Tôhoku Math. J. (2) **10** (1958), 338–354; ibid. (2) **14** (1962), 146–155. MR **22** #3007; **26** #2987.

8. S. Tanno, *Infinitesimal isometry of tangent bundles with Sasaki metric* (to appear).

9. K. Yano and S. Kobayashi, *Prolongations of tensor fields and connections to tangent bundles*. I,II, J. Math. Soc. Japan **18** (1966), 194–210, 236–246. MR **33** #1814; **34** #743.

University of Illinois, Chicago

Universidade Federal de Minas Gerais

Belo Horizonte, Brazil

Proceedings of Symposia in Pure Mathematics
Volume 27, 1975

CURVATURE AND CRITICAL RIEMANNIAN METRIC*

YOSIO MUTŌ

Let M be a compact orientable C^∞ manifold of dimension n and g be a C^∞ Riemannian metric on M satisfying

(1) $$\int_M dV_g = 1$$

where dV_g is the volume element of M measured by g. We denote the set of all such metrics by $\mathscr{M}(M)$ or $\mathscr{M}$. When g is fixed we have a Riemannian manifold (M, g).

Let us take a covering $\{U\}$ of M by coordinate neighborhoods and denote the local coordinates by x^h. In each neighborhood U we use the natural frame. Then the components of the curvature tensor of (M, g) in U are given by

$$\{ = K \{ -\partial\, \partial_j \left\{ {h \atop ki} \right\} + \left\{ {\ \atop kp} \right\} \left\{ {p \atop ji} \right\} - \left\{ {h \atop jp} \right\} \left\{ {p \atop ki} \right\}$$

where $\left\{ {h \atop ji} \right\}$ are the Christoffel symbols derived from the components g_{ji} of g, Latin indices run the range $\{1, \cdots, n\}$, and the summation convention is adopted. The Ricci tensor and the scalar curvature are given respectively by

$$K_{ji} = K_{pji}^{p}, \qquad K = g^{ji} K_{ji}$$

where g^{ji} are defined by $g_{ik}\, g^{kh} = \delta_i^h$. Similarly all tensors will be expressed in terms of their components.

AMS (MOS) subject classifications (1970). Primary 53C20; Secondary 53C25.

*Full notes will be published elsewhere; see [3], [4].

In a Riemannian manifold (M, g) indices can be raised and lowered by g^{ji} and g_{ji} so that, for example, $K^{kjih} = K_{dcb}^{h} g^{dk} g^{cj} g^{bi}$ are the contravariant components of the curvature tensor. Thus $K_{kjih} K^{kjih}$ is a scalar. Considering this at each point of M we get a scalar field.

Let us consider the integral

$$I[g] = \int_M K_{kjih} K^{kjih} \, dV_g$$

over the Riemannian manifold (M, g). When g moves this gives a mapping $I: \mathcal{M} \to \boldsymbol{R}$. The integral $I[g]$ has a remarkable property that $I[g]$ is nonnegative and moreover that, if M does not admit a locally flat metric, then $I[g]$ is strictly positive.

If η is a diffeomorphism of M and η^* its pullback, we have $I[g] = I[\eta^*(g)]$. Hence we can deduce a mapping $\tilde{I}: \mathcal{M}/\mathcal{D} \to \boldsymbol{R}$ from the mapping $I: \mathcal{M} \to \boldsymbol{R}$ where $\mathcal{D}$ is the diffeomorphism group of M and $\mathcal{M}/\mathcal{D}$ is the space of orbits generated by $\mathcal{D}$ of Riemannian metrics [2].

If $\bar{g}$ is a critical point of I, then the orbit of $\bar{g}$ by $\mathcal{D}$ is a critical point of $\tilde{I}$ and vice versa. In this case let us say that $\bar{g}$ is a critical point of $\tilde{I}$ for convenience sake. This convention is useful since there can exist no local minimum, in the narrow sense, for the mapping I but $\tilde{I}$ may possibly have a local minimum and the present paper concerns this. $\bar{g}$ is also called a critical Riemannian metric.

Assume M does not admit a locally flat metric. Then we have many questions about $\tilde{I}$ or $I[g]$, for example, the following.

Has $\tilde{I}$ a local minimum?

How many critical points has $\tilde{I}$?

Has $\{I[g], g \in \mathcal{M}\}$ the least value?

Is the equation $\inf I[g] = 0$ valid?

The answer will depend on M.

M. Berger has obtained the differential equations of critical points [1] which we write in our notation in the form

$$\begin{aligned} 2\nabla^j\nabla^i K - 4\nabla_p\nabla^p K^{ji} + 4K^j_{\ p}K^{pi} - 4K^{ji}_{\ \ qp}\, K^{qp} \\ - 2K^{rqpj} K^i_{\ rqp} + \tfrac{1}{2} K_{dcba} K^{dcba} g^{ji} = c\, g^{ji} \end{aligned}$$

where c is a constant in M and must satisfy

$$c = -(2/n)\, \nabla_p \nabla^p K + (1/2 - 2/n) K_{dcba} K^{dcba}.$$

The following theorem is easily proved.

THEOREM 1. *Let M be a C^∞ manifold diffeomorphic to S^n. Then $\tilde{I}$ has a critical point at the Riemannian metric $\bar{g}$ of positive constant curvature.*

Now the following is our main theorem.

THEOREM 2. *Let M be as in Theorem 1. Then $\tilde{I}$ has a local minimum at the metric $\bar{g}$ of Theorem 1.*

In order to prove the main theorem, we first calculate the second derivative $(I''[g(t)])_0$, namely,

$$(d^2 I[g(t)]/dt^2)_{t=0}$$

for an arbitrary curve $g(t)$ of $\mathcal{M}$ where $g(0) = \bar{g}$.

The resulting formula is very long and complicated if $\bar{g}$ is an arbitrary critical Riemannian metric. But in our case $\bar{g}$ is the metric of constant curvature. Moreover, we can use a theorem of Berger and Ebin in **[2]** and consider the space $\mathcal{M}/\mathcal{D}$ rather than $\mathcal{M}$ itself. Thus it is sufficient to consider only curves $g(t)$ of $\mathcal{M}$ such that D_{ji} defined by

$$D_{ji} = (\partial g_{ji}(x, t)/\partial t)_{t=0}$$

satisfies $\nabla^j D_{ji} = 0$ **[1]**. The second derivatives $(\partial^2 g_{ji}/\partial t^2)_{t=0}$ do not appear in the final formula because of (1).

Thus we get

$$\begin{aligned} I''[g(0)] = \int_M \Big[& 2\, \nabla_q \nabla^q D_{ji} \nabla_p \nabla^p D^{ji} \\ & - \frac{2K}{n-1} (D_{ji} \nabla_p \nabla^p D^{ji} - n^{-1} D^r_r \nabla_q \nabla^q D^p_p) \\ & + \frac{4(n-2)}{n^2(n-1)^2} K^2 D_{ji} D^{ji} - \frac{2(n-3)}{n^2(n-1)^2} K^2 (D^p_p)^2 \Big] d V_{g(0)}. \end{aligned}$$

Putting

$$D_{ji} = H_{ji} + (H/n) g_{ji}, \qquad H = D^p_p,$$

we get $H^p_p = 0$ and $I''[g(0)]$ can be written as the sum, $I''[g(0)] = I_1 + I_2$, where I_1 and I_2 are integrals obtained when D_{ij} is replaced with H_{ji} and $(H/n)g_{ji}$ respectively in $I''[g(0)]$. We easily get $I_1 \geqq 0$ and moreover, $I_1 = 0$ if and only if $H_{ji} = 0$. On the other hand we have

$$I_2 = \frac{2}{n} \int_M [(\Delta H)^2 - \frac{n-4}{n^2(n-1)} K^2 H^2] \, dV.$$

As H satisfies $\int_M H \, dV = 0$ we have

$$\int_M (\Delta H)^2 dV \geqq \frac{K^2}{(n-1)^2} \int_M H^2 dV.$$

Thus we get $I_2 \geqq 0$ and moreover, $I_2 = 0$ if and only if $H = 0$.

Thus we get the main theorem.

We add here some other results **[4]**.

Let M be a manifold diffeomorphic to $S^n \times S^m$ and let us consider a Riemannian metric $g_{12} \in \mathcal{M}(M)$ such that $(M, g_{12}) = (S^n, g_1) \times (S^m, g_2)$ where (S^n, g_1) and (S^m, g_2) are Riemannian manifolds of constant curvature with positive scalar curvature K_1 and K_2 respectively. Then g_{12} is a critical Riemannian metric of $\tilde{I}: \mathcal{M}(M) \to \boldsymbol{R}$ if and only if

$$K_1{}^2/n^2(n-1) = K_2{}^2/m^2(m-1).$$

But this metric $\bar{g}_{12}$ does not give a local minimum of $\tilde{I}$ if $n > 2$ and $m > 2$.

Let M be a manifold diffeomorphic to $S^n \times T^m$ where $n \geqq 2$ and $m \geqq 1$. Then $\inf I[g] = 0$. This is proved as follows. Let g_{12} be a metric on M such that $g_{12} \in \mathcal{M}(M)$ and $(M, g_{12}) = (S^n, g_1) \times (T^m, g_2)$ where (S^n, g_1) is a Riemannian manifold of positive constant curvature and g_2 is a locally flat metric on T^m. Then $g \in \mathcal{M}(M)$ if $(M, g) = (S^n, \alpha^2 g_1) \times (T^m, \beta^2 g_2)$ and $\alpha^n\beta^m = 1$. On the other hand we have $I[g] = \alpha^{-4}\, \alpha^n\, \beta^m\, I[g_{12}]$, hence $I[g] = \alpha^{-4}\, I[g_{12}]$. As α is arbitrary we get $\inf I[g] = 0$.

References

1. M. Berger, *Quelques formules de variation pour une structure riemannienne*, Ann. Sci. École Norm. Sup. (4) **3** (1970), 285–294. MR **43** #3969.

2. M. Berger and D. Ebin, *Some decompositions of the space of symmetric tensors on a Riemannian manifold*, J. Differential Geometry **3** (1969), 379–392. MR **42** #993.

3. Y. Mutō, *Curvature and critical Riemannian*, J. Math. Soc. Japan **26** (1974), no. 4.

4. ———, Kōdai Math. Sem. Rep.

YOKOHAMA NATIONAL UNIVERSITY

Proceedings of Symposia in Pure Mathematics
Volume 27, 1975

AXIOMS FOR THE EULER CHARACTERISTIC

HOWARD OSBORN

Abstract. The Euler characteristic χ is the only real-valued function on the set of closed orientable manifolds which satisfies certain axioms. This characterization of χ provides a brief proof of the Gauss-Bonnet theorem.

Since χ is additive with respect to disjoint sum, and since $\chi(M) = 0$ for any odd-dimensional closed orientable manifold M, we restrict attention to the union $\bigcup_{m\geqq 1} \mathscr{M}_{2m}$, where $\mathscr{M}_{2m}$ is the set of closed connected orientable $2m$-dimensional manifolds in any of the categories DIFF, PL, or TOP. (The category DIFF will be specified later.) If $m > 1$ the *natural sum* $M \natural N$ of $M \in \mathscr{M}_{2m}$ and $N \in \mathscr{M}_{2m}$ is defined by removing an open null-homotopically embedded torus $S^1 \times D^{2m-1}$ from each of M and N and identifying boundary points of the resulting manifolds-with-boundary in the obvious way, an operation which can easily be performed in a fashion such that $M \natural N \in \mathscr{M}_{2m}$. (One loses connectedness in the case $m = 1$.) Since $\chi(S^1 \times D^{2m-1}) = 0$ it follows that $\chi(M \natural N) = \chi(M) + \chi(N)$.

PROPOSITION 1. *The Euler characteristic is the only map* $\varphi : \bigcup_{m\geqq 1} \mathscr{M}_{2m} \to R$ *such that* (i) $\varphi(M \natural N) = \varphi(M) + \varphi(N)$ *for* $M \in \mathscr{M}_{2m}$ *and* $N \in \mathscr{M}_{2m}$ *whenever* $m > 1$, (ii) $\varphi(M \times N) = \varphi(M)\varphi(N)$ *for any* M *and any* N, (iii) *if* $\chi(M) = 0$ *then* $\varphi(M) = 0$, (iv) *if* $M \in \mathscr{M}_2$ *then* $\varphi(M) = \chi(M)$, *and* (v) $\varphi(CP^2) = 3$.

PROOF. For any $k > 0$ one uses products and natural sums of the 2-sphere $S^2 \in \mathscr{M}_2$, the surface $G^2 \in \mathscr{M}_2$ of genus 2, and $CP^2 \in \mathscr{M}_4$ to construct elements $E^{4k}_{\pm} \in \mathscr{M}_{4k}$ and $E^{4k+2}_{\pm} \in \mathscr{M}_{4k+2}$ such that $\chi(E^{4k}_{\pm}) = \pm 1$ and $\chi(E^{4k+2}_{\pm}) = \pm 2$. If $M \in \mathscr{M}_{4k}$ let $\tilde{M} = (\mp \chi(M))\, E^{4k}_{\pm}$, where $\mp$ is chosen so that $\mp \chi(M) \geqq 0$; and similarly if $M \in \mathscr{M}_{4k+2}$ let $\tilde{M} = \mp \frac{1}{2} \chi(M)\, E^{4k+2}_{\pm}$. Then for any $M \in \mathscr{M}_{2m}$ with $m > 1$ the

AMS (MOS) subject classifications (1970). Primary 53C20; Secondary 57D20.
Key words and phrases. Euler characteristic, Gauss-Bonnet theorem.

conditions other than (iii) give $\varphi(\tilde{M}) = -\chi(M)$, and since $\chi(M \natural \tilde{M}) = 0$, condition (iii) gives $\varphi(M) - \chi(M) = \varphi(M \natural \tilde{M}) = 0$. If $M \in \mathcal{M}_2$ then $\varphi(M) = \chi(M)$ by condition (iv).

Proposition 1 was partially motivated by **[2]** and **[3]**. As an example we shall apply it to prove the Gauss-Bonnet theorem, formulated later in terms of the following definitions.

Let $\mathcal{F}$ be the $C^\infty(M)$-module of smooth sections of a smooth orientable $2n$-plane bundle ξ on some $M \in \mathcal{M}_{2m}$ (in DIFF); in particular let $\mathcal{E}$ be the $C^\infty(M)$-module of smooth sections of the tangent bundle of M. Using exterior algebras $\wedge \mathcal{E}$ and $\wedge \mathcal{F}$ over $C^\infty(M)$ one forms a graded $\wedge \mathcal{E}$-module $\wedge^q F = \wedge \mathcal{E} \otimes \wedge^q \mathcal{Y}$ for each $q = 0, \cdots, 2n$, elements of $\wedge^p \mathcal{E} \otimes \wedge^q \mathcal{F}$ being assigned degree $p + q$. Any metric $\mathcal{F} \otimes \mathcal{F} \to C^\infty(M)$ induces a *metric* $\langle\ ,\ \rangle: \wedge^q F \otimes \wedge^q F \to \wedge \mathcal{E}$ in the obvious way, and since $\mathcal{F}$ is orientable there exists an $S \in \wedge^{2n} F$ such that $\langle S, S\rangle = 1$, uniquely determined up to $\pm$. A *Hodge operator* $*: \wedge^n F \to \wedge^n F$ is defined with respect to $\langle\ ,\ \rangle$ (and S) by requiring $\langle s, * t\rangle = \langle s \wedge t, S\rangle$ for any $s, t \in \wedge^n F$.

A *connection in* $\wedge^q F$ is any real linear map $D: \wedge^q F \to \wedge^q F$ of degree $+ 1$ such that $D(\theta s) = d\theta \cdot s + (-1)^p \theta \cdot Ds$ for $\theta \in \wedge^p \mathcal{E}$, and D is *metric* with respect to $\langle\ ,\ \rangle$ if $\langle Ds, t\rangle + (-1)^{p+q} \langle s, Dt\rangle = d\langle s, t\rangle$ whenever $s \in \wedge^q F$ is of degree $p + q$. The *curvature* K of D is the $\wedge \mathcal{E}$-linear map $DD: \wedge^q F \to \wedge^q F$ of degree $+ 2$; in case $q = 1$ the curvature $K : F \to F$ of any connection $D : F \to F$ induces a $\wedge \mathcal{E}$-linear map $\wedge^n K: \wedge^n F \to \wedge^n F$ of degree $2n$. In case K is the curvature of a metric connection $D : F \to F$ the *pfaffian* pf K is the element $(-1)^{n(n-1)/2} \operatorname{tr}(* \circ \wedge^n K) \in \wedge^{2n}\mathcal{E}$.

LEMMA 1. pf K *is a closed element of* $\wedge^{2n}\mathcal{E}$ *whose de Rham cohomology class* $[\text{pf } K] \in H^{2n}(M; R)$ *is independent of the metric and independent of the metric connection.*

PROOF. The Chern-Weil theorem states that pf K is closed and that [pf K] is independent of the metric connection, for a given metric. An explicit proof that [pf K] is also independent of the metric appears in **[1]**.

PROPOSITION 2 (GAUSS, BONNET, HOPF, ALLENDOERFER, CHERN). *Let* $\mathcal{E}$ *be the* $C^\infty(M)$*-module of sections of the tangent bundle of any smooth* $M \in \mathcal{M}_{2m}$ *and let* $\mathcal{F} = \mathcal{E}$; *then* $(4\pi)^{-m} \int_M \text{pf } K = \chi(M)$ *for any metric connection* D *in the tangent bundle.*

In order to obtain Proposition 2 from Proposition 1 we formulate two familiar properties of pf K; brief proofs are indicated for the sake of completeness.

LEMMA 2. $\text{pf}(K \oplus K') = (\text{pf } K)(\text{pf } K')$ *for the curvatures* K, K' *of any metric connections* D, D' *in bundles* ξ, ξ' *and* $\xi \oplus \xi'$.

PROOF. This result is in fact valid for any $\wedge\mathcal{E}$-linear maps K and K' of even degree, as in **[1]**.

LEMMA 3. *If an orientable* $2n$*-plane bundle* ξ *admits a nowhere-vanishing section then* pf $K = 0$ *for the curvature* K *of some metric connection* $D : F \to F$.

PROOF. Let $s \in F$ be the nowhere-vanishing section, which may be chosen to satisfy $\langle s, s\rangle = 1$, so that $F = F_1 \oplus F_2$ where F_2 is the free $\wedge \mathscr{E}$-module spanned by s. For any metric connection $D_0 : F \to F$ the map $F_1 \to F_1$ given by (projection) $\circ D_0 \circ$ (injection) is a metric connection D_1 in F_1, there is a unique metric connection D_2 in F_2 such that $D_2 s = 0$, and hence there is a new metric connection $D = D_1 \oplus D_2$ in F. But $\wedge^n F = G_1 \oplus G_2$, where G_2 consists of all elements with the factor s, and the Hodge operator $*$ interchanges G_1 and G_2. Since $Ds = 0$ the curvature K of D also annihilates s, which implies $\operatorname{im}(* \circ \wedge^n K) \subset G_2$, and the restriction of $* \circ \wedge^n K$ to G_2 vanishes for the same reason. Hence $\operatorname{tr}(* \circ \wedge^n K) = 0$.

PROOF OF PROPOSITION 2. Set $\varphi(M) = (4\pi)^{-m} \int_M \operatorname{pf} K$, which depends only on M by Lemma 1; we shall verify that $\varphi : \bigcup_{m \geqq 1} \mathscr{M}_{2m} \to R$ satisfies conditions (i)—(v) of Proposition 1. If $m > 1$ then for any $M, N \in \mathscr{M}_{2m}$ one chooses open balls $U \subset M$ and $V \subset N$ and metrics which are flat on U and V, respectively; the incisions used to construct $M \natural N$ can then be made in U and in V in such a way that the metrics on M and N induce a metric on $M \natural N$, and since the curvatures of the Levi-Civita connections vanish on U and V, respectively, condition (i) follows. Condition (ii) is an immediate consequence of the product formula of Lemma 2. If $\chi(M) = 0$ the Hopf theorem provides a nowhere-vanishing section of the tangent bundle, so that Lemma 3 implies condition (iii). Condition (iv) is the elementary Gauss-Bonnet theorem for surfaces, and condition (v) is a simple verification.

REFERENCES

1. H.Osborn, *Representations of Euler classes*, Proc. Amer. Math. Soc. **31** (1972), 340–346. MR **45** #9349.

2. B.L.Reinhart, *Cobordism and the Euler number*, Topology **2** (1963), 173–177. MR **27** #2990.

3. C.E.Watts, *On the Euler characteristic of polyhedra*, Proc. Amer. Math. Soc. **13** (1962), 304–306. MR **25** #565.

UNIVERSITY OF ILLINOIS, URBANA

Proceedings of Symposia in Pure Mathematics
Volume 27, 1975

RIEMANNIAN MANIFOLDS WITHOUT CONJUGATE POINTS

JOHN J. O'SULLIVAN*

A connected riemannian manifold M is said to have no conjugate points if there is a unique geodesic segment joining each pair of points in its universal covering space $\tilde{M}$. M is said to have no focal points if, for each geodesic c in $\tilde{M}$ and each point p not on c, there is a unique geodesic passing through p which intersects c perpendicularly. It is well known that manifolds without focal points have no conjugate points and that manifolds of nonpositive curvature have no focal points.

We now summarize some results about manifolds without conjugate points which we have recently obtained. A fuller account is to appear in [3].

THEOREM 1. *Let M be a complete manifold without conjugate points. Let X be a Killing vector field on M whose length is bounded and attains its least upper bound. Then X is parallel.*

COROLLARY. (i) *A left-invariant metric on a nonabelian nilpotent Lie group has conjugate points.*

(ii) *If M is compact, then $I^0(M)$ is a torus of dimension $r \leqq$ rank Z where Z is the center of $\pi_1(M)$. Further, the isotropy subgroup at any point is finite.*

(iii) *A compact homogeneous riemannian manifold without conjugate points is flat.*

THEOREM 2. *Let M be a compact riemannian manifold without focal points. Then:*

(i) *The center Z of $\pi_1(M)$ is isomorphic to $\mathbf{Z} \oplus \mathbf{Z} \oplus \cdots \oplus \mathbf{Z}$ (k times) where $k \leqq \dim(M)$.*

AMS (MOS) subject classifications (1970). Primary 53C20; Secondary 53C70.
*Supported in part by National Foundation grant GP-36418X1.

(ii) $\tilde{M}$, *the universal covering space of* M, *is isometric to* $\boldsymbol{R}^k \times M^*$ *where* $\boldsymbol{R}^k$ *is* k-*dimensional Euclidean space and* M^* *is a simply connected riemannian manifold without focal points. Furthermore* Z *acts on* $\boldsymbol{R}^k$ *by translations.*

(iii) M *is foliated by flat totally geodesic* k-*dimensional tori.*

(iv) *Let* $n = \dim(M)$. *If* $\operatorname{rank} Z \geqq n - 1$, M *is a flat manifold. Further,* M *is a flat torus if and only if* $\pi_1(M)$ *is abelian.*

The fact that M is a flat torus when $\pi_1(M)$ is abelian was also obtained by Avez [**1**].

COROLLARY. *Let* M *be a compact manifold without focal points. Then*:

(i) $\dim I^0(M) = \operatorname{rank} Z$ *and the orbits of* $I^0(M)$ *coincide with the flat totally geodesic tori determined by* Z.

(ii) *Any isometry of* M *which is homotopic to the identity is in* $I^0(M)$.

It would be interesting to know whether part (i) of the previous corollary holds when we just assume that M has no conjugate points. If this were so, it would follow that the only metric without conjugate points on a torus of any dimension is the flat metric. This result was proved in two dimensions by E. Hopf [**2**].

REFERENCES

1. A. Avez, *Variétés riemanniennes sans points focaux*, C.R. Acad. Sci. Paris Sér. A-B **270** (1970), A188–A191. MR **41** #961.

2. E. Hopf, *Closed surfaces without conjugate points*, Proc. Nat. Acad. Sci. U.S.A. **34** (1948), 47–51. MR **9**, 378.

3. J. J. O'Sullivan, *Manifolds without conjugate points*, Math. Ann. (to appear).

INSTITUTE FOR ADVANCED STUDY

Proceedings of Symposia in Pure Mathematics
Volume 27, 1975

LOCAL AND GLOBAL PROPERTIES OF CONVEX SETS IN RIEMANNIAN SPACES

ROLF WALTER

First, generalizing some results of Cheeger and Gromoll [1], it is shown [5] that for every closed locally convex set C of a Riemannian space M, the relative boundary $\mathrm{bd}(C)$ is a locally Lipschitz submanifold and that almost all points of $\mathrm{bd}(C)$ have a unique supporting half-space (see [2] for the flat case). Introducing the metric projection f onto C as the map which, to a point q, associates the point $f(q) \in C$ closest to q, the following is true: f is well defined and locally Lipschitz in an open neighborhood U of C, hence differentiable a.e. in U. This implies e.g. that the distance function ρ_C from C is of class $\mathscr{C}^1$ and twice differentiable a.e. in $U \backslash C$. For details see [3]. Associated with f there is the map which gives, for every $q \in U \backslash C$, the unit initial vector $F(q)$ of the geodesic segment $f(q)\, q$; F is again locally Lipschitz. If C is compact and f, F are restricted to the boundary ∂C^r of the outer parallel set C^r (r small) then a generalized second fundamental form II is defined a.e. by $\mathrm{II} = \langle df \otimes DF \rangle$, and it is symmetric and positive semidefinite a.e.

The second part [4] deals with a formula of the Allendoerffer-Weil type which is valid for arbitrary compact locally convex sets $C \subset M$. Under various assumptions on the curvature along ∂C one can deduce from this formula a Cohn-Vossen type inequality for such C which, in dimensions < 7, only uses sectional curvature. There are applications on the existence and boundedness of total curvature of complete manifolds of nonnegative curvature which, after an earlier announcement of the author (1971), have been proved by a different method in the Stony Brook

AMS (MOS) subject classifications (1970). Primary 53—02, 53B20, 53C20.

thesis of W.A. Poor (1973). There are also corollaries for isometric immersions of codimension $\leqq 2$ [**4**].

LITERATURE

1. J. Cheeger and D. Gromoll, *On the structure of complete manifolds of nonnegative curvature*, Ann. of Math. (2) **96** (1972), 413–443. MR **46** #8121.

2. K. Reidemeister, *Über die singulären Randpunkte eines konvexen Körpers*, Math. Ann. **83** (1921), 116–118.

3. R. Walter, *On the metric projection onto convex sets in Riemannian spaces*, Arch. Math. **25** (1974), 91–98.

4. ———, *A generalized Allendoerffer-Weil formula and an inequality of the Cohn-Vossen type*, J. Differential Geometry **10** (1975), 167–180.

5. ———, *Some analytical properties of geodesically convex sets* (to appear).

UNIVERSITÄT DORTMUND, FEDERAL REPUBLIC OF GERMANY

Proceedings of Symposia in Pure Mathematics.
Volume 27, 1975

ON THE VOLUME OF MANIFOLDS ALL OF WHOSE GEODESICS ARE CLOSED

ALAN WEINSTEIN*

A riemannian manifold (M, g) is called a C_L-manifold if all the geodesics on M are closed and have length $2\pi L$. For example, the symmetric spaces of rank 1 are C_L-manifolds for suitably chosen L. In particular the standard sphere (S^n, can) is a C_1-manifold. Zoll (1903) constructed surfaces of revolution with nonconstant curvature which are C_1-manifolds, and Funk (1913) observed that such surfaces must have area 4π. In this paper we prove that, for any n-dimensional C_L-manifold (M, g), the ratio

$$i(M, g) = \frac{\mathrm{vol}(M, g)}{L^n \,\mathrm{vol}(S^n, can)}$$

is an integer. In fact, $i(M, g)$ is a characteristic number of the circle bundle determined by the geodesic flow on the unit tangent bundle of (M, g). In case M is an even-dimensional sphere, one can show that $i(M, g)$ must equal 1.

AMS (MOS) subject classifications (1970). Primary 53C20; Secondary 53C15, 57D20.
*Paper appears in J. Differential Geometry **9** (1974), 513–517.

SUBMANIFOLDS

Proceedings of Symposia in Pure Mathematics
Volume 27, 1975

ON A GENERALIZATION OF THE CATENOID

DAVID E. BLAIR

It is a classical result that the only surface of revolution in Euclidean space E^3 which is minimal is the catenoid. Of course the surface is conformally flat, but if M^n, $n \geqq 4$, is a conformally flat hypersurface of Euclidean space E^{n+1}, then M^n admits a distinguished direction ("tangent to the meridians"). Thus we seek to characterize conformally flat hypersurfaces of E^{n+1} which are minimal. Specifically we prove the following

THEOREM. *Let M^n, $n \geqq 4$, be a conformally flat, minimal hypersurface immersed in E^{n+1}. Then M^n is either a hypersurface of revolution $S^{n-1} \times M^1$ where S^{n-1} is a Euclidean sphere and M^1 is a plane curve whose curvature κ as a function of arc length s is given by $\kappa = -(n-1)\alpha$, $\alpha = -1/\nu^n$ and*

$$s = \int \frac{\nu^{n-1}\, d\nu}{(-1 + A\nu^{2n-2})^{1/2}}$$

where A is a constant, or M^n is totally geodesic.

AMS (MOS) subject classifications (1970). Primary 53C40; Secondary 53B25.

ON A GENERALIZATION OF THE CATENOID

Proceedings of Symposia in Pure Mathematics
Volume 27, 1975

GEOMETRIC APPLICATIONS OF CRITICAL POINT THEORY TO SUBMANIFOLDS OF COMPLEX PROJECTIVE SPACE AND HYPERBOLIC SPACE

THOMAS E. CECIL

I. Introduction. In a recent paper [1], Nomizu and Rodriguez found a geometric characterization of umbilical submanifolds $M^n \subset R^{n+p}$ in terms of the critical point behavior of a certain class of functions L_p, $p \in R^{n+p}$, on M^n. In that case, if $p \in R^{n+p}$, $x \in M^n$, then $L_p(x) = (d(x, p))^2$, where d is the Euclidean distance function.

Their result can be expressed as follows. Let M^n ($n \geqq 2$) be a connected, complete Riemannian manifold isometrically immersed in R^{n+p}. Suppose there exists a dense subset D of R^{n+p} such that every function of the form L_p, $p \in D$, has index 0 or n at any of its nondegenerate critical points. Then M^n is an umbilical submanifold, that is M^n is embedded in R^{n+p} as a Euclidean subspace R^n or a Euclidean n-sphere, S^n.

Since the set of all points $p \in R^{n+p}$ such that L_p is a Morse function is a dense subset of R^{n+p}, the above theorem could also have been stated in terms of Morse functions of the form L_p.

A corollary of the above theorem is the following. Let M^n be a connected, compact Riemannian manifold isometrically immersed in R^{n+p}. If every Morse function of the form L_p, $p \in R^{n+p}$, has exactly 2 critical points, then $M^n = S^n$.

In this paper, we prove results analogous to those of Nomizu and Rodriguez for submanifolds of complex projective space, $P^m(C)$, endowed with the standard

AMS (MOS) subject classifications (1970). Primary 53B25, 53A35, 53C40.

Fubini-Study metric, and hyperbolic space, H^m, the real space-form of constant sectional curvature -1.

II. Submanifolds of complex projective space. Let M^n be a complex n-dimensional submanifold of $P^{n+p}(\boldsymbol{C})$. For $p \in P^{n+p}(\boldsymbol{C})$, $x \in M^n$, the function $L_p(x)$ which we define is essentially the distance in $P^{n+p}(\boldsymbol{C})$ from p to x. We define the concept of a focal point of (M^n, x) and prove an Index Theorem for L_p which states that the index of L_p at a nondegenerate critical point x is equal to the number of focal points of (M^n, x) on the geodesic in $P^{n+p}(\boldsymbol{C})$ from x to p. In the process, we find that if $L_p(x) = \pi/2$, then L_p has a degenerate critical point at x. Because of this, it is impossible to state our main result in terms of Morse functions of the form L_p. The result is:

THEOREM 1. *Let M^n $(n \geqq 2)$ be a connected, complete, complex n-dimensional Kählerian manifold which is holomorphically and isometrically immersed in $P^{n+p}(\boldsymbol{C})$. Assume there exists a dense subset D of $P^{n+p}(\boldsymbol{C})$ such that every function of the form L_p, $p \in D$, has index* 0 *or n at any of its nondegenerate critical points. Then M^n is either $P^n(\boldsymbol{C})$ or Q^n.*

Here $P^n(\boldsymbol{C})$ denotes a totally geodesic submanifold of $P^{n+p}(\boldsymbol{C})$, and Q^n is the standard complex quadric hypersurface of a totally geodesic $P^{n+1}(\boldsymbol{C}) \subset P^{n+p}(\boldsymbol{C})$. To begin the proof, one shows that $A_\beta^2 = \lambda^2 I$ for any normal β to M^n at any point $x \in M^n$. It is then shown that this implies that M^n is a hypersurface of a totally geodesic $P^{n+1}(\boldsymbol{C})$. Then one remarks that $A_\beta^2 = \lambda^2 I$ implies that M^n is Einstein. The result then follows from a theorem of Smyth **[2]** which states that the only hypersurfaces of $P^{n+1}(\boldsymbol{C})$ which are both complete and Einstein are $P^n(\boldsymbol{C})$ and Q^n. Finally, we study the interesting special case $Q^n \subset P^{n+1}(\boldsymbol{C})$ and find that the set of focal points is $P^{n+1}(R)$, a real $(n+1)$-dimensional projective space naturally embedded in $P^{n+1}(\boldsymbol{C})$.

III. Submanifolds of hyperbolic space. Let M^n be an n-dimensional differentiable manifold immersed in H^{n+p}. For $p \in H^{n+p}$, $x \in M^n$, define $L_p(x) = d(x, p)$, the distance in H^{n+p} from p to x. We define the concept of a focal point of (M^n, x) and prove an Index Theorem for L_p similar to the one mentioned in §II. The main result of this section can then be stated as follows:

THEOREM 2. *Let M^n be a connected, compact, Riemannian manifold isometrically immersed in H^{n+p}. If every Morse function of the form L_p, $p \in H^{n+p}$, has exactly* 2 *critical points, then M^n is embedded as a geometric sphere $S^n \subset H^{n+1} \subset H^{n+p}$, where H^{n+1} is a totally geodesic submanifold of H^{n+p}.*

Finally, we give an example which shows that a result analogous to that of Nomizu and Rodriguez for the noncompact case cannot be proven. More explicitly we exhibit a complete surface $M^2 \subset H^3$ which is not umbilic on which every Morse function of the form L_p has index 0 at any of its critical points. The results of this section will appear in the Tôhoku Math. J. **26** (1974) under the title, *A characterization of metric spheres in hyperbolic space by Morse theory.*

BIBLIOGRAPHY

1. K. Nomizu and L. Rodriguez, *Umbilical submanifolds and Morse functions*, Nagoya Math. J. **48** (1972), 197–201.

2. B. Smyth, *Differential geometry of complex hypersurfaces*, Ann. of Math. (2) **85** (1967), 246–266. MR **39** #6697.

VASSAR COLLEGE

Proceedings of Symposia in Pure Mathematics
Volume 27, 1975

MEAN CURVATURE VECTOR OF A SUBMANIFOLD

BANG-YEN CHEN*

1. Introduction. In the classical theory of surfaces in an ordinary space E^3, the two most important curvatures are the Gauss curvature and the mean curvature. It is well known that the Gauss curvature is intrinsic and the mean curvature is extrinsic. The integral of Gauss curvature gives the beautiful Gauss-Bonnet formula. On the other hand, concerning surfaces with constant mean curvature we have the following results due to Hopf and Klotz-Osserman.

THEOREM A (HOPF [1]). *A closed surface of genus zero in E^3 with constant mean curvature is a sphere.*

THEOREM B (KLOTZ AND OSSERMAN [1]). *A complete surface in E^3 with Gauss curvature $G \geqq 0$ (or $G \leqq 0$) and constant mean curvature is either a minimal surface, a sphere, or a right circular cylinder.*

It is clear that Theorem B gives a generalization of Liebmann's result which says that the only ovaloid with constant mean curvature in E^3 is a sphere.

The main purpose of this note is to give a survey of some recent developments for submanifolds with parallel mean curvature and for total mean curvature.

2. Preliminaries. Let $x = M \to R^m$ be an isometrical immersion of an n-dimensional manifold in an m-dimensional Riemannian manifold R^m and let ∇ and ∇' be the covariant differentiations of M and R^m respectively. Let X and Y be two vector fields on M. Then the second fundamental form h is given by

---

AMS (MOS) subject classifications (1970). Primary 53–02, 53A05, 53B25, 53C40; Secondary 53A10, 53C65.

*This work was partially supported by NSF grant GP-36684.

(2.1) $$\nabla'_X Y = \nabla_X Y + h(X, Y).$$

It is well known that $h(X, Y)$ is a normal vector field on M and it is symmetric on X and Y. Let ξ be a normal vector field on M, we write

(2.2) $$\nabla'_X \xi = -A_\xi(X) + D_X\xi,$$

where $-A_\xi(X)$ and $D_X\xi$ denote the tangential and normal components of $\nabla'_X\xi$. Then we have

(2.3) $$\langle A_\xi(X), Y\rangle = \langle h(X, Y), \xi\rangle,$$

where $\langle\ ,\ \rangle$ denotes the scalar product in R^m. A normal vector field ξ is said to be *parallel* if $D_X\xi = 0$ for all tangent vectors X. *The mean curvature vector* H is defined by

(2.4) $$H = n^{-1} \text{ trace } h.$$

It is clear that a hypersurface has parallel mean curvature (vector) H if and only if H has constant length and a curve in E^m has parallel mean curvature if and only if it is either an open piece of a straight line or an open piece of a plane circle.

In Ruh and Vilms [1], they proved that an immersion of M into a euclidean m-space E^m has parallel mean curvature if and only if the Gauss map associated to the immersion is harmonic in the sense of Eells and Sampson.

3. Submanifolds with parallel mean curvature. It is easy to see that a minimal submanifold of E^m, a minimal submanifold of a hypersphere of E^m, product submanifolds of some submanifolds with parallel mean curvature and some other submanifolds in E^m have parallel mean curvature. Hence it seems to be interesting to ask the following.

PROBLEM I. What kind of n-dimensional submanifolds in a euclidean m-space E^m have parallel mean curvature?

When n is two, this problem has the following complete answer for the local case.

THEOREM 1 (CHEN [1], YAU [1]). *Let M be a surface isometrically immersed in a euclidean space E^m with parallel mean curvature. Then M is one of the following surfaces*:

(a) *minimal surface of E^m*,

(b) *minimal surface of a hypersphere of E^m*,

(c) *surface in a 4-plane of E^m and is locally given by surface $M(\Phi, \alpha, \beta)$, for some analytic quadratic differential Φ and two constants $\alpha > 0$ and β, where $M(\Phi, \alpha, \beta)$ are the surfaces constructed by Hoffman* [1], [2].

If M is a closed surface of genus zero, then we have the following generalization of Theorem A.

THEOREM 2 (FERUS [1], HOFFMAN [1], [2], RUH [1], SMYTH [1]). *The minimal surfaces of a hypersphere of E^m are the only closed surfaces of genus zero in E^m with parallel mean curvature.*

It is still an open problem whether the 2-sphere is the only closed surface in E^3 with parallel mean curvature.

If the dimension of submanifold n is greater than 2, Problem I appears to be quite difficult. However, if we put some additional assumptions on sectional curvatures, then we have the following generalizations of Liebmann's Theorem.

THEOREM 3 (ERBACHER [1], YANO AND ISHIHARA [1]). *Let M be an n-dimensional complete submanifold of nonnegative sectional curvature in E^m. If the mean curvature vector is parallel, the normal bundle is flat, and if either M is closed or M has constant scalar curvature, then M is the product submanifold of some spheres (of certain dimensions) and possibly with one linear subspace.*

THEOREM 4 (SMYTH [1]). *Let M be a closed submanifold of nonnegative sectional curvature in E^m with parallel mean curvature. Then M is a product submanifold of some minimal submanifolds of some hyperspheres of linear subspaces of E^m.*

If the codimension is one, these two theorems are due to Nomizu and Smyth [1]. If the codimension is one and M is convex, they are due to Voss.

4. Total mean curvature. As we mentioned in the introduction, the Gauss curvature and the mean curvature are two of the most important geometric invariants. The integral of Gauss curvature gives the important Gauss-Bonnet formula. Thus it seems to be interesting to study the integral of mean curvature.

For the mean curvature vector we have the following integral formula

$$\int_M H dV = 0 \tag{4.1}$$

for any closed submanifold in E^m. On the other hand, since the integral of the mean curvature, $|H|$, is not a global invariant under similarity transformations, we shall consider the integral of nth power of the mean curvature

$$TMC(x) \equiv \int_M |H|^n \, dV \tag{4.2}$$

instead of the integral of mean curvature. We shall call this integral invariant (4.2) the *total mean curvature* of the immersion x. In Chen [2] (see also Willmore [1]) we have

THEOREM 5. *Let M be an n-dimensional closed submanifold of E^m. Then we have*

$$\int_M |H|^n \, dV \geqq c_n \tag{4.3}$$

where c_n is the volume of a unit n-sphere. The equality sign holds when and only when M is an n-sphere in E^m.

In view of inequality (4.3) and Gauss-Bonnet's formula, it is natural to pose the following problems.

PROBLEM II. Let M be an n-dimensional closed submanifold of E^m. What are the relations between the topological structure of M and total mean curvature? In particular, if the homology groups of M are large, is the total mean curvature large?

PROBLEM III. Let M be an n-dimensional closed Riemannian manifold and let $I(M, m)$ be the set of all isometric immersions of M in E^m. What is

$$\tau(M, m) \equiv \inf_{X \in I(M,m)} TMC(x)? \tag{4.4}$$

If M has nonnegative scalar curvature, then the answer to the second part of Problem II is in the affirmative, in fact, we have the following:

THEOREM 6. *Let M be an n-dimensional closed submanifold of E^m with nonnegative scalar curvature. Then we have*

$$\int_M |H|^n \, dV > a\, \beta(M), \tag{4.5}$$

where $\beta(M) = \max \{\sum_{i=0}^{n} \dim H_i(M: F); F \text{ fields}\}$, $H_i(M, F)$ the ith homology group of M over F, and

$$\begin{aligned} a &= (4n^n)^{-1/2} c_n, && \textit{if } n \textit{ is even},\\ &= (2n^n c_{m-n-1}\, c_{m+n-1})^{-1/2} (c_{2n})^{1/2} c_{m-1}, && \textit{if } n \textit{ is odd}. \end{aligned}$$

For surfaces in E^4 we have the following partial answers to Problem III.

THEOREM 7. *Let M be a real projective plane with nonnegative Gauss curvature. Then we have*

$$\tau(M, 4) > (2 + \pi)\, \pi. \tag{4.6}$$

THEOREM 8. *Let M be a closed surface with nonpositive Gauss curvature. Then we have*

$$\tau(M, 4) \geqq 2\pi^2. \tag{4.7}$$

Related to Problem III, we have the following result due to Shiohama and Takagi [**1**] and Willmore [**2**].

THEOREM 9. *Let M be a surface in E^3 such that M is generated by carrying a small circle around a closed space curve so that the center moves along the curve and the plane of the circle is in the normal plane to the curve at each point. Then we have*

$$\int_M |H|^2 \, dV \geqq 2\pi^2. \tag{4.8}$$

The equality sign holds when and only when M is congruent to the anchor ring given by

$$\begin{aligned} x_1 &= (2^{1/2} c + c \cos \mu) \cos \nu,\\ x_2 &= (2^{1/2} c + c \cos \mu) \sin \nu,\\ x_3 &= c \sin \mu, \end{aligned}$$

where c is a positive constant.

REFERENCES

Chen, B.-Y.

1. *On the surfaces with parallel mean curvature vector*, Indiana Univ. Math. J. **22** (1973), 655–666.
2. *On the total curvature of immersed manifolds*. I, II, III, Amer. J. Math. **93** (1971), 148–162; **94** (1972), 799–809; **95** (1973), 636–642. MR **43** #3971.
3. *Geometry of submanifolds*, Dekker, New York, 1973.

Chen, B.-Y and Ludden, G.D.

1. *Surfaces with mean curvature vector parallel in the normal bundle*, Nagoya Math. J. **47** (1972), 161–167.

Erbacher, J.A.

1. *Isometric immersions with constant mean curvature and triviality of the normal bundle*, Nagoya Math. J. **45** (1972), 139–165.

Ferus, D.

1. *The torsion form of submanifolds in* E^N, Math. Ann. **193** (1971), 114–120. MR **44** #4697.

Hoffman, D.A.

1. *Surfaces in constant curvature manifolds with parallel mean curvature vector field*, Bull. Amer. Math. Soc. **78** (1972), 247–250. MR **45** #7653.
2. *Surfaces of constant mean curvature in constant curvature manifold*, J. Differential Geometry **8** (1973), 161–176.

Hopf, H.

1. *Über Flachen mit einer Relation zwischen den Hauptkrümmungen*, Math. Nachr. **4** (1951), 232–249. MR **2**, 634.

Klotz, T. and Osserman, R.

1. *Complete surfaces in* E^3 *with constant mean curvature*, Comment. Math. Helv. **41** (1966/67), 313–318. MR **35** #2213.

Nomizu, K. and Smith, B.

1. *A formula of Simons' type and hypersurfaces with constant mean curvature*, J. Differential Geometry **3** (1969), 367–377. MR **42** #1018.

Ruh, E.A.

1. *Minimal immersions of 2-spheres in* S^4, Proc. Amer. Math. Soc. **28** (1971), 219–222. MR **42** #6761.

Ruh, E.A. and Vilms, J.

1. *The tension of the Gauss map*, Trans. Amer. Math. Soc. **149** (1970), 569–573. MR **41** #4400.

Shiohama, K. and Takagi, R.

1. *A characterization of a standard torus in* E^3, J. Differential Geometry **4** (1970), 477–485. MR **43** #2646.

Smyth, B.

1. *Submanifolds of constant mean curvature*, Ann. of Math. (2) **205** (1973), 265–280.

Willmore, T.J.

1. *Mean curvature of immersed surfaces*, An. Şti. Univ. "Al. I. Cuza" Iaşi Secţ. I a Mat. **14** (1968), 99–103. MR **38** #6496.
2. *Mean curvature of Riemannian immersions*, J. London Math. Soc. **3** (1971), 307–310.

Yano, K. and Ishihara, S.

1. *Submanifolds with parallel mean curvature vector*, J. Differential Geometry **6** (1971), 95–118.

Yau, S.-T.

1. *Submanifolds with constant mean curvature*. I (to appear).

MICHIGAN STATE UNIVERSITY

Proceedings of Symposia in Pure Mathematics
Volume 27, 1975

GEOMETRY OF SUBMANIFOLDS OF EUCLIDEAN SPACES

ROBERT B. GARDNER*

Many mathematicians agree that the subject of differential geometry was born with Gauss' famous memoire in 1827, *Disquitones generales circa superficies curvas*. The natural generalization of this pioneering work is the analysis of submanifolds

$$X\colon M_m \to R^{m+p}$$

in arbitrary codimension euclidean spaces. Although this setting is natural and the problems studied have important applications to other fields, there is much work left to be done.

We will be considering two basic geometric problems, the imbedding problem and the rigidity problem. The imbedding problem is the question of whether a given abstract Riemannian metric is isometric to a submanifold of euclidean space, the rigidity problem is the question of determining when two isometric submanifolds of euclidean space differ by a euclidean motion.

We will use the word global to mean a manifold which is compact, connected, oriented, and without boundary, and we will restrict this survey to the geometry of global manifolds in arbitrary codimension euclidean spaces.

The induced metric. Let $X\colon M_m \to R^{m+p}$ be an immersion; this induces a Riemannian metric $\mathrm{I} = dX \cdot dX$ where

AMS (MOS) subject classifications (1970). Primary 53–02; Secondary 53C40.

Key words and phrases. Imbedding problem, rigidity problem, induced metric, vector valued second fundamental forms, Laplacian, mean curvature vector, Gauss map, Lipschitz-Killing curvature, Gauss equations, Riemannian product, volume preserving immersions.

*Research partially supported by National Science Foundation grant GP-38419.

$$\mathrm{I}([\alpha], [\beta]) = \left.\frac{dX\circ\alpha}{dt}\right|_{t=0} \cdot \left.\frac{dX\circ\beta}{dt}\right|_{t=0}.$$

If I is given locally by

$$\mathrm{I}_U = \sum g_{\alpha\beta}\, du^\alpha\, du^\beta$$

then there is a unique connection matrix

$$\theta_\alpha^\beta = \sum \Gamma_{\alpha\tau}^\beta\, du^\tau$$

characterized by the two sets of conditions

$$dg_{\alpha\beta} = \sum \theta_\alpha^\tau g_{\tau\beta} + g_{\alpha\sigma}\theta_\beta^\sigma \quad \text{and} \quad \Gamma_{\alpha\tau}^\beta = \Gamma_{\tau\alpha}^\beta.$$

This connection allows us to introduce a covariant derivative D defined on all tensor fields. If $f\colon M \to R$ is a real valued function then the Hessian of f is the symmetric quadratic differential form defined by

$$H_{\mathrm{I}}(f) = D(df).$$

The local expression for $H_{\mathrm{I}}(f)$ is $\sum f_{\alpha;\beta}\, du^\alpha\, du^\beta$ where

$$f_{\alpha;\beta} = \partial^2 f/\partial u^\alpha \partial u^\beta - \sum \partial f/\partial u^\tau \Gamma_{\alpha\beta}^\tau.$$

The Laplace operator Δ is the second order partial differential operator defined by the metric trace of the Hessian. Thus locally

$$\Delta f = \sum g^{\alpha\beta} f_{\alpha;\beta}$$

where $g^{\alpha\beta}$ is the matrix inverse of $g_{\alpha\beta}$. The fundamental property of this operator is given by the Bochner lemma.

LEMMA. *Let M be a global manifold, then $\Delta f = 0$ implies that f is constant.*

The vector valued second fundamental form. If we let $a \in R^{m+p}$ then the position vector of an immersion $X\colon M \to R^{m+p}$ allows us to define a real valued function by $X \cdot a$. This function is called the height function in the direction a. The Hessians of these functions characterize a normal vector valued quadratic differential form II via

$$H_{\mathrm{I}}(X \cdot a) = \mathrm{II} \cdot a \quad \text{for all} \quad a \in R^{m+p}.$$

Intuitively II controls the convexity properties of the submanifold. More precisely let $\alpha\colon R \to M$ be a curve parameterized by arc length having tangent vector T on M, and let

$$X\circ\alpha\colon R \to R^{m+p}$$

have first normal N and first curvature κ, then

$$\kappa N = \frac{d^2 X\circ\alpha}{ds^2} = X_*(\nabla_T T) + \mathrm{II}(T, T).$$

The quantity $\nabla_T T = 0$ if and only if α is a geodesic, hence the range of II is the set of vectors of the form $d^2X \circ \alpha / ds^2$ where α is a geodesic.

In particular if $O_U^1(M)$ denotes the first osculating space of M at U, that is the vector space of first normals to curves lying in M, then

$$\dim O_U^1(M) = \dim T_U(M) + \dim \{\mathrm{II} \cdot a(U) \mid a \in R^{m+p}\}.$$

The integer

$$r = \dim \{\mathrm{II} \cdot a(U) \mid a \in R^{m+p}\}$$

is called the rank of II at U.

If II $\equiv 0$, then the first curvature of every geodesic is zero and as a result every geodesic is a straight line. This implies that the image $X(M)$ is a piece of an m-plane.

Another natural condition on a vector of quadratic differential forms is semi-definiteness of its components. The following result due to do Carmo and Lima **[1]** gives the best global result in this direction.

THEOREM. *Let $X: M_m \to R^{m+p}$ be an immersion of a global m-manifold with* $\mathrm{II} \cdot a$ *pointwise semidefinite for all $a \in R^{m+p}$ and* $\mathrm{II} \cdot a_0$ *definite for some point U and direction a_0, then the image $X(M_m)$ lies in a linear subspace of dimension $m + 1$ as the boundary of a convex body.*

We remark that the condition $\mathrm{II} \cdot a$ positive semidefinite for all $a \in R^{m+p}$ does not locally imply that the image is contained in a linear subspace of dimension $m + 1$ as is shown by

$$X(t, u) = (\sin t, \cos t, t, u).$$

The mean curvature vector. The Laplace operator on height functions characterizes a normal vector field h called the mean curvature vector via

$$\Delta(X \cdot a) = h \cdot a \quad \text{for all} \quad a \in R^{m+p}.$$

A good geometric interpretation of h is obtained as follows, let

$$F: M \times R \to R^{m+p}$$

be a deformation of the given immersion with deformation vector $B = F_* \partial / \partial t$, then if $A(t)$ = area of the image of $F(\ ,t)$ then

$$\frac{dA}{dt}(0) = \int_M B \cdot h \, dA$$

hence the maximal change in area is achieved by taking a deformation in the direction of h. (For details see Chern **[4]**.)

The immersions with $h \equiv 0$ are called minimal and the Bochner lemma immediately implies the nonexistence of global minimal immersions in euclidean spaces since

$$h = 0 \quad \text{implies} \quad \Delta(X \cdot a) = 0$$

which implies that X is constant.

This last result is a special case of the observation that the range of the mean curvature vector controls the range of a global immersion since

$$h \cdot a = 0 \quad \text{implies} \quad \Delta(X \cdot a) = 0$$

which implies that $X \cdot a =$ constant and hence that $X(M)$ lies in a hyperplane perpendicular to a.

Another immediate consequence of the Bochner lemma is a reasonable condition guaranteeing global rigidity.

THEOREM. *Let*

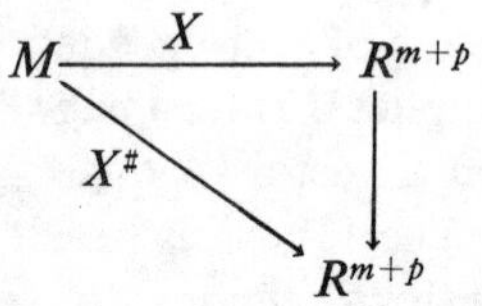

be two isometric immersions of a global manifold M, then X and $X^{\#}$ differ by a translation if and only if they have the same mean curvature vectors.

PROOF. For all $a \in R^{m+p}$,

$$\Delta(X - X^{\#}) \cdot a = (h - h^{\#}) \cdot a = 0;$$

hence

$$X = X^{\#} + a_0.$$

A natural geometric problem is the classification of those immersions having a parallel mean curvature vector field, that is those with $Dh = 0$. The only satisfactory results in this direction are for surfaces in R^4 (see Hoffman [8]).

The normal mappings. The Gauss or normal map associated to an immersed hypersurface

$$X \colon M_m \to R^{m+1}$$

is the map

$$g \colon M_m \to S^m$$

defined by translating the unit normal at a point to the origin.

There are two natural generalizations of this notion to arbitrary codimension immersions

$$X \colon M_m \to R^{m+p}.$$

The first is the map into the Grassmannian of m-planes in $m + p$ space

$$g \colon M_m \to G(m, m + p)$$

given by translating the tangent plane at a point to the origin. This mapping has considerable topological significance, but we will defer to the survey article of Chern in [3].

The number $s(p) = \text{rank } g_*|_u$ is known as the dual dimension. This number s measures the number of parameters needed to locally describe the family of tangent spaces. We will study some properties of this number later.

The second generalization of the normal map is due to Chern and Lashof **[5]**, and is the mapping defined on the unit normal bundle

$$G: N(M) \to S^{m+p-1}$$

defined by translating the given unit normal to the origin. This mapping defines a local degree via

$$\underbrace{G^* \, d\Sigma_{m+p-1}|_a}_{\text{area } S^{m+p-1}} = \det \text{II} \cdot a \; \underbrace{d\Sigma_{p=1}}_{\text{area fiber}} \; \underbrace{dA.}_{\text{area base}}$$

This local degree $\det \text{II} \cdot a$ is called the Lipschitz-Killing curvature.

The integral

$$\int_{N(M)} |G^* \, d\Sigma_{m+p-1}|$$

is called the total curvature of the immersion. This integral has an obvious lower bound which comes from the observation that G at least double covers the sphere. In order to see this let $a \in S^{m+p-1}$ and consider the height function $X \cdot a$. Since M is global there is a maximum and a minimum at which

$$0 = d(X \cdot a) = dX \cdot a$$

which implies that a is normal at these points.

As a result

$$\int |G^* \, d\Sigma_{m+p-1}| = \int_{G(N(M) - \text{crit } G)} d\Sigma_{m+p-1}$$

but by Sard's theorem $G(\text{crit } G)$ has measure zero and hence

$$\int |G^* \, d\Sigma_{m+p-1}| \geqq 2 \text{ Area } S^{m+p-1}.$$

If

$$\int |G^* \, d\Sigma_{m+p-1}| < 3 \text{ Area } S^{m+p-1},$$

then there exists a set of positive measure of $a \in S^{m+p-1}$ such that $X \cdot a$ has precisely two critical points and such that $G(n(p)) \neq 0$ with the translate of $n(p)$ equal to a. This means that the two critical points are nondegenerate and a theorem of Reeb (see Milnor **[11]**) which follows from a simple analysis of gradient flows implies that M_m is homeomorphic to a sphere.

The critical case of the inequality

$$\int |G^* \, d\Sigma_{m+p-1}| = 2 \text{ Area } S^{m+p-1}$$

can be shown to imply that $X(M_m)$ lies in an $(m + 1)$-linear space.

The proof of the theorem of do Carmo and Lima referred to in the second section is proved using similar ideas and a more detailed analysis of gradient flows.

The Gauss equations. One of the basic ideas initiated by Gauss was the distinction between the properties which could be derived from the metric alone which are called intrinsic, and the properties which depend essentially on both the metric and the second fundamental forms which are called extrinsic.

The question of which properties of a submanifold are intrinsic is not as simple as it might appear, for example if

$$X\colon M_m \to R^{m+1}$$

there is a single second fundamental form and

$$\det(\mathrm{I} + t\mathrm{II}) = \sum \binom{m}{\alpha} \sigma_\alpha t^\alpha$$

defines what appears to be a family of extrinsic invariants, but if α is even it can be shown that they are actually intrinsic.

One of the basic sets of relations between the metric and the second fundamental forms is the Gauss equations. If we restrict ranges of indicies to $1 \leqq \alpha, \beta, \sigma \leqq m$ and $m + 1 \leqq a \leqq m + p$ and let

$$\mathrm{I} = \sum (\tau^\sigma)^2 \quad \text{and} \quad \mathrm{II} = \sum \phi^a_\alpha \tau^\alpha \otimes e_a$$

then the Levi-Civita connection is defined by

$$d\tau^\alpha = \sum \tau^\sigma \wedge \phi^\alpha_\sigma \quad \text{with } \phi^\alpha_\sigma = -\phi^\sigma_\alpha$$

and the curvature matrix is defined by

$$\Theta^\beta_\alpha = d\phi^\beta_\alpha - \sum \phi^\tau_\alpha \wedge \phi^\beta_\tau.$$

As such the Gauss equations may be written

$$\Theta^\beta_\alpha = -\sum_a \phi^a_\alpha \wedge \phi^a_\beta.$$

Matters being so, we may introduce arithmetic invariants as follows:

$h =$ minimal number of 1-forms needed to express all Θ^β_α,
$s = \dim \{\phi^a_\alpha\}$,
$r = \operatorname{rank} \mathrm{II}$.

The integer h is a measure of flatness, s is easily seen to be the rank of the tangent map $g\colon M_m \to G(m, m + p)$ which we saw was a measure of the number of parameters needed to locally describe the family of tangent spaces, and r is the codimension of the image in its osculating space. The Gauss equation forces a relation between these three numbers and results in the inequality

$$h \leqq s \leqq h + r.$$

The case $h = 0$ is due to E. Cartan [2] and the inequality with r replaced by the codimension p is due to Chern and Kuiper [5]. The inequality is purely algebraic and may be reduced to showing that the exterior equations

$$\sum_{i=1}^{q} y_\alpha^i \wedge y_\beta^i = 0, \qquad 1 \leqq \alpha, \beta \leqq m,$$

implies dim $\{y_\alpha^i\} \leqq q$.

We say that II is definite at a point if

$$\mathrm{II}(Z, Z) = 0 \text{ implies } Z = 0.$$

If II is definite at a point u, then $s(u) = m$, for otherwise there is a nonzero element $Z \in \{\phi_\alpha^a\}^\perp$ and

$$\mathrm{II}(Z, Z) = \sum \langle Z, \phi_\alpha^a \rangle \langle Z, \tau^\alpha \rangle \otimes e_a = 0.$$

LEMMA. *Let $X: M_m \to R^{m+p}$ be an immersion of a global manifold. Then there is at least one point at which* II *is definite.*

PROOF. Choose the origin of R^{m+p} to be interior to $X(M_m)$, and consider the function $(X \cdot X)/2$. An easy calculation shows

$$H_{\mathrm{I}}(X \cdot X/2) = \mathrm{I} + X \cdot \mathrm{II}$$

and at a maximum value

$$H_{\mathrm{I}}(X \cdot X/2) < 0$$

but $\mathrm{I} > 0$, hence $X \cdot \mathrm{II} < 0$.

As a result of these observations we have the following theorem of Chern and Kuiper [5].

THEOREM. *Let M_m be a global Riemannian manifold with $h \leqq n$. Then it cannot be isometrically immersed in R^{2m-n-1}.*

PROOF. At a point where $s = m$ we have

$$m \leqq h + r \leqq n + r \leqq n + p$$

and hence

$$2m - n \leqq m + p.$$

As a corollary we have the following important theorem of Tompkins [13].

THEOREM. *Let $X: M_m \to R^{m+p}$ be a global immersion inducing a flat Riemannian metric (i.e. $\Theta_\alpha^\beta = 0$). Then $p \geqq m$.*

PROOF. The condition $\Theta_\alpha^\beta = 0$ forces $h = 0$.

E. Cartan studied the geometry of the critical case of a global flat immersion (see [2])

$$X: M_m \to R^{2m}$$

and established the following algebraic theorem.

THEOREM. *If* $\mathrm{II} = \sum \phi_\alpha^a \tau^\alpha \otimes e_a$ *is a definite pencil of symmetric bilinear forms on an m-dimensional space which satisfy*

$$\sum \phi_\alpha^a \wedge \phi_\beta^a = 0, \qquad 1 \leqq \alpha, \beta \leqq m,$$

then there exists an orthogonal change of basis such that

$$\mathrm{II} = \sum b_a w^a w^a \otimes e_a.$$

J. D. Moore [9] has recently used this theorem in a new way to prove the following theorem.

THEOREM. *Let* M_i *be global Riemannian manifolds with* $\dim M_i = n_i \geqq 2$ *for* $1 \leqq i \leqq p$, *and let* $M = M_1 X \cdots XM_p$ *be the Riemannian product, then any isometric immersion*

$$X: M \to R^{n_1+\cdots+n_p+p}$$

is a product of hypersurface immersions.

This coupled with classical theorems on surfaces yields the interesting example that

$$X: \underbrace{S^2X\cdots XS^2}_{p} \to R^{3p}$$

is rigid. This is a genuine global theorem since local isometric deformations exist.

Global isometric imbedding problem. The question of whether a global Riemannian manifold can be isometrically immersed in a euclidean space was solved affirmatively by J. Nash (see [12]).

THEOREM. *Every global Riemannian manifold* M_m *can be isometrically immersed in* $R^{m(3m+11)/2}$.

Unfortunately his proof does not give information on the geometric nature of isometric immersions. There is, however, a formulation of this question in terms of a variational problem which brings more of the geometry into focus.

Thus let $\mathrm{I}^\#$ be a Riemannian metric on a global manifold M_m. If there is any immersion

$$X: M_m \to R^{m+p}$$

then by a theorem of J. Moser [10] there is a volume preserving immersion obtained by preceding X by a diffeomorphism and following X by a constant homothety. The problem then is to characterize isometric immersions among the volume preserving immersions.

Thus let $X: M_m \to R^{m+p}$ be a volume preserving immersion. The abstract metric induces new invariants via the consideration of

$$H_{\mathrm{I}^\#}(X \cdot a) \quad \text{and} \quad \Delta_\#(X \cdot a).$$

The difference of the Hessians relative to the abstract metric $I^\#$ and the induced metric $I = dX \cdot dX$ characterizes a tangential valued quadratic differential form called the difference tensor via

$$H_{I^\#}(X \cdot a) - H_I(X \cdot a) = D(I, I^\#) \cdot a \quad \text{for all } a \in R^{m+p}.$$

The vector field h^* characterized by

$$\Delta_\#(X \cdot a) = h^* \cdot a$$

is called the tension field of X relative to $I^\#$, and by the above

$$h^* = \operatorname{trace}_{I^\#} II + \operatorname{trace}_{I^\#} D(I, I^\#).$$

Since X is volume preserving we have

$$dA^\# = dA \text{ and } \det_{I^\#} I = 1.$$

Matters being so we have

$$0 = \int \Delta_\#\left(\frac{X \cdot X}{2}\right) - \Delta\left(\frac{X \cdot X}{2}\right) dA$$

which by the calculation (see [7])

$$\Delta_{I^\#} \frac{X \cdot X}{2} = \operatorname{trace}_{I^\#} I + X \cdot h^*$$

gives

$$0 = \int \operatorname{trace}_{I^\#} I - m + X \cdot (h^* - h)\, dA,$$

but by Newton's inequality

$$\operatorname{trace}_{I^\#} I \geqq m(\det_{I^\#} I)^{1/m} = m$$

and hence

$$-\, m \text{ Area} = \int X \cdot h\, dA \geqq \int X \cdot h^*\, dA.$$

Thus for any volume preserving immersion

$$m \text{ Area} \leqq -\int X \cdot h^*\, dA$$

with equality occurring if and only if

$$\operatorname{trace}_{I^\#} I = m$$

but this implies that $I = I^\#$ and hence that X is an isometric immersion. The global isometric imbedding problem is thus reduced to studying the minima of

$$-\int X \cdot h^*\, dA$$

which has the lower bound m Area.

REFERENCES

1. M. do Carmo and E. Lima, *Isometric immersions with semi-definite second fundamental forms*, Arch. Math. **20** (1969), 173–175.

2. E. Cartan, *Sur les variétés de courbure constante d'un espace euclidien à non-euclidien*, Bull. Soc. Math. France **47** (1919), 125–160.

3. S.-S. Chern, *La géométrie des sous-variétés d'un espace euclidien à plusieurs dimensions*, Enseignement Math. **40**, 26–46 (1954). MR **16**, 856.

4. ———, *Minimal submanifolds*, Lecture Notes, University of Kansas, 1968. MR **40** #1899.

5. S.-S. Chern and N. Kuiper, *Some theorems on the isometric imbedding of compact Riemann manifolds in euclidean space*, Ann. of Math. (2) **56** (1952), 422–430. MR **14**, 408.

6. S.-S. Chern and R. Lashof, *On the total curvature of immersed manifolds*, Amer. J. Math. **79** (1957), 306–318. MR **18**, 927.

7. R. Gardner, *An integral formula for immersions in euclidean space*, J. Differential Geometry, **3** (1969), 245–252. MR **40** #7992.

8. D. Hoffman, *Surfaces in constant curvature manifolds with parallel mean curvature vector field*, Bull. Amer. Math. Soc. **78** (1972), 247–250. MR **45** #7653.

9. J. D. Moore, *Isometric immersions of riemannian products*, J. Differential Geometry **5** (1971), 159–168. MR **46** #6249.

10. J. Moser, *On the volume elements on a manifold*, Trans. Amer. Math. Soc. **120** (1965), 286–294. MR **32** #409.

11. J. Milnor, *On manifolds homeomorphic to the 7-sphere*, Ann. of Math. (2) **64** (1956), 399–405. MR **18**, 498.

12. J. Nash, *The imbedding problem for Riemannian manifolds*. Ann. of Math. (2) **63** (1956), 20–63. MR **17**, 782.

13. C. Tompkins, *Isometric embedding of flat manifolds in euclidean space*, Duke Math. J. **5** (1939), 58–61.

UNIVERSITY OF NORTH CAROLINA AT CHAPEL HILL

Proceedings of Symposia in Pure Mathematics
Volume 27, 1975

THE HOPF CONJECTURE CONCERNING SURFACES IN E^3

S. I. GOLDBERG[1]

The Gauss map of a surface of constant mean curvature μ immersed in E^3 is harmonic. This fact is used to show that a closed orientable surface S in E^3 with $\mu = \text{const}$ such that $K^2 \geqq C/2\mu^2$, where C is a nonnegative scalar invariant of the Gauss map and K is the Gaussian curvature, is a sphere. (C vanishes if S is a sphere.) This invariant is apparently new and can be given a geometrical interpretation. For surfaces of constant mean curvature it is also shown that $\log K^2$ ($K \neq 0$) is a superharmonic function. The latter seems to be a nontrivial result, but we are unable to draw a geometrical conclusion from it. Some information on the symmetry of S may be useful in settling the conjecture.

UNIVERSITY OF ILLINOIS, URBANA

AMS (MOS) subject classifications (1970). Primary 53A99, 53C40, 53C99.

[1]This is a portion of a paper with S.-S. Chern to appear in Amer. J. Math. entitled *On the volume decreasing property of a class of real harmonic mappings.*

Proceedings of Symposia in Pure Mathematics
Volume 27, 1975

RELATIVE CHERN-LASHOF THEOREMS

NATHANIEL GROSSMAN

We assign to each compact connected manifold with boundary smoothly embedded in a Euclidean space a number, the total absolute curvature, which is positive and at least 1. We prove that the total absolute curvature can equal 1 only when the embedding is onto a closed convex body of a Euclidean subspace of the same dimension. If the total absolute curvature is less than 2, we show that the manifold is contractible and its boundary is a homology sphere. The manifold is diffeomorphic to a closed disc, if its dimension is at least 6, and it as well as its boundary is simply connected. We obtain restrictions on the topology, if the total absolute curvature is less than 5/2. Finally we consider a surface in three-space, and relate the total absolute curvature to the knottedness of the boundary.

(This is the abstract of a paper that has appeared in the Journal of Differential Geometry **7** (1972), 611–618.)

UNIVERSITY OF CALIFORNIA, LOS ANGELES

AMS (MOS) subject classifications (1970). Primary 53C40.

Proceedings of Symposia in Pure Mathematics
Volume 27, 1975

A SOBOLEV INEQUALITY FOR RIEMANNIAN SUBMANIFOLDS

DAVID HOFFMAN AND JOEL SPRUCK

Let $M^m \to \bar{M}^n$ be an isometric immersion of Riemannian manifolds of dimension m and n respectively.

$\bar{K}$ = sectional curvature in $\bar{M}$.

H = mean curvature vector field of the immersion.

$\bar{R}$ = injectivity radius of $\bar{M}$ = minimum distance to the cut locus for all points of $\bar{M}$.

U = a neighborhood of M^m in $\bar{M}$.

ω_m = volume of the unit ball in E^m.

We assume throughout that $\bar{K}_\pi \leqq b^2$ where b is either real or pure imaginary.

1. THEOREM (SOBOLEV INEQUALITY). *Let h be a nonnegative $C^1(U)$ function with compact support. Then*

$$\left(\int_M h^{m/m-1}\, dV_M\right)^{m-1/m} \leqq c(m,b) \int_M [|\nabla h| + h|H|]dV_M,$$

(*where $c(m, b)$ is a universal constant*) *provided both of the following conditions are satisfied*:

(i) $$b^2(\omega_m^{-1}\, 2\, \mathrm{Vol}_M(\mathrm{supp}\ h))^{2/m} < \tfrac{1}{2};$$

(ii) $$\begin{aligned} (2\,\mathrm{Vol}_M(\mathrm{supp}\ h)\,\omega_m^{-1})^{1/m} &\leqq |b|^{-1}\sin(|b|\bar{R}/4), && \text{if } b^2 \geqq 0,\\ &\leqq \bar{R}/4, && \text{if } b^2 \leqq 0. \end{aligned}$$

AMS (MOS) subject classifications (1970). Primary 53C40; Secondary 53A10, 49F10.

*Partially supported by NSF contract 010185.

In the case where $\bar{M}^n = E^n$, this theorem was proved by Michael and Simon [1]. In E^n, $\bar{K}_\pi \equiv 0$ and $\bar{R} = \infty$, hence there are no volume restrictions on the support of h. Similarly

2. COROLLARY. *If $\bar{M}$ is a simply connected, complete, negatively curved ($\bar{K}_\pi \leqq b^2 \leqq 0$) Riemannian manifold then Theorem 1 is true without volume restrictions* (i) *and* (ii).

Theorem 1 has a companion *isoperimetric inequality*:

3. THEOREM. *Let M^m be a compact manifold with boundary ∂M isometrically immersed in $\bar{M}$. Then*

$$(\mathrm{Vol}\,(M))^{m-1/m} \leqq c(m,b)\left(\mathrm{Vol}\,(\partial M) + \int_M |H|\,dV_M\right)$$

provided

(i) $$b^2\,(\omega_m^{-1}\cdot 2\,\mathrm{Vol}\,(M))^{2/m} < \tfrac{1}{2}$$

and

(ii) $$(2\,\mathrm{Vol}\,(M)\cdot\omega_m^{-1})^{1/m} \leqq \begin{cases} |b|^{-1}\sin(|b|\bar{R}/4), & b^2 \geqq 0,\\ \leqq \bar{R}/4, & b^2 \leqq 0.\end{cases}$$

From Theorem 3, it is easily seen that some sort of restriction on the volume of M, or the support of h, is necessary in order to make the inequality possible. For example, there are compact minimal submanifolds without boundary in both positively and negatively curved spaces. These must violate either condition (i) or (ii). Hence we have an a priori lower bound on their volume; this lower bound depends only on b, $\bar{R}$ and m.

More generally:

4. THEOREM. *Let M be a compact manifold without boundary which admits an isometric immersion into $\bar{M}$ satisfying $|H| < \Lambda$. Then there exists a real positive constant $V_0 = V_0(m, b, \bar{R})$ such that*

(a) $$\mathrm{Vol}\,(M) < V_0 \Rightarrow \mathrm{Vol}\,(M) > 1/\Lambda c(m, b),$$

(b) $$\Lambda < 1/c(m, b)\, V_0^{1/m} \Rightarrow \mathrm{Vol}\,(M) > V_0.$$

5. COROLLARY. *A compact manifold M^m without boundary which can be minimally immersed in $\bar{M}$ must satisfy*

$$\mathrm{Vol}\,(M) > V_0.$$

In proving Theorem 1, two basic lemmas are used. For $\zeta \in M$, λ a nondecreasing C^1 function on $\boldsymbol{R}$ with $\lambda(t) = 0$ for $t \leqq 0$, and h as in Theorem 1, we define the following:

$$r(x) = \text{geodesic distance in } \bar{M} \text{ from } \zeta \text{ to } x.$$
$$B_\rho(\zeta) = \text{geodesic ball in } \bar{M} \text{ with center } \zeta \text{ and radius } \rho.$$

$$S_\rho(\zeta) = M \cap B_\rho(\zeta).$$
$$\varphi_\zeta(\rho) = \int_M \lambda(\rho - r)h \, dV_M.$$
$$\psi_\zeta(\rho) = \int_M \lambda(\rho - r)[|\nabla h| + h|\mathrm{H}|] \, dV_M.$$
$$\bar{\varphi}_\zeta(\rho) = \int_{s_\rho(\zeta)} h \, dV_M.$$
$$\bar{\psi}_\zeta(\rho) = \int_{s_\rho(\zeta)} [|\nabla h| + h|H|] \, dV_M.$$
$$\rho_0 = |b|^{-1} \sin^{-1}\left\{|b|\left(2\omega_m^{-1}\int_M h \, dV_M\right)^{1/m}\right\}, \text{ if } b \text{ real},$$
$$= \left(2\omega_m^{-1}\int_M h \, dV_M\right)^{1/m}, \qquad \text{if } b \text{ imaginary}.$$

6. LEMMA. *For $\rho < \inf\{\bar{R}, \pi/b\}$ if b real, and $\rho < \bar{R}$ if b imaginary,*

$$-\frac{d}{d\rho}\left[\frac{\varphi_\zeta(\rho)}{\sinh(|b|_\rho)^m}\right] \leqq \frac{\psi_\zeta(\rho)}{\sin(|b|\rho)^m} \quad \textit{if } b \textit{ real},$$
$$-\frac{d}{d\rho}\left[\frac{\varphi_\zeta(\rho)}{\rho^m}\right] \leqq \frac{\psi_\zeta(\rho)}{\rho^m} \quad \textit{if } b \textit{ pure imaginary}.$$

7. LEMMA. *Suppose $h(\xi) \geqq 1$. There is a ρ, $0 < \rho < \rho_0$, such that $\bar{\varphi}_\zeta(4\rho) \leqq c(m)\,\rho_0\,\bar{\psi}_\zeta(\rho)$ provided*

(i) $$b^2\left(2\omega_m^{-1}\int_M h \, dV_M\right)^{1/m} < \tfrac{1}{2},$$

(ii) $$\rho_0 < \bar{R}.$$

Detailed proofs and other related results will appear elsewhere.

REFERENCES

1. J. H. Michael and L. M. Simon, *Sobolev and meanvalue inequalities on generalized submanifolds of R^n*, Comm. Pure Appl. Math. **26** (1973), 361–379.

UNIVERSITY OF MICHIGAN

COURANT INSTITUTE OF MATHEMATICAL SCIENCES

Proceedings of Symposia in Pure Mathematics
Volume 27, 1975

MINIMAL VARIETIES*

H. BLAINE LAWSON, JR.

The subject of minimal varieties has in the last decade flowered into one of the very active areas of differential geometry. In designing a set of expository lectures on this subject, I was faced, because of time considerations, with having to select certain topics for emphasis. I decided, for example, largely to ignore the classical theory of minimal surfaces in euclidean space because of the existence of an excellent literature in this area. Generally, I have tried to touch upon the results which are more fundamental to geometry as a whole and to give the lectures some internal coherence and natural evolution.

The discussion begins with a derivation of the variational formulas for the area of general, p-dimensional objects in a manifold and leads to some general conclusions about minimal varieties and minimal submanifolds. Minimal varieties in euclidean space are then discussed, with an emphasis on the Plateau problem and various generalizations of Bernstein's theorem. The study of the singularity structure of area-minimizing varieties in this case motivates an exposition of results on minimal submanifolds of spheres. Minimal representatives for integral homology classes are then treated and some applications are made of a generalized method of Synge.

I have finished with a section on minimal varieties in Kähler manifolds which includes a survey of known structure theorems for rectifiable (p, p)-currents, a generalization of Wirtinger's inequality due to Harvey and Knapp, and the formulation of a conjecture related to the Hodge conjecture.

1. The variational formulas. Let S be a p-dimensional submanifold of finite volume in a riemannian manifold M, and let E be a smooth vector field on M with

AMS (MOS) subject classifications (1970). Primary 49F10, 49F20, 49F22, 53C65, 53A10; Secondary 53C40, 32C10, 32C30.

*General lecture given at the Institute.

associated flow φ_t. The main purpose of this section is to compute the derivatives of the function $\mathscr{V}(t) = \mathrm{vol}(\varphi_t(S))$ in terms of E and certain differential-geometric invariants of S. Of course, whenever S is an object which has a well-defined, p-dimensional volume, such as a p-dimensional analytic subvariety or a C^1 image of a polyhedron, $\mathscr{V}(t)$ is still a smooth function, and it is of value to have variational formulas that apply to these cases. It turns out that, in fact, the derivation of the formulas is most transparent in the most general case, namely that of a p-dimensional varifold as introduced by Almgren [**4**]. So, for a while, we shall work in this general context.

Let $G_p(M) \to^{\pi} M$ be the bundle of oriented tangent p-planes on M, and note that $G_p(M) \cong \{\xi \in \Lambda^p TM : \xi \text{ is a simple vector of unit length}\}$. Then $G_p'(M) = G_p'(M)/\mathbf{Z}_2$, where $\mathbf{Z}_2$ acts on $G_p(M)$ by sending ξ to $-\xi$, is the bundle of unoriented tangent p-planes. We shall abuse notation slightly and indicate elements of $G_p'(M)$ also by simple vectors in $\Lambda^p TM$.

DEFINITION 1.1. A *p-dimensional varifold* on M is a Radon measure on $G_p'(M)$. The set of all such varifolds is denoted by $\boldsymbol{V}_p(M)$.

For convenience we shall assume our varifolds to have compact support. (Our discussion will always apply to general varifolds by first restricting them to compact subsets.) We can then consider them as continuous linear functionals on $C(G_p'(M))$. Given a varifold $\mathfrak{S} \in \boldsymbol{V}_p(M)$, we define its *mass* (or "weighted volume") to be $\boldsymbol{M}(\mathfrak{S}) = \mathfrak{S}(1)$, i.e., the $\mathfrak{S}$-measure of the total space.

EXAMPLE 1.2. Let S be a p-dimensional submanifold of finite volume embedded in M. For $x \in S$, let $S_x^{\leftrightarrow} \in G_p'(M)$ denote the tangent plane $T_x S \subset T_x M$. Then S determines a varifold $|S| \in \boldsymbol{V}_p(M)$ by setting

$$|S|(f) = \int_S f(S_x^{\leftrightarrow})\, d\mathscr{H}^p(x) \tag{1.1}$$

for $f \in C(G_p'(M))$ where $\mathscr{H}^p$ denotes Hausdorff p-measure on M. If S is a "manifold with singularities," such as a polyhedral object, $|S|$ is defined in a similar way by restricting to the regular set. Note that $\boldsymbol{M}(|S|) = \mathscr{H}^p(S)$.

The theory of varifolds is based principally on this example. This is particularly clear in the following law for the transformation of varifolds under smooth mappings of the basic space. Let $\varphi : M \to N$ be a C^1 map between manifolds. Then for each $\xi \in G_p'(M)$ we have the associated "Jacobian determinant" $j_\varphi(\xi) = \|\varphi_* \xi\|$, and we define a map $\varphi_* : \boldsymbol{V}_p(M) \to \boldsymbol{V}_p(N)$ by

$$(\varphi_* \mathfrak{S})(f) = \mathfrak{S}((f \circ \varphi) j_\varphi) \tag{1.2}$$

for $f \in C(G_p'(N))$. In particular we have that

$$\boldsymbol{M}(\varphi_* \mathfrak{S}) = \mathfrak{S}(j_\varphi). \tag{1.3}$$

Note that if $\mathfrak{S} = |S|$ where S is a submanifold as in Example 1.2, then $\varphi_* |S| = |\varphi(S)|$.

DEFINITION 1.3. A varifold $\mathfrak{S} \in \boldsymbol{V}_p(M)$ is called a *minimal* (or *stationary*) *variety* if for all vector fields $E \in \mathfrak{X}_M$ we have

(1.4) $$\frac{d}{dt}\boldsymbol{M}(\varphi_{t*}\mathfrak{S})\Big|_{t=0} = 0$$

where φ_t is the flow generated by E. $\mathfrak{S}$ is called a *stable variety* if, for each such E, there is an $\varepsilon > 0$ such that $\boldsymbol{M}(\varphi_{t*}\mathfrak{S}) \geqq \boldsymbol{M}(\mathfrak{S})$ for all $|t| < \varepsilon$. In this case we also have

(1.5) $$\frac{d^2}{dt^2}\boldsymbol{M}(\varphi_{t*}\mathfrak{S})\Big|_{t=0} \geqq 0.$$

These definitions carry over to general varifolds by considering only vector fields E with compact support and replacing the mass by the measure of $\pi^{-1}(K)$ where $\pi : G_p'(M) \to M$ is projection and $K = \operatorname{supp}(E)$.

Whenever a minimal varifold is determined by a submanifold S of M as in Example 1.2, we shall also use the term minimal variety to refer to the closure of S in M.

It is important to consider the closure of this set. For example, let $e_1, \cdots, e_k$ be unit vectors at the origin in $\boldsymbol{R}^n$ and consider $S = \bigcup_k \{te_k : t > 0\}$. S determines a minimal variety in $\boldsymbol{R}^n - \{0\}$, but this variety is minimal in $\boldsymbol{R}^n$ if and only if $\sum_k e_k = 0$.

The following is a list of some general classes of minimal varieties:

(i) Totally geodesic submanifolds.

(ii) Singularities and extremal orbits of compact group actions. (Let G be a compact Lie group acting by isometries on a riemannian manifold M, and let $H \subset G$ be a compact subgroup. We define $M_{(H)}$ to be the set of all points $x \in M$ such that $G_x = \{g \in G : g(x) = x\}$ is conjugate to H. Then $M_{(H)}$ is a submanifold of M with $\mathrm{Cl}(M_{(H)}) = \bigcup \{M_{(H')} : H' \subset H\}$, and $\bar{M}_{(H)}$ is a minimal variety. Furthermore, if we consider in $M_{(H)}$ the function v which assigns to each point the volume of the orbit through that point, then the orbit through each critical point of v on $M_{(H)}$ is a minimal variety. See **[33]** and **[35]** for details.)

(iii) Complex analytic subvarieties of Kähler manifolds (cf. § 5).

(iv) Integral currents of least mass in a homology class with coefficients in $\boldsymbol{Z}$ or $\boldsymbol{Z}_\nu$ (cf. § 4).

Let us at this point justify the term minimal variety for such very general objects and at the same time introduce a notion that will be important in subsequent discussion. Recall that the Hausdorff p-measure of a set $S \subset M$ is defined as

$$\mathscr{H}^p(S) = \liminf_{\varepsilon \downarrow 0} \Big\{ \sum_{\mathcal{O} \in C} \alpha_p (\operatorname{diam}(\mathcal{O})/2)^p : C \in \mathscr{C}_\varepsilon(S) \Big\}$$

where $\mathscr{C}_\varepsilon(S)$ is the collection of countable coverings of S by sets of diameter $\leqq \varepsilon$, and α_p is the volume of the unit ball in $\boldsymbol{R}^p$.

DEFINITION 1.4. An $\mathscr{H}^p$-measurable set $S \subset M$ with $\mathscr{H}^p(S) < \infty$ is called *p-rectifiable* if for all $\varepsilon > 0$ there is an embedded C^1 submanifold S' of dimension p such that $\mathscr{H}^p(S \,\Delta\, S') < \varepsilon$.

Federer has shown that every $\mathscr{H}^p$-measurable set S with $\mathscr{H}^p(S) < \infty$ can be decomposed $\mathscr{H}^p$-uniquely as a disjoint union $S = S_0 \cup S_1$ where S_0 is p-rectifiable and S_1 contains no p-rectifiable subsets of positive measure. Furthermore, if S is p-rectifiable, then at $\mathscr{H}^p$-a.a. $x \in S$, S has an approximate tangent p-plane $S_x^{\leftrightarrow} \in G_p(M)$. (See [**20**, 9.6] or [**22**, Chapter 3].) Therefore, if S is p-rectifiable, we can define a varifold $|S| \in \boldsymbol{V}_p(M)$ as in Example 1.2, by setting

$$|S|(f) = \int_S f(S_x^{\leftrightarrow})\, d\mathscr{H}^p(x) \tag{1.6}$$

for $f \in C(G_p'(M))$. This definition is again natural with respect to transformations; namely, if $\varphi : M \to N$ is a C^1 map between manifolds, then $\varphi_*(|S|) = |\varphi(S)|$ (cf. [**22**, 3.2.20]).

A varifold $\mathfrak{S} \in \boldsymbol{V}_p(M)$ is called *rectifiable* (resp. *integral*) if $\mathfrak{S} = \sum_{n=1}^{\infty} r_n |S_n|$ where $\{S_n\}_{n=1}^{\infty}$ is a disjoint sequence of p-rectifiable sets and $\{r_n\}_{n=1}^{\infty}$ is a sequence of positive real numbers (resp. integers) such that $\sum r_n \mathscr{H}^p(S_n)\ (= \boldsymbol{M}(\mathfrak{S})) < \infty$ and $\mathrm{Cl}(\bigcup S_n)$ is compact.

Given such a varifold, we define a measure $\|\mathfrak{S}\|$ on M by $\|\mathfrak{S}\|(S) = \sum r_n \mathscr{H}^p(S \cap S_n)$. ($\|\mathfrak{S}\| = \pi_*' \mathfrak{S}$ where $\pi' : G_p'(M) \to M$ is the natural projection, and $\operatorname{supp} \|\mathfrak{S}\| = \mathrm{Cl}(\bigcup S_n)$.) We also define a field $\mathfrak{S}^{\leftrightarrow}$ of p-planes by setting $\mathfrak{S}_x^{\leftrightarrow} = (S_n^{\leftrightarrow})_x$ for $x \in S_n$. Then

$$\mathfrak{S}(f) = \int_M f(\mathfrak{S}_x^{\leftrightarrow})\, d\|\mathfrak{S}\|(x)$$

in analogy with formula (1.1) for submanifolds.

Special cases of the fundamental work of Allard and Almgren are the following deep results [**2**].

THEOREM 1.5. *Every minimal variety is rectifiable.*

THEOREM 1.6. *For each $C > 0$ and each compact set $K \subset M$, the set $\{\mathfrak{S} \in \boldsymbol{V}_p(M) : \boldsymbol{M}(\mathfrak{S}) \leqq C$, $\operatorname{supp} \|\mathfrak{S}\| \subset K$ and $\mathfrak{S}$ is rectifiable$\}$ is compact in the weak topology.*

The corresponding theorem holds for integral varifolds.

If $\mathfrak{S} \in \boldsymbol{V}_p(M)$ is a minimal variety, then $\mathscr{H}^p(\operatorname{supp} \|\mathfrak{S}\|) < \infty$. Moreover, we have

THEOREM 1.7 (ALLARD [**2**]). *If $\mathfrak{S}$ is a minimal variety, then there is a relatively open, dense subset R of* supp $\|\mathfrak{S}\|$ *which is a regular (minimal) submanifold of dimension p in M.*

Each point of R has a neighborhood in which $\mathfrak{S} = c\,|R|$ for some $c > 0$. Finally, we remark that in analogy with the existence of closed geodesics the following has been proved.

THEOREM 1.8 (ALMGREN [**4**]). *Let M be a compact riemannian n-manifold. Then for each p, $1 \leqq p \leqq n - 1$, there exists at least one minimal variety of dimension p in M.*

John Pitts, a student of Almgren, has recently proven a stronger regularity theorem for minimal varieties that arise in this way.

We shall now derive the general variational formulas. Fix $\mathfrak{S} \in \boldsymbol{V}_p(M)$ and let E

$\in \mathfrak{X}_M$ be a vector field with associated flow φ_t. Then from (1.3) we have for each k that

$$\frac{d^k}{dt^k} \boldsymbol{M}(\varphi_{t*}\mathfrak{S})\Big|_{t=0} = \mathfrak{S}\Big(\frac{d^k}{dt^k} j_{\varphi_t}\Big|_{t=0}\Big).$$

To compute the function in the argument of $\mathfrak{S}$ on the right we first consider

$$\frac{d^k}{dt^k} \|\varphi_{t*}\xi\|^2\Big|_{t=0} = \frac{d^k}{dt^k} (\varphi_t{}^*g)(\xi, \xi)\Big|_{t=0} = (\mathscr{L}_E{}^k g)(\xi, \xi)$$

where g denotes the metric tensor and $\mathscr{L}_E$ Lie differentiation with respect to E. Let ∇ denote the riemannian connection on M. Then at each point $x \in M$, E gives rise to an endomorphism $\mathscr{A}^E : T_xM \to T_xM$ by setting $\mathscr{A}^E(X) = \nabla_X E$. $\mathscr{A}^E$ extends uniquely to an endomorphism of the full tensor algebra of T_xM as a derivation which commutes with contractions. Since, for all $X, E \in \mathfrak{X}_M$ we have

$$\nabla_E X - \nabla_X E = [E, X] \tag{1.7}$$

we get that

$$\nabla_E - \mathscr{A}^E = \mathscr{L}_E. \tag{1.8}$$

This identity together with the fact that $\nabla g = 0$ easily gives the following (cf. [**45**] or [**44**]).

THEOREM 1.9 (THE FORMULAS OF FIRST AND SECOND VARIATION). *For a varifold $\mathfrak{S} \in \boldsymbol{V}_p(M)$ and a vector field $E \in \mathfrak{X}_M$ with associated flow φ_t, the derivatives of $\boldsymbol{M}(\varphi_{t*}\mathfrak{S})$ are given by*

$$\frac{d^k}{dt^k} \boldsymbol{M}(\varphi_{t*}\mathfrak{S})\Big|_{t=0} = \mathfrak{S}(j_\varphi{}^{(k)})$$

where

$$j_\varphi'(\xi) = \langle \mathscr{A}^E\xi, \xi\rangle, \tag{1.9}$$

$$j_\varphi''(\xi) = \|\mathscr{A}^E\xi\|^2 - \langle \mathscr{A}^E \xi, \xi\rangle^2 + \langle \mathscr{A}^E\mathscr{A}^E\xi, \xi\rangle + \langle (\nabla_E\mathscr{A}^E)(\xi), \xi\rangle. \tag{1.10}$$

In these formulas ξ is viewed as a unit simple vector in Λ^pT_xM determined up to sign. Thus, for example, if $\xi = e_1 \wedge \cdots \wedge e_p$ for orthonormal vectors $e_1, \cdots, e_p \in T_xM$, then

$$\begin{aligned} \langle \mathscr{A}^E \xi, \xi\rangle &= \sum_{j=1}^p \langle e_1 \wedge \cdots \wedge \mathscr{A}^E e_j \wedge \cdots \wedge e_p, e_1 \wedge \cdots \wedge e_p\rangle \\ &= \sum_{j=1}^p \langle \mathscr{A}^E e_j, e_j\rangle. \end{aligned} \tag{1.11}$$

Consider now the case that $\mathfrak{S} = |S|$ where S is a smooth p-dimensional submanifold. Recall that the *second fundamental form* of S at a point $x \in S$ is a bilinear form B on T_xS with values in the normal space N_xS, given by

$$B(X, Y) = (\nabla_X \tilde{Y})^N \tag{1.12}$$

where $\tilde{Y}$ is any extension of Y to a local tangent vector field on S, and $(\cdot)^N$ denotes orthogonal projection onto N_xS. The definition is independent of the choice of the extension $\tilde{Y}$ and is symmetric in X and Y because of equation (1.7). We can then define a normal vector field $H = \text{trace}(B)$, called the *mean curvature vector field*, which is an invariant of the riemannian geometry of the pair (M, S).

For any $E \in \mathfrak{X}_M$ we have a splitting $E|S = E^T + E^N$ of the restriction of E to S into tangential and normal components. It follows immediately from the computation (1.11) that, for any $x \in S$,

$$\begin{aligned}\langle \mathscr{A}^E(S_x), S_x\rangle &= \langle \mathscr{A}^{E^T}(S_x), S_x\rangle + \langle \mathscr{A}^{E^N}(S_x), S_x\rangle \\ &= \text{div}_S(E^T) - \langle E, H\rangle.\end{aligned} \tag{1.13}$$

Note that if p were equal to n, this term would be the divergence of E on M. However, for $p < n$, this trace-like expression splits as a tangential term which is a divergence and a normal term which involves the mean curvature field. From equations (1.1), (1.9) and (1.13) we have that for a compact submanifold S with boundary

$$\frac{d}{dt}\text{vol}(\varphi_t(S))\Big|_{t=0} = \int_{\partial S}\langle E, \nu\rangle * 1 - \int_S \langle E, H\rangle * 1$$

where ν is the exterior unit normal field to ∂S in S.

THEOREM 1.10. *Let S be a p-dimensional submanifold of M with (possibly empty) boundary ∂S. Then the corresponding varifold $|S|$ is a minimal variety in $M - \partial S$ iff the mean curvature $H \equiv 0$.*

Submanifolds with $H \equiv 0$ are classically called *minimal submanifolds*. An immersion of a manifold into M such that the image has $H \equiv 0$ is called a *minimal immersion.*

Suppose now that $\psi : S \to \mathbf{R}^n$ is an isometric immersion of a riemannian manifold S into $\mathbf{R}^n$ equipped with the standard flat metric. Let ∇^2 denote the Laplace-Beltrami operator on S. Then we have the well-known identity

$$\nabla^2 \psi \cong H \tag{1.14}$$

(see [**43**], for example) where "$\cong$" means "is a parallel translate of" and where $\mathbf{R}^n$ has been identified with its tangent space at the origin.

COROLLARY 1.11. *An immersion $\psi : S \to \mathbf{R}^n$ is minimal if and only if $\nabla^2\psi = 0$, i.e., if and only if the coordinate functions are harmonic in the induced metric.*

Let $S^n = \{x \in \mathbf{R}^{n+1} : \|x\| = 1\}$ be the euclidean sphere considered as a subset of $\mathbf{R}^{n+1}$.

COROLLARY 1.12. *An immersion $\psi : S \to S^n$ is minimal in S^n if and only if*

$$\nabla^2 \psi = -p\psi \tag{1.15}$$

where $p = \dim(S)$ and where ψ is considered as an $\mathbf{R}^{n+1}$-valued function.

PROOF. Let H be the mean curvature vector of S in $\boldsymbol{R}^{n+1}$. If S is minimal in S^n, then H is parallel to a multiple to ψ at each point. Thus, by (1.14), $\nabla^2\psi = f\psi$ for some $f \in C^\infty(S)$. However, since $\|\psi\|^2 \equiv 1$, we have $0 = \langle\psi, \nabla^2\psi\rangle + \|\nabla\psi\|^2 = \langle\psi, f\psi\rangle + p$. This completes the proof.

For minimal submanifolds which are stable we have the inequality (1.5), and it is useful to have the second variational formula expressed in terms of invariants of the normal bundle. In particular, we define a natural riemannian connection ∇' in the normal bundle by setting $\nabla_X'\nu = (\nabla_X\nu)^N$ for $\nu \in C^\infty(N(S))$ and X a tangent vector to S. We then let $R_{X,Y} = [\nabla_X, \nabla_Y] - \nabla_{[X,Y]}$ denote the riemannian curvature tensor of M, and for each $x \in S$ we define a transformation $\mathscr{R} : N_xS \to N_xS$ by

$$\mathscr{R}(\nu) = \sum_{j=1}^{p} (R_{e_j,\nu}e_j)^N$$

where $e_1, \cdots, e_p$ is an orthonormal basis of T_xS. A straightforward computation (cf. [**44**, Chapter I]) gives the following.

THEOREM 1.13. *Let S be a p-dimensional minimal submanifold of a riemannian manifold M, and let $E \in \mathfrak{X}_M$ be a vector field such that $E|S$ is normal to S and $E|\partial S = 0$. Then*

$$\frac{d^2}{dt^2}\operatorname{vol}(\varphi_t(S))\bigg|_{t=0} = \int_S \{\|\nabla' E\|^2 - \|\langle B, E\rangle\|^2 + \langle\mathscr{R}(E), E\rangle\}\, d\mathscr{H}^p.$$

2. Minimal submanifolds of $\boldsymbol{R}^n$. This section will be concerned primarily with minimal varieties in euclidean space. It is easy to see that no such varieties exist with compact support. (Suppose $\mathfrak{S} \in \boldsymbol{V}_p(\boldsymbol{R}^n)$ were minimal and consider the radial vector field $E(x) = x$. Then for any tangent vector e in $\boldsymbol{R}^n$, we have $\nabla_e E = e$, and so for any $\xi \in G_p(\boldsymbol{R}^n)$, we have $\langle\mathscr{A}^E\xi, \xi\rangle = p$. It now follows from the first variational formula applied to E that $\boldsymbol{M}(\mathfrak{S}) = 0$, and so $\mathfrak{S} = 0$.) Consequently, the only geometrically interesting minimal submanifolds are those which are compact with boundary and those noncompact ones which are complete in the induced metric. The principal results in these respective areas concern solutions to Plateau's problem and generalizations of Bernstein's theorem. In the interests of time we shall confine our discussion principally to these questions. The most striking theorems concerning these questions have been proved in dimension two and in codimension one. We begin with the case of surfaces.

It is well known that in 1930 J. Douglas and T. Radó gave (independently) the first general solution to the classical Plateau problem for surfaces. For completeness sake we shall state their theorem here. Let $\varDelta = \{(x, y) : x^2 + y^2 < 1\}$, and for each Jordan curve Γ in $\boldsymbol{R}^n$ we consider the space $\mathscr{C}_\Gamma$ of continuous functions $F : \bar{\varDelta} \to \boldsymbol{R}^n$ having first derivatives in L_2, which map $\partial\varDelta$ onto Γ. For each $F \in \mathscr{C}_\Gamma$ we define

$$A(F) = \iint_\varDelta (|F_x|^2|F_y|^2 - \langle F_x, F_y\rangle^2)^{1/2} dx\, dy$$

and

$$D(F) = \iint_{\Delta} (|F_x|^2 + |F_y|^2)\, dx\, dy$$

and set $a_\Gamma = \inf \{A(F) : F \in \mathscr{C}_\Gamma\}$ and $d_\Gamma = \inf \{D(F) : F \in \mathscr{C}_\Gamma\}$.

THEOREM 2.1 (DOUGLAS **[19]** AND RADÓ **[55]**). *Given a Jordan curve $\Gamma \subset \boldsymbol{R}^n$ there exists $F \in \mathscr{C}_\Gamma$ such that:*

(i) $F \mid \Delta \in C^\omega(\Delta)$.

(ii) *F is almost conformal, i.e., $\langle F_x, F_y \rangle = 0$ and $|F_x| = |F_y|$ in Δ with $|dF| > 0$ except at isolated (branch) points.*

(iii) *$F|\partial\Delta : \partial\Delta \to \Gamma$ is a homeomorphism.*

(iv) *$A(F) = a_\Gamma$ and $D(F) = d_\Gamma$.*

Note that it is not necessary that a_Γ be finite. When it is, it follows from (ii) that $2a_\Gamma = d_\Gamma$.

This theorem has been generalized in many ways. (See **[15]**, **[52]** or **[43]**, or the book of Nitsche which will appear very soon.) In fact, in recent years there has been a renaissance of interest in this work, and a number of results have been proved concerning it, including a basic boundary regularity theorem **[32]** and a proof of the nonexistence of branch points **[54]**, **[28]** for $n = 3$.

One of the most seminal results in the geometric theory of minimal submanifolds has been the classical theorem of S. Bernstein proved in 1915.

THE BERNSTEIN THEOREM. *Let f: $\boldsymbol{R} \to \boldsymbol{R}$ be a function whose graph is a minimal surface, then f is linear.*

Note that the graph of a function has the property that its image under the Gauss map lies in the upper hemisphere of S^2. As such, the Bernstein theorem turns out to be a rather special case of the following striking result **[49]**.

THE OSSERMAN THEOREM. *Let Σ be a complete (regular) minimal surface in $\boldsymbol{R}^3$, which is not a plane. Then the Gaussian image of Σ is dense in S^2. (In fact, the complement of the image has zero logarithmic capacity.) If, moreover, the total curvature of Σ is finite, then the Gaussian image omits at most three points.*

This theorem carries over to surfaces in $\boldsymbol{R}^n$. The key to studying such surfaces is the observation that the generalized Gauss map is (anti-) holomorphic. To make this statement precise, recall from the general existence of isothermal parameters in dimension two that we can assume any surface in $\boldsymbol{R}^n$ to be presented as a conformal immersion $\psi : R \to \boldsymbol{R}^n$ where R is a Riemann surface. Let $z = x + iy$ be a local parameter on R and consider the local $\boldsymbol{C}^n$-valued function

$$\varphi = \partial\psi/\partial z. \tag{2.1}$$

Setting $\varphi^2 = \sum_k \varphi_k^2 = \frac{1}{4}(|\psi_x|^2 - |\psi_y|^2 - 2i\langle \psi_x, \psi_y \rangle)$, we have that the conformality of ψ is equivalent to

$$\varphi^2 = 0. \tag{2.2}$$

The induced metric on R has the form $ds^2 = 2F|dz|^2$ where $2F = |\psi_x|^2 = |\psi_y|^2$

$= 2|\varphi|^2$. Hence, the laplacian $\nabla^2 = (1/2F)(\partial/\partial z)(\partial/\partial \bar{z})$, and so the minimal surface equation (cf. Corollary 1.10) is equivalent to

$$\partial\varphi/\partial\bar{z} = 0. \tag{2.3}$$

We can now describe the generalized Gauss map. Let $P^{n-1}(\boldsymbol{C})$ denote complex projective $(n-1)$-space with the standard Fubini-Study metric of holomorphic curvature 2, and let $\boldsymbol{Q}_{n-2} \subset P^{n-1}(\boldsymbol{C})$ be the complex hypersurface defined in homogeneous coordinates by the equation $\sum_k Z_k^2 = 0$. The functions (2.1) describe a map $\Phi : R \to \boldsymbol{Q}_{n-2}$ which by (2.3) is holomorphic. In fact, $\boldsymbol{Q}_{n-2} = \mathrm{SO}(n)/(\mathrm{SO}(2) \times \mathrm{SO}(n-2))$ = the Grassmannian of oriented 2-planes in $\boldsymbol{R}^n$. To see this directly, suppose $P \subset \boldsymbol{R}^n$ to be an oriented plane and let (v_1, v_2) be an oriented, orthogonal basis of P with $|v_1| = |v_2|$. Consider $Z(P) = v_1 + iv_2 \in \boldsymbol{C}^n$. If (v_1', v') is another such basis of P, then $Z'(P) = \alpha Z(P)$ for some $\alpha \in \boldsymbol{C}^*$. Hence, P determines a point $[Z(P)] \in \boldsymbol{Q}_{n-2}$. The map $P \to [Z(P)]$ is a bijection.

NOTE. Under this standard correspondence the Gauss map should be given by $\bar{\Phi}$, i.e., by $\partial\psi/\partial\bar{z}$ locally, and would thus be antiholomorphic. (This is classical in $\boldsymbol{R}^3$.) For convenience of language we choose to discuss Φ.

NOTE. When $n = 3$, $\boldsymbol{Q}_1$ is isometric to the *unit* 2-sphere. This is the reason for the unusual normalization of the metric on $P^{n-1}(\boldsymbol{C})$.

The metric induced on R by Φ has the form

$$ds_\Phi^2 = 2(|\varphi \wedge \varphi'|^2/|\varphi|^4)|dz|^2 \tag{2.4}$$

and it is not difficult to check that

$$ds_\Phi^2/ds^2 = -K \tag{2.5}$$

where K is the Gauss curvature of ds^2. It follows from (2.5) that $K \leqq 0$, and since Φ is holomorphic, the zeros of K are isolated unless $K \equiv 0$. If $K \equiv 0$, then by (2.4), $\varphi \wedge \varphi' \equiv 0$, and it follows easily that $\psi(R)$ is a plane.

The main idea now is to study the geometry of Φ.

Let $\nu = (\nu_1, \cdots, \nu_n) \in \boldsymbol{R}^n$ be a unit vector. Then ν is normal to ψ at $p \in R$ iff $\langle \psi(p), \nu \rangle = \langle \psi_x(p), \nu \rangle - i\langle \psi_y(p), \nu \rangle = 0$ (in some local coordinate $z = x + iy$ at p), i.e., iff $\psi(p)$ is contained in the hyperplane defined by ν in $P^{n-1}(\boldsymbol{C})$. The generalized Bernstein Theorem says the following.

THEOREM 2.2 (CHERN [11], OSSERMAN [51]). *The Gauss map of a complete minimal surface in $\boldsymbol{R}^n$ which is not a plane must meet a dense set of hyperplanes in $P^{n-1}(\boldsymbol{C})$.*

In particular, if the normals to a complete minimal surface omit a neighborhood of some direction, then the surface is a plane. (Here the hyperplane is determined by a linear function with real coefficients.)

PROOF. The fundamental observation in the proof of Theorem 2.6 is the following lemma of Osserman. By passing to the universal covering surface, we may assume that either $R = \Delta$ or $R = \boldsymbol{C}$.

LEMMA 2.3. *Let $\psi : \Delta \to \boldsymbol{R}^n$ be a minimal immersion and set $\varphi = \partial\psi/\partial z$. Suppose* *that for some unit vector* $\nu \in \boldsymbol{C}^n$, *there is an* $\varepsilon > 0$ *such that*

$$|\langle \varphi/|\varphi|, \nu\rangle| \geqq \varepsilon. \tag{2.6}$$

Then Δ *is not complete in the induced metric.*

PROOF. Consider the mapping $w(z) = \int_0^z \langle \varphi(\zeta), \nu\rangle\, d\zeta$. This map has a local inverse $z = F(w)$ with $F(0) = 0$, defined in a disk about $w = 0$. Let ρ be the radius of the largest such disk. Since F is bounded and holomorphic, $\rho < \infty$. Note that there must be some w_0 with $|w_0| = \rho$ such that the curve $\gamma = F[\{tw_0 : 0 \leqq t < 1\}]$ leaves every compact set in Δ. We claim that γ has finite length in the metric $ds^2 = |\varphi|^2|dz|^2$ on Δ. In fact, using (2.6) we have

$$\text{length}(\gamma) = \int_\gamma |\varphi||dz| \leqq \frac{1}{\varepsilon}\int_\gamma |\langle\varphi, \nu\rangle||dz| = \frac{1}{\varepsilon}\int_{F^{-1}(\gamma)} |dw| = \rho/\varepsilon < \infty.$$

Thus, Δ is not complete in the metric ds^2.

It follows from this lemma that $R = \boldsymbol{C}$. The inequality (2.6) now implies that each holomorphic function $\varphi_k/\langle\varphi, \nu\rangle$ (where $\varphi = (\varphi_1, \cdots, \varphi_n)$) is bounded and thus a constant c_k. Consequently, $\varphi(z) = f(z)c$ where $c = (c_1, \cdots, c_n)$ and $f(z) = \langle\varphi(z), \nu\rangle$. Since $\psi(z) = \mathrm{Re}\{\int\varphi(\zeta)d\zeta\}$, we have that $\psi(\boldsymbol{C})$ is a plane. This proves Theorem 2.2.

There are a number of other beautiful, global theorems proved by Osserman for $n = 3$ and by Chern and Osserman for $n > 3$. Many of them involve the total curvature $\int\int|K|dA = -\int\int K\, dA$ of the surface.

THEOREM 2.4. *Let* $\psi : R \to \boldsymbol{R}^n$ *be a complete minimal surface of finite total curvature. Then:*

(i) *R is conformally equivalent to a compact Riemann surface R' punctured at a finite number of points.*

(ii) $\Phi : R \to \boldsymbol{Q}_{n-2}$ *extends to a holomorphic map of* R'.

(iii) $\int\int KdA = -2\pi N$ *for some* $N \in \boldsymbol{Z}^+$ $(-4\pi N$ *if* $n = 3)$.

(iv) $\int\int KdA \leqq 2\pi(\chi - e)$ *where* χ *is the Euler characteristic of* R, *and* e *is the number of ends.*

See **[14]**, **[50]**, **[52]** or **[44]** for details.

We shall now consider varieties in codimension one. For this case it is necessary to enlist the forces of geometric measure theory. We begin by recalling the notion of a rectifiable current.

Let M be a riemannian manifold and $S \subset M$ a p-rectifiable subset with tangent plane field $S^{\leftrightarrow}$. (See § 1.) An *orientation* for S is a choice of orientations for the planes $S_x^{\leftrightarrow}$ so that the resulting functions $S \to G_p(M)|S$, which we denote $x \mapsto \vec{S_x}$, are $\mathscr{H}^p$ -measurable. Recall that $G_p(M) \subset \Lambda^p T(M)$. Thus, given such an oriented,p-rectifiable set, we can define a continuous, linear functional $\mathscr{S}$ on the space of C^∞ p-forms $\mathscr{E}^p(M)$ on M, by

$$\mathscr{S}(\omega) = \int_S \omega(\vec{S_x})\, d\mathscr{H}^p(x) \tag{2.7}$$

for $\omega \in \mathscr{E}^p(M)$. This embeds the oriented p-rectifiable sets (modulo sets of measure zero) into $\mathscr{E}_p(M) = (\mathscr{E}^p(M))'$, the space of de Rham currents on M. We look at the group generated by these and take its closure in a strong topology. Namely, we say a current $\mathscr{S}$ is *rectifiable* if $\mathscr{S} = \sum_{n=1}^{\infty} n\,\mathscr{S}_n$ where $\mathscr{S}_n = (S_n, \vec{S_n})$, $n = 1, 2, 3, \cdots$, is a sequence of mutually disjoint, oriented p-rectifiable sets with

$$\boldsymbol{M}(\mathscr{S}) = \sum_{n=1}^{\infty} n\,\mathscr{H}^p(S_n) \quad \text{and} \quad \operatorname{supp}(\mathscr{S}) = \operatorname{Cl}\Big(\bigcup_n S_n\Big) \text{ compact.}$$

Associated to $\mathscr{S}$ is a measure $\|\mathscr{S}\|$ given on $A \subset M$ by $\|\mathscr{S}\|(A) = \sum n\mathscr{H}^p(S_n \cap A)$. For $x \in S_n$ the plane $(\vec{S_n})_x$ is called the *oriented tangent plane* of $\mathscr{S}$ and is denoted $\vec{\mathscr{S}_x}$. For any $\omega \in \mathscr{E}^p(M)$ we can then write

$$\mathscr{S}(\omega) = \int \omega(\vec{\mathscr{S}_x})\, d\|\mathscr{S}\|(x). \tag{2.8}$$

Note. To any rectifiable current $\mathscr{S}$ there corresponds an underlying rectifiable (in fact, integral) varifold $|\mathscr{S}|$ obtained by neglecting orientations. $\mathscr{S}$ is *minimal*, by definition, if and only if $|\mathscr{S}|$ is minimal.

We denote by $\mathscr{R}_p(M)$ the group of rectifiable p-currents on M and observe that since $\mathscr{R}_p(M) \subset \mathscr{E}_p(M)$ we can take boundaries of elements in $\mathscr{R}_p(M)$ as de Rham currents. That is, for $\mathscr{S} \in \mathscr{E}_p(M)$ we define $\partial\mathscr{S} \in \mathscr{E}_{p-1}(M)$ by setting

$$(\partial\mathscr{S})(\omega) = \mathscr{S}(d\omega)$$

for $\omega \in \mathscr{E}^{p-1}(M)$. In general, $\partial[\mathscr{R}_p(M)] \not\subset \mathscr{R}_{p-1}(M)$, and the space

$$\mathscr{I}_p(M) = \{\mathscr{S} \in \mathscr{R}_p(M) : \partial\,\mathscr{S} \in \mathscr{R}_{p-1}(M)\}$$

is called the set of *integral p-currents* on M. One of the fundamental properties of the space of integral currents is the following.

THEOREM 2.5 (FEDERER AND FLEMING [25]). *Given a compact subset $K \subset M$ and a number $c > 0$, the set*

$$\mathscr{I}_{p,K,c}(M) = \{\mathscr{S} \in \mathscr{I}_p(M) : \operatorname{supp}\mathscr{S} \subset K \text{ and } \boldsymbol{M}(\mathscr{S}) + \boldsymbol{M}(\partial\,\mathscr{S}) \leqq c\}$$

is sequentially compact in the weak topology.

One of the elementary properties of the mass function $\boldsymbol{M}(\cdot)$ is that it is lower semicontinuous in the weak topology. Therefore, an immediate consequence of this theorem is that we can always solve Plateau's problem in the space of integral currents. The varifold underlying such a solution is a minimal variety, and the regularity theorem (Theorem 1.7) applies. However, in codimension one a much stronger theorem is true. From the work of Fleming [27], Almgren [3], deGiorgi [16], Simons [58], Federer [23], and Allard [1] we have

THEOREM 2.6. *Let B be an oriented, closed, $(n-2)$-dimensional submanifold of $\boldsymbol{R}^n$, and suppose $\mathscr{S}_0 \in \mathscr{I}_{n-1}(\boldsymbol{R}^n)$ is any current with $\partial\mathscr{S}_0 = B$ and*

$$\boldsymbol{M}(\mathscr{S}_0) = \inf\{\boldsymbol{M}(\mathscr{S}) : \mathscr{S} \in \mathscr{I}_{n-1}(\boldsymbol{R}^n) \text{ and } \partial\mathscr{S} = B\}.$$

(Such a current exists by Theorem 2.4.) Then $\operatorname{supp}(\mathscr{S}_0) - B$ *is a real analytic manifold with singularities of codimension* 8, *i.e., there is a compact set* $\Sigma \subset \operatorname{supp}(\mathscr{S}_0)$ *with* $\mathscr{H}^\alpha(\Sigma) = 0$ *for all* $\alpha > n - 8$ *and* $\operatorname{supp}(\mathscr{S}_0) - \Sigma - B$ *is a real analytic submanifold of dimension* $n - 1$. *Furthermore, each point of B which admits a support plane has a neighborhood in which* $\operatorname{supp}(\mathscr{S}_0)$ *is a regular manifold with B as boundary.*

In particular, if $n \leqq 7$, then $\mathscr{S}_0 = |S|$ where S is an oriented, real analytic submanifold properly embedded in $\boldsymbol{R}^n - B$.

Thus, while in $\boldsymbol{R}^3$ the Douglas solution to the Plateau problem is always an immersed surface, it may have self-intersections (if, for example, the boundary curve is knotted). However, by Theorem 2.6 there is always an *embedded* analytic surface which solves the Plateau problem among surfaces of all topological types.

A similar regularity theorem holds in any codimension for the unoriented Plateau problem **[23]**. Here the singular set has codimension 2. In general one might conjecture that the singular set of any mass-minimizing integral p-current in $\boldsymbol{R}^n$ would have $\mathscr{H}^{p-1}$-measure zero. This would be nice, since then the regular points would define the current. However, this result is at the moment far from proven.

The codimension 8 restriction in Theorem 2.6 is necessary. The counterexamples are constructed as follows. For any set $S \subset S^{n-1} \subset \boldsymbol{R}^n$ we define the cone on S to be $C(S) = \{tx : x \in S, 0 \leqq t \leqq 1\}$. If S is an oriented p-rectifiable set, then $C(S)$ is naturally an oriented $(p + 1)$-rectifiable set. More generally, if $\mathscr{S} \in \mathscr{I}_{p-1}(S^{n-1})$ has $\partial \mathscr{S} = 0$ and $p > 1$, there is a natural cone $C(\mathscr{S}) \in \mathscr{I}_p(\boldsymbol{R}^n)$ with $\partial C(\mathscr{S}) = \mathscr{S}$ and $\operatorname{supp} C(\mathscr{S}) = C(\operatorname{supp} \mathscr{S})$. Furthermore, $\mathscr{S}$ is minimal if and only if $C(\mathscr{S})$ is minimal. Similar statements apply to varifolds. For positive integers p, q and positive real numbers r, s we define

$$S^p(r) \times S^q(s) = \{(x, y) \in \boldsymbol{R}^{p+1} \times \boldsymbol{R}^{q+1} : \|x\| = r \text{ and } \|y\| = s\} \subset \boldsymbol{R}^{p+q+2}.$$

It is easy to check that

$$\Sigma_{p,q} = S^p\left(\left(\frac{p}{p+q}\right)^{1/2}\right) \times S^q\left(\left(\frac{q}{p+q}\right)^{1/2}\right) \subset S^{p+q+1} \tag{2.9}$$

is minimal in S^{p+q+1}.

THEOREM 2.7. *For* $p + q > 6$ *or for* $p + q = 6$ *and* $|p - q| \leqq 2$, *the cone* $C(\Sigma_{p,q})$ *represents the unique solution to the Plateau problem for the boundary* $\Sigma_{p,q}$.

This was first proven for $p = q \geqq 3$ by Bombieri, deGiorgi and Giusti **[8]**, and the general statement for $p + q > 6$ by the author **[42]**. (Examples of other topological types were also given here.) Recently P. Simões **[64]** proved the result for $\Sigma_{2,4}$ and has proven it to be false for $\Sigma_{1,5}$.

We shall now outline the key geometric ideas in the proof of Theorem 2.6. We begin with some fundamental properties of minimal varieties in $\boldsymbol{R}^n$.

LEMMA 2.8 (MONOTONICITY). *Let* $\mathfrak{S} \in \boldsymbol{V}_p(\boldsymbol{R}^n)$ *be a minimal variety with* $0 \in \operatorname{supp} \|\mathfrak{S}\|$. *Set* $v(r) = \boldsymbol{M}(\mathfrak{S} \cap B_r)$ *where* $B_r = \{x \in \boldsymbol{R}^n : |x| \leqq r\}$. *Then the function*

(2.10) $$\phi(r) = \nu(r)/r^p$$

is monotone nondecreasing for $r > 0$.

PROOF. Consider a vector field $E(x) = f(|x|)x$ for $x \in \boldsymbol{R}^n$, where f approximates the characteristic function of $[-r, r]$ and apply the first variational formula.

Given a minimal variety $\mathfrak{S} \in \boldsymbol{V}_p(\boldsymbol{R}^n)$ we can define a *density function* which at $x \in \boldsymbol{R}^n$ is given as

(2.11) $$\Theta(x, \mathfrak{S}) = \lim_{r \downarrow 0} \frac{\boldsymbol{M}(\mathfrak{S} \cap B_r(x))}{\alpha_p r^p}$$

where α_p is the measure of the unit p-disk and $B_r(x) = x + B_r$.

COROLLARY 2.9. *Let* $\mathfrak{S} \in \boldsymbol{V}_p(\boldsymbol{R}^n)$ *be a minimal variety. Then the function* $x \to \Theta(x, \mathfrak{S})$ *is upper semicontinuous.*

If $\mathfrak{S}$ corresponds to a minimal integral current $\mathscr{S}$, then $\Theta \geqq 1$, $\|\mathscr{S}\|$-a.e. (cf. [**22**, 4.1.28]). Hence, $\mathscr{H}^p(\operatorname{supp} \mathfrak{S}) < \infty$.

We now wish to consider a current which minimizes mass as in Theorem 2.6. Using the compactness theorem, it is possible to get a first order approximation to such a current at (the possibly singular) points of its support.

PROPOSITION 2.10 (CF. [**22**], [**2**] OR [**44**]). *Let* $\mathscr{S}_0 \in \mathscr{I}_p(B_1)$ *be a mass-minimizing current with* $0 \in \operatorname{supp} \mathscr{S}_0 - \operatorname{supp} \partial \mathscr{S}_0$. *Then the weak adherence of the set of dilations*

$$\mathscr{S}(r) = \delta_{r*}(\mathscr{S} \cap B_r)$$

as $r \downarrow 0$ (*where* $\delta_r(x) = (1/r)x$) *is a set of mass-minimizing cones, all of mass* $\Theta(0, |\mathscr{S}|)\, \alpha_p$.

A similar statement holds for general minimal varieties.

A crucial step in the proof of this proposition is the derivation of the following formula. Let $\phi(r)$ be defined as in (2.10). Then for $0 < a < b$,

(2.12) $$\phi(b) - \phi(a) = \int_{B_b - B_a} \frac{1}{|x|^p}\left(1 - \frac{|x^T|^2}{|x|^2}\right) d\|\mathscr{S}_0\|(x)$$

where x^T denotes the orthogonal projection of x onto (the plane parallel to) the tangent plane to $\mathscr{S}_0$ at x.

We shall begin by assuming the following basic regularity result due essentially to Reifenberg. (See Fleming [**27**].)

THEOREM 2.11. *Let* $\mathscr{S}_0 \in \mathscr{I}_p(\boldsymbol{R}^n)$ *be a current which minimizes mass among all currents* $\mathscr{S} \in \mathscr{I}_p(\boldsymbol{R}^n)$ *with* $\partial \mathscr{S} = \partial \mathscr{S}_0$. *Suppose* $0 \in \operatorname{supp} \mathscr{S}_0 - \operatorname{supp} \partial \mathscr{S}_0$ *and* $\Theta(0, \mathscr{S}_0) = 1$. *Then there is a neighborhood* U *of* O *in* $\boldsymbol{R}^n$ *such that* $\operatorname{supp} \mathscr{S}_0 \cap U$ *is a p-dimensional, embedded, analytic submanifold.*

The idea is to show that the tangent cones, given by Proposition 2.10, are all flat disks. To do this we must use Fleming's decomposition lemma [**27**].

LEMMA 2.12. *Let* $\mathscr{S} \in \mathscr{I}_{n-1}(B_1^n)$ *be a mass-minimizing integral current with* $\partial\mathscr{S} \in \mathscr{I}_{n-2}(S^{n-1})$. *Then* $\mathscr{S}$ *can be written as a locally finite sum* $\mathscr{S} = \sum \mathscr{S}_j$ *where each* $\mathscr{S}_j \in \mathscr{I}_{n-1}(B_1^n)$ *is mass-minimizing with* $\partial\mathscr{S}_j \in \mathscr{I}_{n-2}(S^{n-1})$ *and* $\Theta(x, |\mathscr{S}_j|) = 1$ *for* $\|\mathscr{S}_j\|$ *-a.e. x.*

We now proceed by induction on n.

ASSERTION n. An area minimizing cone in $\mathscr{I}_{n-1}(B_1^n)$ having density 1 almost everywhere is a flat disk.

Suppose the assertion true for $n - 1$. Let $\mathscr{S}_\infty$ be such a cone in B_1^n. Choose $x \in \operatorname{supp} \mathscr{S}_\infty - 0$. A tangent cone to $\mathscr{S}_\infty$ at x is a product $\boldsymbol{R} \times \hat{\mathscr{S}}_\infty$ where $\hat{\mathscr{S}}_\infty$ is a mass-minimizing cone in $\boldsymbol{R}^{n-1}$ of density 1 a.e. (That is, it is a product near the vertex of the cone.) Hence, $\hat{\mathscr{S}}_\infty$ is a flat disk. Then by Theorem 2.11, $\mathscr{S}_\infty$ is a regular submanifold outside the origin, i.e., $\mathscr{S}_\infty$ is a cone on a regular minimal submanifold of the sphere. We then apply the following geometric result.

THEOREM 2.13 (ALMGREN [3], $n = 4$, SIMONS [58], $n \geqq 4$). *Let* $M^{n-2} \subset S^{n-1}$ *be a regular (closed) minimal* $(n - 2)$*-dimensional submanifold such that the cone* $C(M^{n-2})$ *is stable with respect to variations supported in* $\operatorname{int}(B_1^n)$. *Then, if* $n \leqq 7$, M^{n-2} *is a totally geodesic* $(n - 2)$*-sphere.*

This theorem fails for $n = 8$. The counterexample $\Sigma_{3,3}$ was first observed by Simons.

The outlines of the proof should now be clear. Of course, for $n \geqq 8$, the inductive assumption must be changed. However, the use of tangent cones is still essential.

The above discussion makes the study of minimizing cones of particular interest. There are at present a number of important, outstanding questions. A basic one is whether there is only one tangent cone to a minimizing current at a given point (cf. Proposition 2.10). Using the compactness theorem, one can show that up to diffeomorphisms of $\boldsymbol{R}^8$, there is a finite number of minimizing cones of dimension 7 (cf. author's private notes). What is this number? One might conjecture that every minimizing current in $\boldsymbol{R}^n$ is supported on a real analytic subvariety. If this were true, then every minimizing cone in $\boldsymbol{R}^n$ would be supported on a homogeneous algebraic subvariety. Is this latter statement true?

Using the above, we can now discuss the following codimension one generalization of the Bernstein theorem, whose proof is due to the same collection of people responsible for Theorem 2.6.

THEOREM 2.14. *Let* $f: \boldsymbol{R}^n \to \boldsymbol{R}$ *be a function whose graph is a minimal submanifold. If* $n \leqq 7$, *then* f *is linear.*

Counterexamples have been found by Bombieri, deGiorgi and Giusti for $n \geqq 8$ [8].

PROOF. Let $G_f = \{(x_1, \cdots, x_{n+1}) \in \boldsymbol{R}^{n+1} : x_{n+1} = f(x_1, \cdots, x_n)\}$ and let $\mathscr{S}$ be the current defined by a continuous orientation of G_f. We first observe that for any r, $\mathscr{S} \cap B_r$ is mass-minimizing for its boundary. To see this we foliate $\boldsymbol{R}^{n+1}$ by the x_{n+1}-translates of G_f, and let ν denote the field of unit normal vectors to this foliation. We choose ν so that $*\nu$ corresponds to $\mathscr{S}^{\rightarrow}$. A direct computation shows that, if Ω is the n-form such that

$$\Omega(\xi) = \langle *\nu, \xi \rangle,$$

then the minimal surface equation is equivalent to the fact that $d\Omega = 0$. It follows that any $\mathscr{S}' \in \mathscr{I}_n(\boldsymbol{R}^{n+1})$ with $\partial \mathscr{S}' = \partial(\mathscr{S} \cap B_r)$ has $\mathscr{S}'(\Omega) = (\mathscr{S} \cap B_r)(\Omega) = \boldsymbol{M}(\mathscr{S} \cap B_r)$. However,

$$\mathscr{S}'(\Omega) = \int \Omega(\mathscr{S}_x'^{\rightarrow}) d\|\mathscr{S}'\|(x) \leqq \int d\|\mathscr{S}'\| = \boldsymbol{M}(\mathscr{S}').$$

We now define $\mathscr{S}(r) \in \mathscr{I}_n(B_1)$ by

$$\mathscr{S}(r) = \delta_{r*}(\mathscr{S} \cap B_r)$$

where $\delta_r(x) = (1/r)x$. $\mathscr{S}(r)$ is mass-minimizing by the above. Moreover, for almost all r, $\partial \mathscr{S}(r)$ is a compact embedded submanifold of S^n (apply Sard's theorem to the function $\|x\|^2$ on G_f) which must separate S^n into two domains one of which has volume $\leqq \frac{1}{2}\operatorname{vol}(S^n)$. It follows that the function $v(r) = M(\mathscr{S}(r))$, which is nondecreasing by Lemma 2.7, satisfies $v(r) \leqq \frac{1}{2}\operatorname{vol}(S^n)$. It is not difficult to see that we also have the inequality

$$\int_0^1 t^{n-1} \boldsymbol{M}(\partial \mathscr{S}(tr))\, dt \leqq \boldsymbol{M}(\mathscr{S}(r)). \tag{2.13}$$

We now choose a sequence $r_j \to \infty$ such that

(i) $\lim_{j\to\infty} \boldsymbol{M}(\partial \mathscr{S}(r_j)) = \liminf_{r\to\infty} \boldsymbol{M}(\partial \mathscr{S}(r))$,

(ii) $\mathscr{S}_j \to \mathscr{S}_\infty \in \mathscr{I}_n(B_1)$.

(Part (ii) is guaranteed by Theorem 2.4.) We observe that $\mathscr{S}_\infty$ is also mass-minimizing for its boundary. To see this it is sufficient to know that $\lim_{j\to\infty}(\inf\{\boldsymbol{M}(\mathscr{S}') : \mathscr{S}' \in \mathscr{I}_n(\boldsymbol{R}^{n+1})$ and $\partial\mathscr{S}' = \partial \mathscr{S}(r_{j+1}) - \partial \mathscr{S}(r_j)\}) = 0$. This follows from the equivalence of the weak topology and the Whitney flat topology on sets of type $\mathscr{I}_{p,K,c}(\boldsymbol{R}^{n+1})$. (See [25].)

We now claim that $\boldsymbol{M}(\mathscr{S}_\infty) = (1/n)\boldsymbol{M}(\partial\mathscr{S}_\infty)$ ($= \boldsymbol{M}$(cone over $\partial\mathscr{S}_\infty$)). The inequality "$\leqq$" follows from the fact that $\mathscr{S}_\infty$ is mass-minimizing. For the reverse inequality, let $\beta = \lim_{j\to\infty} \boldsymbol{M}(\mathscr{S}(r_j))$ and set $\dot\beta = \liminf_{j\to\infty} \boldsymbol{M}(\partial \mathscr{S}(r_j))$. Since the mass function is lower semicontinuous on currents and since $\mathscr{S}_\infty$ is mass-minimizing, we have $\beta = \boldsymbol{M}(\mathscr{S}_\infty)$. Thus, equation (2.13) implies $\dot\beta/n \leqq \beta$. We conclude that *the cone over* $\partial \mathscr{S}_\infty$ *is mass-minimizing*.

Now, arguing as above, we see that interior regularity in dimension n implies that $C(\partial \mathscr{S}_\infty)$ is the cone on a regular, compact, minimal submanifold of S^n.

THEOREM 2.15 (DEGIORGI, SIMONS [58]). *Let $M^{n-1} \subset S^n$ be a compact, oriented minimal submanifold with second fundamental form B. Let ν be the field of unit normal vectors to M^{n-1} in S^n. Then with ν considered as an $\boldsymbol{R}^{n+1}$-valued function on M, we have*

$$\nabla^2\nu = -\|B\|^2\nu. \tag{2.14}$$

Since $\mathscr{S}(r)$ is the graph of a function for each r, the component $\nu_{n+1} \geqq 0$ on $\partial \mathscr{S}_\infty$. Thus, by (2.14), $\nu_{n+1} \equiv 0$ or $\|B\| \equiv 0$. If $B \equiv 0$, $C(\partial \mathscr{S}_\infty) = B_1^n$. If $\nu_{n+1} \equiv 0$, $C(\partial \mathscr{S}_\infty)$ is a product near 0 of a line with a minimizing cone $\hat{\mathscr{S}}_\infty$ in $\boldsymbol{R}^n$. By interior regularity, $\hat{\mathscr{S}}_\infty = B_1^{n-1}$ and so $\mathscr{S}_\infty = B_1^n$. Hence, $\nu(r) \equiv \alpha_n$. The proof now follows by applying formula (2.12). (See [**27**, Lemma 2.3].)

3. Minimal submanifolds of S^n. The existence of tangent cones to minimal varieties in a riemannian manifold M (cf. [**2**]) makes the study of minimal varieties in spheres of fundamental importance. (All such cones are cones on minimal varieties in the unit sphere of the tangent space of M.) From this point of view it is important to have theorems which characterize those compact minimal submanifolds which are totally geodesic, and to understand the geometric structure of those which are not. Such shall be the main concern of this section.

Our discussion will deal principally with smooth submanifolds. However, we begin with observations about general minimal varieties of spheres.

LEMMA 3.1. *Let $\mathfrak{S} \in \boldsymbol{V}_p(S^n)$ be a minimal variety. Then if $\mathfrak{S}$ is carried in a closed hemisphere H of S^n, it is actually carried in the geodesic subsphere ∂H. In particular, there are no minimal varieties in an open hemisphere of S^n.*

PROOF. Consider $S^n = \{x \in \boldsymbol{R}^{n+1} : \|x\| = 1\}$ and let $e \in \boldsymbol{R}^{n+1}$ be the unit vector such that $H = \{x \in S^n : \langle e, x\rangle \geqq 0\}$. Let $E \in \mathfrak{X}_{S^n}$ be the vector field given by $E_x = e - \langle e, x\rangle$. Then for $v \in T_x(S^n)$ we have $\nabla_v E = -\langle e, x\rangle v$ where ∇ is the riemannian connection on S^n. (Recall that $\nabla_X Y = (\overline{\nabla}_X Y)^T$ where $\overline{\nabla}$ is the euclidean connection on $\boldsymbol{R}^{n+1}$.) Therefore, if $\xi = e_1 \wedge \cdots \wedge e_p$ where $e_1, \cdots, e_p \in T_x S^n$ are orthonormal, we have

$$\langle \mathscr{A}^E(\xi), \xi\rangle = \sum_{i=1}^{p} \langle \nabla_{e_i} E, e_i\rangle = -p\langle e, x\rangle.$$

The result now follows directly from the first variational formula.

LEMMA 3.2. *Let $\mathfrak{S} \in \boldsymbol{V}_p(S^n)$ be an integral varifold which is a minimal variety. Then*

$$\boldsymbol{M}(\mathfrak{S}) \geqq \mathrm{vol}\,(S^p)$$

where S^p denotes the totally geodesic p-sphere, and equality holds iff $\mathfrak{S} = |S^p|$.

PROOF. Let $C(\mathfrak{S}) \in \boldsymbol{V}_{p+1}(\boldsymbol{R}^{n+1})$ be the cone on $\mathfrak{S}$. Since $\mathfrak{S}$ is minimal and integral so is $C(\mathfrak{S})$. Since $C(\mathfrak{S})$ is minimal its density is upper semicontinuous by Corollary 2.15. Since $C(\mathfrak{S})$ is integral, the density is $\geqq 1$, $\|C(\mathfrak{S})\|$-a.e. (See [**22**, 3.2.19]. Actually this is trivial if $C(\mathfrak{S})$ has a dense set of manifold points.) It follows that the density is $\geqq 1$ at $0 \in \boldsymbol{R}^{n+1}$. However, since $C(\mathfrak{S})$ is a cone,

$$\Theta(0, C(\mathfrak{S})) = \boldsymbol{M}(C(\mathfrak{S}))/\mathrm{vol}\,(B_1^{p+1}) = \boldsymbol{M}(S)/\mathrm{vol}(S^p).$$

This proves the first part.

If we now suppose $\boldsymbol{M}(\mathfrak{S}) = \mathrm{vol}\,(S^p)$, then by the above, $\Theta(0, C(\mathfrak{S})) = 1$. Moreover, by upper semicontinuity $\Theta \equiv 1$ on the support of $\|C(\mathfrak{S})\|$. A regularity result of Allard [2], similar to Theorem 2.17, now shows that $C(\mathfrak{S})$ is a manifold at 0, and so $C(\mathfrak{S}) = |C(S^p)|$. When $\mathfrak{S}$ corresponds to an immersed manifold, an elementary argument can be given as follows. Continue the cone to an infinite cone by extending all rays. For any point x on the cone, let $v_x(r)$ be the mass of the cone in $B_r(x) = \{x' \in \boldsymbol{R}^{n+1} : \|x' - x\| \leqq r\}$. By formula (2.12), we have $v_x(r)/\alpha_p r^p \geqq \Theta(x, C(\mathfrak{S})) = 1$; but $v_x(r)/\alpha_p r^p$ increases monotonically and its limit is 1 by comparison since $v_0(r) = \alpha_p r^p$. Thus, the integrand in (2.12) is identically zero, and $C(\mathfrak{S})$ is also a cone at x. It follows that $C(\mathfrak{S})$ is a plane.

Note that if $\mathfrak{S}$ corresponds to an immersion of a closed manifold, $\psi : M^p \to S^n$, and if $x \in S^n$ is a double point of the immersion, then $\Theta(x, |\psi(M)|) = 2\alpha_p$. The above upper semicontinuity argument applied to the cone then shows the following.

LEMMA 3.3. *Let $\psi : M^p \to S^n$ be a minimal immersion of a closed manifold M^p. Then, if the induced metric satisfies* vol $(M^p) < 2$ vol (S^p), *the map ψ is an embedding.*

By using the full strength of the regularity theory for minimal varieties it is possible to prove the following.

THEOREM 3.4 (ALLARD [2]). *There exists an $\varepsilon > 0$ depending only on n and p such that if $\mathfrak{S} \in V_p(S^n)$ is an integral minimal variety with $M(\mathfrak{S}) <$* vol $(S^p) + \varepsilon$, *then $\mathfrak{S} = |S^p|$, for some totally geodesic $S^p \subset S^n$.*

We shall now enter a discussion of the more delicate geometry of regular minimal submanifolds of S^n. As in §2 we begin with the two-dimensional case.

Let $\psi : R \to S^n \subset \boldsymbol{R}^{n+1}$ be a conformal immersion of a Riemann surface with second fundamental form B. (As noted in §2, every surface can be presented this way.) The induced metric on R has the form $ds^2 = 2F|dz|^2$ where $F = |\partial\psi/\partial z|^2$. The minimal surface equation (Corollary 1.11) has the form

$$\frac{\partial}{\partial z}\frac{\partial}{\partial \bar{z}}\psi = -F\psi \tag{3.1}$$

where ψ is considered as an $\boldsymbol{R}^{n+1}$-valued function with $\|\psi\|^2 = 1$. If K is the curvature of the induced metric, the classical Gauss curvature equation says:

$$2(1 - K) = \|B\|^2 \tag{3.2}$$

and the Mainardi-Codazzi equations are:

$$(\nabla_X B)(Y, Z) = (\nabla_Y B)(X, Z) \tag{3.3}$$

for $X, Y, Z \in \mathfrak{X}_R$.

In S^3 the most elementary examples of compact minimal surfaces are the geodesic 2-sphere and the Clifford torus $\Sigma_{1,1}$, given by (2.9). In fact, the collection of minimal surfaces is quite rich.

THEOREM 3.5 (LAWSON [38]). *For each integer $g \geqq 0$, there is a compact minimal surface of genus g embedded in S^3. If g is not prime, the embedding is not unique, in fact, there are at least n noncongruent embeddings where n is the number of primes* *in the prime factorization of g. Moreover, for each $k \geqq 0$ there is a minimal immersion of the compact, nonorientable surface of Euler characteristics $-k$.*

The outline of the proof is as follows. Let $S^3 = \{(z_1, z_2) \in C^2 : |z_1|^2 + |z_2|^2 = 1\}$ and let $S_k^1 = S^3 \cap \{z_k = 0\}$. Fix integers $m_1, m_2 \geqq 0$ and choose pairs of points $P_k = \{p_k, q_k\} \subset S_k{}^1$ with $\text{dist}(p_k, q_k) = 2\pi/(m_k + 1)$ where distance is taken in S^3. We form a piecewise smooth simple closed curve $\Gamma_{m_1, m_2} \subset S^3$ by taking the geodesic join of P_1 and P_2. We then solve the Plateau problem for Γ_{m_1, m_2} and reflect the solution about the smooth boundary arcs until eventually a closed regular surface, Σ_{m_1, m_2} is formed. This surface is embedded and has genus $m_1 \cdot m_2$.

These examples give a very pretty illustration of the following general theorem. Let $\psi : R \to S^3$ be as above, and let $z = x_1 + ix_2$ be a local complex parameter on R. Since $B_{11} = -B_{22}$ we can form the quadratic differential form

$$\begin{aligned}\omega &= \tfrac{1}{2}(B_{11} - iB_{12})\, dz^2 \\ &= \frac{1}{iF}\left(\psi \wedge \frac{\partial\psi}{\partial z} \wedge \frac{\partial\psi}{\partial\bar{z}} \wedge \frac{\partial^2\psi}{\partial z^2}\right) dz^2\end{aligned}$$

on R. The Mainardi-Codazzi equations (3.3) imply that ω is holomorphic, and by equation (3.2) we have $\|\omega\|^2 = \frac{1}{2}(1 - K)$. Hence, $K \leqq 1$, and $K = 1$ precisely at the zeros of the holomorphic quadratic differential ω. If the genus g of R is zero, then $\omega \equiv 0$. Otherwise ω has $4g - 4$ zeros to multiplicity. The zeros of this form can be interpreted geometrically as follows.

DEFINITION 3.6. For $p \in R$, let O_p denote the order of contact of the totally geodesic S^2 which is tangent to $\psi(R)$ at p $(O_p \geqq 1)$. The *degree of spherical flatness* of ψ at p is defined as $d_p = O_p - 1$.

THEOREM 3.7 (LAWSON [38]). *Let $\psi : R \to S^3$ be a minimal immersion where R is a compact Riemann surface of genus g. Then:*

(a) *(Almgren) If $g = 0$, ψ is totally geodesic.*

(b) *If $g > 0$, then $\sum_{p \in R} d_p = 4g - 4$.*

COROLLARY 3.8. *If $\psi : R \to S^3$ is a minimal surface of nonpositive curvature, then $\psi(R)$ is the Clifford torus.*

It can be proven (cf. [38] and [39]) that the spherically flat points on Σ_{m_1, m_2} correspond to p_i, q_i and their reflected images, and that $d_{p_1} = d_{q_1} = m_2 - 1$ and $d_{p_2} = d_{q_2} = m_1 - 1$. Thus, $4g - 4 = (2m_1 + 2)(m_2 - 1) + (2m_2 + 2)(m_1 - 1)$, and $g = m_1 m_2$, as can be proved by other means.

An interesting open question is to what extent the embeddings given in Theorem 3.5 are unique. Many more immersions of compact surfaces into S^3 have been constructed. (See [38] and [35].) However, for embeddings there is the following topological uniqueness (or "unknottedness") result.

THEOREM 3.9 (LAWSON [41]). *Let $\Sigma \subset S^3$ be a compact, embedded minimal surface of genus g. Then there exists a diffeomorphism $f : S^3 \to S^3$ such that $f(\Sigma)$*

$= \hat{\Sigma}_g$ *where* $\hat{\Sigma}_g$ *is a standardly embedded surface in* S^3.

Thus no minimal embedding of a torus can be carried by a diffeomorphism of S^3 to the boundary of a tubular neighborhood of a knot.

We shall now consider the case of compact minimal surfaces in S^n. One of the most profound contributions to this subject is the theory of Calabi, which has been elaborated by Barbosa and Chern. To describe the theory we need to recall the geometry of isotropic subspaces. Consider $\boldsymbol{C}^n = \boldsymbol{R}^n \otimes \boldsymbol{C}$ and let $(\cdot, \cdot)$ denote the $\boldsymbol{C}$-bilinear extension of the inner product on $\boldsymbol{R}^n$ to $\boldsymbol{C}^n$. A vector $v \in \boldsymbol{C}^n$ is called *isotropic* if $(v, v) = 0$. A $\boldsymbol{C}$-linear subspace $V \subset \boldsymbol{C}^n$ is *totally isotropic* if every $v \in V$ is isotropic. A holomorphic curve $\varphi : R \to \boldsymbol{C}^n$ is *totally isotropic* if for each $z \in R$, the space spanned by φ and all its derivatives at z is totally isotropic. These definitions carry over in an obvious way to $\boldsymbol{P}^{n-1}(\boldsymbol{C})$.

Given a holomorphic curve $\Phi : R \to \boldsymbol{P}^{n-1}(\boldsymbol{C})$, we denote by $\Phi_l : R \to \boldsymbol{P}^{N_l-1}(\boldsymbol{C})$, $N_l = n!/l!(n-l)!$, its lth associated curve. Φ_l is given locally in homogeneous coordinates $\boldsymbol{C}^{N_l} \cong \Lambda^l \boldsymbol{C}^n$ for projective space by $\varphi_l(z) = \varphi(z) \wedge \varphi^1(z) \wedge \cdots \wedge \varphi^{(l-1)}(z)$ where $z \to \varphi(z) \in \boldsymbol{C}^n$ is a local, homogeneous coordinate representation for Φ. Note that $\Phi_l(R)$ is contained in the Grassmannian of complex l-planes in $\boldsymbol{C}^n$. If Φ is totally isotropic, then $\Phi_l(R)$ is contained in the subvariety of totally isotropic l-planes.

For a better statement of the result we shall consider the general class of *branched minimal immersions* into S^n, i.e., C^ω maps $\Psi : R \to S^n$ which are conformal outside a set of isolated points and satisfy equation (3.1).

The following theorem is due fundamentally to Calabi ([9] and [10]) and has undergone some additions from Chern [12] and Barbosa [6].

THEOREM 3.10. *Let* $\Psi : S^2 \to S^n$ *be a branched minimal immersion whose image lies in no proper geodesic subsphere of* S^n. *Then* $n = 2k$. *Furthermore*, Area (Ψ) *is a multiple of* 4π *and is* $\geqq 2\pi k(k+1)$.

The set of all such branched immersions is in natural one-to-one correspondence with the set of totally isotropic holomorphic maps $\Phi : S^2 \to \boldsymbol{P}^{2k}(\boldsymbol{C})$. *In fact,* Ψ *is isometric to the* $(k-1)$*st osculating curve* $\Phi_{k-1} : S^2 \to \mathscr{H}_k$ *where* $\mathscr{H}_k \cong \mathrm{SO}(2k+1)/U(k)$ *is the space of totally isotropic k-planes in* $\boldsymbol{C}^{2k+1}$. *Furthermore,* Ψ *can be recaptured as* $\Psi = \pi \circ \Phi_{k-1}$ *where* $\pi : \mathscr{H}_k \to S^{2k}$ *is given by the projection* $\mathrm{SO}(2k+1)/U(k) \to \mathrm{SO}(2k+1)/\mathrm{SO}(2k)$ *where the inclusions* $U(k) \subset \mathrm{SO}(2k) \subset \mathrm{SO}(2k+1)$ *are standard.*

Note that the statement that $n = 2k$ includes part (a) of Theorem 3.7.

We can now bring the general theory of algebraic curves (in particular the Plücker formulas and the Calabi rigidity theory) to bear on the subject. One result of such a study is the following

THEOREM 3.11 (BARBOSA [6]). *Any two isometric (branched) minimal immersions of* S^2 *into* S^n *are congruent.*

Such a result is false for noncompact surfaces (cf. [38]) and for compact surfaces of higher genus it is unknown. It should be remarked that for minimal surfaces in $\boldsymbol{R}^n$ there is a detailed theory of noncongruent, isometric minimal surfaces. The work is due to Calabi (cf. [**10**] or [**43**, Chapter IV]) who shows that in general the space of noncongruent minimal surfaces isometric to a given one in $\boldsymbol{R}^n$ are in natural 1-1 correspondence with points in a certain algebraic subvariety of the generalized Siegel upper half-plane.

Theorem 3.11 has the following corollary proved originally by do Carmo and Wallach [**17**].

COROLLARY 3.12. *Let* $\Psi : S^2 \to S^n$ *be a minimal immersion such that* S^2 *has constant curvature c in the induced metric. Then* $c = 2/k(k+1)$ *for some integer* $k > 0$, *and* Ψ *is congruent to the immersion* $\Psi_k = (\phi_1, \cdots, \phi_{N_k})$ *where* $\{\phi_j\}$ *form an orthonormal basis of the spherical harmonics of degree k on* S^2 *with respect to an* SO(3)-*invariant metric.*

PROOF. Since $\nabla^2\Psi = -2\Psi$ and since ∇^2 has spectrum $\lambda_k = -ck(k+1)$ for $k \in \boldsymbol{Z}^+$, it follows that $c = 2/k(k+1)$ for some k. Each Ψ_k is an isometric immersion of a constantly curved 2-sphere, because Ψ_k is an equivariant mapping with respect to the transitive action of SO(3). Let c_k be the curvature of the metric induced by Ψ_k. Since $\|\Psi_k\|^2 = 1$, we have

$$\langle \nabla^2\Psi_k, \Psi_k\rangle + \|\nabla\Psi_k\|^2 = -c_k k(k+1) + 2 = 0.$$

Hence $c_k = 2/k(k+1)$ and so Ψ_k is minimal. The corollary now follows directly from Theorem 3.11.

Given the above results, it is natural to ask for explicit spaces of moduli for minimal immersions of S^2 into S^{2k}. Barbosa has a number of results in this line, the most striking of which is the following [**6**].

THEOREM 3.13. *The space of minimal immersions* $\Psi : S^2 \to S^{2k}$ *of area* $2\pi k(k+1)$ *is naturally diffeomorphic to* $\mathrm{SO}(2k+1, \boldsymbol{C})/\mathrm{SO}(2k+1, \boldsymbol{R})$.

We note that for higher genus minimal surfaces, no such theory exists. Of course, given any totally isotropic holomorphic curve $\Psi : R \to \boldsymbol{P}^{2m}(\boldsymbol{C})$, there corresponds a minimal surface $\Phi = \pi \circ \Psi_{m-1}$ in S^{2m}. Such immersions are characterized by the fact that $(\partial^i\psi/\partial z^i, \partial^j\psi/\partial z^j) \equiv 0$ for all $i + j > 0$. However, minimal surfaces in S^3 are not of this type.

We now turn our attention to minimal submanifolds of higher dimension in S^n. Our first remark is that there is a considerable number of such varieties. By using large symmetry groups, many families of minimal immersions of compact manifolds in codimension one were constructed in [**35**]. Passing to arbitrary codimension, the following strong result can be proved. (See [**35**] for more details of proof.)

THEOREM 3.14 (HSIANG [**33**]). *Every compact homogeneous space admits a minimal (equivariant) immersion into some sphere.*

If one restricts to isotropy irreducible homogeneous spaces, the result is more elementary and was first proven by Takahashi [**59**] as follows. Let $M = G/H$

be such a space where G is assumed compact. Since the isotropy representation of H on the tangent spaces of M is irreducible, there is only one G-invariant metric (up to scalar multiples) on M. Let E be a nontrivial eigenspace for the laplacian (on functions) for this metric. Let $\{\psi_1, \cdots, \psi_N\}$ be an orthonormal basis for E with respect to a G-invariant inner product on E. (G acts on E by sending $\psi \to g^*\psi$ where $g^*\psi(x) = \psi(gx)$.)

We now define a map $\Psi : M \to \boldsymbol{R}^N$ by $\Psi = (\psi_1, \cdots, \psi_N)$. Since $\{\psi_1, \cdots, \psi_N\}$ is an orthonormal basis, there is a homomorphism $G \to O(N)$ sending $g \to \mathcal{O}_g$ such that $\Psi(gx) = (g^*\Psi)(x) = \mathcal{O}_g\Psi(x)$ for all x. If we renormalize Ψ so that $\|\Psi(x_0)\|^2 = 1$ for a fixed $x_0 \in M$, then $\|\Psi\|^2 \equiv 1$ since G is transitive on M. Thus, $\Psi : M \to S^{N-1}$.

Clearly Ψ induces a G-invariant metric on M, and so by changing the metric on M by a constant scalar, Ψ becomes an isometry. Then since $\|\Psi\|^2 = 1$, we have $\langle\nabla^2\Psi, \Psi\rangle + \|\nabla\Psi\|^2 = \lambda + p = 0$ where $p = \dim(M)$ and $\nabla^2\Psi = -\lambda\Psi$. Hence, Ψ is a minimal immersion.

This construction introduces representation theory into the study of minimal immersions of irreducible symmetric spaces (and conversely). Extensive use has been made of this by N. Wallach (cf. **[61]**).

In particular, for the sphere S^p of constant curvature there is a sequence of minimal immersions $\{\Psi_k\}$ given for each k by choosing an orthonormal basis of the generalized spherical harmonics of degree k. It was conjectured for a while that, as in Corollary 3.12, these immersions were rigid. Indeed, if $\Psi : S^p \to S^n$ is a minimal immersion inducing a metric of constant curvature c, then the argument of Corollary 3.12 shows that $c = c_k$, the curvature induced by some Ψ_k. It follows that $\Psi = A\Psi_n$ for some (constant) linear map A. However, in general A will not be orthogonal.

THEOREM 3.15 (DO CARMO AND WALLACH **[18]**). *The minimal immersions of a sphere S^p of constant curvature c_k are rigid for $p = 2$ or $k \leqq 3$. In general, the space of such immersions, identified up to congruence, is homeomorphic to a nontrivial convex body (explicitly described) in some vector space.*

A second rich class of minimal submanifolds of codimension $\geqq 2$ is the following. Recall that a manifold $M^p \subset S^n$ is minimal if and only if the cone $C(M^p)$ on M^p is minimal in $\boldsymbol{R}^{n+1}$. Suppose now that $V \subset \boldsymbol{C}^{n+1}$ is an algebraic variety defined by homogeneous polynomials. Then for all $z \in V$ and all $\alpha \in \boldsymbol{C}$, we have $\alpha \cdot z \in V$. If V is regular outside the origin, then V is the cone on a submanifold $M \subset S^{2n+1}$. Since V is minimal (cf. § 5), so is M^p. Each such manifold is the total space of a circle bundle over a compact, complex submanifold of $\boldsymbol{P}^n(\boldsymbol{C})$. This construction gives a large collection of beautiful minimal submanifolds of spheres.

It is still an interesting open question whether there exists a smooth manifold which cannot be minimally immersed into some sphere.

Some of the most important work on minimal varieties in spheres is that of J. Simons, who characterized the totally geodesic subspheres in a strong way within the class of minimal submanifolds. To best state the results we should examine his fundamental equation.

Let M^p be a minimal submanifold of S^n with second fundamental form B. B is a section of the bundle Hom(S, N) where S is the bundle of symmetric transformations of $T(M)$. We then have the transposed section B^t in Hom(N, S). Using this, we can define a section $\tilde{B}$ of Hom (N, N) by

$$\tilde{B} = B \circ B^t$$

and a section $\underset{\sim}{B}$ of Hom (S, S) by

$$\underset{\sim}{B} = \operatorname{tr}(\operatorname{ad}_{B^t} \circ \operatorname{ad}_{B^t})$$

where the trace is taken over N.

THEOREM 3.16 (SIMONS [58]). *Let B be the second fundamental form of a minimal submanifold of dimension p in S^n. Then*

$$\nabla^2 B = pB - \tilde{B} \circ B - B \circ \underset{\sim}{B} \tag{3.3}$$

where ∇^2 is the connection laplacian on Hom (S, N). *If $p = n - 1$, then* (3.3) *becomes*

$$\nabla^2 B = pB - \|B\|^2 B. \tag{3.4}$$

COROLLARY 3.17. *Let M^p be a compact minimal submanifold of dimension p in S^n with second fundamental form B satisfying $\|B\|^2 \leqq p/q$ where $q = 2 - 1/(n - p)$. Then M^p is a totally geodesic S^p or $\|B\|^2 \equiv p/q$.*

It has been proven by Chern, do Carmo and Kobayashi [13] (see also [37]) that in the second case either $p = 2$ and M^2 is the Veronese surface ($\mathscr{V}_2(S^2)$ in the above terminology) or $p = n - 1$ and $M^p = \Sigma_{k,l}$ (cf. equation (2.9)) for some k, l. For each k, l with $k + l = n - 1$, $\Sigma_{k,l}$ has $\|B\|^2 \equiv n - 1$.

It is a direct consequence of the general Gauss curvature equation that the normalized scalar curvature $\mathscr{K}$ of a minimal M^p in S^n satisfies

$$1 - \mathscr{K} = \frac{1}{p(p-1)} \|B\|^2. \tag{3.5}$$

Consequently we have the following.

COROLLARY 3.18. *Let M^p be a compact minimal submanifold of dimension p in S^n with normalized scalar curvature $\mathscr{K}$. Then $\mathscr{K} \leqq 1$, and if*

$$\mathscr{K} > 1 - \frac{1}{q(p-1)}$$

where $q = 2 - 1/(n - p)$, then M^p is a totally geodesic p-sphere.

A theorem similar to this has been proven by Lawson [40] in complex and quaternionic projective spaces, and a number of interesting theorems of a similar nature have been proved for positively curved, minimal $M^p \subset S^n$ by Yau [63].

It should be noted that equation (3.4) is an important part of proving Theorem 2.19.

4. Homology theory. Note that for any riemannian manifold M^n, the integral currents

$$0 \to \mathscr{I}_n(M^n) \xrightarrow{\partial} \mathscr{I}_{n-1}(M^n) \xrightarrow{\partial} \cdots \xrightarrow{\partial} \mathscr{I}_0(M^n) \to 0 \tag{4.1}$$

form a chain complex ($\partial^2 = 0$ since $d^2 = 0$), and therefore give rise to a sequence of homology groups. One of the fundamental results in homological integration theory is the following.

THEOREM 4.1 (FEDERER AND FLEMING [25] OR [22, 4.4]). *There is a natural isomorphism $H_*(\mathscr{I}_*(M^n)) \cong H_*(M^n; \boldsymbol{Z})$ of the homology of the complex* (4.1) *with the integral singular homology groups of M^n.*

This theorem is a special case of the corresponding general result for pairs of spaces (A, B) where A and B are local Lipschitz neighborhood retracts in some $\boldsymbol{R}^n$. In particular, it applies to globally embeddable analytic spaces, to polyhedral objects, etc. It is also possible to devise a similar theory of "integral currents mod ν" which gives homology groups isomorphic to singular homology with coefficients in $\boldsymbol{Z}/\nu\boldsymbol{Z}$. (See [22, 4.4].) For the case of real coefficients it is possible to use the full complex of currents $\mathscr{E}_*(M^n)$ (de Rham [57]) or the subcomplex of currents (with boundary also) representable by integration (Federer and Fleming [25]).

Theorem 4.1 together with the compactness theorem, 2.4, and the lower semicontinuity of the mass function, has the following important consequence.

COROLLARY 4.2. *Let M be a compact riemannian manifold. Then each class $\alpha \in H_p(\mathscr{I}_*(M)) \cong H_p(M; \boldsymbol{Z})$ contains an integral current of least mass among all integral currents in α.*

This corollary remains true when working with coefficients $\boldsymbol{Z}/\nu\boldsymbol{Z}$ or $\boldsymbol{R}$.

Using the codimension one regularity proof of § 2 and a basic regularity result for general elliptic integrands due to Almgren [5] the following can be proven.

THEOREM 4.3. *Let M^n be a compact riemannian manifold of dimension $n \leqq 7$. Then for each $\alpha \in H_{n-1}(M^n; \boldsymbol{Z})$, there is a finite collection of mutually disjoint, compact, oriented minimal submanifolds $N_1, \cdots, N_k$ of dimension $n-1$ embedded in M^n, and integers $n_1, \cdots, n_k$, such that $\mathscr{S} = \sum_{j=1}^k n_j N_j \in \alpha$ and $M(\mathscr{S}) \leqq M(\mathscr{S}')$ for all integral currents $\mathscr{S}' \in \alpha$. Furthermore, every $\mathscr{S}' \in \alpha$ with $M(\mathscr{S}') = M(\mathscr{S})$ has this smooth structure.*

For $n \geqq 8$, the currents of least mass in codimension one can have singularities, even though the homology classes always contain embedded manifolds. However, one point that I want to emphasize in these lectures is that, in general, it is not necessary to have a smooth manifold structure on minimizing currents in order to use them to prove global theorems. Every rectifiable current has enough of the general structure of a manifold that many proofs can be carried over from the smooth case without change. Moreover, it is always possible to apply the general variational formulas of § 1. For example, we made use of the general formula for the first variation in the beginning of § 2 and in Lemma 3.1. Somewhat more profound results can be obtained from the second variation.

Note that any integral current (or integral current mod ν) of least mass in its homology class carries an underlying stable varifold. The inequality (1.5) together with the formula for the second variation gives a great deal of geometric information about any such varifold.

As an example, consider the case where $M = S^n \subset \boldsymbol{R}^{n+1}$. Let $\mathscr{V} \subset \mathfrak{X}_{S^n}$ be the space of gradients of linear functions in $\boldsymbol{R}^{n+1}$ restricted to S^n. (Note that $\mathscr{V}$ is canonically isomorphic to $\boldsymbol{R}^{n+1}$.) A straightforward calculation using (1.10) gives the following.

PROPOSITION 4.4 (LAWSON AND SIMONS [45]). *Let $\mathfrak{S} \in V_p(S^n)$ be any varifold and associate to $\mathfrak{S}$ a quadratic form $Q_{\mathfrak{S}}$ on $\mathscr{V}$ by setting*

$$Q_{\mathfrak{S}}(E) = \frac{d^2}{dt^2} M(\varphi_{t*}\mathfrak{S})\Big|_{t=0}$$

for $E \in \mathscr{V}$, where φ_t is the flow generated by E. Then

$$\operatorname{tr}(Q_{\mathfrak{S}}) = -\,p(n-p)M(\mathfrak{S}).$$

COROLLARY 4.5. *There are no stable varifolds of dimension p, $1 \leqq p \leqq n-1$, on S^n.*

This result contrasts sharply with the large number of minimal varifolds on S^n discussed in § 3. This result not only implies that no compact minimal submanifolds in S^n are stable, it also shows that configurations such as an elaborate "fish net" on S^2 are never stable.

As a result of Corollaries 4.2 and 4.5 we can deduce the rather well-established fact that $H_p(S^n; \boldsymbol{Z}) = H_p(S^n; \boldsymbol{Z}/\nu\boldsymbol{Z}) = 0$ for $p = 1, \cdots, n-1$. However, the analogous argument can be applied to submanifolds of spheres to prove the following.

THEOREM 4.6 (LAWSON AND SIMONS [45]). *Let M^n be a compact n-manifold immersed in S^N with second fundamental form B satisfying $\|B\|^2 < \min\{pq, 2(pq)^{1/2}\}$ where $p + q = n$ and $pq > 0$. Then $H_p(M^n; A) = H_q(M^n; A) = 0$ for any finitely generated, abelian group A.*

It should be noted that Theorem 4.6 is sharp. That is, for $r^2 = p^{1/2}/(p^{1/2} + q^{1/2})$ and $s^2 = q^{1/2}/(p^{1/2} + q^{1/2})$ the product $S^p(r) \times S^q(s) \subset S^{p+q+1}$ (cf. § 2) has $\|B\|^2 \equiv 2(pq)^{1/2}$.

5. Kähler manifolds. Minimal varieties bear a relation to complex analytic subvarieties not unlike that between harmonic and pluriharmonic (or holomorphic) functions. The subject of complex subvarieties of a Kähler space forms a beautiful, special topic in the study of the problem of least area. In this section we shall try to present some of the major aspects of the known theory.

Let M^n be a Kähler manifold, that is, a complex hermitian manifold whose almost complex structure J satisfies

$$\nabla_X J = 0 \tag{5.1}$$

for all $X \in \mathfrak{X}_M$. Then the second fundamental form B of any complex submanifold

S of M is actually a *complex bilinear* form, that is,

$$B(JX, Y) = JB(X, Y) = B(X, JY) \tag{5.2}$$

for all $X, Y \in \mathfrak{X}_S$. To see this note that $(\nabla_X JY)^N = (J\nabla_X Y)^N = J(\nabla_X Y)^N$ and use the symmetry of B. Choosing an orthonormal basis of type $e_1, Je_1, \cdots, e_p, Je_p$ in each tangent space of S and using (5.2) immediately proves the following.

LEMMA 5.1. *Any complex submanifold of a Kähler manifold is minimal.*

In fact something much stronger is true. For this it is necessary to consider the fundamental 2-form ω defined by

$$\omega(X, Y) = \langle JX, Y\rangle \quad \text{for } X, Y \in \mathfrak{X}_M. \tag{5.3}$$

Condition (5.1) is equivalent to the fact that

$$d\omega = 0 \tag{5.4}$$

(cf. [**43**]). The fundamental result which relates complex subvarieties to the general Plateau problem is the following linear algebraic proposition.

THEOREM 5.2 (THE WIRTINGER INEQUALITY). *Let M be any almost complex manifold with a hermitian metric and a fundamental 2-form ω given by* (5.3). *Fix $x \in M$ and let $\xi \in \Lambda^{2p}T_xM$ be any simple $2p$-vector (where the exterior product is taken over $\boldsymbol{R}$). Then*

$$(\omega^p/p!)(\xi) \leqq \|\xi\| \tag{5.5}$$

and equality holds if and only if ξ corresponds to a canonically oriented, complex subspace of $T_x M$.

For a brief elegant proof of this due to Federer see [**22**, 1.8.2] or [**44**].

COROLLARY 5.3. *A canonically oriented, compact complex submanifold (with possibly empty boundary) of a Kähler manifold M defines a current $\mathscr{S} \in \mathscr{I}_{2p}(M)$ of least mass among all currents $\mathscr{S}' \in \mathscr{I}_{2p}(M)$ homologous to $\mathscr{S}$.*

PROOF. By (5.4) we have $d\omega^p = p\omega^{p-1} \wedge d\omega = 0$. Therefore, since $\mathscr{S} - \mathscr{S}' = \partial\mathscr{T}$ for some $\mathscr{T} \in \mathscr{I}_{2p+1}(M)$, we have $\mathscr{S}(\omega^p) - \mathscr{S}'(\omega^p) = 0$. Since the tangent planes to $\mathscr{S}$ are all complex subspaces, Theorem 5.2 shows that

$$\mathscr{S}(\omega^p) = \int \omega^p(\vec{\mathscr{S}}_x)\, d\|\mathscr{S}\|(x) = (p!) \int d\|\mathscr{S}\| = (p!)\boldsymbol{M}(\mathscr{S}).$$

Similarly, the inequality (5.5) shows that $\mathscr{S}'(\omega^p) \leqq (p!)M(\mathscr{S}')$. This completes the proof.

It follows from this corollary that any compact complex submanifold of $\boldsymbol{C}^n$ (with boundary) represents a solution to the Plateau problem in the strong sense. Moreover, any compact, complex submanifold of $\boldsymbol{P}^n(\boldsymbol{C})$ represents a current of least mass in its homology class. Note that the minimal representatives in this case are by no means unique. For $n > 2$ and $m > 1$, the class m in $H_2(\boldsymbol{P}^n(\boldsymbol{C})\,;\,\boldsymbol{Z}) \cong \boldsymbol{Z}$ contains complex curves of differing genus.

We now observe that in the proof of Corollary 5.3 the only fact about $\mathscr{S}$ we used was that its tangent planes $\vec{\mathscr{S}}_x$ were complex and canonically oriented almost everywhere. Moreover, from the last statement of Theorem 5.2 it follows that the current $\mathscr{S}'$ in Corollary 5.3 has $\boldsymbol{M}(\mathscr{S}') = \boldsymbol{M}(\mathscr{S})$ if and only if its tangent planes also have this property. These observations lead us to make the following definition.

DEFINITION 5.4. Let M be a complex manifold. Then a current $\mathscr{S} \in \mathscr{R}_{2p}(M)$ is said to be of *type* (p, p) if for $\|\mathscr{S}\|$-a.e. x, $\vec{\mathscr{S}}_x$ represents a complex subspace of $T_x M$. If, in addition, $\vec{\mathscr{S}}_x$ is canonically oriented $\|\mathscr{S}\|$-almost everywhere, then $\mathscr{S}$ is called *positive*.

From our remarks above we have the following strengthened version of Corollary 5.3.

COROLLARY 5.5 (FEDERER [21]). *Let $\mathscr{S} \in \mathscr{R}_{2p}(M)$ be a positive (p, p)-current in a Kähler manifold M. Then*

$$\boldsymbol{M}(\mathscr{S}) \leqq \boldsymbol{M}(\mathscr{S}') \tag{5.6}$$

for all $\mathscr{S}' \in \mathscr{R}_{2p}(M)$ such that $\partial\mathscr{S} = \partial\mathscr{S}'$ and $\mathscr{S} - \mathscr{S}'$ is homologous to zero. Furthermore, equality holds in (5.6) *if and only if $\mathscr{S}'$ is also a positive (p, p)-current.*

Of course any complex submanifold of M determines a positive (p, p)-current. More generally, however, suppose V is any complex analytic subvariety of pure (complex) dimension p in M. Then $\mathscr{H}^{2p}(V)$ is locally finite (cf. [44]), and so V determines a current $[V] \in \mathscr{R}_{2p}^{\mathrm{loc}}(M)$ by integration over the canonically oriented, regular set $R(V)$ of V. Since $\operatorname{supp}(\partial[V]) \subset V - R(V)$ and $V - R(V)$ is a subvariety of complex dimension $\leqq p - 1$, it follows from [22, 4.1.20] that $\partial[V] = 0$.

Highly relevant to Corollary 5.5 above is the following beautiful structure theorem.

THEOREM 5.6 (J. KING [36]). *Let $\mathscr{S} \in \mathscr{R}_{2p}^{\mathrm{loc}}(M)$ be a positive (p, p)-current with $\partial\mathscr{S} = 0$ in a complex manifold M. Then $\mathscr{S}$ is a locally finite sum, $\mathscr{S} = \sum n_i[V_i]$, where for each i, V_i is an irreducible, p-dimensional, complex analytic subvariety of M and n_i is a positive integer.*

REMARK. If V is a p-dimensional, complex subvariety of $\boldsymbol{C}^n$, then by the above, the current $\mathscr{S} = [V \cap B_1^{2n}]$ is a solution to the Plateau problem for the boundary $\partial\mathscr{S}$. Hence, in real codimension $\geqq 2$, mass-minimizing p-currents can have singularities of dimension $p - 2$. (Compare with Theorem 2.12.) In spite of the singularities, complex analytic subvarieties are particularly nice examples of minimal varieties. For example, at any point p on such a variety V, the tangent cone to V is the homogeneous algebraic variety whose ideal is generated by the first nonzero, homogeneous polynomial terms in the power series expansions at p of functions in the ideal of V at p. This is the Whitney tangent cone (cf. [63, p. 210]). This homogeneous algebraic variety corresponds to a projective variety whose volume, by the Wirtinger inequality, is an integral multiple of the volume of a linear subspace of the same dimension. The multiple is the homology degree. It follows that the density of $[V]$ at p is an integer, called the *Lelong number* of V at p.

Note that if $\mathscr{B} \in \mathscr{I}_{2p-1}(\boldsymbol{C}^n)$ is the boundary of an analytic variety as above, then the tangent spaces to $\mathscr{B}$ contain a complex subspace of maximal possible dimension $\|\mathscr{B}\|$-a.e. A strong converse to this has recently been proved by R. Harvey and the author [**30**].

The appearance of Theorem 5.6 immediately raised the question of the analogous result for general (p, p)-currents. However, circumventing the assumption of positivity proved to be a difficult task, and an entirely new approach was required to prove the following.

THEOREM 5.7 (R. HARVEY AND B. SHIFFMAN [**31**]). *Let $\mathscr{S} \in \mathscr{R}_{2p}^{\mathrm{loc}}(M)$ be a (p, p)-current with $\partial\mathscr{S} = 0$ and $\mathscr{H}^{2p+2}(\operatorname{supp}\mathscr{S}) = 0$, in a complex manifold M. Then $\mathscr{S}$ is a locally finite sum*

(5.7) $$\mathscr{S} = \sum n_i[V_i]$$

where, for each i, V_i is an irreducible, p-dimensional complex analytic subvariety of M and $n_i \in \boldsymbol{Z}$.

Note that this result generalizes Theorem 5.6, since by Corollary 5.5 any positive (p, p)-current $\mathscr{S}$ is a minimal variety (in the euclidean metric on any local coordinate system), and so by the discussion following Corollary 2.15, $\mathscr{H}^{2p}(\operatorname{supp}\mathscr{S})$ is locally finite.

It is not difficult to see that, in general, complex currents are not area minimizing. (Consider $\mathscr{S} = [\{(z, 0) \in \boldsymbol{C}^2 : |z| \leqq 1\}] - [\{(z, \varepsilon) \in \boldsymbol{C}^2 : |z| \leqq 1\}]$ with $0 < \varepsilon \ll 1$, for example.) However, every current of type (5.7) is certainly stable, and in projective space we shall see that the converse is true.

Let J be the almost complex structure on a (p, p)-manifold M. At each point $x \in M$, J can be extended uniquely to a derivation of the algebra $\Lambda^* T_x M$. It is then straightforward to check that a simple vector $\xi \in \Lambda^{2p} T_x M$ corresponds to a complex subspace of $T_x M$ if and only if $J\xi = 0$.

THEOREM 5.8 (LAWSON AND SIMONS [**45**]). *Let $\mathscr{V}$ denote the space of holomorphic vector fields on $\boldsymbol{P}^n(\boldsymbol{C})$ with the standard inner product. Then to each current $\mathscr{S} \in \mathscr{R}_p(\boldsymbol{P}_n(\boldsymbol{C}))$ (p not necessarily even) we associate a quadratic form $Q_{\mathscr{S}}$ on $\mathscr{V}$ as follows. For $E \in \mathscr{V}$, let φ_t be the flow generated by E and set*

$$Q_{\mathscr{S}}(E) = \frac{d^2}{dt^2} M(\varphi_{t*}\mathscr{S})\bigg|_{t=0}.$$

Then

(5.8) $$\operatorname{tr} Q_{\mathscr{S}} = -\int \|J\vec{\mathscr{S}}\|^2 \, d\|\mathscr{S}\|.$$

It follows that a submanifold of finite volume in $\boldsymbol{P}^n(\boldsymbol{C})$ defines a stable current if and only if it is a complex submanifold. More generally, using Theorem 5.7 and the theorem of Chow (cf. [**46**]) we have the following.

COROLLARY 5.9. *A closed current $\mathscr{S} \in \mathscr{R}_{2p}(\boldsymbol{P}^n(\boldsymbol{C}))$ is stable if and only if it is an algebraic cycle.*

Of course, it is necessary to observe that stable currents automatically have support of the appropriate dimension (cf. Corollary 2.9 ff.).

Note that if p is odd, then $\|J\xi\|^2 \geqq 1$. Thus, there are no stable currents, in fact, no stable varifolds of odd dimension on $\boldsymbol{P}^n(\boldsymbol{C})$.

As the final topic in this section we shall prove a generalization of the Wirtinger inequality for arbitrary vectors in $\Lambda^{2p}T_xM$ and examine some of its consequences. For this generalization we must introduce the notion of positive (p, p)-vectors.

DEFINITION 5.10. Let V be a real vector space with an almost complex structure J and extend J to Λ^*V as a derivation. Then a vector $\xi \in \Lambda^{2p}V$ is said to be of *type* (p, p) iff $J\xi = 0$. The space of such vectors is denoted $\Lambda^{(p,p)}V$. A vector $\xi \in \Lambda^{(p,p)}V$ is then said to be *positive* iff $\xi = \sum_{i=1}^m \xi_i$ where each ξ_i is a simple vector corresponding to a canonically oriented complex subspace.

REMARK 5.11. Let $P^{(p,p)} \subset \Lambda^{(p,p)}V$ denote the convex cone of positive vectors, and suppose V has a hermitian inner product. Then we can define a vector $\omega \in \Lambda^{(1,1)}V$ by

$$\omega = e_1 \wedge Je_1 + \cdots + e_n \wedge Je_n$$

where $\{e_1, Je_1, \cdots, e_n, Je_n\}$ is an orthonormal basis of V, and it can be shown that $\omega^p \in \text{interior}(P^{(p,p)})$ for $1 \leqq p \leqq n$. (See Harvey-Knapp [**29**, Corollary 1.10].)

Recall that any inner product on a vector space V extends naturally to an inner product on all of Λ^*V, and thus defines a norm $\|\cdot\|$ on this space. Given such an "L_2" norm on Λ^pV, we introduce a second, so-called *mass-norm* $|\cdot|$ by defining its unit ball to be the convex hull of $\boldsymbol{S} = \{\xi \in \Lambda^pV : \xi \text{ is a simple vector with } \|\xi\| = 1\}$. Clearly, $|\cdot| \geqq \|\cdot\|$ and equality holds precisely for simple vectors. This norm has a second useful description, derived as follows. For any p-vector $\xi \neq 0$, we have $\xi/|\xi| = \sum c_i\xi_i$ where $\xi_i \in \boldsymbol{S}$, $c_i > 0$ and $\sum c_i \leqq 1$. Since $|\xi_i| = \|\xi_i\| = 1$ for all i, the triangle inequality gives $|\xi| \leqq \sum c_i|\xi| \leqq |\xi|$. Thus, $\sum c_i = 1$. Furthermore, if we set $\eta_i = c_i|\xi|\xi_i$, then we have proved that ξ can be expressed as a finite sum of simple vectors $\xi = \sum \eta_i$ such that $|\xi| = \sum \|\eta_i\|$. In particular,

$$|\xi| = \inf\left\{\sum \|\eta_i\| : \xi = \sum \eta_i \text{ and each } \eta_i \text{ is a simple vector}\right\}.$$

REMARK 5.12. It should be pointed out that this norm is quite natural in the following sense. Let M be a riemannian manifold, and let $|\cdot|'$ be any norm in the fibers of Λ^pTM. This determines a dual norm $|\cdot|'$ in the fibers of $\Lambda^pT^*M = (\Lambda^pTM)^*$. For $\varphi \in \mathscr{E}^p(M)$, let $M'(\varphi) = \sup\{|\varphi_x|' : x \in M\}$. Then for any current $\mathscr{S} \in \mathscr{E}_p(M)$ we consider

$$\boldsymbol{M}'(\mathscr{S}) = \sup\{\mathscr{S}(\varphi) : \boldsymbol{M}'(\varphi) \leqq 1\}. \tag{5.9}$$

Those currents for which $\boldsymbol{M}'(\mathscr{S}) < \infty$ are said to be *representable by integration*. This class of currents is independent of the norm used. However, if $|\xi|' = \|\xi\|$ on simple vectors ξ, then $\boldsymbol{M}'(\mathscr{S}) = \boldsymbol{M}(\mathscr{S})$ for rectifiable currents, so let us assume $|\cdot|'$ has this property. Then there is a theorem (cf. Federer [**24**]) that every current on M representable by integration can be approximated in the $\boldsymbol{M}'$-norm by real polyhedral chains of class C^1 if and only if $|\cdot|'$ is a multiple of the mass norm at each point.

The result of central interest here is the following theorem of Harvey and Knapp **[29]**.

THEOREM 5.13 (THE GENERALIZED WIRTINGER INEQUALITY). *Let M and ω be as in Theorem 5.2. Fix $x \in M$ and choose any $\xi \in \Lambda^{2p}T_xM$. Then $(\omega^p/p!)(\xi) \leqq |\xi|$ and equality holds if and only if ξ is of type (p, p) and positive.*

PROOF. We proved above that for a given $\xi \in \Lambda^p T_x M$ there exist simple p-vectors $\xi_1, \cdots, \xi_m$ such that $\xi = \sum \xi_i$ and $|\xi| = \sum \|\xi_i\|$. Now by Theorem 5.2,

$$\left(\frac{\omega^p}{p!}\right)(\xi) = \sum_{i=1}^{m}\left(\frac{\omega^p}{p!}\right)(\xi_i) \leqq \sum_{i=1}^{m}\|\xi_i\| = |\xi|$$

and equality holds iff each ξ_i corresponds to a canonically oriented complex subspace, i.e., iff ξ is of type (p, p) and positive.

To discuss the consequences of this theorem we need to generalize the notion of positive currents. Let M^n be a complex hermitian manifold, and let $|\cdot|$ denote the mass norm in the fibers of $\Lambda^p TM^n$. Then we can introduce a notion of mass on general currents in $\mathscr{E}_p(M)$ as in Remark 5.12. It follows from the Riesz representation theorem that to any $\mathscr{S} \in \mathscr{E}_p(M^n)$ with $M(\mathscr{S}) < \infty$ there is associated a Borel measure $\|\mathscr{S}\|$ on M^n and a measurable field of p-vectors $\vec{\mathscr{S}_x}$ having $|\mathscr{S}^{\rightarrow}| = 1$ $\|\mathscr{S}\|$-a.e., such that

$$\mathscr{S}(\omega) = \int_M \varphi(\mathscr{S}^{\rightarrow})d\|\mathscr{S}\|$$

for all $\varphi \in \mathscr{E}^p(M)$. If $\mathscr{S} \in \mathscr{E}_{2p}(M)$ satisfies $\mathbf{M}(\mathscr{S}) < \infty$, and if for $\|\mathscr{S}\|$-a.a. $x \in M^n$ the vector $\vec{\mathscr{S}_x}$ is of type (p, p) and positive, then $\mathscr{S}$ is said to be a *positive (p, p)-current.*

COROLLARY 5.14. *Let $\mathscr{S} \in \mathscr{E}_{2p}(M)$ be any positive (p, p)-current in a Kähler manifold M^n. Then $\mathbf{M}(\mathscr{S}) \leqq \mathbf{M}(\mathscr{S}')$ for any $\mathscr{S}' \in \mathscr{E}_{2p}(M)$ for which $\partial\mathscr{S} = \partial\mathscr{S}'$ and $\mathscr{S} - \mathscr{S}'$ is homologous to zero as a de Rham current.*

PROOF. The proof is exactly the same as the proof of Corollary 5.3, except that the Wirtinger inequality is used in generalized form.

A second important class of such currents arises from the following. A (real) differential form $\varphi \in \mathscr{E}^{2p}(M)$ is said to be a *positive (p, p)-form* if for all $x \in M$, φ_x is of type (p, p) and positive. Note that φ is a positive (p, p)-form iff $*\varphi$ is a positive $(n-p, n-p)$-form. If ω is the fundamental 2-form of the hermitian metric, then ω^p is a positive (p, p)-form for all p. In fact, it follows from Remark 5.11 that if M^n is compact, then for any (p, p)-form φ, there is a number $c > 0$ such that $\varphi + c\omega^p$ is positive. Of course, if M^n is a Hodge manifold, then whenever φ is closed and has intergral periods, c can be chosen so that $\varphi + c\omega^p$ is also closed with integral periods.

Now given any form $\varphi \in \mathscr{E}^p(M)$ with compact support, we can define a current $\mathscr{S}_\varphi \in \mathscr{E}_{2n-p}(M)$ by setting

$$\mathscr{S}_\varphi(\alpha) = (\varphi, *\alpha) = \int_M \varphi \wedge \alpha,$$

where M is assumed to be canonically oriented. It is straightforward to check that $\boldsymbol{M}(\mathscr{S}_\varphi) < \infty$, in fact, $\boldsymbol{M}(\mathscr{S}_\varphi) = \int_M |\varphi| *1$ (where $|\cdot|$ is the mass norm in the fibers of $\Lambda^p T^*M$, not the dual norm). Clearly, $\mathscr{S}_\varphi$ is a positive $(n-p, n-p)$-current iff φ is a positive (p,p)-form. If, moreover, $d\varphi = 0$, then $\partial\mathscr{S}_\varphi = 0$ and $\mathscr{S}_\varphi$ belongs to a homology class of de Rham currents $[\mathscr{S}_\varphi] \in H_{2n-2p}(M; \boldsymbol{R})$. This is the dual class to $[\varphi] \in H^{2p}(M; \boldsymbol{R})$. Using the mass function $\boldsymbol{M}(\cdot)$ we can formulate a Plateau problem on $[\mathscr{S}_\varphi]$, and there is the compactness necessary in $\mathscr{E}_*(M)$ to guarantee solutions on compact manifolds. However, it follows from Corollary 5.14 that if φ is a positive (p,p)-form, then $\mathscr{S}_\varphi$ is automatically a current of least mass in $[\mathscr{S}_\varphi]$.

This discussion leads us to the following conjecture. Let M be a compact Kähler manifold and fix a homology class of integral currents $\alpha \in H_{2p}(M; \boldsymbol{Z})$. Let $\alpha^{\boldsymbol{R}} \subset \mathscr{E}_{2p}(M)$ denote the homology class of de Rham currents containing α, and define

$$|\alpha| = \inf\{\boldsymbol{M}(\mathscr{S}) : \mathscr{S} \in \alpha\}, \qquad |\alpha|^{\boldsymbol{R}} = \inf\{\boldsymbol{M}(\mathscr{S}) : \mathscr{S} \in \alpha^{\boldsymbol{R}}\}.$$

Clearly $|\alpha|^{\boldsymbol{R}} \leqq |\alpha|$. Furthermore, it may happen that $|\alpha|^{\boldsymbol{R}} \neq |\alpha|$ (if α is a torsion class, for example). However, one may hope for the following.

CONJECTURE 5.15. *There exists an integer $m \geqq 1$ such that $|\alpha|^{\boldsymbol{R}} = (1/m)|m \cdot \alpha|$.*

Using the polyhedral approximation property, Federer **[24]** has proven that, in general riemannian manifolds,

$$|\alpha|^{\boldsymbol{R}} = \lim_{m\to\infty} \frac{1}{m} |m \cdot \alpha|.$$

It follows easily from our discussion above that for integral classes α on a Hodge manifold M, where $\alpha^{\boldsymbol{R}}$ is dual to a positive, closed (p,p)-form, Conjecture 5.15 implies the Hodge conjecture.

For general (possibly real analytic) riemannian manifolds, the conjecture is false. (See **[24]** for a nice counterexample due to Almgren.) For flat tori, the conjecture has been shown recently by the author to be true for homology classes of dimension $\leqq 2$ but to fail in higher dimensions. In fact, these examples show that the conjecture fails on an abelian variety of complex dimension four, for a class dual to a positive, harmonic $(2, 2)$-form. However, this form lies on the boundary of the positive harmonic forms, and we can still formulate the following.

CONJECTURE. *Let $\alpha \in H_{2p}(M; \boldsymbol{Z})$ be a class dual to an $(n-p, n-p)$-form on a Hodge manifold M, and let $\lambda_p \in H_{2p}(M; \boldsymbol{Z})$ be the algebraic class dual to the $(n-p)$th power of the Kähler form. Then there exists a $K > 0$ such that Conjecture 5.15 holds for all classes of the form $\alpha + k\lambda_p$ for all $k \geqq K$.*

This conjecture can be shown to be equivalent to the Hodge conjecture.

REFERENCES

1. W. K. Allard, *On boundary regularity for Plateau's problem*, Bull. Amer. Math. Soc. **75** (1969), 522–523. MR **39** #3140.

2. ———, *On the first variation of a varifold*, Ann. of Math. (2) **95** (1972), 417–491.

3. F. J. Almgren, Jr., *Some interior regularity theorems for minimal surfaces and an extension of Bernstein's theorem*, Ann. of Math. (2) **84** (1966), 277–292. MR **34** #702.

4. ———, *The theory of varifolds*, Princeton mimeographed notes, 1965.

5. ———, *Existence and regularity almost everywhere of solutions to elliptic variational problems among surfaces of varying topological type and singularity structure*, Ann. of Math. (2) **87** (1968), 321–391. MR **37** #837.

6. J. L. Barbosa, *On minimal immersions of S^2 into S^{2m}*, Thesis, University of Calif., Berkeley, Calif., 1972.

7. S. N. Bernšteĭn, *Sur un théorème de géométrie et son application aux équations aux dérivées partielles du type elliptique*, Soobšč. Har'kov. Mat. Obšč. **15** (1915), 38–45.

8. E. Bombieri, E. de Giorgi and E. Giusti, *Minimal cones and the Bernstein problem*, Invent. Math. **7** (1969), 243–268. MR **40** #3445.

9. E. Calabi, *Minimal immersions of surfaces in Euclidean spheres*, J. Differential Geometry **1** (1967), 111–125. MR **38** #1616.

10. ———, *Quelques applications de l'analyse complexe aux surfaces d'aire minima* (together with *Topics in complex manifolds* by H. Rossi), Les Presses de l'Université de Montréal, 1968.

11. S. S. Chern, *Minimal surfaces in an Euclidean space of N-dimensions*, Differential and Combinatorial Topology, Princeton Univ. Press, Princeton, N. J., 1965, pp. 187–198. MR **31** #5156.

12. ———, *On minimal immersions of the 2-sphere in a space of constant curvature*, Problems in Analysis, Princeton, 1970, pp. 27–40.

13. S. S. Chern, M. do Carmo and S. Kobayashi, *On minimal submanifolds of a sphere with second fundamental form of constant length*, Functional Analysis and Related Fields, Springer, New York, 1970, pp. 59–75. MR **42** #8424.

14. S. S. Chern and R. Osserman, *Complete minimal surfaces in euclidean n-space*, J. Analyse Math. **19** (1967), 15–34. MR **37** #2103.

15. R. Courant, *Dirichlet's principle, conformal mapping and minimal surfaces*, Interscience, New York, 1950. MR **12**, 90.

16. E. de Giorgi, *Una extensione del theorema di Bernstein*, Ann. Scuola Norm. Sup. Pisa **19** (1965), 79–80.

17. M. P. do Carmo and N. R. Wallach, *Representations of compact groups and minimal immersions into spheres*, J. Differential Geometry **4** (1970), 91–104. MR **42** #1013.

18. ———, *Minimal immersions of spheres into spheres*, Ann. of Math. (2) **93** (1971), 43–62. MR **43** #4048.

19. J. Douglas, *Solution of the problem of Plateau*, Trans. Amer. Math. Soc. **33** (1931), 263–321.

20. H. Federer, *The (ϕ, k) rectifiable subsets of n-space*, Trans. Amer. Math. Soc. **62** (1947), 114–192. MR **9**, 231.

21. ———, *Some theorems on integral currents*, Trans. Amer. Math. Soc. **117** (1965), 43–67. MR **29** #5984.

22. ———, *Geometric measure theory*, Die Grundlehren der math. Wissenschaften, Band 153, Springer-Verlag, New York, 1969. MR **41** #1976.

23. ———, *The singular sets of area minimizing rectifiable currents with codimension one and of area minimizing flat chains modulo two with arbitrary codimension*, Bull. Amer. Math. Soc. **76** (1970), 767–771. MR **41** #5601.

24. ———, *Real flat chains, cochains and variational problems* (to appear).

25. H. Federer and W. H. Fleming, *Normal and integral currents*, Ann. of Math. (2) **72** (1960), 458–520. MR **23** #A588.

26. W. H. Fleming, *Flat chains over a finite coefficient group*, Trans. Amer. Math. Soc. **121** (1966), 160–186. MR **32** #2554.

27. ———, *On the oriented Plateau problem*, Rend. Circ. Mat. Palermo (2) **11** (1962), 69–90. MR **28** #499.

28. R. Gulliver, *Regularity of minimizing surfaces of prescribed mean curvature*, Thesis, Stanford University, Stanford, Calif., 1971.

29. R. Harvey and A. Knapp, *Positive (p, p)-forms, Wirtinger's inequality and currents*, Proc. of Tulane conference on complex analysis, Springer Lecture Notes (to appear).

30. R. Harvey and H. B. Lawson, Jr., *Boundaries of complex analytic varieties*, Bull. Amer. Math. Soc. **80** (1974), 180–183.

31. R. Harvey and B. Shiffman, *A characterization of holomorphic chains*, Ann. of Math. (to appear).

32. S. Hildebrandt, *Boundary behavior of minimal surfaces*, Arch. Rational Mech. Anal. **35** (1969), 47–82. MR **40** #1901.

33. Wu-yi Hsiang, *On the compact homogeneous minimal submanifolds*, Proc. Nat. Acad. Sci. U.S.A. **56** (1966), 5–6. MR **34** #5037.

34. ———, *Remarks on closed minimal submanifolds in the standard Riemannian m-sphere*, J. Differential Geometry **1** (1967), 257–267. MR **37** #838.

35. Wu-yi Hsiang and H. B. Lawson, Jr., *Minimal submanifolds of low cohomogeneity*, J. Differential Geometry **5** (1970), 1–37.

36. J. King, *The currents defined by analytic varieties*, Acta Math. **127** (1971), 185–220.

37. H. B. Lawson, Jr., *Local rigidity theorems for minimal hypersurfaces*, Ann. of Math. (2) **89** (1969), 187–197. MR **38** #6505.

38. ———, *Complete minimal surfaces in S^3*, Ann. of Math. (2) **92** (1970), 335–374. MR **42** # 5170.

39. ———, *The global behavior of minimal surfaces in S^n*, Ann. of Math. (2) **92** (1970), 224–237. MR **42** #5169.

40. ———, *Rigidity theorems in rank-*1 *symmetric spaces*, J. Differential Geometry **4** (1970), 349–357. MR **42** #2394.

41. ———, *The unknottedness of minimal embeddings*, Invent. Math. **11** (1970), 183–187. MR **44** #4651.

42. ———, *The equivariant Plateau problem and interior regularity*, Trans. Amer. Math. Soc. **173** (1972), 231–250.

43. ———, *Lectures on minimal submanifolds*, I. M. P. A., Rua Liuz de Camões 68, Rio de Janeiro, 1973.

44. ———, *Minimal varieties in real and complex geometry*, Univ. of Montréal Press, Montréal, 1973.

45. H. B. Lawson, Jr. and J. Simons, *On stable currents and their application to global problems in real and complex geometry*, Ann. of Math. (2) **98** (1973), 427–450.

46. R. Narasimhan, *Introduction to the theory of analytic spaces*, Lecture Notes in Math., no. 25, Springer-Verlag, Berlin and New York, 1966. MR **36** #428.

47. J. C. C. Nitsche, *On new results in the theory of minimal surfaces*, Bull. Amer. Math. Soc. **71** (1965), 195–270. MR **30** #4200.

48. H. Omari, *Isometric immersions of riemannian manifolds*, J. Math. Soc. Japan **19** (1967), 205–214.

49. R. Osserman, *Proof of a conjecture of Nirenberg*, Comm. Pure Appl. Math. **12** (1959), 229–232. MR **21** #4436.

50. ———, *On complete minimal surfaces*, Arch. Rational Mech. Anal. **13** (1963), 392–404. MR **27** #1888.

51. ———, *Global properties of minimal surfaces in E^3 and E^n*, Ann. of Math. (2) **80** (1964), 340–364. MR **31** #3946.

52. ———, *A survey of minimal surfaces*, Van Nostrand Reinhold, New York, 1969. MR **41** 934.

53. ———, *Minimal varieties*, Bull. Amer. Math. Soc. **75** (1969), 1092–1120. MR **43** #2615.

54. ———, *A proof of the regularity everywhere of the classical solution to Plateau's problem*, Ann. of Math. (2) **91** (1970), 550–569. MR **42** #979.

55. T. Radó, *On Plateau's problem*, Ann. of Math. (2) **31** (1930), 457–469.

56. E. R. Reifenberg, *Solution of the Plateau problem for m-dimensional surfaces of varying topological type*, Acta Math. **104** (1960), 1–92. MR **22** #4972.

57. G. de Rham, *Variétés différentiables. Formes courants, formes harmoniques*, Actualités Sci. Indust., no. 1222, Hermann, Paris, 1955. MR **16**, 957.

58. J. Simons, *Minimal varieties in riemannian manifolds*, Ann. of Math. (2) **88** (1968), 62–105. MR **38** #1617.

59. T. Takahashi, *Minimal immersions of Riemannian manifolds*, J. Math. Soc. Japan **18** (1966), 380–385. MR **33** #6551.

60. N. Wallach, *Minimal immersions of symmetric spaces into spheres*, Symmetric spaces, Ed. Boothby and Weiss, Dekker, New York, 1972.

61. W. Whitney, *Complex analytic varieties*, Addison-Wesley, Reading, Mass., 1972.

62. S. T. Yau, *Submanifolds with constant mean curvature*. I, II, Amer. J. Math. (to appear).

63. P. Simões, Thesis, University of California, Berkeley, Calif., 1973.

UNIVERSITY OF CALIFORNIA, BERKELEY

Proceedings of Symposia in Pure Mathematics
Volume 27, 1975

ON A HOLOMORPHIC ANALOGUE OF VANISHING NORMAL SCALAR CURVATURE

GERALD D. LUDDEN* AND KOICHI OGIUE**

Let M be a submanifold immersed in $\tilde{M}$. In generalizing the results of Nomizu and Smyth [3] on hypersurfaces of nonnegative sectional curvature and constant mean curvature immersed in real space forms to general submanifolds of real space forms Yano and Ishihara [8] and Erbacher [2] found that it was necessary to impose the condition that the scalar normal curvature K_N vanished. Let σ be the second fundamental form of the immersion. Then σ is a normal bundle-valued symmetric bilinear form. If M is an m-dimensional Riemannian submanifold of $\tilde{M}$, the vanishing of K_N is equivalent to:

(#) at each point x of M, there exists an orthonormal basis $e_1, \cdots, e_m$ of $T_x(M)$ such that $\sigma(e_i, e_j) = 0$ for $1 \leqq i \neq j \leqq m$.

Here $T_x(M)$ denotes the tangent space to M at x. It was shown in [1] that if M is a complex submanifold of the Kaehler manifold $\tilde{M}$ then K_N vanishes identically if and only if M is totally geodesic, i.e. $\sigma = 0$. Therefore (#) does not play an important role in the theory of Kaehler submanifolds.

Let M be an n-dimensional Kaehler submanifold of the $(n + p)$-dimensional complex space form $\tilde{M}_{n+p}(\tilde{c})$ (or simply $\tilde{M}$), i.e. $\tilde{M}_{n+p}(\tilde{c})$ is a Kaehler manifold of constant holomorphic sectional curvature $\tilde{c}$. Consider the following condition:

(*) At each point x of M, there exists an orthonormal frame $e_1, \cdots, e_n, Je_1, \cdots, Je_n$ such that $\sigma(e_a, e_b) = 0$ for $1 \leqq a \neq b \leqq n$, where J denotes the complex structure of M.

AMS (MOS) subject classifications (1970). Primary 53B35, 53B25.

*Partially supported by NSF grant GP-36684.
**Partially supported by the Matsunaga Science Foundation.

Condition (*) may be considered as a complex analogue of (#). The purpose of this paper is to prove the following theorems.

THEOREM 1. *Let M be an n-dimensional Kaehler submanifold immersed in* $\tilde{M}_{n+p}(\tilde{c})$. *If M is a complex space form and if the immersion satisfies* (*), *then M is totally geodesic in* $\tilde{M}$.

THEOREM 2. *Let M be an n-dimensional complete Kaehler submanifold immersed in* $\tilde{M}_{n+p}(\tilde{c})$, $\tilde{c} > 0$. *If every holomorphic sectional curvature of M is greater than* $\tilde{c}/2$ *and if the immersion satisfies* (*), *then M is totally geodesic in* $\tilde{M}$.

REMARK. It is clear that if M is a Riemann surface or if M is a hypersurface of $\tilde{M}$ then condition (*) holds.

1. Preliminaries. Let $\tilde{M}_{n+p}(\tilde{c})$ (or simply $\tilde{M}$) be an $(n+p)$-dimensional complex space form of constant holomorphic sectional curvature $\tilde{c}$ and let M be an n-dimensional Kaehler submanifold immersed in $\tilde{M}$. Let J (resp. $\tilde{J}$) be the almost complex structure of M (resp. $\tilde{M}$) and let g (resp. $\tilde{g}$) be the Kaehler metric of M (resp. $\tilde{M}$). We denote by ∇ (resp. $\tilde{\nabla}$) the covariant differentiation operator with respect to g (resp. $\tilde{g}$). Then the second fundamental form σ of the immersion is given by

$$\sigma(X, Y) = \tilde{\nabla}_X Y - \nabla_X Y$$

and it satisfies

$$\sigma(X, Y) = \sigma(Y, X), \quad \sigma(JX, Y) = \sigma(X, JY) = \tilde{J}\sigma(X, Y). \tag{1.1}$$

We choose a local field of orthonormal frames

$$e_1, \cdots, e_n, e_{1^*} = \tilde{J}e_1, \cdots, e_{n^*} = \tilde{J}e_n, \xi_1, \cdots, \xi_p, \xi_{1^*} = \tilde{J}\xi_1, \cdots, \xi_{p^*} = \tilde{J}\xi_p$$

in $\tilde{M}$ in such a way that restricted to M, $e_1, \cdots, e_n, e_{1^*}, \cdots, e_{n^*}$ are tangent to M.[1] If we set

$$g(A_\lambda X, Y) = g(\sigma(X, Y), \xi_\lambda), \tag{1.2}$$

or

$$\sigma(X, Y) = \sum_\lambda g(A_\lambda X, Y)\,\xi_\lambda, \tag{1.2$'$}$$

then $A_1, \cdots, A_p, A_{1^*}, \cdots, A_{p^*}$ are local fields of symmetric linear transformations. We can easily see from (1.1) that

$$A_{\alpha^*} = JA_\alpha \quad \text{and} \quad JA_\alpha = -A_\alpha J.$$

Thus tr $A_\lambda = 0$ for all λ and hence M is a minimal submanifold of $\tilde{M}$.

The condition (*) is equivalent to the following:

(*′) at each point of M, there exists an orthonormal frame $e_1, \cdots, e_n, Je_1, \cdots, Je_n$ with respect to which the matrices of the A_α's are of the form

---

[1]We use the following convention on the range of indices: $a, b \cdots = 1, \cdots, n, \alpha, \beta, \cdots = 1, \cdots, p$, $\lambda, \mu, \cdots = 1, \cdots, p, 1^*, \cdots, p^*$.

(1.3) 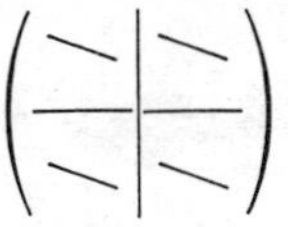

or

(*″) at each point of M, there exists an orthogonal matrix which simultaneously transforms all of the A_α's into the form (1.3).

The sectional curvature K of M determined by orthonormal vectors X and Y is given by

$$(1.4)\quad K(X, Y) = \tilde{c}/4\,\{1 + 3g(X, JY)^2\} + \tilde{g}(\sigma(X, X), \sigma(Y,Y)) - \|\sigma(X, Y)\|^2.$$

The holomorphic sectional curvature H of M determined by a unit vector X is given by

$$(1.5)\qquad H(X) = \tilde{c} - 2\,\|\sigma(X, X)\|^2 = \tilde{c} - 2\sum_\lambda g(A_\lambda X, X)^2.$$

Let ρ be the scalar curvature of M. Then we have

$$(1.6)\qquad \rho = n(n+1)\,\tilde{c} - \|\sigma\|^2,$$

where $\|\sigma\|$ is the length of the second fundamental form σ of the immersion so that

$$(1.7)\qquad \|\sigma\|^2 = \sum_\lambda \operatorname{tr} A_\lambda^2 = 2\sum_\alpha \operatorname{tr} A_\alpha^2.$$

The second fundamental form satisfies a differential equation.

PROPOSITION 1.1 [4]. *Let M be an n-dimensional Kaehler submanifold immersed in $\tilde{M}_{n+p}(\tilde{c})$. Then*

$$\frac{1}{2}\Delta\|\sigma\|^2 = \|\nabla'\sigma\|^2 - 8\operatorname{tr}\Big(\sum_\alpha A_\alpha^2\Big)^2 - \sum_{\lambda,\mu}(\operatorname{tr} A_\lambda A_\mu)^2 + \frac{n+2}{2}\tilde{c}\,\|\sigma\|^2,$$

where Δ denotes the Laplacian and ∇' the covariant differentiation with respect to the connection in (tangent bundle) $\oplus$ (normal bundle).

REMARK. Condition (*) implies that the A_α^2 are simultaneously diagonalizable. However the immersion of space forms of holomorphic sectional curvature $\tilde{c}/2$ in $\tilde{M}_{n+n(n+1)/2}(\tilde{c})$ shows that the converse is not true. The condition that the A_α^2 are simultaneously diagonalizable can be expressed as the vanishing of a scalar on M and hence it may be of interest to study this condition.

2. Proof of Theorem 1. Let M be an n-dimensional complex space form of constant holomorphic sectional curvature c which satisfies the assumption of Theorem 1. In consideration of a result of B. O'Neill (Theorem 3 in [6]), we may assume that $p \geqq n(n+1)/2$.

First we note that $c \leqq \tilde{c}$. If $c = \tilde{c}$ then M is totally geodesic in $\tilde{M}$. We may therefore assume that $c \neq \tilde{c}$.

Since $H = c$, from (2.4), we have

$$\|\sigma(X, X)\|^2 = (\tilde{c} - c)/2 \tag{2.1}$$

for every unit vector X.

On the other hand, every antiholomorphic sectional curvature of a complex space form of constant holomorphic sectional curvature c is $c/4$. Therefore

$$K(X, Y) = K(X, JY) = c/4.$$

provided that X, Y and JY are orthonormal. This, together with (2.1) and (2.3), implies

$$\|\sigma(X, Y)\|^2 = (\tilde{c} - c)/4 \tag{2.2}$$

for orthonormal X, Y and JY.

Let $e_1, \cdots, e_n, e_{1^*}, \cdots, e_{n^*}$ be local fields of orthonormal vectors on M as in §1. Then we have the following.

LEMMA 2.1 [6]. *The $n(n+1)$ local fields of vectors $\sigma(e_a, e_b)$, $\tilde{J}\sigma(e_a, e_b)$, $1 \leqq a \leqq b \leqq n$, are orthogonal.*

This, together with (2.1) and (2. 2), implies that $\sigma(e_a, e_b)$, $\tilde{J}\sigma(e_a, e_b)$, $1 < a < b < n$, are linearly independent at each point.

We choose local fields of orthonormal vectors $\xi_1, \cdots, \xi_p, \xi_{1^*}, \cdots, \xi_{p^*}$ normal to M (i.e., a local flied of normal frames) in such a way that

$$\xi_a = \frac{2^{1/2}}{(\tilde{c}-c)^{1/2}}\,\sigma(e_a, e_a), \qquad \xi_{(a,b)} = \frac{2}{(\tilde{c}-c)^{1/2}}\,\sigma(e_a, e_b),$$

where

$$(a, b) = \operatorname{Min}\{a, b\} + \frac{|a-b|(2n+1-|a-b|)}{2}$$

for $a \neq b$. From (1.2) we can see that, with respect to the local field of normal frames chosen above, the matrices of A_α's have the following form:

$$A_a = \left(\begin{array}{ccc|ccc} & \overset{a}{\vdots} & & & & \\ a\ \cdots & \# & \cdots & & 0 & \\ & \vdots & & & & \\ \hline & & & & \vdots & \\ & 0 & & \cdots & -\# & \cdots \ n+a \\ & & & & \vdots & \\ & & & & n+a & \end{array}\right) \tag{2.3}$$

$$A_{(a,b)} = \left(\begin{array}{cccc|cccc} & \overset{a}{\vdots} & \overset{b}{\vdots} & & & & & \\ a\ \cdots & \cdot & * & \cdots & & & & \\ b\ \cdots & * & \cdot & \cdots & & 0 & & \\ & \vdots & \vdots & & & & & \\ \hline & & & & & \vdots & \vdots & \\ & & & & \cdots & \cdot & * & \cdots \ n+a \\ & & 0 & & & \vdots & \vdots & \\ & & & & \cdots & -* & \cdot & \cdots \ n+b \\ & & & & & \vdots & \vdots & \\ & & & & & n+a & n+b & \end{array}\right)$$

$$A_\alpha = 0 \text{ for } \alpha > n(n+1)/2,$$

where $\# = (\tilde{c} - c)^{1/2}/2^{1/2}$ and $* = (\tilde{c} - c)^{1/2}/2$.

We can easily see that the A_a's in (2. 3) do not satisfy (*)' or (*)''. Therefore we have proved that a complex space form which is a Kaehler submanifold of $\tilde{M}_{n+p}(\tilde{c})$ cannot satisfy (*) unless it is totally geodesic. This completes the proof of Theorem 1.

3. Proof of Theorem 2. First we note that by a theorem of Tsukamoto [7] M is compact.

By the assumption (*) or equivalently (*)', at each point of M we can choose an orthonormal frame $e_1, \cdots, e_n, e_{1^*}, \cdots, e_{n^*}$ with respect to which the matrix of A_α is of the form

$$\left(\begin{array}{cc|cc} \lambda_1^\alpha & 0 & \mu_1^\alpha & 0 \\ & \ddots & & \ddots \\ 0 & \lambda_n^\alpha & 0 & \mu_n^\alpha \\ \hline \mu_1^\alpha & 0 & -\lambda_1^\alpha & 0 \\ & \ddots & & \ddots \\ 0 & \mu_n^\alpha & 0 & -\lambda_n^\alpha \end{array}\right).$$

For each a, from (1. 5), we have

$$H\left(\frac{e_a + e_{a^*}}{2^{1/2}}\right) = \tilde{c} - 2\sum_\alpha \{(\lambda_a^\alpha)^2 + (\mu_a^\alpha)^2\}.$$

This, together with the assumption that $H > \tilde{c}/2$, implies that

$$\sum_\alpha \{(\lambda_a^\alpha)^2 + (\mu_a^\alpha)^2\} < \tilde{c}/4,$$

from which we obtain

$$\text{(3.1)} \qquad \operatorname{tr}(\sum A_\alpha^2)^2 \leqq \frac{\tilde{c}}{4} \operatorname{tr} \sum A_\alpha^2 = \frac{\tilde{c}}{8} \|\sigma\|^2.$$

On the other hand, we have the following:

LEMMA 3.1. *Let M be an n-dimensional Kaehler submanifold immersed in $\tilde{M}_{n+p}(\tilde{c})$ (not necessarily assumed to satisfy (*)). If $H > \tilde{c}/2$, then* $\sum (\operatorname{tr} A_\lambda A_\mu)^2 < (n/2)\tilde{c}\|\sigma\|^2$.

PROOF. Let $\Lambda = (\operatorname{tr} A_\lambda A_\mu)$. Then Λ is a symmetric $(2p, 2p)$-matrix and it is covariant for the change of $\xi_1, \cdots, \xi_p, \xi_{1^*}, \cdots, \xi_{p^*}$. In other words, let $\Lambda' = (\operatorname{tr} A'_\lambda A'_\mu)$ be the corresponding matrix with respect to $\xi'_1, \cdots, \xi'_p, \xi'_{1^*}, \cdots, \xi'_{p^*}$ and let $U = (U_{\lambda\mu})$ be the real representation of the unitary matrix given by $\xi'_\mu = \sum \xi_\lambda U_{\lambda\mu}$. Then $\Lambda' = {}^tU \wedge U$.

Moreover $\Lambda' = {}^tU \wedge U$ implies that Λ can be diagonalized for a suitable choice of $\xi_1, \cdots, \xi_p, \xi_{1^*}, \cdots, \xi_{p^*}$, that is,

$$
{}^tU \wedge U = \begin{pmatrix} \operatorname{tr} A_1'^2 & & & & & 0 \\ & \ddots & & & & \\ & & \operatorname{tr} A_p'^2 & & & \\ & & & \operatorname{tr} A_1'^2 & & \\ & & & & \ddots & \\ 0 & & & & & \operatorname{tr} A_p'^2 \end{pmatrix}
$$

for some U. Therefore we have

$$
\sum (\operatorname{tr} A_\lambda A_\mu)^2 = 2 \sum (\operatorname{tr} A_\alpha'^2)^2 \leqq 4n \sum \operatorname{tr} A_\alpha'^4, \tag{3.2}
$$

where we use the general fact that a symmetric $(2n, 2n)$-matrix A satisfies $(\operatorname{tr} A^2)^2 \leqq 2n \operatorname{tr} A^4$.

On the other hand, from (1.5) we can see that if $H > \tilde{c}/2$, then the square of every eigenvalue of A_λ must be smaller than $\tilde{c}/4$. Therefore we have

$$
\operatorname{tr} A_\lambda^2 A_\mu^2 \leqq \frac{\tilde{c}}{4} \operatorname{tr} A_\lambda^2
$$

for all λ and μ. This, together with (3.2), implies

$$
\sum (\operatorname{tr} A_\lambda A_\mu)^2 \leqq n\tilde{c} \sum \operatorname{tr} A_\alpha'^2 = \frac{n}{2} \tilde{c} \, \| \sigma \|^2. \quad \text{Q.E.D.}
$$

Proposition 1.1, (3.1) and Lemma 3.1 imply $\Delta \|\sigma\|^2 \geqq 0$. Therefore, by a well-known theorem of E. Hopf, $\| \sigma \|^2$ is constant so that ρ is constant. Hence Theorem 2 follows from Theorem 1 in [5].

REMARK. If we replace the condition $H > \tilde{c}/2$ in Theorem 2 by $H \geqq \tilde{c}/2$ then we still have $\Delta \| \sigma \|^2 \geqq 0$ and so $\| \sigma \|^2$ is constant. It can then be shown that M is Einstein and $\nabla' \sigma = 0$. Hence, it is reasonable to conjecture that under these conditions M is totally geodesic or a complex quadric.

REFERENCES

1. B.-Y. Chen and G.D. Ludden, *Riemann surfaces in complex projective spaces*, Proc. Amer. Math. Soc. **32** (1972), 561–566. MR **44** #7446.

2. J.A. Erbacher, *Isometric immersions with constant mean curvature and triviality of the normal bundle*, Nagoya Math. J. **45** (1972), 139–165.

3. K. Nomizu and B. Smyth, *A formula of Simons' type and hypersurfaces with constant mean curvature*, J. Differential Geometry **3** (1969), 367–377. MR **42** #1018.

4. K. Ogiue, *Complex submanifolds of the complex projective space with second fundamental form of constant length*, Kōdai Math. Sem. Rep. **21** (1969), 252–254. MR **40** #1902.

5. ———, *Positively curved complex submanifolds immersed in a complex projective space*, J. Differential Geometry **7** (1972), 603–606.

6. B. O'Neill, *Isotropic and Kaehler immersions*, Canad. J. Math. **17** (1965), 907–915. MR **32** #1654.

7. Y. Tsukamoto, *On Kählerian manifolds with positive holomorphic sectional curvature*, Proc. Japan Acad. **33** (1957), 333–335. MR **19**, 880.

8. K. Yano and S. Ishihara, *Submanifolds with parallel mean curvature vector*, J. Differential Geometry **6** (1971/72), 95–118. MR **45** #7650.

MICHIGAN STATE UNIVERSITY

Proceedings of Symposia in Pure Mathematics
Volume 27, 1975

COMPLETE, OPEN SURFACES IN E^3

TILLA KLOTZ MILNOR

This article outlines a small assortment of results and questions about complete, open surfaces in E^3. The discussion will center upon recent work of N. V. Efimov, and some related conjectures due to John Milnor.

In 1964, Efimov proved that no complete surface can be C^2-immersed in E^3 so as to have Gauss curvature $K \leqq$ constant < 0 (see **[2]** or **[6]**). This generalized Hilbert's classical result which stated that no sufficiently smooth immersion exists with $K \equiv -1$. In 1968, Efimov obtained the following improvement **[4]** of his 1964 result.

THEOREM (N. V. EFIMOV). *On a complete surface C^2-immersed in E^3 with $K < 0$, one cannot have*

$$\left| \frac{1}{|K(p_2)|^{1/2}} - \frac{1}{|K(p_1)|^{1/2}} \right| \leqq C_1\, d(p_1, p_2) + C_2 \tag{1}$$

if C_1 and C_2 are positive constants, and $d(p_1, p_2)$ the distance determined between any two points p_1 and p_2 on the surface by the Riemannian metric.

On a surface with $K \leqq$ constant < 0, (1) can be satisfied with $C_1 = 0$. To check that the result above covers cases not handled by Efimov's earlier theorem, it is handy to use the following **[4]**.

COROLLARY TO EFIMOV'S THEOREM. *A complete surface with $K < 0$ cannot be C^2-immersed in E^3 if for some constant q,*

$$|\,\text{gradient}\,|K|^{-1/2}| \leqq q < \infty. \tag{2}$$

AMS (MOS) subject classifications (1970). Primary 53A05, 53C20.

Using this corollary, we see that the plane with the complete metric $ds^2 = dx^2 + (1 + x^2)^2\, dy^2$ cannot be C^2-immersed in E^3. Here

$$K = -2/(1 + x^2),$$

so that K is certainly not bounded away from zero.

In [3], Efimov had obtained a weaker version of the corollary above, using $q \leqq 1/3^{1/2}$. In the same paper, one finds the following observation (see Lemma IV in [3]).

LEMMA (N. V. EFIMOV). *If* (2) *holds on a complete surface with* $K < 0$, *then the metric* $|K|\, ds^2$ *is also complete.*

The metric $|K|\, ds^2$ plays a central role throughout [3] and [**4**]. Indeed, on any surface with $K \neq 0$, $|K|\, ds^2$ is bound to be interesting, since it is conformally equivalent to ds^2, while at the same time having total area equal to the absolute value of the total curvature of the surface. For surfaces immersed in E^3, $|K|\, ds^2$ is equiareal with the third fundamental form III (which is obtained by pulling back the metric on the unit sphere, using the Gauss spherical image mapping). For minimal surfaces in E^3, $|K|\, ds^2$ is identical with III.

It was Efimov's guess (see p. 52 in [3]) that a complete surface with $K < 0$ cannot be C^2-immersed in E^3 if $K\, ds^2$ is complete, with area greater than 2π. While this guess is not correct (see Remark 1 below), it does suggest an inquiry into the possible completeness of metrics of the form $|K|^\alpha\, ds^2$ on complete, open surfaces in E^3 with $K \neq 0$. Thus, for any real constant α, we consider the following possible assertion.

STATEMENT$_\alpha$. *A complete, open surface with* $K \neq 0$ *cannot be* C^2-*immersed in* E^3 *if* $|K|^\alpha\, ds^2$ *is complete.*

Pogorelov has shown that all complete metrics with $K > 0$ are imbeddable in E^3, subject to certain restrictions on smoothness (see Theorems 2 and 3, p. 104 of [**10**]). To this extent, Statement$_\alpha$ places a restriction on the complete metrics with $K > 0$ definable on an abstract, noncompact surface. We can now report the following guess.

CONJECTURE (JOHN MILNOR). *Statement*$_\alpha$ *is true if* $1 < \alpha \leqq 2$.

REMARK 1. Statement$_\alpha$ is false if $\alpha \leqq 1$ or if $\alpha > 2$, both for surfaces on which $K > 0$, and for surfaces on which $K < 0$. Certain surfaces of revolution pointed out to us by John Milnor serve as counterexamples. To give one sample argument, consider a plane curve with arc length s and distance $y(s)$ from the x-axis. Here $y(s)$ can be any twice differentiable function with

$$y(s) > 0, \qquad |\dot{y}(s)| < 1,$$

where the dot indicates differentiation with respect to s. Rotating about the x-axis, one obtains a surface with Gauss curvature $K = -\ddot{y}/y$. It is not difficult to construct a complete surface with $K < 0$ in this way, using a function $y(s)$ which satisfies

$$y(s) = y(-s) > 0, \quad \ddot{y} > 0$$

for all real s, and

$$\dot{y}(s) = 1 - 1/(\log s)$$

for all sufficiently large $s > 0$. Then, when s is large,

$$\ddot{y} = 1/(s \log^2 s).$$

But y is clearly asymptotic to s, so that

$$K \sim -1/(s^2 \log^2 s),$$

as $s \to \infty$. Thus, for large enough $c > 0$, $\int_c^\infty |K|^{\alpha/2}\, ds$ is approximately $\int_c^\infty ds/s^\alpha (\log s)^\alpha$. For $\alpha \leqq 1$, this is infinite, so that one easily concludes that the metric $|K|^\alpha ds^2$ is complete on the surface constructed.

REMARK 2. It is not hard to check that Statement$_\alpha$ is true for surfaces of revolution if $K < 0$ and $1 < \alpha \leqq 2$, or if $K > 0$ and $\alpha = 2$. The arguments are elementary.

Since the known counterexamples to Statement$_\alpha$ for $\alpha > 2$ all involve arbitrarily large values of K occurring on increasingly tiny sets farther and farther out on the surface, it seems natural to consider the following modified assertion.

STATEMENT$'_\alpha$. *A complete, open surface with $K \neq 0$ cannot be C^2-immersed in E^3 if* $\min(1, |K|^\alpha)\, ds^2$ *is complete.*

CONJECTURE$'$ (JOHN MILNOR). *Statement$'_\alpha$ is true if* $1 < \alpha$.

REMARK 3. If $|K|$ is bounded, Statement$_\alpha$ and Statement$'_\alpha$ are equivalent. For all $\beta > \alpha$, Statement$_\alpha$ implies Statement$'_\alpha$ which in turn implies Statement$'_\beta$. Thus a counterexample to Statement$'_\beta$ for any $\beta > 2$ would disprove both the conjectures just stated. Finally, Statement$'_\beta$ implies Efimov's 1964 Theorem for any $\beta > 0$.

We note in passing that there is an easy way of producing complete, nonstandard metrics of geometric interest on any complete surface with $K \neq 0$ which is C^2-immersed in E^3. Specifically, if H denotes mean curvature, one can take any function $f(H, K) > 0$, and form the complete metric $I + f(H, K)\,\mathrm{III}$. Particularly interesting examples result if one takes $f = 1/K$ for $K > 0$, or $f = -1/K$ for $K < 0$. On a complete surface with $K > 0$, one obtains the complete metric $(H/K)\,\mathrm{II}$. On a complete surface with $K < 0$, one obtains the complete metric $(H'/|K|)\,\mathrm{II}'$, where II' is given by $H'\,\mathrm{II}' = H\,\mathrm{II} - K\mathrm{I}$, and $H' = (H^2 - K)^{1/2}$. See [8] for results involving these and similar metrics definable on surfaces immersed in arbitrary Riemannian manifolds once some unit normal vector field is fixed.

There is yet another conjecture due to John Milnor which would generalize Efimov's 1964 Theorem (see [5] or [6]). It states that the sum of the absolute values of the principal curvatures cannot be bounded away from zero on a complete, umbilic-free surface, C^2-immersed in E^3, unless K changes sign, or else vanishes identically. A review of progress toward verification of this conjecture can be found in [7]. A more recent result with some bearing on the conjecture is the following [1].

THEOREM (D. BLEEKER). *Consider a complete, finitely-connected, umbilic-free surface, C^2-immersed in E^3 with K locally constant outside some compact set. If the sum of the absolute values of the principal curvatures is bounded away from zero, then total curvature is zero.*

This last theorem checks the Milnor conjecture for all finitely-connected surfaces on which K is locally constant outside some compact set. The next result checks the conjecture for all "$\mathscr{Q}\,dz^2$-bounded" surfaces on which $K \leqq 0$. Before stating the theorem, some definitions are in order.

If x, y are isothermal coordinates on a surface S which is C^2-immersed in E^3, then the second fundamental form is given by

$$\begin{aligned} \mathrm{II} &= L dx^2 + 2M dx dy + N dy^2 \\ &= \mathscr{Q}\,dz^2 + 2\mathscr{P} dz d\bar{z} + \bar{\mathscr{Q}} d\bar{z}^2 \end{aligned}$$

where $4\mathscr{Q} = L - N - 2iM$, and $z = x + iy$. Under change of isothermal parameters, $\mathscr{Q}dz^2$ behaves like a quadratic differential. We call S $\mathscr{Q}dz^2$-bounded if there exist global isothermal coordinates on the universal covering surface $\tilde{S}$ of S in terms of which $\log \mid \mathscr{Q} \mid < u$, for some harmonic function u on $\tilde{S}$. Thus, if S is $\mathscr{Q}dz^2$-bounded, $\tilde{S}$ cannot be conformally a sphere, for no global isothermal coordinates exist on the sphere. However, it is easy to check that $\mathscr{Q}\,dz^2$-boundedness is independent of the particular choice of global isothermal parameters on $\tilde{S}$. As an example, any nonumbilic surface of constant mean curvature is $\mathscr{Q}dz^2$-bounded, since $\mathscr{Q}dz^2 \neq 0$ is holomorphic, and $\log \mid \mathscr{Q} \mid$ itself is harmonic on S.

THEOREM (TILLA MILNOR). *If S is a complete, $\mathscr{Q}dz^2$-bounded surface, C^2-immersed in E^3 with $K \leqq 0$ and $H' = (H^2 - K)^{1/2} \geqq$ constant > 0, then $K \equiv 0$ on S.*

The result is an improvement of Theorem 1 in [7] only to the extent that its hypotheses may be more easily verified and understood. For it is not hard to show that S is $\mathscr{Q}dz^2$-bounded if and only if S is $\mathscr{Q}$-bounded in the sense of [7]. Similar improvements are available for nearly all results in [7], using natural arguments based upon a lemma due to Osserman (see p. 78 of [9]). See [8] for details and related results.

REFERENCES

1. D. Bleeker, Dissertation, University of California, Berkeley, Calif., 1973.

2. N. V. Efimov, *Appearance of singularities on surfaces of negative curvature*, Mat. Sb. **64 (106)** (1964), 286–320; English transl., Amer. Math. Soc. Transl. (2) **66** (1968), 154–190. MR **29** #5203.

3. ———, *Surfaces with slowly varying negative curvature*, Uspehi Mat. Nauk **21** (1966), no. 5 (131), 3-58 = Russian Math. Surveys **21** (1966), no. 5, 11-55. MR **34** #1966.

4. ———, *Differential criteria for homeomorphism of certain mappings with applications to the theory of surfaces*, Mat. Sb. **76 (118)** (1968), 499–512 = Math. USSR Sb. **5** (1968), 475–488. MR **37** #5821.

5. T. Klotz and R. Ossermann, *Complete surfaces in E^3 with constant mean curvature*, Comment. Math. Helv. **41** (1966/67), 313–318. MR **35** #2213.

6. T. K. Milnor, *Efimov's theorem about complete immersed surfaces of negative curvature*, Advances in Math. **8**, 300–369.

7. ———, *Some restrictions on the smooth immersion of complete surfaces in E^3*, *Contributions to analysis, a collection of papers dedicated to Lipman Bers*, Academic Press, New York, 1974, pp. 313–323.

8. ———, *Restrictions on the curvatures of Φ-bounded surfaces*, J. Differential Geometry (to appear).

9. R. Osserman, *A survey of minimal surfaces*, Van Nostrand Reinhold, New York, 1969. MR 41 #934.

10. A. V. Pogorelov, *Extrinsic geometry of convex surfaces*, "Nauka", Moscow, 1969; English transl., Transl. Math. Monographs, vol. 35, Amer. Math. Soc., Providence, R.I., 1973. MR **39** #6222.

RUTGERS, THE STATE UNIVERSITY

Proceedings of Symposia in Pure Mathematics
Volume 27, 1975

MINIMAL VARIETIES IN TORI

TADASHI NAGANO AND BRIAN SMYTH*

There are no compact minimal submanifolds in Euclidean space; this situation changes radically if we take a flat torus as ambient space, and because of absolute parallelism in the torus the Gauss map survives intact. Indeed the Gauss map is then harmonic in the sense of Eells and Sampson [**1**] (see [**4**]). From the point of view of minimal submanifolds in Euclidean space invariant by a lattice, it is natural to consider compact minimal submanifolds in a flat torus. A more compelling influence is the fact that a compact Kähler manifold M has a canonical holomorphic map a into a canonical flat complex torus $A(M)$ (known respectively, as the Albanese map and Albanese torus of M). When this map is an immersion, as happens, it is a minimal immersion.

Indeed the Albanese map turns out to be the most important object associated with a minimal variety in a torus. Here, it should be said, that the Albanese map $a: M \to A(M)$ is defined when M is any compact orientable Riemannian manifold. $A(M)$ is a flat torus of dimension $b_1(M)$ and a is a harmonic map. The remarkable property of a is its universality, i.e. for each harmonic map f of M into a flat torus T there is a unique affine map $g: A(M) \to T$ such that $f = g \circ a$. It can be shown that a map $f: M \to T$ is harmonic if and only if the pull-back of every harmonic-form is harmonic. While much can be said of harmonic maps into a torus we will be content here to outline the results obtained for minimal submanifolds.

Let M^n be a compact orientable Riemannian manifold and $f: M^n \to T^{n+r}$ a minimal isometric immersion of M in a flat torus.

(i) Any minimal isometric immersion of M in T which is homotopic to f differs from f by a translation (rigidity).

AMS (MOS) subject classifications (1970). Primary 53A10, 53C20; Secondary 53C40, 57D40, 32C10.

*Partially supported by NSF grant GP-29662.

(ii) The identity component $I_0(M)$ of the isometry group of M is a torus which acts freely on M.

(iii) The isometry group of M is finite only if the Ricci curvature is negative at some point (since the Ricci curvature of M is nonpositive a theorem of Bochner asserts the converse).

(iv) $I_0(M) \to M \to M_1 = M/I_0(M)$ is a principal torus bundle and the base M_1, with the projected metric, has a natural isometric immersion as a minimal submanifold in a flat torus, also with codimension r.

(v) For compact complex submanifolds in complex tori the obvious complex analogue of each of the above results holds.

(vi) For compact complex hypersurfaces in a complex torus the following are equivalent.

(a) the Euler number is nonzero,

(b) $\mathrm{Aut}_0(M)$ is the identity,

(c) the Ricci curvature is negative definite on an open dense set,

(d) the Gauss map $\pi : M \to P^n(C)$ is a branched covering.

This last result (vi) was also obtained by Matsushima [2] this past winter by a very different approach using theta functions.

A special class of examples is provided by nonsingular hyperplane sections V of an abelian variety imbedded in complex projective space. Our work shows these V satisfy the properties in (vi) above and for the homotopy groups $\pi_1(V) = \bigoplus^{2n+2} Z$, $\pi_i(V) \neq 0$ for some $i > 1$; here n denotes the complex dimension of V. In particular, the universal covering of V is not a cell. Clearly these V are useful test spaces for relationships between the curvature and topology. It follows, for example, that the fundamental group of a compact Riemannian manifold with negative Ricci curvature need not have exponential growth and can even be free abelian. This answers a question raised by Milnor [3].

A fuller account of this work is to appear.

REFERENCES

1. J. Eells, Jr. and J. H. Sampson, *Harmonic mappings of Riemannian manifolds*, Amer. J. Math. **86** (1964), 109–160. MR **29** #1603.

2. Y. Matsushima, *Holomorphic immersions of a compact Kähler manifold into complex tori*, University of Notre Dame, Notre Dame, Ind. (preprint).

3. J. Milnor, *A note on curvature and fundamental group*, J. Differential Geometry **2** (1968), 1–7. MR **38** #636.

4. E.A. Ruh and J. Vilms, *The tension field of the Gauss map*, Trans. Amer. Math. Soc. **149** (1970), 569–573. MR **41** #4400.

UNIVERSITY OF NOTRE DAME

Proceedings of Symposia in Pure Mathematics
Volume 27, 1975

ELIE CARTAN'S WORK ON ISOPARAMETRIC FAMILIES OF HYPERSURFACES

KATSUMI NOMIZU

In a series of papers **[1]**—**[4]** E. Cartan studied isoparametric families of hypersurfaces. It seems that this work, which, incidentally, was done when he was about 70 years old, has not received due attention in spite of its rich content as well as a wide range of interesting problems arising from it.

In this lecture, I should like to outline what this work of Cartan is and discuss a few problems and recent results in **[5]** and **[7]**. I wish to acknowledge helpful conversations I had with D. S. Leung while I was giving a series of lectures on this subject in the spring of 1972.

1. Parallel hypersurfaces. A geometric definition of an isoparametric family of hypersurfaces can be given by means of a parallel family of hypersurfaces. Let $M^{n+1}(c)$ be a simply connected, complete Riemannian manifold of dimension $n+1$ with constant sectional curvature c. According as $c = 1, 0, -1$, $M^{n+1}(c)$ is thus the sphere S^{n+1}, the Euclidean space R^{n+1}, the hyperbolic space H^{n+1}, respectively. Let M^n be a hypersurface immersed in $M^{n+1}(c)$ and let f be the immersion. The following considerations are essentially local. Let $x \to \xi_x$ be a field of unit normals to the hypersurface M^n. For each $t > 0$, let $f_t(x)$, $x \in M^n$, be the point of $M^{n+1}(c)$ on the geodesic from $f(x)$ starting in the direction ξ_x which has geodesic distance t from $f(x)$. In the Euclidean case ($c = 0$), we have

$$f_t(x) = f(x) + t\,\xi_x.$$

In the spherical case ($c = 1$), we have

$$f_t(x) = (\cos t) f(x) + (\sin t)\,\xi_x$$

AMS (MOS) subject classifications (1970). Primary 53C40; Secondary 57D40, 58C05.

by considering S^{n+1} as the unit sphere in R^{n+2}.

In all cases, f_t is an immersion of M^n into $M^{n+1}(c)$, provided t is small enough. Of course, f_0 is the original immersion f. Let g, h, d be the first, second, and third fundamental forms of the immersion f, and let g_t and h_t be the first and second fundamental forms for the immersion f_t (thus they are all defined on M^n).

In the euclidean case, we have

$$g_t = g - 2th + t^2 d, \tag{1}$$

$$h_t = h - td. \tag{2}$$

In terms of the so-called shape operator A (i.e. the symmetric endomorphism such that $h(X, Y) = g(AX, Y)$ for tangent vectors X, Y at each point of M^n), we have from (1) and (2)

$$A_t = (I - tA)^{-1}A, \tag{3}$$

where I denotes the identity transformation.

At a point x of M^n, let $a_1, \cdots, a_n$ be the principal curvatures for the immersion f with principal vectors $e_1, \cdots, e_n$ which are orthonormal relative to g:

$$Ae_k = a_k e_k, \qquad 1 \leqq k \leqq n. \tag{4}$$

Then

$$A_t(\bar{e}_k) = \frac{a_k}{1 - ta_k} \bar{e}_k \tag{5}$$

where

$$\bar{e}_k = e_k / (1 - ta_k) \tag{6}$$

are orthonormal relative to the metric g_t. Thus $a_k/(1 - ta_k)$ are the principal curvatures for the immersion f_t. Note that (5) and (6) are valid for $t < 1/\text{Max}\,|a_k|$.

PROPOSITION 1. *f_t has constant mean curvature for each t if and only if f has constant principal curvatures.*

PROOF. If f has constant principal curvatures, that is, $a_1, \cdots, a_n$ are constant, then the principal curvatures $a_k/(1 - ta_k)$ are constant on M^n for each t, and thus f_t has constant mean curvature. Conversely, assume that f_t has constant mean curvature for each t so that

$$\sum_{k=1}^{n} \frac{a_k(x)}{1 - ta_k(x)} = \varphi(t), \tag{7}$$

where we have indicated the principal curvatures $a_1, \cdots, a_n$ for f as functions on M^n; the right-hand side is simply a function of t. Evaluating $\varphi(t)$, $d\varphi/dt$, $d^2\varphi/dt^2$, $\cdots$, $d^n\varphi/dt^n$ at $t = 0$, we obtain

$$\Sigma\, a_k(x) = c_1,\ \Sigma\, a_k^2(x) = c_2, \cdots, \Sigma\, a_k^n(x) = c_n, \tag{8}$$

where $c_1, c_2, \cdots, c_n$ are constants. These relations imply that each of the functions $a_k(x)$ on M^n is a constant.

In the spherical case ($c = 1$), the formula corresponding to (3) takes the following form

$$A_t = (\cos tI - \sin tA)^{-1} (\cos tA + \sin tI) \tag{9}$$
$$= (\cot tI - A)^{-1} (\cot A + I).$$

Again, if $a_1, \cdots, a_n$ are the principal curvatures of f, the principal curvatures for the immersion f_t are given by

$$(\cos ta_k + 1)/(\cot t - a_k) = \cot (\theta_k - t), \tag{10}$$

where $a_k = \cot \theta_k$, $1 \leqq k \leqq n$.

Proposition 1 is still valid for a parallel family of hypersurfaces in S^{n+1} (with a similar proof). The situation for a parallel family of hypersurfaces in the hyperbolic space H^{n+1} can be discussed in a similar manner and Proposition 1 remains valid.

A geometric definition of an isoparametric family of hypersurfaces is then the following. *In $M^{n+1}(c)$, an isoparametric family of hypersurfaces is a family of parallel hypersurfaces $f_t : M^n \to M^{n+1}(c)$ obtained from a hypersurface $f : M^n \to M^{n+1}(c)$ with constant principal curvatures.*

2. Level hypersurfaces. An analytic definition of an isoparametric family of hypersurfaces is now explained. Let F be a differentiable function on a Riemannian manifold, say, $M^{n+1}(c)$, for our purpose. Using the classical notation, let

$$\Delta_1 F = \| dF \|^2 = \| \operatorname{Grad} F \|^2,$$

where $\| \quad \|$ denotes the length of the 1-form dF and the gradient vector field Grad F relative to the metric on $M^{n+1}(c)$, and let $\Delta_2 F =$ Laplacian of F. Let

$$M_s = \{ x \in M^{n+1}(c); F(x) = s \}, \tag{11}$$

which is a nonsingular hypersurface of $M^{n+1}(c)$ provided $\Delta_1 F$ never vanishes on M_s. Under this assumption,

$$\xi = \operatorname{Grad} F/(\Delta_1 F)^{1/2} \tag{12}$$

is a field of unit normals to the hypersurface M_s. The second fundamental form h for M_s is given by

$$h(X, Y) = - H_F(X, Y)/(\Delta_1 F)^{1/2}, \tag{13}$$

where H_F denotes the Hessian of the function F on $M^{n+1}(c)$. The mean curvature (= trace of h) is given by

$$((\operatorname{Grad} F)(\Delta_1 F)^{1/2} - (\Delta_1 F)^{1/2} \Delta_2 F)/\Delta_1 F. \tag{14}$$

Now suppose that $\Delta_1 F$ and $\Delta_2 F$ are functions of F, that is, there are functions Φ_1 and Φ_2 of a real variable such that

$$(\Delta_1 F)(x) = \Phi_1(F(x)) \text{ and } (\Delta_2 F)(x) = \Phi_2(F(x)) \tag{15}$$

for all $x \in M^{n+1}(c)$.

Under this assumption, (14) implies that each level hypersurface M_s has constant mean curvature. We shall give some simple examples.

EXAMPLE 1. On R^{n+1}, let

$$F(x_1, \cdots, x_{n+1}) = x_1^2 + \cdots + x_{k+1}^2,$$

where k is a fixed integer $0 \leqq k \leqq n$. We have

$$\Delta_1 F = 4F \quad \text{and} \quad \Delta_2 F = 2(k+1).$$

The level hypersurface, for $s > 0$,

$$M_s = \{ (x_1, \cdots, x_{n+1}) \in R^{n+1}; x_1^2 + \cdots + x_{k+1}^2 = s\}$$

is of the form $S^k(s^{1/2}) \times R^{n-k}$, where $S^k(s^{1/2})$ is the sphere of radius $s^{1/2}$ in the euclidean subspace R^{k+1} perpendicular to the Euclidean subspace R^{n-k}.

EXAMPLE 2. On S^{n+1} (which we consider as the unit sphere in R^{n+2} with rectangular coordinates $x_0, x_1, \cdots, x_{n+1}$), let F be the restriction of x_0 to S^{n+1}. Then

$$\Delta_1 F = 1 - F_2 \quad \text{and} \quad \Delta_2 F = -(n+1)F.$$

The level hypersurfaces

$$M_s = \{ x \in S^{n+1}; F(x) = s\}, \qquad -1 < s < 1,$$

are small n-spheres in S^{n+1} except for $s = 0$ which gives the great n-sphere.

EXAMPLE 3. Again on S^{n+1}, let

$$F(x) = x_0^2 + x_1^2 + \cdots + x_k^2, \qquad x \in S^{n+1},$$

where k is a fixed integer, $1 \leqq k \leqq n-1$. We have

$$\Delta_1 F = 4F(1-F) \quad \text{and} \quad \Delta_2 F = 2(k+1) - 2(n+2)F.$$

For each s, $0 < s < 1$, the level hypersurface

$$M_s = \{ x \in S^{n+1}; F(x) = s\}$$

is what is known as the product $S^k(s^{1/2}) \times S^{n-k}((1-s)^{1/2})$ imbedded as a hypersurface of S^{n+1}.

Going back to the general case, let F be a function satisfying condition (15). The vector field ξ in (12) is actually defined on $M^{n+1}(c)$ (that is, on the open subset on which $\Delta_1 F \neq 0$). We have

PROPOSITION 2. *Each integral curve of ξ is a geodesic in $M^{n+1}(c)$.*

PROOF. Denoting the metric and covariant differentiation on $M^{n+1}(c)$ by g and $\tilde{\nabla}$, we prove $\tilde{\nabla}_\xi \xi = 0$. Let X be a vector field perpendicular to ξ everywhere. Thus X is tangent to each level hypersurface so that $XF = 0$ and $\xi(XF) = 0$ everywhere. Since $\xi F = dF(\xi) = (\Delta_1 F)^{1/2}$ is a function of F, ξF is constant on each level hypersurface. Thus $X(\xi F) = 0$. We have hence $[X, \xi]F = 0$. Using $[X, \xi] = \tilde{\nabla}_X \xi - \tilde{\nabla}_\xi X$, we obtain

$$g(\tilde{\nabla}_X \xi, \xi) - g(\tilde{\nabla}_\xi X, \xi) = g([X, \xi], \xi) = 0.$$

But $g(\tilde{\nabla}_X \xi, \xi) = 0$ because $g(\xi, \xi) = 1$. Therefore we get $g(\tilde{\nabla}_\xi X, \xi) = 0$. On the other hand, $g(X, \xi) = 0$ implies $g(\tilde{\nabla}_\xi X, \xi) + g(X, \tilde{\nabla}_\xi \xi) = 0$. It follows that $g(\tilde{\nabla}_\xi \xi, X) = 0$. This shows that $\tilde{\nabla}_\xi \xi$ is perpendicular to every vector field X perpendicular to ξ. Because $g(\tilde{\nabla}_\xi \xi, \xi) = 0$, we conclude that $\tilde{\nabla}_\xi \xi = 0$.

By virtue of Proposition 2 we see that a family of level hypersurfaces defined by a function F satisfying condition (15) is essentially (that is, except for a parametrization) a parallel family of hypersurfaces obtained from a hypersurface with constant principal curvatures. In the special case where $\Delta_1 F = 1$, the level hypersurface M_{s+t} is obtained from M_s by moving every point by distance t on the geodesic in the direction of ξ.

Conversely, suppose M_t is a family of parallel hypersurfaces obtained from a hypersurface M_0 with constant principal curvatures. We may consider t as a differentiable function (on a certain open subset of $M^{n+1}(c)$) such that its gradient vector field is a unit vector field with the property that $\tilde{\nabla}_\xi \xi = 0$. Furthermore, since the mean curvature of each hypersurface M_t is constant together with its principal curvatures, the formula (14) shows that $\Delta_2 t$ is constant on each M_t, that is, $\Delta_2 t$ is a function of t. We have thus shown that the geometric and analytic definitions of isoparametric families coincide.

3. Cartan's results and problems. Let M^n be a hypersurface immersed in $M^{n+1}(c)$ and assume that it has constant principal curvatures. Let $a_1, \cdots, a_p$ be all the distinct constant principal curvatures with multiplicities $\nu_1, \cdots, \nu_p$, respectively.

In R^{n+1} the hypersurfaces in Example 1 of § 2 have two distinct principal curvatures except for the case where $k = 0$ (hyperplane) or $k = n$ (sphere). *There are no other hypersurfaces in R^{n+1} with constant principal curvatures.* This was proved by Levi-Civita for $n = 2$ and by B. Segre for an arbitrary dimension n. In the hyperbolic space H^{n+1}, Cartan showed that p is one or two. For the sphere S^{n+1}, however, he found far more examples, all geometrically interesting.

The starting point of Cartan's work is the following basic identity concerning $a_1, \cdots, a_p$, when $p \geqq 2$.

BASIC IDENTITY. *For each i, $1 \leqq i \leqq p$,*

$$\sum_{j \neq i} \nu_j \frac{c + a_i a_j}{a_i - a_j} = 0.$$

In the Euclidean case ($c = 0$), we can show that p is at most 2. Let a_1 be the smallest positive number among $a_1, \cdots, a_p$ and write down the identity above for $i = 1$. If $a_j > 0$ or $a_j < 0$, the summand is negative. Thus the only other principal curvature, if any, has to be 0. Thus $p \leqq 2$. If $p = 2$, one of the principal curvatures must be 0. This argument can be modified to yield the result $p \leqq 2$ (and $c + a_1 a_2 = 0$ if $p = 2$) in the case $c < 0$.

In S^{n+1} Cartan has shown the following results: If $p = 3$, then $n = 3, 6, 12$, or 24 and the multiplicities are the same in each case. If $p = 4$ and if the multiplicities are equal, then $n = 4$ or 8. For each case he gave an example of a hypersurface with constant principal curvatures.

In fact, one major part of Cartan's work is to give an algebraic method of finding in S^{n+1} an isoparametric family of hypersurfaces with p distinct principal curvatures with the same multiplicity ν (so $n = p\nu$). The result is that such a family is defined by

$$M_t = \{x \in S^{n+1};\ F(x) = \cos pt\},$$

where F is a harmonic homogeneous polynomial of degree p on R^{n+2} satisfying

$$\nabla_1 F = p^2 (x_0^2 + \cdots + x_{n+1}^2)^{p-1},$$

where $(x_0, \cdots, x_{n+1})$ are the rectangular coordinates on R^{n+2} and $\nabla_1 F = \| dF \|^2$ relative to the Euclidean space R^{n+2}.

An interesting corollary is that a piece of hypersurface with constant principal curvatures of the same multiplicity can be extended to a compact hypersurface with the same property, a fact which is far from obvious.

Cartan lists the following problems.

PROBLEM 1. *For each positive integer p, does there exist an isoparametric family with p distinct principal curvatures of the same multiplicity?*

PROBLEM 2. *Does there exist an isoparametric family with more than* 3 *distinct principal curvatures not of the same multiplicity?*

PROBLEM 3. *Does every isoparametric family of hypersurfaces admit a transitive group of isometries?*

The last problem seems to be a central one. All the compact hypersurfaces with constant principal curvatures in S^{n+1} given by Cartan are homogeneous; indeed, each of them is the orbit of a certain point by an appropriate closed subgroup of $SO(n + 2)$ acting on S^{n+1}. (It is obvious that such an orbit hypersurface has constant principal curvatures.) In a recent paper [7] T. Takahashi and R. Takagi determined all the orbit hypersurfaces in S^{n+1}. From their results we get the following.

For Problem 1, the answer is yes for $p = 6$ (in addition to $p = 1, 2, 3$ and 4 as Cartan knew). In fact, both S^7 and S^{13} have an isoparametric family with 6 distinct principal curvatures of the same multiplicity 1 and 2, respectively.

For Problem 2, the answer is affirmative. The list in [7] includes 5 different classes of orbit hypersurfaces each with 4 distinct principal curvatures not of the same multiplicity.

Before [7] came to my attention, I had an example for the affirmative answer to Problem 2. In the next section I shall describe a class of hypersurfaces with constant principal curvatures from the analytic point of view given in §2.

4. Isoparametric family M_t^{2n} in S^{2n+1}. Let R^{n+1} be the $(n + 1)$-dimensional real vector space with usual inner product $\langle x, y\rangle$. If we consider the $(n + 1)$-dimensional complex vector space C^{n+1} as a real vector space, we have $C^{n+1} = R^{n+1} + iR^{n+1}$. The real inner product in C^{n+1} is given by

$$\langle z, w\rangle = \langle x, u\rangle + \langle y, v\rangle$$

for $z = x + iy$, $w = u + iv$, where $x, y, u, v \in R^{n+1}$. Let

$$S^{2n+1} = \{z \in C^{n+1};\ \| z \| = 1\}$$

and consider a function

$$F(z) = (\| x \|^2 - \| y \|^2)^2 + 4 \langle x, y\rangle^2 \quad \text{for } z = x + iy$$

on S^{2n+1}. We find

$$\Delta_1 F = 16F(1 - F) \quad \text{and} \quad \Delta_2 F = 16(1 - 2F),$$

that is, F on S^{2n+1} satisfies condition (15) in §2. For each t, $0 < t < \pi / 4$, let

$$M_t^{2n} = \{z \in S^{2n+1};\ F(z) = \cos^2 2t\}.$$

According to a general result stated in §2, M_t^{2n} is an isoparametric family of hypersurfaces in S^{2n+1}. The principal curvatures of the hypersurface M_t^{2n} for a fixed t can be computed; they are $\cot(-t)$ of multiplicity $n - 1$, $(1 + \sin 2t)/\cos 2t = \cot(-t + \pi/4)$ of multiplicity 1, $\cot(-t + \pi/2)$ of multiplicity $n - 1$, and $(-1 + \sin 2t)/\cos 2t = \cot(-t + 3\pi/4)$ of multiplicity 1.

The computation of the principal curvatures is based on the following observations. Let S^1 be the unit circle $\{e^{i\theta}\}$ and let $S_{n+1,2}$ be the Stiefel manifold consisting of all orthonormal pairs of vectors (x, y) in R^{n+1}. Let f be the mapping of $S^1 \times S_{n+1,2}$ into S^{2n+1} defined by

$$f(e^{i\theta}, (x,y)) = e^{i\theta/2} (\cos tx + i \sin ty),$$

where t is fixed. It is easily verified that f is an imbedding and its image is precisely the hypersurface M_t^{2n} in S^{2n+1}.

Let us note that the hypersurface M_t^4 in S^5 (i.e. $n = 2$) is one of the examples given by Cartan.

5. Focal varieties. Another interesting notion that appears in Cartan's work is that of focal varieties for an isoparametric family. For an arbitrary hypersurface M^n in S^{n+1}, let ξ be a field of unit normal vectors. A point y_0 of S^{n+1} is called a focal point of (M, x_0), where $x_0 \in M^n$, if

$$y_0 = \cos t_0\, x_0 + \sin t_0\, \xi_{x_0} \tag{16}$$

for some t_0 and if the differential of the mapping

$$(x, t) \to y = \cos tx + \sin t\xi_x$$

is singular at (x_0, t_0). A point y_0 is called a focal point of M if it is a focal point of (M, x_0) for some $x_0 \in M^n$.

It is easily verified that (16) is a focal point of (M^n, x_0) if and only if $\cot t_0$ is one of the principal curvatures (for the field ξ). This fact is related to the expression (10).

Now let us assume that M^n has distinct constant principal curvatures $a_1, \cdots, a_p$ with multiplicities $\nu_1, \cdots, \nu_p$ respectively. For each i, $1 \leqq i \leqq p$, let

(17) $$T_i(x) = \{X \in T_x(M^n);\ AX = a_i X\}, \qquad x \in M^n.$$

The distribution T_i: $x \to T_i(x)$ of dimension ν_i is integrable and the maximal integral manifold $M_i(x)$ of T_i through x is totally geodesic in M^n and umbilical as a submanifold of S^{n+1}. (If M^n is connected and compact, $M_i(x)$ is thus a ν_i-dimensional small sphere of S^{n+1}.)

We define a differentiable mapping $f_i : M^n \to S^{n+1}$ by

(18) $$f_i(x) = \cos\theta_i x + \sin\theta_i \xi x,$$

where $a_i = \cot\theta_i$ as in (10). The differential of f_i at x has $T_i(x)$ as the null space and is injective on the subspace $\sum_{k\neq i} T_k(x)$ of $T_x(M)$. It follows that for each point x there is a local coordinate system $u^1, \cdots, u^n$ with origin x such that f_i is a one-to-one immersion of the slice $u^1 = \cdots = u^{\nu_i} = 0$ into S^{n+1}, the image being contained in $f_i(M^n)$. In this way we may consider $f_i(M^n)$ as a submanifold of dimension $n - \nu_i$ in a neighborhood of $f_i(x)$. For each i, $1 \leqq i \leqq p$, $f_i(M^n)$ is called a *focal variety* of M^n. Geometrically, the mapping f_i maps $M_i(x)$ into one single point $f_i(x)$ and $f_i(M^n)$ collapses by dimension ν_i.

If M_t^n is the parallel family arising from the hypersurface $M^n = M_0^n$, then we may speak of the focal varieties of this isoparametric family M_t^n in the following sense. Let $x = x_0$ be a point of M^n and let x_t be the geodesic starting from x_0 in the direction normal to M^n (so the point x_{t_1} is a point of $M_{t_1}^n$). If $a_i = \cot\theta_i$ is one of the principal curvatures of M^n at x, then the corresponding principal curvature of $M_{t_1}^n$ at the point x_{t_1} is $\cot(\theta_i - t_1)$ as we know from formula (10). The geodesic distance from x_0 to $f_i(x_0)$ is θ_i and the geodesic distance from x_{t_1} to $f_i(x_0)$ is $\theta_i - t_1$. This means that the focal variety of $M_{t_1}^n$ corresponding to the principal curvature $\cot(\theta_i - t_1)$ coincides with the focal variety of M^n corresponding to $\cot\theta_i$.

The discussions above are local. If M^n is compact and connected, one can indeed give a global submanifold structure to each focal variety $f_i(M^n)$. The foliation T_i is regular in the sense of Palais [**6**]. The space of leaves, say, V_i, admits a structure of differentiable manifold of dimension $n - \nu_i$ by a result in [**6**]. From this one can state [**5**]

THEOREM. *Let M^n be a connected compact hypersurface in S^{n+1} with constant principal curvatures. Each focal variety $f_i(M^n)$ is a submanifold of S^{n+1} in the following sense: there exist an $(n - \nu_i)$-dimensional differentiable manifold V_i, a differentiable mapping π_i of M onto V_i, and a differentiable imbedding g_i of V_i into S^{n+1} such that $f_i = g_i \pi_i$ and thus $f_i(M^n) = g_i(V_i)$.*

Let us examine the focal varieties of the isoparametric family M_t^{2n} in S^{2n+1} given in §4. We know that for a fixed t, $0 < t < \pi/4$, the hypersurface M_t^{2n} consists of all points of S^{2n+1} of the form

(19) $$x_t = e^{i\theta}(\cos tx + i\sin ty),$$

where θ is arbitrary and $(x, y) \in S_{n+1,2}$.

We fix θ, x and y and consider x_t in (19) as a curve as t varies. The curve x_t is a geodesic in S^{2n+1} normal to each hypersurface M_t^{2n} for $0 < t < \pi/4$. If we take $t = 0$, the point $x_0 = e^{i\theta}x$ is of distance $-t$ from the point x_t. Thus the focal variety corresponding to the principal curvature $\cot(-t)$ of multiplicity $n - 1$ consists of all points in S^{2n+1} of the form $e^{i\theta}x$, where θ is arbitrary and $x \in S^n$ (the sphere consisting of real unit vectors in R^{n+1}). We also find that this focal variety of dimension $n + 1$ coincides with

(20) $$M_0 = \{z \in S^{2n+1}; F(z) = 1\}$$

and that it is doubly covered by $S^1 \times S^n$.

If we now let $t = \pi/4$ in (19), we get the point $x_{\pi/4} = e^{i\theta}(x + iy)/2^{1/2}$ of distance $\pi/4 - t$ from the point x_t. Thus the focal variety corresponding to the principal curvature $\cot(-t + \pi/4)$ of multiplicity 1 consists of all points of the form $e^{i\theta}(x + iy)/2^{1/2}$, where θ is arbitrary and $(x, y) \in S_{n+1,2}$. Actually, it is the set of all points of the form $(x + iy)/2^{1/2}$, where $(x, y) \in S_{n+1,2}$ and coincides with

(21) $$M_{\pi/4} = \{z \in S^{2n+1}; F(z) = 0\}.$$

This focal variety of dimension $2n - 1$ is doubly covered by the Stiefel manifold $S_{n+1,2}$.

The focal varieties for the principal curvatures $\cot(-t + \pi/2)$ and $\cot(-t + 3\pi/4)$ coincide with (20) and (21), respectively.

6. Minimality of focal varieties and other questions. In [5] I announced the following result with a brief indication of the proof.

THEOREM. *Each focal variety of an isoparametric family M_t^n in S^{n+1} is a minimal submanifold in S^{n+1}.*

It seems that the question of minimality was not touched upon by Cartan. The Veronese surface in S^4, which is a notable example of minimal submanifolds, appears as a focal variety of Cartan's example M_t^3 in S^4 which is defined by the equation

$$\cos 3t = x_5^3 + \frac{3}{2}(x_1^2 + x_2^2)x_5 - 3(x_3^2 + x_4^2)x_5 + \frac{3\cdot 3^{1/2}}{2}(x_1^2 - x_2^2)x_4 + 3\cdot 3^{1/2}x_1x_2x_3,$$

where $(x_1, x_2, x_3, x_4, x_5)$ are rectangular coordinates in R^5. The Veronese surface is parametrized by

$$x_1 = 3^{1/2}vw, \quad x_2 = 3^{1/2}wu, \quad x_3 = 3^{1/2}uv,$$
$$x_4 = \frac{3^{1/2}}{2}(u^2 - v^2), \quad x_5 = w^2 - \frac{u^2 + v^2}{2},$$

where $(u, v, w) \in S^2$.

In the proof of the theorem, the basic identity of Cartan in §3 plays an important role. In fact, this identity is equivalent to the vanishing of the trace of the second fundamental form of each focal variety in the normal direction given by the geodesic $\cos t\, x + \sin t\, \xi_x$, $x \in M^n$, coming into the focal variety.

As for the proof of the basic identity, it is essentially based on the equations of Codazzi and Gauss, although it is somewhat intricate. Cartan posed the following question.

PROBLEM 4. *Is there a geometric proof of the basic identity*?

In view of the remark above, one might ask if minimality of focal varieties can be proved without appealing to the basic identity.

Note (added on October 11, 1973). After the lecture was given at Stanford, the following recent work has come to my attention:

(1) H. F. Münzner (*Isoparametrische Hyperflächen in Sphären*, to appear in Math. Ann.) has proved that the number of distinct principal curvatures of an isoparametric family can be only 1, 2, 3, 4 or 6. Thus Problem 1 has been answered negatively (even without assuming that the multiplicities are equal).

(2) Using an idea due to Münzner, one can now prove minimality of focal varieties and Cartan's basic identity at the same time. Thus Problem 4 is solved in a certain sense.

(3) R. Takagi has communicated to me that the hypersurface M_t^{2n} in §4 can be characterized as a connected, compact, orientable hypersurface in S^{2n+1} having 4 distinct constant principal curvatures with multiplicities $1, n-1, 1$ and $n-1$.

(4) J. Eells, Ted Smith and R. Wood have pointed out a close relationship of Cartan's work to the problem of polynomial and harmonic maps from S^n into S^r.

REFERENCES

1. E. Cartan, *Familles de surfaces isoparamétriques dans les espaces à courbure constante*, Ann. Mat. **17** (1938), 177–191.

2. ———, *Sur des familles remarquables d'hypersurfaces isoparamétriques dans les espaces sphériques*, Math. Z. **45** (1939), 335–367. MR **1**, 28.

3. ———, *Sur quelques familles remarquables d'hypersurfaces*, C. R. Congrès Math. Liège, 1939, 30–41.

4. ———, *Sur des familles d'hypersurfaces isoparamétriques des espaces sphériques à 5 et à 9 dimensions*, Univ. Nac. Tucumán Revista A **1** (1940), 5–22. MR **3**, 18.

5. K. Nomizu, *Some results in E. Cartan's theory of isoparametric families of hypersurfaces*, Bull. Amer. Math. Soc. **79** (1973), 1184–1188.

6. R.S. Palais, *A global formulation of the Lie theory of transformation groups*, Mem. Amer. Math. Soc. No. 22 (1957). MR **22** #12162.

7. R. Takagi and T. Takahashi, *On the principal curvatures of homogeneous hypersurfaces in a sphere*, Differential Geometry, in Honor of K. Yano, Kinokuniya, Tokyo, 1972, pp. 469–481.

BROWN UNIVERSITY

UNIVERSITÄT BONN

Proceedings of Symposia in Pure Mathematics
Volume 27, 1975

ON KAEHLER SUBMANIFOLDS

KOICHI OGIUE

A Kaehler manifold of constant holomorphic sectional curvature is called a *complex space form*. There are three types of complex space forms: elliptic, hyperbolic or flat according as the holomorphic sectional curvature is positive, negative or zero.

Let $P_m(C)$ be an m-dimensional complex projective space with the Fubini-Study metric of constant holomorphic sectional curvature 1. Then $P_m(C)$ is a complete and simply connected elliptic complex space form.

Complex Euclidean space C^m with the usual Hermitian metric is a complete and simply connected flat complex space form.

Let D_m be the open unit ball in C^m with the natural complex structure and the Bergman metric of constant holomorphic sectional curvature -1. Then D_m is a complete and simply connected hyperbolic complex space form.

Any m-dimensional complex space form is (after multiplying the metric by a suitable constant) locally complex analytically isometric to $P_m(C)$, C^m or D_m, according as the holomorphic sectional curvature is positive, zero or negative.

By a *Kaehler submanifold* we mean a complex submanifold with the induced Kaehler structure.

In this report we consider some properties of Kaehler submanifolds of complex space forms.

We use the following notation:

$M_n(c)$: n-dimensional complex space form of constant holomorphic sectional curvature c (no topological conditions assumed unless otherwise stated),

K: the sectional curvature of submanifold,

AMS (MOS) subject classifications (1970). Primary 53C40, 53C55; Secondary 14M10.

H: the holomorphic sectional curvature of submanifold,
S: the Ricci tensor of submanifold,
p: the scalar curvature of submanifold,
g: the Kaehler metric of submanifold.

1. Problems. There have been many problems in the theory of Kaehler submanifolds, among which we confine our attention to the following.

(I) Let M be a complete Kaehler submanifold immersed in $P_m(C)$. If $H > \frac{1}{2}$, is M totally geodesic in $P_m(C)$?

(II) Let M_n be an n-dimensional complete Kaehler submanifold immersed in $P_{n+p}(C)$. If $K > 0$ and if $p < n(n+1)/2$, is M_n totally geodesic in $P_{n+p}(C)$?

(III) Let M_n be an n-dimensional complete Kaehler submanifold immersed in $P_m(C)$. If every Ricci curvature is $> n/2$, is M_n totally geodesic in $P_m(C)$?

(IV) Let M_n be an n-dimensional compact Kaehler submanifold immersed in $P_m(C)$. If $\rho > n^2$, is M_n totally geodesic in $P_m(C)$?

(V) Let $M_n(c)$ be a Kaehler submanifold immersed in $M_m(\tilde{c})$. If the immersion is full (i.e., $M_n(c)$ cannot be immersed in any proper totally geodesic submanifold of $M_m(\tilde{c})$), is the following true?

(i) $\tilde{c} = \nu c$,

(ii) $m = \binom{n+\nu}{\nu} - 1$,

for some positive integer ν.

2. Examples in this direction. (1) If $P_n(C)$ is imbedded in $P_m(C)$ as a linear subspace, then $P_n(C)$ is totally geodesic in $P_m(C)$ as a Kaehler submanifold.

(2) Let $z_0, z_1, \cdots, z_{n+1}$ be a homogeneous coordinate in $P_{n+1}(C)$ and let $Q_n(C) = \{(z_0, z_1, \cdots, z_{n+1}) \in P_{n+1}(C) \mid \sum z_i^2 = 0\}$. Then, as a Kaehler submanifold, $Q_n(C)$ satisfies $0 \leqq K \leqq 1$, $\frac{1}{2} \leqq H \leqq 1$, $S = (n/2)g$, $\rho = n^2$.

(3) An n-dimensional complex projective space of constant holomorphic sectional curvature c can be imbedded as a Kaehler submanifold into an $\{\binom{n+\nu}{\nu} - 1\}$-dimensional complex projective space of constant holomorphic sectional curvature νc.

(4) $P_n(C) \times P_m(C)$ can be imbedded as a Kaehler submanifold into $P_{n+m+nm}(C)$ and it satisfies $0 \leqq K \leqq 1$, $\frac{1}{2} \leqq H \leqq 1$, $\rho = n(n+1) + m(m+1)$.

3. Results in this direction. Using the vanishing theorem of Kodaira, Kobayashi and Ochiai [2] proved an outstanding result, as a corollary to which we first state the following.

THEOREM 1 [2]. *Let M_n be an n-dimensional compact Kaehler submanifold imbedded in $P_m(C)$ as a complete intersection. If $H > \frac{1}{2}$ or if $K > 0$ and $n \geqq 2$, then M_n is totally geodesic in $P_m(C)$.*

Theorem 1 can be considered as partial solutions to Problems (I) and (II).

For a general Kaehler submanifold, we have the following result as a partial solution to (I).

THEOREM 2 [5]. *Let M_n be an n-dimensional complete Kaehler submanifold im-*

mersed in $P_{n+p}(C)$. *If* $H > 1 - (n+2)/2(n+2p)$, *then* M_n *is totally geodesic in* $P_{n+p}(C)$.

Theorem 2 is best possible in the case $p = 1$.

The lower bound for H in Theorem 2 can be improved by imposing some additional assumption, for example, the following can also be considered as a partial solution to (I).

THEOREM 3 **[5]**. *Let* M *be a complete Kaehler submanifold immersed in* $P_m(C)$. *If* $H > \frac{1}{2}$ *and if* ρ *is constant, then* M *is totally geodesic in* $P_m(C)$.

The following is a result of the same type as Theorem 3 and is a partial solution to (II).

THEOREM 4 **[5]**. *Let* M_n *be an n-dimensional complete Kaehler submanifold immersed in* $P_{n+p}(C)$. *If* $K > 0$, ρ *is constant and if* $p < n(n+1)/2$, *then* M_n *is totally geodesic in* $P_{n+p}(C)$.

In the case of hypersurfaces, as a matter of course, the result is simpler, that is,

THEOREM 5 **[5]**. *Let* M *be a complete Kaehler hypersurface immersed in* $P_{n+1}(C)$. *If* $K > 0$ *and if* $n \geqq 4$, *then* M *is totally geodesic in* $P_{n+1}(C)$.

As for problem (III) we have a complete solution.

THEOREM 6 **[5]**. *Let* M_n *be an n-dimensional complete Kaehler submanifold immersed in* $P_m(C)$. *If* $S > (n/2)g$, *then* M_n *is totally geodesic in* $P_m(C)$.

As an immediate consequence of Theorem 6 we have the following result which gives the best possible solution to (I) in the case of complex curves.

COROLLARY 7 **[5]**. *Let* M *be a complete complex curve immersed in* $P_m(C)$. *If* $K > \frac{1}{2}$, *then* M *is totally geodesic in* $P_m(C)$ *(i.e.,* M *is a projective line).*

The special case of this result where $m = 2$ and M is nonsingular (or imbedded) was proved by Nomizu and Smyth **[4]**.

The theorem of Gauss-Bonnet gives a relation between curvature (differential geometric invariant) and the Euler number (topological invariant). The following result is of Gauss-Bonnet type in the sense that it provides a relation between differential geometric invariant and more primitive invariant.

THEOREM 8 **[5]**. *Let* M_n *be an n-dimensional compact Kaehler submanifold imbedded in* $P_{n+p}(C)$. *If* M_n *is a complete intersection of p hypersurfaces of degree* $a_1, \cdots, a_p$ *in* $P_{n+p}(C)$, *then*

$$\begin{aligned}\int_M \rho * 1 &= n(n+p+1-\textstyle\sum a_\alpha)\displaystyle\int_M * 1\\ &= n(n+p+1-\textstyle\sum a_\alpha)(\Pi a_\alpha)\dfrac{(4\pi)^n}{n!}.\end{aligned}$$

This result can be proved by using the properties of the 1st Chern class and some elementary facts in harmonic integral theory.

The following result is an immediate consequence of Theorem 8, which provides a partial solution to (IV).

COROLLARY 9 [5]. *Let M_n be an n-dimensional compact Kaehler submanifold imbedded in $P_m(C)$. If M_n is a complete intersection and if $\rho > n^2$, then M_n is totally geodesic in $P_m(C)$.*

Theorem 8 implies that the integral of the scalar curvature depends only on the degree of M. But the scalar curvature itself depends wholly on the equation defining M. In fact,

THEOREM 10 [5]. *Let M be a compact Kaehler hypersurface of $P_{n+1}(C)$ defined by a homogeneous equation $F(z_0, z_1, \cdots, z_{n+1}) = 0$. Then*

$$\rho = n(n+1) - \left(\sum z_i \bar{z}_i\right)\left\{\frac{\operatorname{tr} A\bar{A}}{{}^t\bar{\mathfrak{U}}\mathfrak{U}} - 2\frac{{}^t(\bar{A}\mathfrak{U})(A\bar{\mathfrak{U}})}{({}^t\bar{\mathfrak{U}}\mathfrak{U})^2} + \frac{({}^t\bar{\mathfrak{U}}A\bar{\mathfrak{U}})({}^t\mathfrak{U}\bar{A}\mathfrak{U})}{({}^t\bar{\mathfrak{U}}\mathfrak{U})^3}\right\},$$

where $\mathfrak{U} = (\partial F/\partial z_i)$ and $A = (\partial^2 F/\partial z_i \partial z_j)$.

EXAMPLE. Let $M = \{(z_0, z_1, \cdots, z_{n+1}) \in P_{n+1}(C) \mid z_0^2 + z_1^2 + \cdots + z_n^2 + az_{n+1}^2 = 0\}$. Then Theorem 8 and Theorem 10 imply

$$n^2 + 1 - a^2 \leqq \rho \leqq n(n+1) - n/a \qquad (a \geqq 1),$$
$$n(n+1) - n/a \leqq \rho \leqq n^2 + 1 - a^2 \qquad (0 < a \leqq 1),$$

$$\int_M \rho * 1 = n^2 \int_M * 1 = n^2 \frac{2(4\pi)^n}{n!} \quad \textit{(which does not depend on a)}.$$

Corollary 9 states that problem (IV) is affirmative for some special class of Kaehler submanifolds. The following result due to S. Tanno gives a partial solution to (IV) in general case.

THEOREM 11 [6]. *Let M_n be a compact Kaehler submanifold immersed in $P_m(C)$. If $\rho > n(n+1) - (n+2)/3$, then M_n is totally geodesic in $P_m(C)$.*

Concerning problem (V), there is an outstanding result due to E. Calabi.

THEOREM 12 [1]. *Let $M_n(c)$ and $M_m(\tilde{c})$ be complete and simply connected complex space forms. Then $M_n(c)$ is imbedded as a Kaehler submanifold in $M_m(\tilde{c})$ and the imbedding is full if and only if*

(i) $\tilde{c} = \nu c$,

(ii) $m = \binom{n+\nu}{\nu} - 1$,

for some positive integer ν.

Using a purely differential-geometric method, Nakagawa and Ogiue obtained the local version of Theorem 12, which provides a complete solution to problem (V).

THEOREM 13 [3]. *If $M_n(c)$ is a Kaehler submanifold immersed in $M_m(\tilde{c})$ and if the immersion is full, then*

(i) $\tilde{c} = \nu c$,

(ii) $m = \binom{n+\nu}{\nu} - 1$,

for some positive integer ν.

Bibliography

1. E. Calabi, *Isometric imbedding of complex manifolds*, Ann. of Math. (2) **58** (1953), 1-23. MR **15**, 160.

2. S. Kobayashi and T. Ochiai, *On complex manifolds with positive tangent bundles*, J. Math. Soc. Japan **22** (1970), 499–525. MR **43** #1231.

3. H. Nakagawa and K. Ogiue, *Complex space forms immersed in complex space forms*, Trans. Amer. Math Soc. (to appear).

4. K. Nomizu and B. Smyth, *Differential geometry of complex hypersurfaces*. II, J. Math. Soc. Japan **20** (1968), 498–521. MR **37** #5827.

5. K. Ogiue, *Differential geometry of Kaehler submanifolds*, Advances in Math. **13** (1974), 73–114.

6. S. Tanno, *Compact complex submanifolds immersed in complex projective spaces*, J. Differential Geometry **8** (1973), 629–641.

MICHIGAN STATE UNIVERSITY

Proceedings of Symposia in Pure Mathematics
Volume 27, 1975

ISOPERIMETRIC AND RELATED INEQUALITIES

ROBERT OSSERMAN

I. Isoperimetric inequalities. For a plane domain D bounded by a curve C, the *isoperimetric inequality* states that the area A of D and the length L of C are related by

$$L^2 \geqq 4\pi A, \tag{1}$$

with equality holding if and only if D is a circular disk.

The isoperimetric inequality has many applications. For example, it allows a partial answer to the question: "Can you hear the shape of a drum?", in the form: "You can hear the shape of a circular drum." Namely, from the set of eigenvalues of the Laplacian on a plane domain D, one can determine both its area A and the length L of its boundary. The domain is circular if and only if $L^2 = 4\pi A$.

The first problem we wish to consider is the following:

Question 1. For what class of domains D on what class of surfaces does the isoperimetric inequality (1) continue to hold?

First of all, let M be a two-dimensional Riemannian manifold, p an arbitrary point of M, D_r the geodesic disk of radius r centered at p, and C_r the geodesic circle bounding D_r. For small r, the area $A(r)$ of D_r and length $L(r)$ of C_r have the asymptotic expansions

$$A(r) = \pi r^2 - \frac{\pi}{12} K r^4 + O(r^5),$$

$$L(r) = 2\pi r - \frac{\pi}{3} K r^3 + O(r^4),$$

AMS (MOS) subject classifications (1970). Primary 53A05, 53A05, 53C40; Secondary 49F20, 49F25, 53C20.

where K is the Gauss curvature of M at p. It follows that

$$L(r)^2 - 4\pi A(r) = -\pi^2 K r^4 + O(r^5)$$

and as a consequence

$$K = -\lim \frac{L(r)^2 - 4\pi A(r)}{(\pi r^2)^2}.$$

We deduce that

$$L^2 \geqq 4\pi A \quad \text{for all } D \subset M \Rightarrow K \leqq 0 \text{ on } M. \tag{2}$$

Thus we have as part of the answer to Question 1, that in order for the isoperimetric inequality (1) to hold for every domain on M, M must have nonpositive Gaussian curvature.

A partial converse to (2) was obtained by Beckenbach and Radó **[5]**.

THEOREM 1.1. *If $K \leqq 0$ on M, then* (1) *holds for every simply-connected domain D bounded by a single curve C.*

To see that some hypothesis on the topology of D is needed, note the following examples.

EXAMPLE 1. If D is a doubly-connected domain on a circular cylinder M, bounded by two parallel circles, then the area of D can be made arbitrarily large without increasing the length of the boundary circles, so that (1) will fail to hold even though $K \equiv 0$ on M.

EXAMPLE 2. If M is a flat torus, and D is the complement of a small disk, the length L of the boundary can be made arbitrarily small, and (1) will fail. This surface can also be realized as the Clifford torus embedded in $\mathbf{R}^4$.

The proof of Beckenbach and Radó uses potential theory, generalizing an earlier result of Carleman **[6]** who proved Theorem 1.1 in the special case that M is a minimal surface. It turns out that for minimal surfaces a stronger result is true.

THEOREM 1.2 (W.T. REID **[17]** FOR $n = 3$; C.C. HSIUNG **[10]** FOR n ARBITRARY). *Let M be a minimal surface in $\mathbf{R}^n$, D a domain on M bounded by a single curve C. Then the isoperimetric inequality* (1) *holds for D.*

Actually the theorem of Reid and Hsiung is much more general, giving an inequality for an arbitrary surface M in $\mathbf{R}^n$ provided the domain D is bounded by a single curve C. Namely, if H is the mean curvature vector of M, and if x is the position vector, then for any point $a \in C$, we have

$$L^2 \geqq 4\pi A + 2\pi \iint_D (x - a) \cdot H \, dA. \tag{3}$$

This formula was later generalized by Kaul **[12]** to surfaces in a Riemannian manifold. We give here a proof adapted from that of Kaul.

PROOF OF THEOREM 1.2. We start with a general formula for area:

(4) $$2A = -\iint_D (x-a)\cdot H\, dA + \int_C (x-a)\cdot \nu\, ds$$

where $a \in \boldsymbol{R}^n$ is arbitrary, and ν is the unit exterior normal to D; that is, ν is a unit tangent vector to M, normal to C, directed outwards from D. This formula is a special case of the formula for the first variation of area, using a radial variation outward from the point a. Clearly (3) follows immediately from (4) if we can show

(5) $$2\pi \int_C (x-a)\cdot \nu\, ds \leqq L^2.$$

To prove this inequality, let $x(s)$ be the parametrization of C with respect to arc length, where $x(0) = x(L) = a$. Next define at each point of C a linear transformation $B : \boldsymbol{R}^n \to \boldsymbol{R}^n$ having the following properties:

1. B is a rotation through $\pi/2$ on the tangent space to M, with $B\nu = x'$, the unit tangent vector to C;
2. B is zero on the normal space to M.

From these two properties follow immediately:

3. $|BV| \leqq |V|$ for any vector V;
4. $U\cdot BV = -V\cdot BU$ for any vectors U, V.

Let $E_1, \cdots, E_n$ be vector fields along C satisfying the linear equation

(6) $$E_k' = (\pi/L)\, BE_k,$$

where the prime denotes differentiation with respect to arc length s, and the initial vectors $E_1(0), \cdots, E_n(0)$ are orthonormal. The skew-symmetry of B, property 4 above, then implies that $E_1, \cdots, E_n$ are orthonormal at each point. If we expand $x-a$ in terms of this basis, say $x - a = \sum c_k E_k$, then differentiating,

$$x' = \sum c_k' E_k + \sum c_k E_k',$$

and using (6), we may compute that

(7) $$\frac{2\pi}{L} x'\cdot B(x-a) = |x'|^2 - \sum\left[(c_k')^2 - \frac{\pi^2}{L^2} c_k^2\right] - [|x-a|^2 - |B(x-a)|^2].$$

Property 1 of B implies that $Bx' = -\nu$, and then by property 4,

$$(x-a)\cdot\nu = -(x-a)\cdot Bx' = x'\cdot B(x-a).$$

Using (7), and the fact that $|x'| \equiv 1$, we find

(8) $$\begin{aligned} 2\pi \int_C (x-a)\cdot\nu\, ds = L^2 &- L\int [|x-a|^2 - |B(x-a)|^2]\, ds \\ &- L\int \sum\left[(c_k')^2 - \frac{\pi^2}{L^2} c_k^2\right] ds. \end{aligned}$$

The fact that $x(0) = a$ gives $c_k(0) = 0$, $k = 1, \cdots, n$. We may then define

$$b_k(s) = c_k(s)/\sin(\pi s/L) \tag{9}$$

and using the identity

$$(c_k')^2 - \frac{\pi^2}{L^2} c_k^2 = (b_k')^2 \sin^2 \frac{\pi s}{L} + \frac{\pi}{L} \frac{d}{ds}\left(c_k^2 \cot \frac{\pi s}{L}\right),$$

we may write (8) in the form

$$\begin{aligned} &L^2 - 2\pi \int_C (x - a) \cdot \nu \, ds \\ &\quad = L\left[\int_0^L \sum (b_k')^2 \sin^2 \frac{\pi s}{L} ds + \int_C [|x - a|^2 - |B(x - a)|^2] \, ds\right]. \end{aligned} \tag{10}$$

Using property 3 of B, we see that both terms on the right are nonnegative; thus (5) is established, proving the theorem.

CONJECTURE 1. The isoperimetric inequality, $L^2 \geqq 4\pi A$, holds for an arbitrary domain on a minimal surface in $\boldsymbol{R}^n$.

We shall review what is known concerning this conjecture.

First of all suppose D is a domain whose boundary C is the union of m curves C_j. Choose points $a_j \in C_j$, and let L_j be the length of C_j. Then equation (10) says

$$L_j^2 - 2\pi \int_{C_j} (x - a_j) \cdot \nu \, ds = M_j \geqq 0, \tag{11}$$

where M_j denotes the corresponding right-hand side of (10). But

$$\int_C x \cdot \nu \, ds = \sum \int_{C_j} (x - a_j) \cdot \nu \, ds + \sum \int_{C_j} a_j \cdot \nu \, ds$$

and combining (4) with (11) gives

$$\sum L_j^2 - 4\pi A = 2\pi \iint_D x \cdot H \, dA + \sum M_j - 2\pi \sum a_j \cdot \int_{C_j} \nu \, ds. \tag{12}$$

Note that $L^2 = (\sum L_j)^2 = \sum L_j^2 + 2 \sum_{i<j} L_i L_j$ so that the inequality

$$\sum L_j^2 \geqq 4\pi A \tag{13}$$

would be strictly stronger than $L^2 \geqq 4\pi A$ for domains with more than one boundary component. In fact, inequality (13) does hold for area-minimizing surfaces. Namely, if the Douglas solution to the Plateau problem for the curve C_j has area A_j, then it follows from Theorem 1.2 that $L_j^2 \geqq 4\pi A_j$. But by the area-minimizing property, $4\pi A \leqq \sum 4\pi A_j \leqq \sum L_j^2$. Thus Conjecture 1 holds in an even stronger form for area-minimizing surfaces.

On the other hand, inequality (13) is not true in general. For example, if one considers a doubly-connected domain on a catenoid bounded by parallel circles, then when the circles are far enough apart, the area A of the catenoid will be greater than the sum of the areas A_j of the disks bounded by the separate circles.

But for each of those disks we have $4\pi A_j = L_j^2$. Hence $4\pi A > 4\pi(A_1 + A_2) = L_1^2 + L_2^2$ and (13) fails to hold.

Returning to equation (12), we see that for minimal surfaces the first term on the right vanishes, and

$$L^2 \geqq 4\pi A \Leftrightarrow 2 \sum_{i<j} L_i L_j + \sum M_j \geqq 2\pi \sum a_j \cdot \int_{C_j} \nu \, ds. \tag{14}$$

Recall that the a_j are arbitrary points on C_j, and the problem would be to choose them so that (14) is satisfied. (M_j also depends on the choice of a_j.)

A useful observation is the following. Applying (4) to a minimal surface with $C = UC_j$ yields

$$a \cdot \int_C \nu \, ds = \int_C x \cdot \nu \, ds - 2A,$$

for any a. Since the right-hand side is independent of a, it follows that $\sum \int_{C_j} \nu \, ds = \int_C \nu \, ds = 0$. In particular, for doubly-connected minimal surfaces, (14) becomes

$$2L_1 L_2 + M_1 + M_2 \geqq 2\pi(a_2 - a_1) \cdot \int_{C_2} \nu \, ds. \tag{15}$$

One can think of various geometric conditions under which the right-hand side of (15) can be made to vanish for suitable choice of $a_j \in C_j$ (for example, if no plane separates C_1 from C_2) in which case $L^2 \geqq 4\pi A$, but so far (15) has not been established in general.

Using similar methods, first Nitsche [**15**], and later Kaul [**13**] proved that there exists a constant $c > 0$, such that the inequality

$$L^2 \geqq cA \tag{16}$$

holds for every doubly-connected minimal surface. Actually an inequality of the form (16) holds for *arbitrary* domains on a minimal surface. This was proved by the methods of geometric measure theory by Almgren [**4**] and Allard [**3**]. It also follows from the work on Sobolev inequalities of Michael and Simon [**14**].

Finally let us note some of the many generalizations of Theorem 1.1.

Shiffman [**18**] considered two cases of singular surfaces. The first is a PL-version of Theorem 1.1, in which D is a domain on a polyhedral surface which satisfies $K \leqq 0$ in the sense that the sum of the angles around each vertex is at least 2π. The second is for harmonic surfaces. If $x(u, v)$ is a map of the disk into $\boldsymbol{R}^n$ with each coordinate function $x_k(u, v)$ harmonic, then the resulting surface must have $K \leqq 0$ at all regular points. (This can be seen, for example, by the fact that a harmonic surface clearly satisfies the convex hull property.) However, harmonic surfaces may be singular at isolated points or along curves, and Theorem 1.1 cannot be applied directly. Shiffman notes that for $\varepsilon > 0$, the surface $y(u, v)$ in $\boldsymbol{R}^{n+2}$ defined by

$$y_k(u, v) = x_k(u, v), \qquad k = 1, \cdots, n,$$
$$y_{n+1} = \varepsilon u,$$
$$y_{n+2} = \varepsilon v,$$

is a regular harmonic surface to which Theorem 1.1 may be applied. Then letting ε tend to zero, we obtain $L^2 \geqq 4\pi A$ on the original surface.

The weaker inequality $L^2 \geqq 4A$ was proved by Courant [7] for harmonic surfaces and used in his treatment of the Plateau problem.

Finally, we mention the result of A. Huber [11] generalizing Beckenbach and Radó's theorem to the case where K may have variable sign. Denoting $K^+ = \max[K, 0]$, Huber's result for a simply-connected domain D is

$$L^2 \geqq 4\pi A - 2A \iint_D K^+ \, dA. \tag{17}$$

II. Related inequalities. We shall use the following notation:

M a 2-manifold immersed in $\boldsymbol{R}^n$,

$B_r = \{x \in \boldsymbol{R}^n : |x| < r\}, \qquad S_r = \{|x| = r\},$

$M_r = M \cap B_r, \qquad C_r = M \cap S_r,$

$A(r) = \text{area } M_r, \qquad L_r = \text{length } C_r.$

We shall assume throughout that M_r has no boundary in B_r.

THEOREM 2.1. *If M is simply-connected, and* $0 \in M$, *then*

$$A(r) \geqq \pi r^2 - \frac{r^2}{2} \iint_{M_r} K^+ \, dA.$$

The proof of this uses Huber's inequality (17) together with the fact that $A(r) \geqq \int_0^r L(t)\, dt$. (See [1].)

COROLLARY. *Under the same hypotheses, if $K \leqq 0$ on M_r, then*

$$A(r) \geqq \pi r^2. \tag{2.1}$$

Note that simple-connectivity is necessary here, since a long thin cylinder through the origin will not satisfy (2.1).

THEOREM 2.2. *If M_r is minimal and* $0 \in M_r$, *then* (2.1) *holds.*

The proof of this follows from the fact that for minimal surfaces $A(t)/\pi t^2$ is monotone increasing (see Fleming [9]).

Note that unlike Theorem 2.1, Theorem 2.2 holds without restriction on the topological type of M_r.

If we drop the assumption that M is minimal in Theorem 2.2, and let $\Lambda = \sup_{M_r} |H|$, then it is not hard to show that $A(r) \geqq \pi r^2/e^{\Lambda r}$. A very general version of this inequality was proved by Allard [3].

Using more delicate reasoning, Michael and Simon [14] obtained the inequality

$$A(r) \geqq \pi r^2/(1 + \Lambda r + \tfrac{1}{2}(\Lambda r)^2).$$

Finally, Trudinger [19] obtained the bound $A(r) \geqq \pi r^2/(1 + \frac{1}{4}(\Lambda r)^2)$ which is asymptotically sharp as $r \to \infty$ when M is a sphere.

We ask next what can be said if one drops the assumption that $0 \in M$. The following results were obtained in joint work of the author with H. Alexander and D. Hoffman [1], [2].

THEOREM 2.3. *Let $a(r, d)$ be the area of the segment of a circle of radius r cut off by a chord at distance d from the origin. Let M_r be simply-connected, and let d be the distance of M_r to the origin. Then*

$$K \leqq 0 \text{ on } M_r \Rightarrow A(r) \geqq \pi(r - d)^2, \tag{2.2}$$

$$M_r \text{ satisfies convex hull property} \Rightarrow A(r) \geqq 2a(r, d), \tag{2.3}$$

$$M_r \text{ minimal} \Rightarrow A(r) \geqq \pi(r^2 - d^2). \tag{2.4}$$

Note that the right-hand sides all coincide when $d = 0$, but for $d > 0$ we have the strict inequalities

$$\pi(r - d)^2 < 2a(r, d) < \pi(r^2 - d^2).$$

The three classes of surfaces considered are related by:

$$\text{minimal} \Rightarrow \text{convex hull property} \Rightarrow K \leqq 0$$

and none of the reverse implications hold except when $n = 3$:

$$\text{convex hull property} \Leftrightarrow K \leqq 0 \text{ in } \boldsymbol{R}^3. \tag{2.5}$$

(See [16].)

THEOREM 2.4. *Equality holds in* (2.4) (*and also in* (2.2) *and* (2.3) *for $d = 0$*) *when M is a plane. For $d > 0$, inequalities* (2.2) *and* (2.3) *are strict, but they are best possible; one cannot replace the right-hand sides by anything larger except in the case of* (2.2) *if one restricts the dimension n. In particular, for $n = 3$,* (2.2) *is superseded by* (2.3), *by virtue of* (2.5).

Note that using cylinders again shows that (2.2) and (2.3) fail if M_r is not simply-connected. However, Theorem 2.2 leads us to believe that (2.4) should hold in general.

Conjecture 2. (2.4) holds for M_r of arbitrary topological type.

The analogy between Theorems k.1, k.2 and Conjecture k, for $k = 1,2$, should be evident. The clincher is the following:

THEOREM 2.5. *Conjecture* 1 *for $C \subset S_r \Rightarrow$ Conjecture* 2; *Theorem* 2.2 $\Rightarrow$ *Conjecture* 1 *for $C \subset S_r$, $0 \in M$.*

In other words, if the isoperimetric inequality holds for general domains (or even for domains whose boundary lies on a sphere) then the area inequality (2.4) also holds. Conversely, the area inequality (2.1) in Theorem 2.2 shows that the isoperimetric inequality holds for any domain on a minimal surface M obtained by intersecting M with a ball centered at a point of M.

This last statement has a two-line proof, which we give in 5 lines.

From equation (4), $2A(r) = \int_{C_r} x \cdot \nu \, ds \leqq rL(r)$ since $|x \cdot \nu| \leqq |x| = r$ on C_r. Thus

$$L(r) \geqq 2A(r)/r. \tag{2.6}$$

By Theorem 2.2, $A(r) \geqq \pi r^2$, so that (2.6) yields

$$L(r) \geqq 2\pi r. \tag{2.7}$$

Multiplying (2.6) and (2.7) gives $L(r)^2 \geqq 4\pi A(r)$.

III. The higher dimensional case. Many of the results discussed above carry over to higher dimensional manifolds. We review the situation briefly.

First of all, let ω_m be the volume of the unit ball in $\boldsymbol{R}^m$. If D is a domain in $\boldsymbol{R}^m$, let V denote its volume, and A the $(m-1)$-dimensional measure of its boundary. Then the isoperimetric inequality states that

$$A^m \geqq m^m \omega_m V^{m-1}, \tag{3.1}$$

with equality only when D is a ball. For a proof that holds in great generality, see Federer [**8**].

Conjecture 3.1. The isoperimetric inequality (3.1) holds when D is an arbitrary domain on an m-dimensional minimal submanifold of $\boldsymbol{R}^n$.

If M is an m-dimensional submanifold of $\boldsymbol{R}^n$, let $V(r)$ be the m-dimensional measure of $M \cap B_r$, $A(r)$ the $(m-1)$-dimensional measure of $M \cap S_r$. Let H be the mean curvature vector of M, and let $\Lambda = \sup_{M_r} |H|$.

THEOREM 3.1 (MICHAEL AND SIMON [**14**]). *If $0 \in M$ and M has no boundary in B_r, then*

$$V(r) \geqq \omega_m r^m/(1 + \Lambda r + \cdots + (\Lambda r)^m/m!). \tag{3.2}$$

COROLLARY (FLEMING [**9**]). *If furthermore M is minimal, then*

$$V(r) \geqq \omega_m r^m. \tag{3.3}$$

Conjecture 3.2. If M is minimal, d is the distance of M to the origin, and M has no boundary in B_r, then

$$V(r) \geqq \omega_m (r^2 - d^2)^{m/2}. \tag{3.4}$$

THEOREM 3.2 (ALEXANDER, HOFFMAN, AND OSSERMAN [**1**]). *If for some constant $c > 0$ the inequality*

$$A^m \geqq c V^{m-1} \tag{3.5}$$

holds for every domain D on an m-dimensional minimal surface in $\boldsymbol{R}^n$, then

$$V(r) \geqq (c/m^m)(r^2 - d^2)^{m/2}. \tag{3.6}$$

COROLLARY. *Conjecture* 3.1 $\Rightarrow$ *Conjecture* 3.2.

Note that the existence of a constant c for which (3.5) holds has been proved by Almgren [4] and Allard [3]. Thus (3.6) holds with the same constant. The question in both cases is whether one can get the best-possible constant, $c = m^m \omega_m$. In that case equality is realized in (3.6) when M is a linear subspace.

In conclusion we note that an analogous argument to that given at the end of the previous section shows that one can deduce from the corollary to Theorem 3.1 that the isoperimetric inequality (3.1) holds for domains on an m-dimensional minimal surface M whose boundary lies on a sphere about some point of M.

REFERENCES

1. H. Alexander, D. Hoffman and R. Osserman, *Area estimates for submanifolds of euclidean space*, INDAM Symposia Mathematica, Vol. 14 (to appear).
2. H. Alexander and R. Osserman, *Area bounds for various classes of surfaces*, Amer. J. Math. (to appear).
3. W. K. Allard, *On the first variation of a varifold*, Ann. of Math. (2) **95** (1972), 417–491. MR **46** #6136.
4. F. J. Almgren, *The theory of varifolds*, Princeton, N. J., 1965 (mimeographed notes).
5. E. F. Beckenbach and T. Radó, *Subharmonic functions and surfaces of negative curvature*, Trans. Amer. Math. Soc. **35** (1933), 662–674.
6. T. Carleman, *Zur Theorie der Minimalflächen*, Math. Z. **9** (1921), 154–160.
7. R. Courant, *Dirichlet's principle, conformal mapping, and minimal surfaces*, Interscience, New York, 1950. MR **12**, 90.
8. H. Federer, *Geometric measure theory*, Die Grundlehren der math. Wissenschaften, Band 153, Springer-Verlag, New York, 1969. MR **41** #1976.
9. W. H. Fleming, *On the oriented Plateau problem*, Rend. Circ. Mat. Palermo (2) **11** (1962), 69–90. MR **28** #499.
10. C. C. Hsiung, *Isoperimetric inequalities for two-dimensional riemannian manifolds with boundary*, Ann. of Math. (2) **73** (1961), 213–220. MR **24** #A497.
11. A. Huber, *On the isoperimetric inequality on surfaces of variable Gaussian curvature*, Ann. of Math. (2) **60** (1954), 237–247. MR **16**, 508.
12. H. Kaul, *Isoperimetrische Ungleichung und Gauss-Bonnet Formel für H-Flächen in Riemannschen Mannigfaltigkeiten*, Arch. Rational Mech. Anal. **45** (1972), 194–221.
13. ———, *Remarks on the isoperimetric inequality for multiply-connected H-surfaces*, Math. Z. **128** (1972), 271–276.
14. J. H. Michael and L. M. Simon, *Sobolev and mean-value inequalities on generalized submanifolds of* R^n, Comm. Pure Appl. Math. **26** (1973), 361–379.
15. J. C. C. Nitsche, *The isoperimetric inequality for multiply-connected minimal surfaces*, Math. Ann. **160** (1965), 370–375. MR **32** #2989.
16. R. Osserman, *The convex hull property of immersed manifolds*, J. Differential Geometry **6** (1971/72), 267–270. MR **45** #7647.
17. W. T. Reid, *The isoperimetric inequality and associated boundary problems*, J. Math. Mech. **8** (1959), 897–905. MR **24** #A483.
18. M. Shiffman, *On the isoperimetric inequality for saddle surfaces with singularities*, Studies and Essays Presented to R. Courant, Interscience, New York, 1948, pp. 383–394. MR **9**, 303.
19. N. Trudinger, *A sharp inequality for subharmonic functions on two dimensional manifolds*, Math. Z. **133** (1973), 75–79.

STANFORD UNIVERSITY

Proceedings of Symposia in Pure Mathematics
Volume 27, 1975

MINIMAL SUBMANIFOLDS WITH m-INDEX 2

TOMINOSUKE OTSUKI

For a submanifold M in a Riemannian manifold $\bar{M}$, the *minimal index* (m-index) at a point of M is by definition the dimension of the linear space of all the 2nd fundamental forms with vanishing trace. The *geodesic codimension* (g-codim) of M in $\bar{M}$ is defined by the minimum of codimensions of M in totally geodesic submanifolds of $\bar{M}$ containing M.

It is clear that

$$m\text{-index} \leqq g\text{-codim}.$$

In this paper we suppose that $\bar{M}$ is of constant curvature and M is a minimal submanifold of $\bar{M}$. If M is of m-index 1 at each point, then its geodesic codimension is one, which we have proved earlier [**5**]. We shall discuss what we can say about M, if it has constant m-index 2 and show some examples of such submanifolds in the space forms, each of which is the locus of points on a moving totally geodesic submanifold intersecting orthogonally a surface W^2 at a moving point. This surface W^2 is called a *base surface* of M. This situation is quite analogous to the case of a right helicoid in E^3 generated by a moving straight line along a base helix.

1. Minimal submanifolds with m-index 2. We summarize here some of the standard results which we shall need. Details can be found in [**8**].

Let $M = M^n$ be an n-dimensional submanifold of an $(n+\nu)$-dimensional Riemannian manifold $\bar{M} = \bar{M}^{n+\nu}$ of constant curvature $\bar{c}$. Let $\bar{\omega}_A, \bar{\omega}_{AB} = -\bar{\omega}_{BA}$, $A, B = 1, 2, \cdots, n+\nu$, be the basic and connection forms of $\bar{M}$ on the orthonormal frame bundle $F(\bar{M})$ over $\bar{M}$, which satisfy the structure equations

$$(1) \qquad d\bar{\omega}_A = \textstyle\sum_B \bar{\omega}_{AB} \wedge \bar{\omega}_B, \; d\bar{\omega}_{AB} = \sum_C \bar{\omega}_{AC} \wedge \bar{\omega}_{CB} - \bar{c}\bar{\omega}_A \wedge \bar{\omega}_B.$$

AMS (MOS) subject classifications (1970). Primary 53A10, 53B25, 53C40.

Let B be the submanifold of $F(\bar{M})$ over M composed of $b = (x, e_1, \cdots, e_{n+\nu})$ such that $(x, e_1, \cdots, e_n) \in F(M)$, where $F(M)$ is the orthonormal frame bundle of M with the induced Riemannian metric from $\bar{M}$. Then deleting the bars of $\bar{\omega}_A$, $\bar{\omega}_{AB}$ on B, we have

$$\omega_\alpha = 0, \qquad \omega_{i\alpha} = \sum_j A_{\alpha ij}\, \omega_j, \qquad A_{\alpha ij} = A_{\alpha ji}, \tag{2}$$
$$\alpha = n+1, \cdots, n+\nu;\ i, j = 1, 2, \cdots, n.$$

At any point $x \in M$, let $\bar{M}_x$, M_x and N_x be the tangent spaces to $\bar{M}$ and M respectively and the normal space to M_x in $\bar{M}_x$. Denoting the space of all symmetric real matrices of order n by S_n, for any $b \in B$ we define a linear mapping $\varphi_b : N_x \to S_n$ by

$$\varphi_b\Big(\sum_\alpha \nu_\alpha e_\alpha\Big) = \sum_\alpha \nu_\alpha A_\alpha, \quad \text{where } A_\alpha = (A_{\alpha ij}). \tag{3}$$

Now, suppose that M is minimal in $\bar{M}$ and of m-index 2 at each point in the following. Then we have

$$\operatorname{trace} A_\alpha = 0, \qquad \alpha = n+1, \cdots, n+\nu, \tag{4}$$

and N_x is decomposed as

$$N_x = N'_x + O'_x, \qquad N'_x \perp O'_x, \tag{5}$$

where $O'_x = \operatorname{kern} \varphi_b$ and

$$\dim N'_x = 2. \tag{6}$$

Let B_1 be the set of $b \in B$ such that $e_{n+1}, e_{n+2} \in N'_x$, then B_1 is a principal bundle over M of structure group $O(n) \times O(2) \times O(\nu - 2)$. On B_1, for $\beta > n+2$ we have

$$A_\beta = 0, \tag{7}$$

$$\omega_{n+1,\beta} \equiv \omega_{n+2,\beta} \equiv 0 \ (\operatorname{mod} \omega_1, \cdots, \omega_n) \tag{8}$$

and

$$\omega_{n+1,\beta} = \omega_{n+2,\beta} = 0 \quad \text{or} \quad \omega_{n+1,\beta} \wedge \omega_{n+2,\beta} \neq 0. \tag{9}$$

By (8), we can define a bilinear mapping $\psi' : M_x \times N'_x \to O'_x$ by

$$\psi'(X, V) = \sum_{\beta > n+2} \langle V, e_{n+1}\, \omega_{n+1,\beta}(X) + e_{n+2}\, \omega_{n+2,\beta}(X)\rangle\, e_\beta, \tag{10}$$

which we call the 1st *torsion operator* of M in $\bar{M}$. (See [9].)

Now, let $\mathfrak{L}_x$ be the space of relative nullity of M in $\bar{M}$ at x, whose dimension is called the index of relative nullity at x. We decompose M_x orthogonally as

$$M_x = \mathfrak{W}_x + \mathfrak{L}_x. \tag{11}$$

LEMMA 1. $\dim \mathfrak{L}_x \leqq n - 2$.

The 1st torsion operator ψ' has an analogous meaning as the torsion for space curves in some sense [8].

THEOREM A. (i) *M is g-codim = 2 if and only if* $\psi' \equiv 0$.

(ii) *If $\psi' \neq 0$ everywhere, then* $\dim \mathfrak{L}_x = n-2$, $\psi'_V(\mathfrak{L}_x) = 0$ *and ψ'_V has a common image $\psi'_V(M_x)$ for any $V \in N'_x$, where* $\psi'_V(X) = \psi'(X, V)$, $V \neq 0$.

Now, we suppose that $\psi' \neq 0$ everywhere. Since $\dim \mathfrak{W}_x = 2$, let B'_1 be the set of $b \in B_1$ such that $e_1, e_2 \in \mathfrak{W}_x$ and $e_3, \cdots, e_n \in \mathfrak{L}_x$ and

$$\begin{aligned} &\omega_{1,n+1} = \lambda\,\omega_1, \quad \omega_{2,n+1} = -\lambda\,\omega_2, \quad \omega_{r,n+1} = 0, \\ &\omega_{1,n+2} = \mu\,\omega_2, \quad \omega_{2,n+2} = \mu\,\omega_1, \quad \omega_{r,n+2} = 0, \\ &r = 3, 4, \cdots, n; \lambda \neq 0, \mu \neq 0. \end{aligned} \tag{12}$$

Then, supposing B'_1 is smooth in B_1, by the above relations we can set on B'_1 as

$$\omega_{1r} + i\omega_{2r} = (p_r + iq_r)(\omega_1 + i\omega_2), 2 < r \leqq n,$$

which define two tangent vector fields $P = \sum_r p_r e_r$ and $Q = \sum_r q_r e_r$ of M called the *principal* and *subprincipal asymptotic vector fields*, respectively. Then, we have the following theorem [8].

THEOREM B. *If the* 1*st torsion operator ψ' is not trivial everywhere and B'_1 is smooth in B_1, then we have:*

(i) *The distribution $\mathfrak{L} = \bigcup \mathfrak{L}_x$ is completely integrable and its integral submanifolds are totally geodesic in $\bar{M}$.*

(ii) *The distribution $\mathfrak{W} = \bigcup \mathfrak{W}_x$ is completely integrable if and only if $Q \equiv 0$.*

(iii) *When $Q \equiv 0$, integral surfaces of $\mathfrak{W}$ are totally umbilic in M.*

(iv) *P and Q are involutive. When $P \neq 0$, the integral curves of P are of* g-codim $\leqq 1$ *and they are geodesics of $\bar{M}$ if and only if $\langle P, Q\rangle = 0$ or $P \wedge Q = 0$.*

Under the conditions of Theorem B and $Q \equiv 0$, on B'_1 we have

$$\{d\log\lambda - \langle P, dx\rangle - i(2\,\omega_{12} - \sigma\,\hat{\omega}_1)\} \wedge (\omega_1 + i\omega_2) = 0, \tag{13}$$

$$\{d\sigma + i(1-\sigma^2)\,\hat{\omega}_1\} \wedge (\omega_1 + i\omega_2) = 0, \tag{14}$$

$$d\omega_{12} = -\,\{\|P\|^2 + \bar{c} - \lambda^2 - \mu^2\}\,\omega_1 \wedge \omega_2, \tag{15}$$

$$d\hat{\omega}_1 = -\frac{1}{\lambda\mu}\{2\lambda^2\mu^2 - \|F\|^2 - \|G\|^2\}\,\omega_1 \wedge \omega_2, \tag{16}$$

where $\sigma = \mu/\lambda$, $\hat{\omega}_1 = \omega_{n+1,n+2}$ and $F = \sum_{\gamma > n+2} f_\gamma e_\gamma$ and $G = \sum_{\gamma > n+2} g_\gamma e_\gamma$ are defined by means of the fact that on B'_1 and for $\gamma > n+2$ we can set as

$$\lambda\omega_{n+1,\gamma} + i\mu\omega_{n+2,\gamma} = (f_\gamma + ig_\gamma)(\omega_1 - i\omega_2). \tag{17}$$

LEMMA 2. *Under the conditions of Theorem* B *and $Q \equiv 0$, $\bar{c} \neq 0$ implies $P \neq 0$.*

If $P \neq 0$ and $Q \equiv 0$, we denote the integral surface of $\mathfrak{M}$ through $x \in M$ by $W^2(x)$. Then we have

LEMMA 3. *The integral curves Γ^1 of P are the orthogonal trajectories of a family of hypersurfaces of M^n containing the integral surfaces W^2 of $\mathfrak{M}$.*

Considering the length of a Jacobi field along Γ^1, we can prove that these geodesics Γ^1 of $\bar{M}$ pass through a common point z_0 or are parallel to each other in case $\bar{c} = 0$. Thus, we have the following theorem which plays a fundamental role in our consideration.

THEOREM C. *Let M^n be a maximal minimal submanifold in a space form $\bar{M}^{n+\nu}$ (of constant curvature $\bar{c}$) which is of m-index 2, has the maximal relative nullity $n-2$, $Q \equiv 0$ and for which B_1' is smooth. Then M^n is a locus of totally geodesic submanifolds $L^{n-2}(y)$ in $\bar{M}^{n+\nu}$ through points y of a base surface W^2 lying in either a Riemannian hypersphere in $\bar{M}^{n+\nu}$ with center z_0 such that*

(i) *$L^{n-2}(y)$ intersects orthogonally W^2 at y and contains the geodesic segment om z_0 to y,*

(ii) *the $(n-3)$-dimensional tangent spaces to the intersection of $L^{n-2}(y)$ and the hypersphere at y are parallel along W^2 in $\bar{M}^{n+\nu}$*

or an $(n+2)$-dimensional linear space in $\bar{M}^{n+\nu} = E^{n+\nu}$ such that M^n is $(n-2)$-dimensionally cylindrical and $L^{n-2}(y)$ are its generators intersecting orthogonally W^2 at y.

In this theorem, denoting the arclength from z_0 to y along the geodesics Γ^1 by v, we have

$$
\begin{aligned}
p = \|P\| &= \bar{c}^{1/2} \cot \bar{c}^{1/2} v && (\bar{c} > 0),\\
&= 1/v && (\bar{c} = 0),\\
&= (-\bar{c})^{1/2} \coth(-\bar{c})^{1/2} v && (\bar{c} < 0).
\end{aligned} \tag{18}
$$

2. Minimal submanifolds with ψ' of rank 1. We shall discuss M^n in $\bar{M}^{n+\nu}$ as in Theorem C under the additional condition that ψ'_V, $V \in N'$, $V \neq 0$, is of rank 1 everywhere, where $N' = \bigcup_x N'_x$ is the vector bundle over M. Let B_2 be the set of $b \in B_1$ such that $e_{n+3} \in \psi'_V(M_x)$, $V \in N'_x$. $V \neq 0$. Then, on $B_2' = B_2 \cap B_1'$, we have

$$F = f e_{n+3}, \qquad G = g e_{n+3}, \qquad f^2 + g^2 \neq 0. \tag{19}$$

THEOREM 1. *Let M and $\bar{M}$ be as in Theorem C and ψ' be of rank 1 everywhere, then the geodesic codimension of M is 3.*

Furthermore, we consider the following conditions **[8]**:

(α_1) *$\sigma = \mu/\lambda$ is constant on W^2,*

(β) *W^2 is of constant curvature c,*

(σ_1) *$\hat{\omega}_1 \neq 0$.*

THEOREM 2. *Let M and $\bar{M}$ be as in Theorem 1. If a base surface W^2 satisfies the conditions (α_1), (β) and (γ_1), then we have $\sigma = 1$ or -1, $2\lambda^2 = p^2 + \bar{c}$ and W^2 is flat. Supposing $\sigma = 1$ and $\omega_{12} = d\theta$ on W^2, we obtain a Frenet formula of W^2 in $\bar{M}: \hat{\omega}_1 = 2d\theta$ and*

$$
\begin{aligned}
dx &= R((e_1^* + ie_2^*)d\bar{z}),\\
\bar{D}(e_1^* + ie_2^*) &= e_3 p dz + (e_{n+1}^* + ie_{n+2}^*)\lambda d\bar{z},\\
\bar{D}e_3 &= -pR((e_1^* + ie_2^*)d\bar{z}),\\
\bar{D}(e_{n+1}^* + ie_{n+2}^*) &= -(e_1^* + ie_2^*)\lambda dz + e_{n+3}2^{1/2}\lambda d\bar{z},\\
\bar{D}e_{n+3} &= -2^{1/2}\lambda R((e_{n+1}^* + ie_{n+2}^*)dz),
\end{aligned} \tag{20}
$$

where z is an isothermal coordinate of W^2 such that $\omega_1 + \omega_2 = \exp(-i\theta)dz$, $e_1^* + ie_2^* = \exp(i\theta)(e_1 + ie_2)$, $e_{n+1}^* + ie_{n+2}^* = \exp(2i\theta)(e_{n+1} + ie_{n+2})$, and $\bar{D}$ is the covariant differential operator of $\bar{M}$.

Now, we shall give an example of W^2 in case $\bar{M}^{n+3}$ is the Euclidean space E^{n+3}. Considering as $\boldsymbol{C}^3 = \boldsymbol{R}^6 = E^6$, we take 3 fixed vectors A_1, A_2 and A_3 in $\boldsymbol{C}^3$ such that

$$A_j \cdot A_j = 0, \qquad A_j \cdot \bar{A}_j = 1/6, \qquad A_j \cdot A_k = A_j \cdot \bar{A}_k = 0, \qquad j, k = 1, 2, 3; j \neq k, \tag{21}$$

then the surface composed of the point x given by

$$\begin{aligned} x = -\nu\Big\{ & A_1 \exp \frac{i(u_1 + 3^{1/2} u_2)}{2^{1/2}\nu} + \bar{A}_1 \exp \frac{-i(u_1 + 3^{1/2} u_2)}{2^{1/2}\nu} \\ & + A_2 \exp \frac{2iu_1}{2^{1/2}\nu} + \bar{A}_2 \exp \frac{-2iu_1}{2^{1/2}\nu} \\ & + A_3 \exp \frac{i(u_1 - 3^{1/2} u_2)}{2^{1/2}\nu} + \bar{A}_3 \exp \frac{-i(u_1 - 3^{1/2} u_2)}{2^{1/2}\nu} \Big\} \end{aligned} \tag{22}$$

is a solution of (20).

By means of Theorem C, we can give an example of M^n in the Euclidean $(n+3)$-space E^{n+3} in Theorem 2 as follows :

Consider $E^{n+3} = \boldsymbol{R}^{n+3} = \boldsymbol{R}^6 \times \boldsymbol{R}^{n-3}$, and take a surface W^2 given by (22) in $\boldsymbol{R}^6$, and the linear subspace $L^{n-2}(y)$ through $y \in W^2$, parallel to e_3 and $\boldsymbol{R}^{n-3}$. Then the locus of the moving $L^{n-2}(y)$ forms a submanifold M^n as stated above.

Making use of the same W^2, we can construct an M^n in case $\bar{M}^{n+3}$ is the unit $(n+3)$-sphere S^{n+3} or the hyperbolic space H^{n+3} of constant curvature -1.

3. Minimal submanifolds with ψ' of rank 2. Now, we shall go back to the situation of Theorem C. We assume that the rank of ψ' is 2, that is $F \wedge G \neq 0$ everywhere. We denote the 2-dimensional normal space spanned by F and G by N''_x and set $N'' = \bigcup_x N''_x$, which is a 2-dimensional vector bundle over M. Then, we decompose N_x as

$$N_x = N'_x + N''_x + O''_x, \qquad O'_x = N''_x + O''_x, \qquad N''_x \perp O''_x. \tag{23}$$

Denote the set of $b \in B_1$ such that e_{n+3}, $e_{n+4} \in N''_x$ by B_2. Then on $B'_2 = B_2 \cap B'_1$ we have

$$\omega_{n+1,\gamma} = \omega_{n+2,\gamma} = 0, \qquad \gamma > n+4, \tag{24}$$

$$\omega_{n+3,\gamma} = \omega_{n+4,\gamma} = 0 \,(\mathrm{mod}\ \omega_1, \omega_2), \qquad \gamma > n+4. \tag{25}$$

Using these facts, we can define a bilinear mapping $\psi'' : M_x \times N''_x \to O''_x$ by

$$\psi''(X, V) = \sum_{\gamma > n+4} \langle V, e_{n+3}\,\omega_{n+3,\gamma}(X) + e_{n+4}\,\omega_{n+4,\gamma}(X) \rangle e_\gamma, \tag{26}$$

which we call the *2nd torsion operator* of M in $\bar{M}$. (See [9].) ψ'' has an analogous property to ψ'.

THEOREM 3. *Let M and $\bar{M}$ be as in Theorem* C *and ψ' be of rank* 2 *everywhere, then M is of* g-codim 4 *if and only if $\psi'' \equiv 0$.*

In the following section, we suppose that $\psi'' \equiv 0$, and so we may put $\nu = 4$ by Theorem 3.

LEMMA 4. *The curvature $d\hat{\omega}_2$ of N'' is not zero everywhere, where $\hat{\omega}_2 = \omega_{n+3,n+4}$ on B_2'.*

PROOF. We get easily

$$\begin{aligned} d\hat{\omega}_2 &= \omega_{n+3,n+1} \wedge \omega_{n+1,n+4} + \omega_{n+3,n+2} \wedge \omega_{n+2,n+4} \\ &= -\frac{\Delta}{\lambda^2}\left((1 + \frac{1}{\sigma^2}\right)\omega_1 \wedge \omega_2, \end{aligned}$$

where $\Delta = f_{n+3}g_{n+4} - f_{n+4}g_{n+3} \neq 0$.

Now, we consider the following conditions as in § 2 :

(α_2) $\sigma = \mu/\lambda$ and $\Phi = (f_{n+4} + ig_{n+4})/(f_{n+3} + ig_{n+3})$ are constants on W^2.

(β) W^2 is of constant curvature c.

(σ_2) $\hat{\omega}_1 \neq 0$, $\hat{\omega}_2 \neq 0$.

THEOREM 4. *Let M and $\bar{M}$ be as in Theorem 3 and $\psi'' \equiv 0$. If a base surface W^2 satisfies the conditions (α_2), (β) and (σ_2), then we have $\sigma = 1$ or -1, $\Phi = i$ or $-i$, $\| F \| = \| G \|$, $\langle F, G \rangle = 0$, $2\lambda^2 = p^2 + \bar{c} - c$ and $2\lambda^2 = 5c$ which implies $c > 0$. Supposing $\sigma = 1$ and $c = 1$, we obtain a Frenet formula of W^2 in $\bar{M}$: $\lambda = \mu = 10^{1/2}/2$, $\|F\| = \|G\| = 15^{1/2}/2$, $\hat{\omega}_1 = 2\omega_{12}$, $\hat{\omega}_2 = 3\omega_{12}$ and*

$$(27)\qquad \begin{aligned} dx &= \frac{1}{h}(\bar{\xi}\, dz + \xi\, d\bar{z}), \quad \bar{D}e_3 = -\frac{p}{h}(\bar{\xi}\, dz + \xi\, d\bar{z}), \\ \bar{D}\xi &= \frac{1}{h}\xi\,(\bar{z}dz - zd\bar{z}) + \frac{2p}{h}e_3\, dz + \frac{10^{1/2}}{h}\eta\, d\bar{z}, \\ \bar{D}\eta &= -\frac{10^{1/2}}{h}\xi\, dz + \frac{2}{h}\eta(\bar{z}dz - zd\bar{z}) + \frac{6^{1/2}}{h}\zeta\, d\bar{z}, \\ \bar{D}\zeta &= -\frac{6^{1/2}}{h}\eta\, dz + \frac{3}{h}\zeta\,(\bar{z}dz - zd\bar{z}), \end{aligned}$$

where z is an isothermal coordinate of W^2 such that $ds^2 = 4dzd\bar{z}/(1 + z\bar{z})^2$, $\xi = e^{i\theta}(e_1 + ie_2)$, $\eta = e^{2i\theta}(e_{n+1} + ie_{n+2})$, $\zeta = e^{3i\theta}(e_{n+3} + ie_{n+4})$, $\omega_1 + i\omega_2 = e^{-i\theta}(2dz/(1 + z\bar{z}))$ and $h = 1 + z\bar{z}$.

Now we shall give a solution of (27) in case $\bar{M}^{n+4}$ is the Euclidean space E^{n+4} which is uniquely determined in a sense. In $\boldsymbol{R}^7 = E^7$, the solution is the surface given by

$$(28)\qquad \begin{aligned} x_1 &= \frac{1 - 3z\bar{z} + z^2\bar{z}^2}{(1 + z\bar{z})^3}(z + \bar{z}), \quad x_2 = -i\frac{1 - 3z\bar{z} + z^2\bar{z}^2}{(1 + z\bar{z})^3}(z - \bar{z}), \\ x_3 &= \frac{5^{1/2}(1 - z\bar{z})}{2^{1/2}(1 + z\bar{z})^3}(z^2 + \bar{z}^2), \quad x_4 = -i\frac{5^{1/2}(1 - z\bar{z})}{2^{1/2}(1 + z\bar{z})^3}(z^2 - \bar{z}^2), \\ x_5 &= \frac{5^{1/2}}{3^{1/2}(1 + z\bar{z})^3}(z^3 + \bar{z}^3), \quad x_6 = -i\frac{5^{1/2}}{3^{1/2}(1 + zz)^3}(z^3 - \bar{z}^3), \\ x_7 &= -\frac{1 - 9z\bar{z} + 9z^2\bar{z}^2 - z^3\bar{z}^3}{6^{1/2}(1 - z\bar{z})^3}. \end{aligned}$$

Making use of the surface (28), we can construct an example of M^n in E^{n+4} in Theorem 4 as follows:

Consider as $E^{n+4} = \boldsymbol{R}^{n-3} \times \boldsymbol{R}^7$ and take a surface W^2 given by (28) in $\boldsymbol{R}^7$, and the linear subspace $L^{n-2}(y)$ through $y \in W^2$ and the origin and parallel to $\boldsymbol{R}^{n-3}$. Then, the locus of the moving $L^{n-2}(y)$ forms a submanifold M^n as stated above.

Making use of the same surface W^2, we can construct M^n in case $\bar{M}^{n+4}$ is the sphere $S^{n+4}(R)$ or the hyperbolic space $H^{n+4}(\bar{c})$, where R is the radius of the sphere and $\bar{c}$ is the curvature.

4. Minimal submanifolds with m-index 2 in space forms. Considering carefully the results in § 3 and § 4, we can prove the following theorem. (See **[10]**.)

THEOREM 5. *Let M^n and $\bar{M}^{n+\nu}$ be as in Theorem* C. *Then, the normal vector bundle N over M^n in $\bar{M}^{n+\nu}$ can be decomposed orthogonally and smoothly in general as follows*:

(i) *If* g-codim *of M^n in $\bar{M}^{n+\nu}$ is even, say* 2*m, then*

$$N = N' + N'' + \cdots + N^{(m)} + O^{(m)}.$$

(ii) *If* g-codim *of M^n in $\bar{M}^{n+\nu}$ is odd, say* 2*m* + 1, *then*

$$N = N' + N'' + \cdots + N^{(m+1)} + O^{(m+1)},$$

where the fibre $N_x^{(t)}$ of $N^{(t)}$ is of 2-*dimension for $t = 1, \cdots, m$ and* 1-*dimension for $t = m + 1$, and $X \in \Gamma(T(M^n))$ and $V \in \Gamma(N^{(t)})$,*

$$\bar{D}_X V \in \Gamma(N^{(t-1)}) + \Gamma(N^{(t)}) + \Gamma(N^{(t+1)})$$

for $t = 1, \cdots, m - 1$ in case (i) *and $t = 1, \cdots, m$ in case* (ii) *and*

$$\bar{D}_X V \in \Gamma(N^{(t-1)}) + \Gamma(N^{(t)})$$

for $t = m$ in case (i) *and $t = m + 1$ in case* (ii), *here $N^{(0)} = \mathfrak{B}$. Furthermore, $N' + N'' + \cdots + N^{(m)}$ or $N' + N'' + \cdots + N^{(m+1)}$ are the normal vector bundles of M^n in the totally geodesic submanifolds of the least dimensions in $\bar{M}^{n+\nu}$ containing M^n.*

We call M^n in this theorem S-type or T-type according to g-codim M^n = even or odd.

Now for $M = M^n$ and $\bar{M} = \bar{M}^{n+\nu}$ as in Theorem 5, we may suppose $\nu = 2m$ when M is of S-type and $\nu = 2m + 1$ when M is of T-type. Let B_t be the set of $b = (x, e_1, \cdots, e_{n+\nu})$ such that

$$e_{n+1}, e_{n+2} \in N'_x; \cdots; e_{n+2t-1}, e_{n+2t} \in N_x^{(t)}.$$

LEMMA 5. *There exist complex valued* (1, 2)-*matrix fields $\Psi_t = (\Psi_{t1}, \Psi_{t2})$ on $B'_t = B_t \cap B'_1$, $t = 1, \cdots, m$, such that*

$$(29_t) \qquad \Psi_t \Omega_t = \Psi_{t+1}(\omega_1 - i\omega_2) \text{ on } B'_{t+1}, \quad t = 1, \cdots, m - 1,$$

and a complex valued function Ψ_{m+1} such that

$$(29_m) \qquad \Psi_m \Omega_m = \Psi_{m+1}(\omega_1 - i\omega_2) \quad \textit{on } B'_{m+1} \ (\textit{in case (ii)}),$$

where

$$\Omega_t := \begin{pmatrix} \omega_{\alpha_1\beta_1} & \omega_{\alpha_1\beta_2} \\ \omega_{\alpha_2\beta_1} & \omega_{\alpha_2\beta_2} \end{pmatrix}, \quad \begin{array}{l} \alpha_1 = n + 2t - 1,\ \alpha_2 = n + 2t, \\ \beta_1 = n + 2t + 1,\ \beta_2 = n + 2t + 2, \\ t = 1, \cdots, m - 1, \end{array}$$

and

$$\Omega_m := \begin{pmatrix} \omega_{\alpha_1\gamma} \\ \omega_{\alpha_2\gamma} \end{pmatrix}, \quad \begin{array}{l} \alpha_1 = n + 2m - 1,\ \alpha_2 = n + 2m, \\ \gamma = n + 2m + 1. \end{array}$$

By this lemma, we can define the following normal vector field $F_t, G_t \in \Gamma(N^{(t)})$ by

$$\Psi_{t1} e_{n+2t-1} + \Psi_{t2} e_{n+2t} = F_t + iG_t$$

(and $\Psi_{m+1} e_{n+2m+1} = F_{m+1} + iG_{m+1}$ in case (ii)).

Now, we consider the following conditions for W^2 as in § 3 and § 4:

(α_m) $\Psi_{t1}/\Psi_{t2} = c_t$ is constant on W^2 for $t = 1, 2, \cdots, m$,

(β) W^2 is of constant curvature c,

(γ_m) $\hat{\omega}_t = \omega_{n+2t-1,n+2t} \neq 0$ on W^2 for $t = 1, 2, \cdots, m$.

THEOREM 6. *Let M^n and $\bar{M}^{n+\nu}$ be as in Theorem* C *and a base surface W^2 satisfies the conditions* (α_m), (β) *and* (γ_m), *then we can find a submanifold $B''_m \subset B'_m$ over W^2 on which* (i) $c_t = i$ *or* $-i$ *for* $t = 1, 2, \cdots, m$; (ii) Ψ_t *is constant for* $t = 1, 2, \cdots, m + 1$; (iii) $\langle F_t, G_t \rangle = 0$ *for* $t = 1, 2, \cdots, m$; (iv) $\hat{\omega}_t = (t + 1)\sigma_t \omega_{12}$, *where* $\sigma_t = c_t/i$, *for* $t = 1, 2, \cdots, m$; *and* (v) $c > 0$ *when M^n is of S-type and* $c = 0$ *when M^n is of T-type.*

5. Determination of W^2 for M^n of S-type in E^{n+2m}. First, we shall show the Frenet formula of W^2 in Theorem 6, in case M^n is of S-type.

In the proof of Theorem 6, we obtain in fact

$$(30) \qquad \Psi_t = a_t(1, i\sigma_t), \quad a_t > 0, \quad \sigma_t = \pm 1 \quad \text{for } t = 1, 2, \cdots, m,$$

$$(31) \qquad a_t^2 = a_{t-1}^2 \left\{ \left(\frac{a_{t-1}}{a_{t-2}} \right)^2 - \frac{t}{2} c \right\} \qquad \text{for } t = 2, \cdots, m,$$

$$(32) \qquad c = \frac{2}{m+1} \left(\frac{a_m}{a_{m-1}} \right)^2,$$

hence $c > 0$ in this case. We may put $c = 1$, then we get

$$b_t = \left(\frac{a_t}{a_{t-1}} \right)^2 = \frac{(m - t + 1)(m + t + 2)}{4}, \qquad t = 1, 2, \cdots, m,$$

$$\lambda^2 = \frac{m(m + 3)}{4}, \qquad \|P\|^2 + \bar{c} = \frac{(m + 1)(m + 2)}{2}.$$

Then, we obtain a Frenet formula of W^2 in $\bar{M}^{n+2m}$:

$$
(33)\qquad
\begin{aligned}
dx &= \frac{1}{h}(\bar{\xi}_0\,dz + \xi_0\,d\bar{z}),\ \bar{D}P = -\frac{\|P\|^2}{h}(\bar{\xi}_0\,dz + \xi_0\,d\bar{z}),\\
\bar{D}\xi_0 &= \frac{1}{h}\xi_0(\bar{z}dz - zd\bar{z}) + \frac{2}{h}Pdz + \frac{2\,b_1{}^{1/2}}{h}\xi_1\,d\bar{z},\\
\bar{D}\xi_1 &= -\frac{2\,b_1^{1/2}}{h}\xi_0\,dz + \frac{2}{h}\xi_1(\bar{z}dz - zd\bar{z}) + \frac{2\,b_2^{1/2}}{h}\xi_2\,d\bar{z},\\
&\cdots\cdots\cdots\cdots\cdots\cdots\cdots\cdots\cdots\cdots\\
\bar{D}\xi_t &= -\frac{2\,b_t^{1/2}}{h}\xi_{t-1}\,dz + \frac{t+1}{h}\xi_t(\bar{z}dz - zd\bar{z}) + \frac{2\,b_{t+1}^{1/2}}{h}\xi_{t+1}\,d\bar{z},\\
&\cdots\cdots\cdots\cdots\cdots\cdots\cdots\cdots\cdots\cdots\\
\bar{D}\xi_m &= \frac{2\,b_m^{1/2}}{h}\xi_{m-1}\,dz + \frac{m+1}{h}\xi_m(\bar{z}dz - zd\bar{z}),
\end{aligned}
$$

where z is an isothermal coordinate of W^2 such that $ds^2 = 4dzd\bar{z}/(1+z\bar{z})^2\ \omega_1 + i\omega_2 = e^{-i\theta}(2\,dz/(1+z\bar{z}))$, $h = 1 + z\bar{z}$ and $\xi_0 = e^{i\theta}(e_1 + ie_2)$, $\xi_t = e^{i(t+1)\theta}\cdot(e_{\alpha_1} + i\sigma_t e_{\alpha_2})$, $\alpha_1 = n + 2t - 1$, $\alpha_2 = n + 2t$, $t = 1, \cdots, m$.

In case $\bar{M}^{n+2m} = E^{n+2m}$, we obtain a solution of (33) which is uniquely determined in a sense and lies in a $(2 + 2m)$-sphere with radius

$$\frac{1}{p} = \left(\frac{2}{(m+1)(m+2)}\right)^{1/2}.$$

We denote this surface by W_m^2 and the one enlarged by the similarity of magnification p by V_m^2. The surface V_m^2 is given by the equations:

$$
(34)\qquad
\begin{aligned}
x_0 &= \frac{1}{(1+z\bar{z})^{m+1}}\sum_{s=0}^{m+1}(-1)^s\binom{m+1}{s}^2(z\,\bar{z})^s\\
x_{2t-1} &= \frac{1}{(1+z\bar{z})^{m+1}}\left(\frac{(m+1)\,m\cdots(m-t+2)}{2(m+t+1)(m+t)\cdots(m+2)}\right)^{1/2}\\
&\quad\times\sum_{s=0}^{m-t+1}(-1)^s\binom{m+t+1}{s+t}\binom{m-t+1}{s}(z\,\bar{z})^s(z^t + \bar{z}^t),\\
x_{2t} &= \frac{-i}{(1+zz)^{m+1}}\left(\frac{(m+1)\,m\cdots(m-t+2)}{2(m+t+1)(m+t)\cdots(m+2)}\right)^{1/2}\\
&\quad\times\sum_{s=0}^{m-t+1}(-1)^s\binom{m+t+1}{s+t}\binom{m-t+1}{s}(z\,\bar{z})^s(z^t - \bar{z}^t), t = 1, 2, \cdots, m+1,
\end{aligned}
$$

and minimal in the unit sphere S^{2m+2}. When $m = 1$, V_1^2 is the Veronese surface in S^4, so V_m^2 may be called the *generalized Veronese surface of index m*.

We can construct an example of M^n of S-type in Theorem 5 by analogous methods stated in § 4.

6. Examples of M^n of T-type. When M^n is of T-type in Theorem 6, for W^2 we obtain in fact : $p^2 = 2\lambda^2 - \bar{c}$ and $\Psi_t = \lambda^t(1, i\sigma_t)$, $t = 1, \cdots, m$; $\Psi_{m+1} =$

$2^{1/2}\lambda^{m+1}e^{-i(m+2)\theta}$, where θ is as in Theorem 2. Setting $\xi_0 = e^{i\theta}(e_1 + ie_2)$, $\xi_t = e^{i(t+1)\theta}(e_{\alpha_1} + i\sigma_t e_{\alpha_2})$, $\alpha_1 = n + 2t - 1$, $\alpha_2 = n + 2t$, $t = 1, \cdots, m$, we obtain a Frenet formula of W^2 in $\bar{M}^{n+2m+1}$:

$$
(35)\qquad
\begin{aligned}
dx &= \tfrac{1}{2}(\bar{\xi}_0\, dz + \xi_0\, d\bar{z}), \qquad \bar{D}\xi_0 = P dz + \lambda\xi_1\, d\bar{z},\\
\bar{D}\xi_t &= -\lambda\xi_{t-1}\, dz + \lambda\xi_{t+1}\, d\bar{z}, \qquad t = 1, 2, \cdots, m-1,\\
\bar{D}\xi_m &= -\lambda\xi_{m-1}\, dz + 2^{1/2}\lambda\, e_{n+2m+1}\, d\bar{z},\\
\bar{D}e_{n+2m+1} &= -\left(\tfrac{1}{2}\right)^{1/2}\lambda(\xi_m\, dz + \bar{\xi}_m\, d\bar{z}),\\
\bar{D}P &= -\frac{\|P\|^2}{2}(\bar{\xi}_0\, dz + \xi_0\, d\bar{z}).
\end{aligned}
$$

Here, we shall give an example of W^2 in case $\bar{M}^{n+2m+1} = E^{n+2m+1}$. Since $p = \|P\| = 2^{1/2}\lambda = 1/v$ in this case, we take the fixed point $x + (1/p^2)P$ as the origin. The point x and the vectors $\xi_0, \xi_1, \cdots, \xi_m, e_{n+2m+1}$ and P satisfy the same equation

$$\partial^2 X / \partial z \partial\bar{z} = -\lambda^2 X.$$

Using this fact, we can obtain a solution surface of (35) given by

$$
(36)\qquad
\begin{aligned}
x = -v\Bigg[\sum_j \Bigg\{ & A_j \exp\frac{i2^{1/2}}{v}\left(u_1 \sin\frac{(2j-1+\varepsilon)\pi}{2(m+2)} + u_2\cos\frac{(2j-1+\varepsilon)\pi}{2(m+2)}\right)\\
&+ \bar{A}_j \exp\frac{-i2^{1/2}}{v}\left(u_1 \sin\frac{(2j-1+\varepsilon)\pi}{2(m+2)} + u_2\cos\frac{(2j-1+\varepsilon)\pi}{2(m+2)}\right)\Bigg\}\Bigg]
\end{aligned}
$$

where $\varepsilon = 0$ or 1 according as m is odd or even and A_j, $j = 1, \cdots, m+2$, are constant complex vectors in $\boldsymbol{C}^{m+2} = \boldsymbol{R}^{2m+4}$ such that

$$A_j\cdot A_j = 0,\quad A_j\cdot\bar{A}_j = \frac{1}{2(m+2)},\quad A_j\cdot A_k = A_j\cdot\bar{A}_k = 0 \qquad (j \neq k).$$

Making use of this surface, we can construct an example of M^n in E^{n+2m+1} and also the sphere S^{n+2m+1} and the hyperbolic space H^{n+2m+1} by analogous methods described in § 2.

References

1. S. S. Chern, M. do Carmo and S. Kobayashi, *Minimal submanifolds of a sphere with second fundamental form of constant length*, Functional Analysis and Related Fields (Proc. Conf. for M. Stone, Univ. Chicago, Chicago, Ill., 1968), Springer, New York, 1970, pp. 59–75. MR **42** #8424.

2. T. Itoh, *Minimal surfaces with M-index 2, T_1-index 2 and T_2-index 2*, Kōdai Math. Sem. Rep. **24** (1972), 1–16. MR **45** #9253.

3. T. Ôtsuki, *A theory of Riemannian submanifolds*, Kōdai Math. Sem. Rep. **20** (1968), 282–295. MR **38** #2707.

4. ———, *Pseudo-umbilical submanifolds with M-index $\leqq 1$ in Euclidean spaces*, Kōdai Math. Sem. Rep. **20** (1968), 296–304. MR **38** #5139.

5. ———, *Minimal hypersurfaces in a Riemannian manifold of constant curvature*, Amer. J. Math. **92** (1970), 145–173. MR **41** #9157.

6. ———, *On principal normal vector fields of submanifolds in a Riemannian manifold of constant curvature*, J. Math. Soc. Japan **22** (1970), 35–46. MR **40** #6457.

7. ———, *Submanifolds with a regular principal normal vector field in a sphere*, J. Differential Geometry **4** (1970), 121–131. MR **42** #1020.

8. T. Ôtsuki, *On minimal submanifolds with M-index* 2, J. Differential Geometry **6** (1971/72), 193–211. MR **46** #833.

9. ———, *Minimal submanifolds with M-index* 2 *in Riemannian manifolds of constant curvature*, Tôhoku Math. J. (2) **23** (1971), 371–402. MR **47** #2515.

10. ———, *Minimal submanifolds with m-index* 2 *and generalized Veronese surfaces*, J. Math. Soc. Japan **24** (1972), 89–122. MR **45** #4327.

TOKYO INSTITUTE OF TECHNOLOGY

Proceedings of Symposia in Pure Mathematics
Volume 27, 1975

A PROBLEM OF ORDNUNGSGEOMETRIE

WILLIAM F. POHL*

R. Thom [4] has given a proof, admittedly incomplete, of the following theorem: a compact real $2k$-dimensional differentiable manifold embedded in CP^{n+k}, the complex projective $(n + k)$-space, which meets a dense set of complex n-planes each in the same number of points must be an algebraic variety. In [3] the author has given a complete and corrected statement and proof of this result. Moreover he proves there a more general result in that not just embeddings are considered but even immersed submanifolds and "submanifolds with singularities". The proof is complicated by the need for elaborate general position lemmas to handle the singularities, and certain algebraic-geometric problems which arise only in higher dimensions. The aim of this seminar is to present the essential steps of the proof of the theorem in the simplest case, so that the main geometric ideas may be brought out, unencumbered by the technical trappings of the general case.

The special case is the following. Let CP^{2*} denote the space of complex lines in CP^2, i.e. the dual projective space.

THEOREM. *Let $M^2 \subset CP^2$ be a compact connected embedded real surface, differentiable of class C^4. Suppose that there exists an everywhere dense subset $U \subset CP^{2*}$ such that if $\lambda \in U$ then $\lambda \cap M^2$ consists of exactly m points, m independent of λ. Then M^2 is either an algebraic curve or, up to a complex projective transformation, the real projective plane with its canonical embedding $RP^2 \subset CP^2$.*

The main considerations of the proof do not involve the main hypothesis of the Theorem, but give information about general real surfaces in CP^2. Let $M \subset C^2 =$

AMS (MOS) subject classifications (1970). Primary 53C75; Secondary 53C40, 53A20.

*Research supported by the National Science Foundation under grant GP-29321.

R^4 be an arbitrary C^4 embedded real surface. (Later M will lie in CP^2.) Let $p \in M$. If the tangent plane to M at p is a complex line in C^2, we say that p is a *point of type C*. Suppose that p is not a point of type C; let t be a real tangent line to M at p, and $K(t)$ the real 3-space in R^4 spanned by the tangent space to M at p and the complex line in $C^2 = R^4$ spanned by t. Let S be a curve on M tangent to t at p. If the curvature vector to S at p is contained in $K(t)$, we say that t is a *direction of type F*. (By Meusnier's theorem this condition depends only on t and not on the choice of S.) If every real tangent line at p is a direction of type F, we say that p is a *point of type F*. If p is not a point of type C or F, then for some real tangent line t at p, the curvature vector of a curve on M tangent to t lies outside $K(t)$, in which case we say that t is an *ordinary direction* and p a *point of type* 0. It may be shown (Lemma 1 below) that these properties of points and tangent directions of M are preserved under complex projective transformation of M in CP^2, so that these properties are well defined for submanifolds $M \subset CP^2$. Having made the classification of points into types 0, F, and C, we can now outline the proof of the Theorem.

PROPOSITION A. *Suppose $M \subset CP^2$ is connected and consists entirely of points of type F. Then a complex projective transform of M lies in $RP^2 \subset CP^2$.*

PROPOSITION B. *Suppose $M \subset CP^2$ is compact and contains a point of type* 0. *Then there exist open sets $U_1, U_2 \subset CP^{2*}$ and a positive integer n such that if $\lambda \in U_1$ then $\lambda \cap M$ consists of exactly n points, and if $\lambda \in U_2$ then $\lambda \cap M$ consists of exactly $n + 2$ points.*

Given Propositions A and B the proof of the Theorem is concluded in the following way. Suppose $M^2 \subset CP^2$ satisfies the hypothesis of the Theorem. The conclusion of Proposition B is clearly incompatible with the hypothesis of the Theorem. Consequently M^2 contains no points of type 0 and hence every point is of type C or F. If M contains a point p of type F, then p cannot be a limit point of points of type C for otherwise the tangent plane at p would also be a complex line. Hence some neighborhood of p in M^2 consists of points of type F, so that the set of points of type F is open. Let M' be the largest connected neighborhood of p in M consisting of points of type F. Then by Proposition A we can apply a complex projective transformation to CP^2 to bring M' into RP^2. Suppose M' has a boundary point q in M^2. If q lies at infinity, apply a complex projective transformation with real coefficients to bring q to a finite point. M' still lies in RP^2. Now the tangent plane to RP^2 at its finite points is constant in C^2. Hence by continuity the tangent plane to M^2 at q must be this same plane, so that the tangent plane to M^2 at q is not a complex line. Since M^2 contains no points of type 0, q must then be a point of type F. But the set of points of type F is open in M^2, contradicting the assumption that q be a boundary point. Consequently $M' = M^2$ and by Proposition A, $M^2 \subset RP^2 \subset CP^2$. Since M^2 is compact, $M^2 = RP^2$. All this comes from the assumption that M^2 contains a point of type F. If M^2 contains a point of type C, then every point of M^2 must be of type C (because if M^2 contained a point of type

F, every point would be of type F, as we have just shown). Then the Cauchy-Riemann equations hold, so that M^2 is an analytic curve, and hence by Chow's theorem an algebraic curve. This completes the proof of the Theorem given Propositions A and B.

The proof of Propositions A and B rests on the following. Let $M \subset CP^2$ be arbitrary as before and let $\pi : T(M) \to M$ denote the bundle of real tangent lines of M (so that $T(M)$ has dimension 3).

LEMMA 1. *Let $l : T(M) \to CP^{2*}$ be the mapping which assigns to each real tangent line t the complex line in CP^2 spanned by t. Then l has rank* 3 *at t if and only if t is an ordinary direction, and rank* 2 *if t is a direction of type F.*

This can be proved in various ways; one is given in [**3**, Proposition 1].

Let us indicate first the proof of Proposition A. We assume that every point of M is of type F. It follows from Lemma 1 that $l : T(M) \to CP^{2*}$ is an immersed surface in CP^{2*}, which we call the *dual surface* of M. Now by a "circle" lying in CP^1 we mean an ordinary circle lying on the Riemann sphere after the latter is canonically identified with CP^1. By a "circle" in a complex projective space we mean a "circle" lying on some projective line lying in the projective space. If we restrict l to a fibre $T_p = \pi^{-1}(p)$ of the bundle $\pi : T(M) \to M$, it may be shown by an elementary analytic-geometric argument that $l(T_p)$ is a "circle" in CP^{2*}. We see therefore that our dual surface contains a two-parameter family of "circles". It may be shown that through every point q of the dual surface there passes a one-parameter family of such "circles" and that the set of points of the dual surface different from q which may be joined to q by such "circles" is open in the dual surface. At such a point q choose two of these "circles" with distinct tangent lines, C_1 and C_2, and choose a point on each "circle", p_1 and p_2, distinct from q. Take as line at infinity in CP^{2*} a complex line containing p_1 and p_2 but not q. The circles C_1 and C_2 are now euclidean straight lines in $R^4 = C^2 \subset CP^{2*}$. We must now use the euclidean geometry of R^4. At q there are now two distinct asymptotic directions. It may be shown by consideration of the second fundamental form that this implies that the curvature vectors at q of all curves on the dual surface through q lie in a single real 3-space, which is spanned by the tangent plane at q and a single normal line. But it is easily shown that the tangent plane at q is not a complex line and hence the complex lines spanned by real tangent lines at q cannot all lie in any single real 3-space. But if the curvature vector of a "circle" is nonzero, that curvature vector together with the tangent line spans the complex line containing that "circle" over the reals. It follows that the curvature vectors of the "circles" on the dual surface through q must vanish, so that all these "circles" are euclidean straight lines, and hence that the (open!) set of points of the dual surface which are joined to q by "circles" lying in the dual surface must lie in the tangent plane at q. From this it follows by easy arguments that the dual surface lies in a real 2-plane and that M^2 lies in a real 2-plane in CP^2 which may be brought to RP^2 by a complex projective transformation. This, in essence, is the proof of Proposition A.

The proof of Proposition B depends on a study of the secant lines, the importance of which I have insisted on for years [2]. Let $S(M) = M \times M - \Delta \cup T(M)$, where $\Delta = \{(x, x) \mid x \in M\}$. This carries a differentiable structure of class C^3 [1], and the map $L : S(M) \to CP^{2*}$ defined by

$$L(x, y) = \text{the complex line in } CP^2 \text{ joining } x \text{ and } y, \qquad x, y \in M,\ x \neq y,$$
$$L(t) = l(t) = \text{the complex line in } CP^2 \text{ spanned by } t, \quad t \in M,$$

is differentiable of class C^3. We call L the *secant map*. The heart of the proof of Proposition B is the following:

LEMMA 2. *Let $t \in T(M)$ be an ordinary direction. Then L is a fold with center $T(M)$ at t.*

Here we say that a map $\varphi : A \to B$ from one 4-dimensional manifold into another is a *fold at x with center C*, $C \subset A$ an embedded submanifold of dimension 3, if $x \in C$ and there exist local coordinate systems $x_1, \cdots, x_4$ and $y_1, \cdots, y_4$ centered at $x \in A$ and $\varphi(x) \in B$, respectively, such that C is defined locally by $x_1 = 0$ and φ takes the form

$$y_1 = x_1^2,$$
$$y_i = x_i, \quad i = 2, 3, 4,$$

in a neighborhood of x. For the proof we refer to [3, Proposition 4].

The proof of Proposition B rests also on some general position arguments. Let $\varphi : N_1 \to N_2$ be a differentiable map of differentiable manifolds. We say that $x \in N_1$ is a *good point* of φ if there exists an open neighborhood U of $\varphi(x)$ in N_2 such that $\varphi^{-1}(U)$ consists of finitely many connected components $V_1, \cdots, V_r$ each of which is mapped diffeomorphically by φ onto $\varphi(V_1)$.

LEMMA 3. *Let $\varphi : N_1 \to N_2$ be differentiable, N_1 compact, and $y \in N_1$ a point at which the Jacobian rank of φ is equal to the dimension of N_1. Then there exists a good point of φ arbitrarily close to y in N_1.*

For the proof we refer to [3, Lemma 7]. Let us proceed to the proof of Proposition B. Let $p \in M$ be a point of type 0, $t \in T_p$ an ordinary direction. Then $l : T(M) \to CP^{2*}$ has rank 3 at t, by Lemma 1. It follows from Lemma 3 that there exists a good point t_0 of l. Let U be the neighborhood of $l(t_0)$ and $V_1, \cdots, V_r$ the neighborhoods whose existence is guaranteed by the definition of "good point". We claim that $r = 1$; for otherwise M admits a "three-parameter family of bitangent complex lines". But the impossibility of this is proved by a kinematic argument using moving frames no different in principle from the argument in the elementary differential geometry that a curve in the ordinary plane has no continuous family of bitangent lines.

Since L is a fold at t_0 with center $T(M)$, by Lemma 2, we can take local coordinates $y_1, \cdots, y_4$ in a neighborhood $U' \subset U$ of $l(t_0)$ in CP^{2*} and local coordinates $x_1, \cdots, x_4$ in a neighborhood of t_0 in $S(M)$ such that $T(M)$ is defined locally by $x_1 = 0$ and the map L takes the form $y_1 = x_1^2$, $y_i = x_i$, $i = 2, 3, 4$. Let $U'_+, U'_- \subset U'$

denote the regions $y_1 > 0$ and $y_1 < 0$, respectively. The situation is now the following: $l(t_0)$ is tangent to M only at p and meets M in, say, n further points, at each of which it is transversal to M. Each $\lambda \in U'_+ \cup U'_-$ meets M transversally. Each $\lambda \in U'_+$ is the image under L of a pair of points "near" to t_0 in $S(M)$; each $\lambda \in U'_-$ is not the image of such a pair of points. After further restriction one can find open sets $U_1 \subset U'_-$ and $U_2 \subset U'_+$ such that if $\lambda \in U_1$ then $\lambda \cap M$ consists of n points and if $\lambda \in U_2$ then $\lambda \cap M$ consists of $n + 2$ points. This completes the proof in outline of Proposition B.

References

1. W.F. Pohl, *Some integral formulas for space curves and their generalization*, Amer. J. Math. **90** (1968), 1321–1345. MR **38** #6523.

2. ———, *The differential geometry of secants*, Differentialgeometrie im Grossen, Berichte aus dem Mathematischen Forschungsinstitut Oberwolfach, vol. 4, 1971.

3. ———, *A theorem of géométrie finie*, J. Differential Geometry (to appear)

4. R. Thom, *Sur les variétés d'ordre fini*, Global Analysis, (Papers in Honor of K. Kodaira), Univ. of Tokyo Press, Tokyo, 1969, pp. 397–401. MR **40** #8068.

University of Minnesota

Proceedings of Symposia in Pure Mathematics
Volume 27, 1975

ON THE HESSIAN OF A FUNCTION AND THE CURVATURES OF ITS GRAPH

ROBERT C. REILLY*

Let $\mathscr{D}$ be a bounded domain in $\boldsymbol{R}^m$ whose boundary is a smooth hypersurface in $\boldsymbol{R}^m$. If f is a smooth function on $\bar{\mathscr{D}}$ we let $S_q(f)$ denote the qth elementary symmetric function of the eigenvalues of the Hessian matrix of f; we set $W^2 = \sum(\partial f/\partial k_j)^2 + 1$.

THEOREM. *If f and g are smooth functions on $\bar{\mathscr{D}}$ such that* grad(f) = grad(g) *on $\partial\mathscr{D}$, then $\int_{\mathscr{D}} S_q(f)\, dx_1 \cdots dx_m = \int_{\mathscr{D}} S_q(g)\, dx_1 \cdots dx_m$.*

THEOREM. *If f is smooth on $\bar{\mathscr{D}}$ and $f|\partial\mathscr{D}$ is constant, and if q is even and $\mathscr{D}$ is strictly convex, then $\int_{\mathscr{D}} S_q(f)\, dx_1 \cdots dx_m \geqq 0$, with equality if and only if the normal derivative of f vanishes on $\partial\mathscr{D}$.*

Now suppose that M is the graph of f. We let $S_q(M)$ and $T_q(M)$ denote, respectively, the qth elementary symmetric function of the principal curvatures and the qth Newton tensor. We denote the shape operator by B and the projection of the vector $A = (0, 0, \cdots 1)$ onto M by A^T.

THEOREM. $S_q(f) = W^q(S_q(M) + W^2\langle (T_{q-1}(M) \cdot B)(A^T), A^T\rangle)$.

REMARK. From this formula we conclude that Chern's Theorem 4 of [3] follows from a similar result in Flanders' paper [4]. Moreover, if we could extend Flanders' result to state that an inequality of the form $|S_q(f)| \geqq CW^{1+\varepsilon} > 0$ implies a bound on the size of the domain, then automatically we could also extend Chern's result.

AMS (MOS) subject classifications (1970). Primary 53B25, 26A57, 26A66.
*This is an abstract of a paper to appear in Michigan Math. J.

Unfortunately, it appears to be quite difficult to extend Flanders' result.

If q is even then we interpret the tensor $T_{q-1}(M) \cdot B$ as a kind of Ricci tensor for the q-sectional curvature of Thorpe [10] and we show how it relates to the generalized Einstein tensor of Lovelock [5].

Finally we derive the following formula.

THEOREM. *If M is the graph of f then*

$$(q+1)\, S_{q+1}(M) = \sum_{ij} \frac{\partial}{\partial x_i}\left(\frac{1}{w} T_q(M)_j{}^i f_j\right).$$

Here $T_q(M)^i_j$ denotes the matrix of $T_q(M)$ relative to the coordinate frame field, and $f_j = \partial f/\partial x_j$ This formula generalizes the well-known formula for the first mean curvature. As an application we have

COROLLARY. *If* $\operatorname{grad} f = 0$ *on* $\partial\mathscr{D}$ *then* $\int_{\mathscr{D}} S_q(M)\, dx_1 \cdots dx_m = 0$.

BIBLIOGRAPHY

1. R. Abraham, *Foundations of mechanics*, Benjamin, New York, 1967. MR **36** #3527.

2. S. Bernstein, *Sur la généralisation due problème de Dirichlet*, Math. Ann. **69** (1910), 82–136.

3. S.-S. Chern, *On the curvatures of a piece of hypersurface in euclidean space*, Abh. Math. Sem. Univ. Hamburg **29** (1965), 77–91. MR **32** #6376.

4. H. Flanders, *Non-parametric hypersurfaces with bounded curvatures*, J. Differential Geometry **2** (1968), 265–277. MR **39** #2097.

5. D. Lovelock, *The Einstein tensor and its generalizations*, J. Mathematical Phys. **12** (1971), 498–501. MR **43** #1588.

6. R. C. Reilly, *Extrinsic rigidity theorems for compact submanifolds of the sphere*, J. Differential Geometry **4** (1970), 487–497. MR **44** #7980.

7. ———, *Variational properties of functions of the mean curvatures for hypersurfaces in space forms*, J. Differential Geometry **8** (1973), 465–477.

8. H. Rund, *Integral formulae on hypersurfaces in Riemannian manifolds*, Ann. Mat. Pura Appl. (4) **88** (1971), 99–122. MR **46** #8136.

9. U. Simon, *Minkowskische Integralformeln und ihre Anwendungen in der Differentialgeometrie im Grossen*. Math. Ann. **173** (1967), 307–321. MR **36** #2094.

10. J. A. Thorpe, *Sectional curvatures and characteristic classes*, Ann. of Math. (2) **80** (1964), 429–443. MR **30** #546.

UNIVERSITY OF CALIFORNIA, IRVINE

Proceedings of Symposia in Pure Mathematics
Volume 27, 1975

PAIRS OF METRICS ON PARALLEL HYPERSURFACES AND OVALOIDS

DONALD H. SINGLEY*

This paper has two purposes: First, we describe a number of general techniques that can be applied to any manifold with a pair of metrics. These techniques should prove useful in extracting information from pairs of metrics whenever such pairs arise in a geometric problem. Second, we give two specific examples of pairs of metrics, parallel hypersurfaces and ovaloids, to show how much of the geometry contained in a given situation is reflected by suitably chosen pairs of metrics. Such a special choice of metrics is actually essential to applications of this theory. Since the space of metrics on any paracompact manifold is quite large, an arbitrary pair of metrics on a manifold yields little information, and the geometric content almost always comes from a very specific choice of metrics, tailored to the problem at hand.

We define an *ovaloid* as a hypersurface M^n of E^{n+1} on which the second fundamental form is everywhere positive definite. Then both the first fundamental form and the second fundamental form define metrics on M^n, and so ovaloids are examples of manifolds admitting a pair of metrics. As a second example, if V is an arbitrary hypersurface in E^{n+1}, we define a *parallel hypersurface of* V, V_r, for r any fixed real number close to zero, by

$$V_r = \{p + rN_p \mid p \in V, N_p = \text{normal to } V \text{ at } p\}.$$

Then the induced Euclidean metrics on the two immersions V and V_r give us a pair

AMS (MOS) subject classifications (1970). Primary 53B25, 53A05, 53B20.

*This work was partially supported by the National Science Foundation under a National Science Foundation Graduate Fellowship and Grant NSF GP-29321.

of metrics in this case, as well. There are many other examples of pairs of metrics—for example, the right and left invariant metrics on a Lie group, or the obvious pair of metrics on a Riemannian manifold W which is diffeomorphic to another Riemannian manifold W'—but we will only deal with the two examples mentioned above in this paper. For the remainder of this paper, we denote the pair of metrics in any of these situations by G and G'.

The most basic of the techniques which can be used on pairs of metrics is proving that the two metrics are either identical or conformal—that is, $G' = f \cdot G$, for some function f. As an example, if the two metrics mentioned above on an ovaloid are conformal, then every point is umbilic, and the ovaloid is a Euclidean sphere. For the case of parallel surfaces in E^3, if the metrics are conformal, the surface is a sphere, a plane, or a surface of constant mean curvature $= -2/r$. [**1**] Clearly, this technique will yield strong results in other geometric situations.

A somewhat weaker, though still useful, structure on any manifold M^n with a pair of metrics G and G' is a set of n functions on M, called the mixed invariants of G and G'. These are defined as follows: Both G and G' give nonsingular linear maps from the tangent space of M at each point p, $T_p(M)$, to the cotangent space at p, $T_p^*(M)$, in the usual way. So, if we compose the inverse of the linear map G with the linear map G', we get a linear transformation on $T_p(M)$, which we denote by $G^{-1}G'(p)$. The characteristic polynomial of this linear transformation, $P(G^{-1}G')(t)$, is a polynomial of degree n. If we set

$$P(G^{-1}G')(t) = \sum_{i=0}^{n} (-1)^i \sigma_i(G^{-1}G')\, t^{n-i},$$

we get a set of n functions, σ_1 to σ_n, defined on M, which we call the *mixed invariants* of G and G'. Thus, for instance, $\sigma_1 = \mathrm{Trace}(G^{-1}G')$ and $\sigma_n = \mathrm{Det}(G^{-1}G')$.

We note that the σ_i's are all C^∞, if the two metrics are C^∞. However, their expression in terms of an arbitrary local basis is complicated, and calculating their derivatives in such a basis is tedious. So, we observe that a particular local basis simplifies the σ_i's enormously—namely, the one which simultaneously orthonormalizes G and orthogonalizes G'. With respect to this basis, each σ_i becomes simply the ith elementary symmetric function in the n real eigenvalues of $G^{-1}G'$, which we will henceforth denote by $g_1, \cdots, g_n$. We have one new difficulty to overcome, however. These eigenvalues are not usually differentiable on all of M. As an example, on a two-dimensional manifold, the eigenvalues are the roots of the quadratic characteristic polynomial and so equal $(\sigma_1 \pm ((\sigma_1)^2 - 4\sigma_2)^{1/2})/2$. But this expression is not necessarily differentiable at conformal points, where $(\sigma_1)^2 - 4\sigma_2 = 0$. However, with some additional work [**2**], we can show that the $\{g_i\}$'s are always C^∞ on an open, dense subset of M, and the corresponding eigenvectors can also always be chosen (locally) to be C^∞ on this same open, dense set. So, if we differentiate a σ_i and write the answer as a sum of terms, each of which has an invariant expression in terms of an arbitrary local basis, we know that the limit of each term exists on the complement of this dense set, and the formula for the derivatives of each σ_i can be extended to all of M, by a limiting argument.

With this difficulty overcome, we may now define the differential invariants of a pair of metrics. The first of these is the *difference tensor* of the two metrics, constructed as follows. Let $\{e_1, \cdots, e_n\}$ be any set of local frame fields around a point p and $\{w^1, \cdots w^n\}$ their dual one-forms. Let $\{w_j^i\}$, i and $j = 1, \cdots, n$, be connection forms for the unique Riemannian connection of the metric G, with respect to this basis, and $w_j'^i$ the connection forms for G', with respect to the same basis. By the first Cartan structure equations, we have

$$dw^j = -\sum_k w_k^j \wedge w^k, \tag{1}$$

$$dw^j = -\sum_k w_k'^j \wedge w^k. \tag{2}$$

We now define the *difference tensor*, K, by the equations

$$K_j^i = w_j'^i - w_j^i \quad \text{and} \quad K_j^i = \sum_k K_{jk}^i w^k.$$

It is well known **[1]** that the K_{jk}^i are the components of a (1, 2) tensor; moreover, by subtracting equations (1) and (2), we get

$$\sum_k K_k^j \wedge w^k = 0.$$

By Cartan's Lemma, this implies that $K_{kl}^j = K_{lk}^j$, for all l and k.

Often, the geometry contained in a pair of metrics yields an additional symmetry of the difference tensor. For example, on an ovaloid M, we choose a basis $\{e_1, \cdots, e_n\}$ which is orthonormal for the first fundamental form (G) and orthogonal for the second fundamental form (G'). The $\{g_i\}$'s then become the principal curvatures of the ovaloid, and the Codazzi equations for M yield, after some computation **[3]**, the symmetry

$$g_i K_j^i = g_j K_i^j \quad \text{for all } i \text{ and } j \text{ (no summation)}. \tag{3}$$

Similarly, if we let G be the induced metric on a hypersurface V, and G' the induced metric on the parallel hypersurface V_r, we again choose $\{e_1, \cdots, e_n\}$ which are orthonormal for G and orthogonal for G'. Then each $g_i = (1 + rk_i)^2$, where $k_i =$ the principal curvature of V in the direction e_i. (Thus, each e_i is a principal vector.) Finally, a computation which uses the Codazzi equations for both V and V_r yields the symmetry

$$g_i^{1/2} K_j^i = g_j^{1/2} K_i^j, \quad \text{all } i \text{ and } j \text{ (no summation)}. \tag{4}$$

An important numerical invariant for the difference tensor K is its rank, defined as follows. Given any tangent vector $Z \in T(M)$, define $Z \cdot K \in T^*(M) \circledcirc T^*(M)$, where $\circledcirc$ is the symmetric product, by

$$Z \cdot K = \sum_a Z_a K_{rs}^a w^r \circledcirc w^s.$$

Next, define the bilinear form B by the equation

$$B(Z, Y) = \operatorname{tr}_{G'}(\{Y \cdot K\}^t \{Z \cdot K\})$$

where $\operatorname{tr}_{G'}$ is trace with respect to G' and ${}^t\{Z \cdot K\}$ is the transpose of $Z \cdot K$. Thus, if the vectors e_i are orthonormal for G',

$$B(e_i, e_j) = \sum_{r,s} K^i_{rs} K^j_{rs}.$$

Notice that $\operatorname{tr}_{G'}(B) = \sum_i B(e_i, e_i) = \sum_{i,j,k} (K^i_{jk})^2 = 0$ if and only if $K \equiv 0$, i.e., if and only if the two metrics have the same connection.

DEFINITION. The rank of K = the rank of B as a bilinear form. Thus, the rank of K is a measure of the extent to which G and G' have the same connection forms. Again, in specific geometric examples, the rank of K has a geometric interpretation, as the following theorem shows:

THEOREM [4]. *If the pair of metrics comes from an ovaloid, and if the rank of K is q on an open set U, then U is $(m - q)$-umbilical (i.e., $m - q$ of the principal curvatures are equal on U), and the principal curvature associated to the $(m - q)$ umbilical directions is constant.*

(The proof is contained in [4].)

Given additional symmetries of the difference tensor, such as equations (3) and (4), we next differentiate them and use the second Cartan structure equations

$$dw^i_j = -\sum_k w^i_k \wedge w^k_j + R^i_j \qquad (R^i_j = \text{curvature form of } G), \tag{5}$$

$$dw'^i_j = -\sum_k w'^i_k \wedge w'^k_j + R'^i_j \qquad (R'^i_j = \text{curvature form of } G') \tag{6}$$

to obtain information on the curvatures of the two metrics. In the case of parallel hypersurfaces, no new information is obtained; however, in the case of ovaloids, differentiation of equation (3) plus a short computation yields the equation

$$g_k R'^k_j - g_j R'^j_k = g_k R^k_j - g_j R^j_k + 2 \sum_s g_s K^s_j \wedge K^s_k, \quad \text{all } j \text{ and } k. \tag{7}$$

Use of the Gauss equation

$$R^j_{iji} = g_i g_j \quad (\text{where } R^j_{iji} = (i, j)\text{-component of } R^j_i) \tag{8}$$

yields a further expression, which in dimension 2 becomes the following:

$$2R_{G'} = H - (1/R) \cdot \{\langle \operatorname{grad} H, M \rangle_G - \sum_{s,i,j} g_s (K^s_{ij})^2\}. \tag{9}$$

Here, $R_{G'}$ = the Gauss curvature in the second fundamental form metric G'; R = the usual Gauss curvature of the ovaloid; $H = \sigma_1 = g_1 + g_2$ = the mean curvature of the ovaloid; M is a vector field on the ovaloid given by $M = \sum_{i,m} K^i_{mm} e_i$, and $\langle\ ,\ \rangle_G$ is the metric given by G. From this we deduce the following:

THEOREM. *On any closed ovaloid, there is a point where $R_{G'} = H/2$.*

PROOF. By (9), at any critical point of H,

$$R_{G'} = \tfrac{1}{2}H + (1/2R)\cdot\Big(\sum_{s,i,j} g_s (K^s_{ij})^2\Big). \tag{10}$$

So, at each critical point, such as max H or min H, $R_{G'} \geqq H/2$.

But on the other hand, $R_{G'}$ cannot be greater than $H/2$ everywhere. This follows from the Gauss-Bonnet Theorem. In the metric G', the element of area $= (g_1 g_2)^{1/2} = R^{1/2}$, since the g_i 's are the principal curvatures. Hence,

$$\int R_{G'}\,(R^{1/2}\,dA) = \int R\ dA = \int R^{1/2}(R^{1/2}\,dA) = 4\pi,$$

by the Gauss-Bonnet Theorem for each metric. In addition, $H - 2R^{1/2} = (g_1^{1/2} - g_2^{1/2})^2 \geqq 0$, with equality only at umbilics. Hence,

$$\int (H - 2R^{1/2})R^{1/2}\,dA = \int (H - 2R_{G'})R^{1/2}\,dA \geqq 0. \tag{11}$$

Hence, $H - 2R_{G'}$ is not everywhere < 0, so $R_{G'} = H/2$ somewhere on the ovaloid.□

In fact, equation (11) proves more:

COROLLARY 1. *If $R_{G'} \geqq H/2$ everywhere, the ovaloid is a Euclidean sphere.*

PROOF. By equation (11), the hypothesis implies that $\int (H - 2R^{1/2})R^{1/2}\,dA = 0$. So, $H - 2R^{1/2} \equiv 0$, and every point is an umbilic. So, the ovaloid is a sphere. □

COROLLARY 2 (R. SCHNEIDER). *If $R_{G'}$ is constant, the ovaloid is a Euclidean sphere.*

PROOF. By equation (10) applied to max H, $R_{G'} \geqq H/2$ everywhere. Corollary 1 then gives the result.□

The last technique we will mention is that of integral formulas derived from Stokes' Theorem. We will henceforth assume that the manifold M is compact, orientable, and without boundary. Both metrics, G and G', give rise to Hodge $*$-operators, $*_G$ and $*_{G'}$, which map 1-forms to $(p-1)$-forms on M. So, we may define two Laplacian operators on M, characterized by the equations

$$d{*_G}d(f) = \Delta_G(f)\,dV_G, \tag{12}$$

$$d{*_{G'}}d(f) = \Delta_{G'}(f)\,dV_{G'}, \tag{13}$$

where f is any smooth function on M, dV_G is the volume element in the metric G, and $dV_{G'}$ is the volume element in the metric G'. By Stokes' Theorem,

$$\int \Delta_G(f)\,dV_G = 0, \tag{14}$$

$$\int \Delta_{G'}(f)\,dV_{G'} = 0. \tag{15}$$

So, for each of the n functions $\sigma_1, \cdots, \sigma_n$, equations (14) and (15) give integral formulas on M.

In order to write down these integral formulas, we need the following definitions: First, $\sigma_p(l)$ is the pth elementary symmetric function in the $(n-1)$-functions $g_1, \cdots, \hat{g}_l, \cdots, g_n$. Similarly, $\sigma_p(l, m)$ is the pth elementary symmetric function in the $(n-2)$-functions $g_1, \cdots, \hat{g}_l, \cdots, \hat{g}_m, \cdots, g_n$. The following formula holds for the $\sigma_p(l, m)$: For $0 \leqq p \leqq n-2$, and $l \neq m$,

$$\sigma_p(l, m) = \sum_{q=0}^{p} (-1)^q \left\{ \sum_{r=0}^{q} (g_l)^r (g_m)^{q-r} \right\} \sigma_{p-q}. \tag{16}$$

We then define the expression $\sigma_p(l, l)$ by setting $m = l$ in (16). Moreover, we set $\sigma_{-1}(l, m) \equiv 0$, $\sigma_{n-1}(l, m) = 0$ for $l \neq m$, and $\sigma_{n-1}(l, l) =$ the expression given by (16) for $p = n-1$ and $m = l$. Then the integral formulas given by (14) become, after a considerable amount of computation [3],

$$\begin{aligned} \Delta(\sigma_p) = {} & \operatorname{div} W_p - 2 \sum_{k,l} g_l \sigma_{p-1}(l) \{R'^l_{klk} - R^l_{klk}\} \\ & + \sum_{l,k,m} g_l g_m \sigma_{p-2}(l, m) \{-2K^m_{ml} K^l_{kk} + 4K^l_{lk} K^m_{mk} - 2K^l_{mk} K^m_{lk}\} \\ & + \sum_{l,k,m} g_l \sigma_{p-1}(l, m) \{-2K^l_{mm} K^l_{kk} + 2K^l_{mk} K^l_{mk}\}. \end{aligned} \tag{17}$$

Here, W_p is the one-form $W_p = 2 \sum_{k,l} g_l \sigma_{p-1}(l) K^l_{kk} w^l$, and div W_p is defined by the formula

$$d *_G W = (\operatorname{div} W)\, dV_G.$$

As before, R'^l_{klk} and R^l_{klk} are the components of the curvature tensors for G and G'. We do not give formulas of the form (15) here. Moreover, additional integral formulas for pairs of metrics can be found in [4].

In specific geometric applications, the additional information present usually allows us to simplify (17) and derive geometric consequences. As an example, for ovaloids the additional information derived from the Codazzi and Gauss equations gives us the following equation for $\Delta(\sigma_2)$:

$$\Delta(\sigma_2) = \operatorname{div} W_2 - \sum_{i>j} g_i g_j (g_i - g_j)^2 + \langle \operatorname{grad} \sigma_1, \operatorname{grad} \sigma_1 \rangle_G - 4 \sum_{i,j,k} g_i^2 (K^i_{jk})^2. \tag{18}$$

We can derive the following geometric consequences from this formula:

THEOREM (HSIUNG ET AL.). *If σ_1 is constant on a closed ovaloid, it is a Euclidean sphere.*

PROOF. grad $\sigma_1 \equiv 0$, and $\int \operatorname{div} W_2 \, dV = 0$, by Stokes' Theorem. Moreover, all the other terms on the right-hand side of (18) are everywhere $\leqq 0$, always. So, integrating both sides of (18) and using Stokes' Theorem again on the left-hand side, we conclude that these two other terms on the right must vanish identically. From the vanishing of the term $\sum g_i g_j (g_i - g_j)^2$, we conclude that every point is umbilic, and so the ovaloid is a Euclidean sphere. □

We also get a generalization of a theorem of Hilbert:

THEOREM. *On an ovaloid, σ_2 cannot have a relative minimum at a point p where σ_1 has a relative maximum, unless the point is an umbilic.*

PROOF. An easy argument from the fact that the Hessian of σ_1 is negative definite at p shows that div $W_2 \leqq 0$ at p. Moreover, grad $\sigma_1 = 0$ at p, and the other two terms on the right-hand side of (18) are everywhere $\leqq 0$. So, the right-hand side of (18) is $\leqq 0$. But, since σ_2 has a minimum at p, $\Delta(\sigma^2)(p) \geqq 0$. So, both sides are $= 0$ at p, and in fact each term in (18) $= 0$ at p. So, $\sum g_i g_j (g_i - g_j)^2 = 0$ at p, and $g_i = g_j$ for all i and j—i.e., the point is an umbilic. □

COROLLARY. *If σ_2 is a monotone decreasing function of σ_1 on an ovaloid, the ovaloid is a Euclidean sphere. In particular, if σ_2 is constant, the ovaloid is a sphere.*

PROOF. $\sigma_2 \leqq ((n-1)/2n)(\sigma_1)^2$, with equality only at umbilics [5]. Let $m =$ the point on the ovaloid where σ_1 has an absolute maximum. Since σ_2 is a monotone decreasing function of σ_1, m is the point where σ_2 has an absolute minimum. So, m is an umbilic point. For any other point q, we have

$$\sigma_2(q) \geqq \sigma_2(m) = ((n-1)/2n)\,\sigma_1^2(m) \geqq ((n-1)/2n)\,\sigma_1^2(q).$$

So, $\sigma_2(q) = ((n-1)/2n)(\sigma_1(q))^2$ for all points q, every point is umbilic, and the ovaloid is a sphere. □

A similar integral formula holds for σ_n:

$$\Delta(\sigma_n) = \operatorname{div} W_n - \sigma_n \sum_{i>j} (g_i - g_j)^2 - 4\sigma_n \Big(\sum_{i,j,k} K^i_{jk} K^j_{ik} - \sum_{i,m,k} K^i_{ik} K^m_{mk} \Big).$$

By arguments similar to those above, we can prove the

THEOREM. *On an ovaloid, σ_n cannot have a relative minimum at a point p where σ_1 has a relative maximum, unless the point is umbilic.*

COROLLARY. *If σ_n is a monotone decreasing function of σ_1 on an ovaloid, the ovaloid is a Euclidean sphere.*

The proof goes exactly as before, using the inequality $\sigma_n \leqq (1/n)^n \sigma_1^n$, with equality only at umbilics [5].

We do not state the theorems for $\Delta(\sigma_3), \cdots, \Delta(\sigma_{n-1})$, since these are more complicated. Similar theorems derived from these integral formulas for the case of parallel hypersurfaces will appear elsewhere.

REFERENCES

1. Noel J. Hicks, *Notes on differential geometry*, Van Nostrand Math. Studies, no. 3, Van Nostrand, Princeton, N.J., 1965. MR **31** #3936.

2. Donald Singley, *Smoothness theorems for the principal curvatures and principal vectors of a hypersurface*, Rocky Mountain J. Math. **5** (1975) (to appear).

3. ———, *Integral formulas for pairs of metrics with applications to convex hypersurfaces* (to appear).

4. Robert B. Gardner, *Subscalar pairs of metrics with applications to rigidity and uniqueness of hypersurfaces with a nondegenerate second fundamental form*, J. Differential Geometry **6** (1972), 437–458.

5. G. H. Hardy, J. E. Littlewood and G. Pólya, *Inequalities*, Cambridge Univ. Press, New York, 1934.

ARKANSAS STATE UNIVERSITY

Proceedings of Symposia in Pure Mathematics
Volume 27, 1975

SOME LEFT-OVER PROBLEMS FROM CLASSICAL DIFFERENTIAL GEOMETRY

MICHAEL SPIVAK

The classification of complete surfaces in $\boldsymbol{R}^3$ of constant curvature K is well known[1] [**17**, Chapter 5].

THEOREM A (HILBERT). *Any immersed complete connected surface in $\boldsymbol{R}^3$ with constant curvature $K > 0$ is a (standard) sphere.*

THEOREM B (HILBERT-HOLMGREN). *There is no complete surface immersed in $\boldsymbol{R}^3$ with constant curvature $K < 0$.*

THEOREM C. *A complete flat surface immersed in $\boldsymbol{R}^3$ is a generalized cylinder.*

Theorem A was actually first proved by Minkowski (see [**4**]); a different proof was given by Liebmann [**11**] in the analytic case, and the result often bears his name. Nowadays the proof is always based on a lemma of Hilbert which allows us to conclude that the surface in Theorem A is all-umbilic, so that we can apply the elementary

PROPOSITION 1. *A connected all-umbilic surface in $\boldsymbol{R}^3$ is part of a plane or sphere.*

Hilbert's original proof [**8**] of Theorem 2 was very complicated and not very convincing, but Holmgren [**9**] replaced the hardest part by a much simpler argument. Bieberbach [**3**] gave a proof, based on Hilbert's ideas, together with an unfounded criticism of Holmgren's proof. Actually, the right approach (implicit in Moore [**13**] and carried out in a more elementary way in do Carmo [**5**]) allows

AMS (MOS) subject classifications (1970). Primary 53-02, 53B25, 53C40.

[1] For all "standard" results, which are actually quite scattered throughout the literature, the reference will be [**17**].

one to give a proof without tears along these lines. Both this proof and Holmgren's are given in [**17**].

We will have no need to consider Theorem C; we merely mention that it is, oddly enough, the most recent (proved, in various formulations, by Pogorelov [**15**], [**16**], Hartman and Nirenberg [**7**], Massey [**12**], and Stoker [**18**]), and that it makes no use of the classical "classification" of flat surfaces.

When we replace $\boldsymbol{R}^3$ by the complete simply-connected 3-manifold of constant curvature $K_0 = 1$ or $K_0 = -1$, the situation is much less familiar, considerably more interesting, and not yet completely elucidated. For $K_0 = 1$ our ambient space is the familiar 3-sphere S^3. For $K_0 = -1$, it can be described in serveral ways, of which three will be important for us [**17**, Chapter 7A]. The "hyperbolic space" H^3, of constant curvature -1, may be defined, first of all, in terms of the "conformal model": this is the open ball of radius 2 with the complete metric

$$\sum_{i=1}^{3} dx^i \otimes dx^i \Big/ [1 - \tfrac{1}{4}\Sigma(x^i)^2]^2,$$

which is conformally equivalent to the usual metric $\Sigma\, dx^i \otimes dx^i$. It can also be described by means of the complete metric

$$\sum_{i=1}^{3} dx^i \otimes dx^i \Big/ (x^3)^2$$

on the upper half-space $x^3 > 0$. On the other hand, we also have the "projective model" of H^3: on the open ball of radius 1 there is a complete metric of constant curvature -1 such that every geodesic is an ordinary line segment (with a different parametrization).

The only important thing about the projective model is that it exists. The properties of the conformal model require more discussion. It turns out that the isometries of the conformal model of H^3 are precisely the usual conformal maps of the ball of radius 2 onto itself, and each of these is a composition of inversions and similarities (at most one of each). [On the other hand, if we compose an inversion through a point on the boundary of the ball with an appropriate orthogonal map, we obtain an isometry onto the upper half-space model.] The totally geodesic surfaces ("planes") are the planes or spheres intersecting the boundary of the ball orthogonally, while the set of points at a fixed distance from a given point ("spheres") is exactly the ordinary spheres completely contained in the open ball. The ordinary spheres intersecting the boundary of the ball in exactly one point are "horospheres"; they are obtained by considering a geodesic ray $\{\gamma(s): 0 \leqq s < \infty\}$ and taking the limit as $s \to \infty$ of the "sphere" around $\gamma(s)$ which passes through $\gamma(0)$. The map of the ball onto the upper half-space described above will take a horosphere to a plane parallel to the (x^1, x^2)-plane; the induced metric is then a constant multiple of $dx^1 \otimes dx^1 + dx^2 \otimes dx^2$, so horospheres are flat. Finally, the ordinary spheres which intersect the boundary of the ball in more than one point, but not orthogonally, are "equidistant surfaces"; they are one component of the set of points at a fixed distance from a "plane".

What happens when we replace $\boldsymbol{R}^3$ by S^3 in Proposition 1? It is easy to see that the surface in question must be part of some sphere in S^3, for one easily shows that an all-umbilic surface in S^3 is all-umbilic in $\boldsymbol{R}^4$, and hence a sphere [**17**, 7D]. For H^3 we need a classical

LEMMA (SCHOUTEN). *A conformal equivalence takes umbilics of hypersurfaces to umbilics on their image.*

The proof is a straightforward calculation [**17**, 7D]. We now see that connected all-umbilic surfaces in H^3 are parts of "planes", "spheres", horosopheres, and equidistant surfaces.

In considering Theorems A and B, we want to distinguish between the extrinsic curvature K_{ext} (the product of the principal curvatures) and the intrinsic curvature K_{int} of the surface as a Riemannian manifold. Gauss' equation gives

$$(*) \qquad \begin{aligned} K_{\text{int}} = K_{\text{ext}} + K_0 &= K_{\text{ext}} + 1 \quad \text{for } S^3, \\ &= K_{\text{ext}} - 1 \quad \text{for } H^3. \end{aligned}$$

The proofs of Theorems A and B use no extrinsic information except for the Codazzi-Mainardi equations, which hold just as well when the ambient space has constant curvature. The only problem is noting when K should be K_{int} and when it should be K_{ext}. It turns out that in both cases one part of the argument involves K_{int}, and one involves K_{ext}. What we obtain is[2]

THEOREM A′. *Any immersed complete connected surface in S^3 or H^3 with constant $K_{\text{int}} > 0$ and $K_{\text{ext}} \geqq 0$ is all-umbilic.*

THEOREM B′. *There is no complete surface immersed in S^3 or H^3 with constant $K_{\text{int}} < 0$ and $K_{\text{ext}} < 0$.*

These results, though they hold for both S^3 and H^3, end up having quite different consequences, because of (∗).

For S^3, we immediately see that

$$\begin{aligned} K_{\text{ext}} > 0 &\Rightarrow \text{surface is a small sphere,} \\ K_{\text{ext}} = 0 &\Rightarrow \text{surface is a totally geodesic "great" } S^2. \\ K_{\text{ext}} < -1 &\Rightarrow \text{surface does not exist.} \end{aligned}$$

A short argument, due to Ferus [**6**], shows that

$$-1 < K_{\text{ext}} < 0 \Rightarrow \text{surface does not exist.}$$

For our surface M would have $K_{\text{int}} > 0$, hence be compact. We can assume M orientable, by considering its double covering. Then M is homeomorphic to S^2, by Gauss-Bonnet. But $K_{\text{ext}} < 0$ means that the principal curvatures are always of different signs, allowing us to pick a continuous field of directions on S^2, which is impossible.

This leaves the isolated possibility $K_{\text{ext}} = -1 \Leftrightarrow K_{\text{int}} = 0$. The product tori in

---

[2] All results not referenced henceforth can be found in [**17**, 7E, F].

S^3 are all examples, and it is an attractive conjecture that these are the only possibilities. But in fact there are many others, all classified by Bianchi [**1**], in a truly beautiful geometric investigation. Although Bianchi did not consider completeness, this is a case, unlike the classification of flat surfaces in $\boldsymbol{R}^3$, where the local classification works just as well as a global classification.

The whole geometry of the situation depends on the fact that S^3 is the group of unit quaternions, and that left and right translations are isometries. (The quaternions never entered the picture for Bianchi, who only knew that there were two kinds of "Clifford translations", isometries $A: S^3 \to S^3$ with the property that $d(x, A(x))$ is constant.)

Let $M \subset S^3$ have $K_{\text{ext}} = -1$. There are 2 distinct asymptotic directions at each point, and the asymptotic lines form a Tchebycheff net (we can parametrize M so that the parameter curves are *arclength* parametrizations of the asymptotic lines); the proof of this is just one of the steps in proving Theorem B$'$. The metric on M then has the form

$$ds \otimes ds + \cos\omega\,[ds \otimes dt + dt \otimes ds] + dt \otimes dt,$$

where ω is the angle between the asymptotic curves. We easily compute that the s-parameter curves, for example, have geodesic curvature $\partial\omega/\partial s$. Since they are asymptotic curves, this is also their curvature as curves in S^3. But a classical computation gives

$$\partial^2\omega/\partial s\partial t = -K_{\text{int}} \cdot \sin\omega = 0, \quad \text{in our case.}$$

So

$$\omega(s, t) = S(s) + T(t) \Rightarrow \partial\omega/\partial s = S', \qquad \partial\omega/\partial t = T'.$$

So all s-parameter curves have the same curvature, and similarly for all t-parameter curves.

Now the classical Beltrami-Enneper theorem [**17**, 7D] says that the torsions of these asymptotic curves (as curves in S^3) are $\pm(-K_{\text{ext}})^{1/2} = \pm 1$, assuming the torsions τ exist. Let us assume this (i.e. assume the curves have nonzero curvature everywhere). Then the s-parameter curves, say, have $\tau = 1$ everywhere, and the t-parameter curves have $\tau = -1$ everywhere. Thus all s- and t-parameter curves are congruent. Moreover, under the 1-parameter family of isometries taking one s-parameter curve into the others, each point moves along an arclength parametrized t-parameter curve. This strongly suggests that all s-parameter curves are translates of each other. In fact,

(**) All s-parameter [t-parameter] curves are right [left] translates of each other.

In other words, our surface is a "translation surface", consisting of right [left] translates of any s-parameter [t-parameter] curve. A straightforward proof of (**) seems fairly hopeless. Instead we use

LEMMA. *A curve in* S^3 *has torsion* 1 $[-1]$ *everywhere if and only if its binormal is left* [*right*] *invariant along it.*

Bianchi proved this by considering $S^3 \subset \boldsymbol{R}^4$, but the nicest proof comes from expressing the Frenet frame of the curve in terms of left-invariant vector fields on S^3, and using the fact that a bi-invariant metric on a Lie group has covariant derivative ∇ given by $\nabla_X Y = \frac{1}{2}[X, Y]$ for left-invariant vector fields X and Y.

It is elementary, but slightly involved, to use the Lemma to deduce (**). But its relevance will be clearly seen when we take the converse situation. Let c and γ be curves of torsions 1 and -1, respectively, with $c(0) = \gamma(0) = 1 \in S^3$, and positioned so that their osculating planes coincide at 0. We consider the surface $M = \{c(s)\cdot\gamma(t)\colon s, t \in \boldsymbol{R}\}$. The Lemma, applied to c, shows that the osculating plane of c at s coincides with the osculating plane of $t \mapsto c(s)\cdot\gamma(t)$ at $t = 0$; hence these osculating planes coincide with the tangent space of M at $c(s)$. Similarly, by considering the curves $t \mapsto c(s)\cdot\gamma(t)$, we find that the same is true at every point of M. Then the Beltrami-Enneper theorem shows that M has $K_{\text{ext}} = -1$. A separate argument shows that the result still holds if one, or both, of c and γ is a geodesic. We thus see that there are many complete flat surfaces in S^3 other than product tori—we just have to choose c and γ as curves of infinite length.

Now consider H^3. We find that

$$K_{\text{ext}} > 1 \Rightarrow \text{the surface is a ``sphere''},$$
$$K_{\text{ext}} < 0 \Rightarrow \text{the surface does not exist.}$$

We will investigate the two extremes of the remaining interval $0 \leqq K_{\text{ext}} \leqq 1$. For $K_{\text{ext}} = 1 \Leftrightarrow K_{\text{int}} = 0$ we can assume that the surface M is simply-connected, and thus isometric to $\boldsymbol{R}^2$ with the standard metric $dx \otimes dx + dy \otimes dy$. The coefficients l_{ij} of the second fundamental form satisfy

$$\text{(1)} \qquad l_{11}l_{22} - (l_{12})^2 = 1 \quad \text{(Gauss)},$$

$$\text{(2)} \qquad \frac{\partial l_{12}}{\partial x} = \frac{\partial l_{11}}{\partial y}, \qquad \frac{\partial l_{22}}{\partial x} = \frac{\partial l_{12}}{\partial y} \quad \text{(Codazzi-Mainardi).}$$

So there is $\varphi\colon \boldsymbol{R}^2 \to \boldsymbol{R}$ with

$$\frac{\partial^2\varphi}{\partial x^2} = l_{11}, \qquad \frac{\partial^2\varphi}{\partial x \partial y} = l_{12}, \qquad \frac{\partial^2\varphi}{\partial y^2} = l_{22},$$

and (1) gives

$$\frac{\partial^2\varphi}{\partial x^2}\frac{\partial^2\varphi}{\partial y^2} - \left(\frac{\partial^2\varphi}{\partial x \partial y}\right)^2 = 1.$$

But a theorem of Jörgens [**10**] (or see Nitsche [**14**]) says that such a φ must be a quadratic polynomial, so the l_{ij} are constants. We can assume $l_{12} = 0$, by an orthogonal transformation of $\boldsymbol{R}^2$. So $l_{11} = k_1$ and $l_{22} = k_2$ are the principal curvatures, with $k_1k_2 = 1$; these constants determine M up to a congruence. For $k_1 = k_2$, M must be a horosphere. For other pairs (k_1, k_2) it turns out that M is given by a right circular cone in the upper half-space model, where the angle θ between the

z-axis and the generators is given by

$$k_1 = \sin\theta, \qquad k_2 = 1/\sin\theta.$$

This set can be described intrinsically as the set of points at a fixed distance from a geodesic. This classification of complete $M \subset H^3$ with $K_{\text{int}} = 0$ is due to Volkov and Vladimirov **[19]**, and independently to Sasaki.

Jörgens' theorem is usually applied to prove Bernstein's theorem about minimal surfaces in $\boldsymbol{R}^3$. Since minimal surfaces are intimately related to holomorphic maps, the following ancient result of Bianchi **[2]** may be of interest in this regard:

Consider a surface M in the upper half-space model of H^3. For $p \in (x, y)$-plane, let γ be a geodesic of H^3 tending to p which intersects M orthogonally, and let $f(p)$ be the other point in the (x, y)-plane towards which γ tends. Then f is holomorphic if and only if M has $K_{\text{int}} = 0$.

Now consider a surface $M \subset H^3$ with $K_{\text{ext}} = 0 \Leftrightarrow K_{\text{int}} = -1$. The same argument which is used for flat surfaces in $\boldsymbol{R}^3$ shows that if the second fundamental form is not 0 at $p \in M$, then p has a neighborhood which is a ruled surface. Synge's inequality **[17**, Corollary 1.6] shows that $K_{\text{int}} \leqq -1$ along any ruling γ, with equality if and only if the tangent space of M is parallel along γ; and this condition holds if and only if M is tangent to a "plane" along γ. This last condition involves only the geodesics of H^3, not its metric, so we are led to the following result:

> Let M be a surface contained in the open unit ball B, and let $\langle\ ,\ \rangle$ be the metric on the unit ball which gives the projective model of H^3. Then $M \subset (B, \langle\ ,\ \rangle)$ has $K_{\text{int}} = -1$ if and only if M is flat as a surface in $\boldsymbol{R}^3$.

Now there are (noncomplete) C^∞ flat surfaces M in $\boldsymbol{R}^3$ of arbitrary genus. For example, by gluing two C^∞ cylinders together along a completely planar region in each, we can obtain a surface homeomorphic to the torus minus a disc.

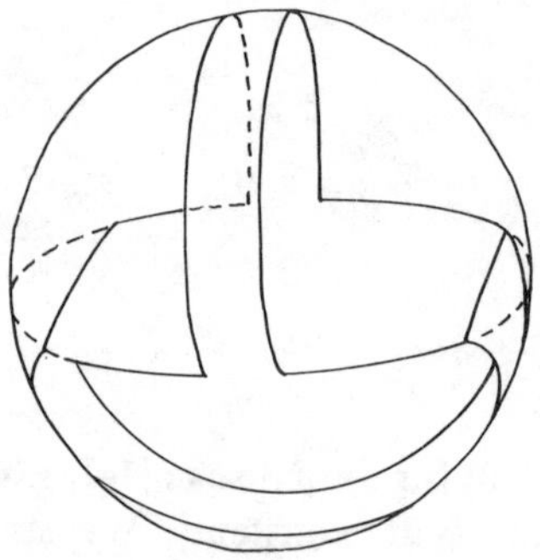

It is easy to arrange for the intersection of this surface with B to be complete in $(B, \langle\ ,\ \rangle)$. So there are C^∞ complete surfaces M in H^3, of arbitrary high genus, with $K_{\text{int}} = -1$. {I am almost certain that there are no *analytic* flat surfaces M in

$\boldsymbol{R}^3$ of genus > 0, but I hesitate to remove the word "almost" because I was burned once before when I found that the classical classification of flat surfaces, as cylinders, cones, and tangent developables, fails even in the analytic case; in fact there is an analytic Möbius strip in $\boldsymbol{R}^3$ (Wunderlich [**20**] and [**17**, Chapter 5]).}

REMARKS. In the lecture I was unable to give examples for $0 < K_{\text{ext}} < 1$, other than the equidistant surfaces. Milnor suggested that there should be infinitely many rotation surfaces for each such K_{ext}, and in fact there is always a l-parameter family of such surfaces, and one additional one [**17**, 7F]. I also raised the question whether there are closed curves in S^3 of torsion 1, and hence flat tori other than the product tori. Milnor pointed out that one can simply take various helices on flat tori; the resulting flat surfaces always seem to be immersed rather than imbedded. (For closed curves of torsion 1 in $\boldsymbol{R}^3$, see the article by J. Weiner.) I was informed that Lawson had observed that the inverse image under the Hopf map of any curve in S^2 is flat. Upon hearing of this, Calabi immediately gave the following proof: If the curve in S^2 is a circle, the inverse image is a product torus; for a point on a general curve, consider its osculating circle (in S^2)—the inverse image surfaces agree up to order 2 along the inverse image of the point, hence have the same curvature there! In these examples of flat imbedded tori, the asymptotic curves of one family are geodesics—it would be interesting to know when the curves in the other family are closed. It would also be interesting to know whether there are one-one imbeddings of $\boldsymbol{R}^2$ as a flat surface in S^3.

BIBLIOGRAPHY

1. L. Bianchi, *Sulle superficie a curvatura nulla in geometria ellittica,* Ann. Mat. **24** (1896), 93–129.

2. ———, *Differential geometrie,* Teubner, Leipzig, 1899.

3. L. Bieberbach, *Hilberts Satz über Flächen konstanter negativer Krümmung,* Acta Math. **48** (1926), 319–327.

4. T. Bonnesen and W. Fenchel, *Theorie der konvexen Körper,* Springer, Berlin, 1934.

5. M. do Carmo,

6. D. Ferus, *Der Satz von Liebmann in Räumen konstanter Krümmung,* Arch. Math. (Basel) **20** (1969), 296–300. MR **39** #6191.

7. P. Hartman and L. Nirenberg, *On spherical image maps whose Jacobians do not change sign,* Amer. J. Math. **81** (1959), 901–920. MR **23** #A 4106.

8. D. Hilbert, *Ueber Flächen von konstanter Gausscher Krümmung,* Trans. Amer. Math. Soc. **2** (1901), 87–99.

9. E. Holmgren, *Sur les surfaces à courbure constant négative,* C. R. Acad. Sci. Paris **134** (1902), 740–743.

10. K. Jörgens, *Über die Lösungen der Differentialgleichung $rt - s^2 = 1$,* Math. Ann. **127** (1954), 130–134. MR **15**, 961.

11. H. Liebmann, *Ueber die Verbiegung der geschlossenen Flächen positiver Krümmung,* Math. Ann. **53** (1900), 81–112.

12. W. S. Massey, *Surfaces of Gaussian curvature zero in Euclidean* 3*-space,* Tôhoku Math. J. (2) **14** (1962), 73–79. MR **25** #2527.

13. J. D. Moore, *Isometric immersions of space forms in space forms,* Pacific J. Math. **40** (1972), 157–166.

14. J. C. C. Nitsche, *Elementary proof of Bernstein's theorem on minimal surfaces*, Ann. of Math. (2) **66** (1957), 543–544. MR **19,** 878.

15. A. W. Pogorelov, *Continuous maps of bounded variations*, Dokl. Akad. Nauk SSSR **111** (1956), 757–759. (Russian) MR **19,** 309.

16. ———, *Extensions of the theorem of Gauss on spherical representation to the case of surfaces of bounded extrinsic curvature*, Dokl. Akad. Nauk SSSR **111** (1956), 945–947. (Russian) MR **19,** 309.

17. M. Spivak, *A comprehensive introduction to differential geometry*. Vol. 3, Publish or Perish, Boston, 1974.

18. J. J. Stoker, *Developable surfaces in the large*, Comm. Pure. Appl. Math. **14** (1961), 627–635. MR **25** #493.

19. Ju. A. Volkov and S. M. Vladimirova, *Isometric immersions of the Euclidean plane in Lobačevskiĭ space*, Mat. Zametki **10** (1971), 327–332 = Math. Notes **10** (1971), 619–622. MR **45** #2624.

20. W. Wunderlich, *Über ein abwickelbares Möbiusband*, Monatsh. Math. **66** (1962), 276–289. MR **26** #680.

UNIVERSITY OF CALIFORNIA, BERKELEY

Proceedings of Symposia in Pure Mathematics
Volume 27, 1975

MANIFOLDS AND SUBMANIFOLDS WITH VANISHING WEYL OR BOCHNER CURVATURE TENSOR

KENTARO YANO

1. Introduction. Let M^n be an n-dimensional Riemannian manifold of class C^∞ covered by a system of coordinate neighborhoods $\{V; \eta^h\}$, where here and in the sequel the indices $h, i, j, k, \cdots$ run over the range $\{1, 2, \cdots, n\}$ and let g_{ji}, ∇_j, $R_{kji}{}^h$, R_{ji} and R be positive definite metric tensor, the operator of covariant differentiation with respect to the Christoffel symbols $\{{h \atop ji}\}$ formed with g_{ji}, the curvature tensor, the Ricci tensor and the scalar curvature of M^n respectively.

If M^n is locally conformal to a Euclidean space, then M^n is said to be conformally flat. It is well known (Schouten [10], Weyl [16]) that a conformally flat Riemannian manifold is characterized, for $n > 3$, by the vanishing of the Weyl conformal curvature tensor

$$C_{kji}{}^h = R_{kji}{}^h + \delta_k^h C_{ji} - \delta_j^h C_{ki} + C_k{}^h g_{ji} - C_j{}^h g_{ki} \tag{1.1}$$

and, for $n = 3$, by the vanishing of the tensor

$$C_{kji} = \nabla_k C_{ji} - \nabla_j C_{ki}, \tag{1.2}$$

where

$$C_{ji} = -\frac{1}{n-2} R_{ji} + \frac{1}{2(n-1)(n-2)} R g_{ji}, \qquad C_k{}^h = C_{kt} g^{th}. \tag{1.3}$$

It is easily seen that a conformally flat space is of constant sectional curvature if and only if it is an Einstein space.

AMS (MOS) subject classifications (1970). Primary 53B20; Secondary 53B25, 53B35.

Bochner (Bochner [1], Yano and Bochner [21]) proved

THEOREM 1.1. *If a compact conformally flat Riemannian manifold has positive definite Ricci curvature, then we have $b_p = 0$ $(0 < p < n)$, where b_p denotes the pth Betti number of the manifold.*

Now let M^n be a real n-dimensional almost Hermitian manifold with almost complex structure $F_i{}^h$ and almost Hermitian metric g_{ji}:

$$F_j{}^i F_t{}^h = -\delta_j^h, \qquad F_j{}^t F_i{}^s g_{ts} = g_{ji}. \tag{1.4}$$

A Kaehlerian manifold is characterized by

$$\nabla_j F_i{}^h = 0. \tag{1.5}$$

Bochner (Bochner [2], Yano and Bochner [21]) introduced in a Kaehlerian manifold a curvature tensor given by

$$\begin{aligned} B^h_{kji} = R^h_{kji} &+ \delta^h_k L_{ji} - \delta^h_j L_{ki} + L^h_k g_{ji} - L^h_j g_{ki} + F^h_k M_{ji} - F^h_j M_{ki} \\ &+ M^h_k F_{ji} - M^h_j F_{ki} - 2(M_{kj} F^h_i + F_{kj} M^h_i), \end{aligned} \tag{1.6}$$

where

$$\begin{aligned} L_{ji} &= -\frac{1}{n+4} R_{ji} + \frac{1}{2(n+2)(n+4)} R g_{ji}, \\ L_k{}^h &= L_{kt} g^{th}, \quad M_{ji} = -L_{jt} F_i{}^t, \quad M_k{}^h = M_{kt} g^{th}, \end{aligned} \tag{1.7}$$

and $F_{ji} = F^t_j g_{ti}$, as a curvature tensor which corresponds to the Weyl conformal curvature tensor in a Riemannian manifold.

We notice here that L_{ji} satisfies

$$L_{ji} = F^t_j F^s_i L_{ts}, \tag{1.8}$$

that is, L_{ji} is a hybrid tensor (see Yano [19]).

Bochner introduced this curvature tensor using a complex coordinate system. The tensor expression (1.6) in a real coordinate system has been given by Tachibana [12]. (See also Tachibana and Liu [13], Takagi and Watanabe [14], Yano [20].)

It is easily seen that a Kaehlerian manifold with vanishing Bochner curvature tensor is of constant holomorphic sectional curvature if and only if the manifold is an Einstein space.

Bochner (Bochner [2], Yano and Bochner [21]) proved

THEOREM 1.2. *If a compact Kaehlerian manifold with vanishing Bochner curvature tensor has positive definite Ricci tensor, then we have $b_{2p} = 1, b_{2p+1} = 0$, where b_i denotes the ith Betti number of the manifold.*

The purpose of the present report is to state some more recent results on manifolds and submanifolds with vanishing Weyl or Bochner curvature tensor and to show the close analogy between Weyl conformal curvature tensor and Bochner curvature tensor.

2. Characterization of manifolds with vanishing Weyl or Bochner curvature tensor. Suppose that M^n is a conformally flat Riemannian manifold of dimension $n > 3$, then we have $C_{kji}{}^h = 0$, from which

$$R_{kjih} = -g_{kh}C_{ji} + g_{jh}C_{ki} - C_{kh}g_{ji} + C_{jh}g_{ki}. \tag{2.1}$$

Using this, we can show that the sectional curvature $K(\sigma)$ of the manifold with respect to a section σ spanned by two mutually orthogonal unit vectors U and V is given by

$$K(\sigma) = -C_{ji}U^jU^i - C_{ji}V^jV^i. \tag{2.2}$$

Conversely suppose that the sectional curvature $K(\sigma)$ of the manifold with respect to a section σ spanned by two mutually orthogonal unit vectors U and V is given by an equation of the form (2.2), then we can prove that the Weyl conformal curvature tensor vanishes. Thus we have

THEOREM 2.1 (CHEN AND YANO [5]). *In order that a Riemannian manifold of dimension $n > 3$ is conformally flat, it is necessary and sufficient that the sectional curvature of the manifold with respect to a section spanned by two mutually orthogonal unit vectors U and V is the sum of a quadratic form of U and that of V, two quadratic forms having the same coefficients.*

Kulkarni [7] also obtained a characterization of a conformally flat space in terms of sectional curvature different from the above.

Suppose that the Bochner curvature tensor of a Kaehlerian manifold vanishes: $B_{kji}{}^h = 0$, then we have

$$\begin{aligned} R_{kjih} = &-g_{kh}L_{ji} + g_{jh}L_{ki} - L_{kh}g_{ji} + L_{jh}g_{ki} - F_{kh}M_{ji} + F_{jh}M_{ki} \\ &- M_{kh}F_{ji} + M_{jh}F_{ki} + 2(M_{kj}F_{ih} + F_{kj}M_{ih}). \end{aligned} \tag{2.3}$$

Using this, we can show that the holomorphic sectional curvature $K(U)$ of the manifold with respect to a holomorphic section spanned by a unit vector U and its transform FU by F is given by

$$K(U) = -8L_{ji}U^jU^i. \tag{2.4}$$

Conversely suppose that the holomorphic sectional curvature $K(U)$ of the manifold with respect to a holomorphic section spanned by a unit vector U and its transform FU by F is given by an equation of the form (2.4), the coefficients L_{ji} defining a hybrid tensor of type (0, 2), then we can prove that the Bochner curvature tensor vanishes. Thus we have

THEOREM 2.2 (CHEN AND YANO [5]). *In order that the Bochner curvature tensor of a Kaehlerian manifold vanishes, it is necessary and sufficient that the holomorphic sectional curvature of the manifold with respect to a holomorphic section spanned by a unit vector U and its transform FU by the complex structure tensor F is given by a quadratic form of U, the coefficients of the quadratic form defining a hybrid tensor of type* (0, 2).

3. Manifolds with constant scalar curvature whose Weyl or Bochner curvature tensor vanishes. Ryan **[9]** proved

THEOREM 3.1. *Let M^n be a compact conformally flat Riemannian manifold of dimension $n \geqq 3$ with constant scalar curvature. If the Ricci tensor is positive semi-definite, then the simply connected Riemannian covering of M^n is one of*

$$S^n(c), \quad R \times S^{n-1}(c) \quad or \quad E^n,$$

the real space forms of curvature c being denoted by $S^n(c)$ or E^n depending on whether c is positive or zero.

For the case in which the Ricci tensor is positive definite we do not have the cases $R \times S^{n-1}(c)$ and E^n (see Goldberg **[6]** and Tani **[15]**).

He first proves that, in a conformally flat Riemannian manifold with constant scalar curvature R, we have

$$\frac{1}{2} \Delta(R_{ih}R^{ih}) = \frac{1}{n-2} P + (\nabla_j R_{ih})(\nabla^j R^{ih}), \tag{3.1}$$

where

$$P = nR_t^{\ s} R_s^{\ r} R_r^{\ t} - \frac{2n-1}{n-1} RR_{ji} R^{ji} + \frac{1}{n-1} R^3 \tag{3.2}$$

and that if we then denote by λ_i the eigenvalues of R_{ji}, we have

$$P = \frac{1}{n-1} \sum_i \sum_{j \neq k} \lambda_i(\lambda_i - \lambda_j)(\lambda_i - \lambda_k). \tag{3.3}$$

He then assumes that the Ricci tensor R_{ji} is positive semidefinite and shows that in this case we have $P \geqq 0$ on M. Thus he obtains $\Delta(R_{ih} R^{ih}) \geqq 0$, from which

$$P = 0 \quad \text{and} \quad \nabla_k R_{ji} = 0. \tag{3.4}$$

From these he obtains the theorem quoted above. The theorem also holds if the assumptions of compactness and constant scalar curvature are replaced by local homogeneity of M^n.

Ishihara and the present author **[22]** proved

THEOREM 3.2. *Let M^n be a Kaehlerian manifold of real dimension n with constant scalar curvature whose Bochner curvature tensor vanishes and whose Ricci tensor is positive semidefinite. If M^n is compact, then the universal covering manifold is a complex projective space $CP^{n/2}$ or a complex space $C^{n/2}$.*

(For the case in which the Ricci curvature is positive definite, see Tachibana **[12]**.)

In this case, we have

$$\frac{1}{2}\Delta(R_{ih}R^{ih}) = \frac{1}{n+4}Q + (\nabla_j R_{ih})(\nabla^j R^{ih}), \tag{3.5}$$

where

$$Q = nR^s_t R^r_s R^t_r - \frac{2(n+1)}{n+2} RR_{ji}R^{ji} + \frac{1}{n+2}R^3, \tag{3.6}$$

and consequently we have

$$Q = P + \frac{3n}{(n-1)(n+2)} R\Big(R_{ji} - \frac{R}{n}g_{ji}\Big)\Big(R^{ji} - \frac{R}{n}g^{ji}\Big). \tag{3.7}$$

Thus if we assume that the Ricci tensor R_{ji} is positive semidefinite, then we have $Q \geqq 0$ on M^n and consequently we can conclude (3.4). From these we obtain Theorem 3.2. The theorem also holds if the assumptions of compactness and constant scalar curvature are replaced by local homogeneity of M^n.

4. Submanifolds with vanishing Weyl or Bochner curvature tensor. In 1921, Schouten [10] proved that the second fundamental tensor of a conformally flat hypersurface of an $(n+1)$-dimensional conformally flat Riemannian manifold has at least $n-1$ equal eigenvalues. Nishikawa and Maeda [8] obtained the same result independently.

Let M^{n+1} be an $(n+1)$-dimensional Riemannian manifold covered by a system of coordinate neighborhoods $\{U;\ \xi^\kappa\}$, where here and in the sequel the indices $\kappa, \lambda, \mu, \nu, \cdots$, run over the range $\{1', 2', \cdots, n', (n+1)'\}$ and let $g_{\mu\lambda}$, ∇_μ, $R_{\nu\mu\lambda}{}^\kappa$, $R_{\mu\lambda}$ and R' be the positive definite metric tensor, the operator of covariant differentiation with respect to the Christoffel symbols $\{{}_\mu{}^\kappa{}_\lambda\}$ formed with $g_{\mu\lambda}$, the curvature tensor, the Ricci tensor and the scalar curvature of M^{n+1} respectively.

We assume that an n-dimensional Riemannian manifold M^n is isometrically immersed as a hypersurface in M^{n+1} by the immersion $i: M^n \to M^{n+1}$. We identify $i(M^n)$ with M^n and represent the immersion locally by

$$\xi^\kappa = \xi^\kappa(\eta^h). \tag{4.1}$$

If we put $B_i{}^\kappa = \partial_i \xi^\kappa$, $\partial_i = \partial/\partial\eta^i$, then we see that the metric tensor $g_{\mu\lambda}$ of M^{n+1} and g_{ji} of M^n are related by

$$g_{ji} = B_{ji}^{\mu\lambda}\, g_{\mu\lambda}, \tag{4.2}$$

where $B_{ji}^{\mu\lambda} = B_j{}^\mu B_i{}^\lambda$.

Denoting by C^κ the unit normal to M^{n+1}, we have equations of Gauss

$$\nabla_j B_i{}^\kappa = H_{ji} C^\kappa, \tag{4.3}$$

where

$$\nabla_j B_i{}^\kappa = \partial_j B_i{}^\kappa + B_{ji}^{\mu\lambda}\left\{ {k \atop \mu\lambda} \right\} - \left\{ {h \atop ji} \right\} B_h{}^\kappa,$$

and H_{ji} is the second fundamental tensor.

Equations of Gauss giving the relation between $R^{\kappa}_{\nu\mu\lambda}$, R^{h}_{kji} and the second fundamental tensor H_{ji} are

$$R_{kji}{}^{h} = B^{\nu\mu\lambda h}_{kji\kappa} R_{\nu\mu\lambda}{}^{\kappa} + H_{k}{}^{h} H_{ji} - H_{j}{}^{h} H_{ki}, \tag{4.4}$$

where $B^{\nu\mu\lambda h}_{kji\kappa} = B_{k}{}^{\nu} B_{j}{}^{\mu} B_{i}{}^{\lambda} B_{\kappa}{}^{h}$ and $B_{\kappa}{}^{h} = B_{i}{}^{\lambda} g^{ih} g_{\lambda\kappa}$.

If we assume that M^{n+1} is locally flat, then we have from (4.4)

$$R^{h}_{kji} = H^{h}_{k} H_{ji} - H^{h}_{j} H_{ki}. \tag{4.5}$$

Now, from (1.1) and (4.5), we have

$$\begin{aligned} H_{h}{}^{k} H^{ji} C_{kji}{}^{h} = \frac{1}{(n-1)(n-2)} \{ & -n(n-1) H^{(4)} + 4(n-1) H^{(1)} H^{(3)} \\ & + (n^2 - 3n + 3)(H^{(2)})^2 - 2n (H^{(1)})^2 H^{(2)} + (H^{(1)})^4 \}, \end{aligned} \tag{4.6}$$

where we have used the notation

$$A^{(p)} = \operatorname{trace}(A^{p})$$

for a square matrix whose elements are $(A_{i}{}^{h})$.

We now denote by λ_i n eigenvalues of a symmetric square matrix A and put

$$\sigma(A) = \sum (\lambda_k - \lambda_j)^2 (\lambda_i - \lambda_h)^2, \tag{4.7}$$

where the summation is taken over all distinct k, j, i, h. Thus if $\sigma(A) = 0$, then at least $n - 1$ of λ's are equal.

We can prove that

LEMMA 4.1 (CHEN AND YANO [3]). *For any real symmetric $n \times n$ matrix A, $n > 3$, we have*

$$\begin{aligned} \tfrac{1}{2}\sigma(A) = & -n(n-1) A^{(4)} + 4(n-1) A^{(1)} A^{(3)} + (n^2 - 3n + 3)(A^{(2)})^2 \\ & - 2n (A^{(1)})^2 A^{(2)} + (A^{(1)})^4. \end{aligned} \tag{4.8}$$

We now assume that $n > 3$ and M^n is conformally flat. Then we have, from (4.6) and (4.8),

$$\sigma(H) = 0 \tag{4.9}$$

and consequently at least $n - 1$ eigenvalues of H are equal. Thus the second fundamental tensor H_{ji} of the hypersurface M^n is of the form

$$H_{ji} = \alpha g_{ji} \quad \text{or} \quad H_{ji} = \alpha g_{ji} + \beta v_j v_i,$$

at a point of M^n, v_j being a covector. If $H_{ji} = \alpha g_{ji}$ at a point, the point is called an umbilical point of M^n and if $H_{ji} = \alpha g_{ji} + \beta v_j v_i$ at a point, then the point is called a quasi-umbilical point of M^n. Thus we have proved

THEOREM 4.1. *A point of a conformally flat hypersurface of a flat Riemannian manifold is either umbilical or quasi-umbilical.*

The property that a point of a hypersurface is umbilical or quasi-umbilical being invariant under conformal change of metric of the ambient manifold, we have, from Theorem 4.1,

THEOREM 4.2. *A point of a conformally flat hypersurface of a conformally flat Riemannian manifold is either umbilical or quasi-umbilical.*

To prove this theorem we can use the conformal equations of Gauss for a hypersurface of a Riemannian manifold (Yano [18]):

$$
\begin{aligned}
C_{kji}^{h} = B_{kji\kappa}^{\nu\mu\lambda h} C_{\nu\mu\lambda}^{\kappa} &- \frac{1}{n-2}\,[\delta_k^h B_\kappa^\nu B_{ji}^{\mu\lambda} C_{\nu\mu\lambda}{}^{\kappa} - \delta_j^h B_\kappa^\nu B_{ki}^{\mu\lambda} C_{\nu\mu\lambda}{}^{\kappa} \\
&\qquad + B_\kappa^\nu B_{kt}^{\mu\lambda} C_{\nu\mu\lambda}{}^{\kappa} g^{th} g_{ji} - B_\kappa^\nu B_{jt}^{\mu\lambda} C_{\nu\mu\lambda}{}^{\kappa} g^{th} g_{ki}] \\
&+ \frac{1}{(n-1)(n-2)} B_\kappa^\nu B^{\mu\lambda} C_{\nu\mu\lambda}{}^{\kappa} (\delta_k^h g_{ji} - \delta_j^h g_{ki}) + M_k{}^h M_{ji} - M_j{}^h M_{ki} \\
&+ \frac{1}{n-2} (\delta_k^h M_j{}^t M_{ti} - \delta_j^h M_k{}^t M_{ti} + M_k{}^t M_t{}^h g_{ji} - M_j{}^t M_t{}^h g_{ki}) \\
&- \frac{1}{(n-1)(n-2)} M_s{}^t M_t{}^s (\delta_k^h g_{ji} - \delta_j^h g_{ki}),
\end{aligned} \tag{4.10}
$$

where $B_\kappa^\nu = B_\kappa{}^t B_t{}^\nu$ $B^{\mu\lambda} = B_{ji}^{\mu\lambda} g^{ji}$ and $M_{ji} = H_{ji} - g^{ts} H_{ts} g_{ji}/n$, $M_i{}^h = M_{it} g^{th}$.

If we assume that the ambient manifold M^{n+1} is conformally flat, then we have using (4.10) and $M_t^t = M^{(1)} = 0$,

$$
M_h{}^k M^{ji} C_{kji}{}^h = \frac{1}{(n-1)(n-2)} \{-n(n-1)M^{(4)} + (n^2 - 3n + 3)(M^{(2)})^2\}. \tag{4.11}
$$

Thus if we assume that the hypersurface M^n is also conformally flat, we have, from (4.11),

$$
-n(n-1)M^{(4)} + (n^2 - 3n + 3)(M^{(2)})^2 = 0
$$

and consequently applying Lemma 4.1 and taking account of $M^{(1)} = 0$, we have $\sigma(M) = 0$ and hence we see that M_{ji} is of the form

$$
M_{ji} = 0 \quad \text{or} \quad M_{ji} = \alpha g_{ji} + \beta v_j v_i,
$$

which proves Theorem 4.2.

For a conformally flat submanifold of codimension 2 of a conformally flat Riemannian manifold, we have

THEOREM 4.3 (CHEN AND YANO [3]). *Let M^n be a conformally flat submanifold of codimension 2 of a conformally flat Riemannian manifold M^{n+2}. If a point of M^n is quasi-umbilical with respect to a normal direction, then the point is also quasi-umbilical with respect to another normal direction orthogonal to the first normal.*

We can prove this theorem using conformal equations of Gauss for a submanifold of codimension 2 of a Riemannian manifold (Yano [18]).

For a further generalization of this theorem to the case of general codimension, see Chen and Yano [4].

To obtain a theorem corresponding to Theorem 4.3 for a complex hypersurface with vanishing Bochner curvature tensor of a Kaehlerian manifold with vanishing Bochner curvature tensor, let M^{n+2} be a real $(n+2)$-dimensional Kaehlerian manifold and let M^n be a real n-dimensional invariant submanifold of M^{n+2}. Let the parametric equations of M^n be given by $\xi^\kappa = \xi^\kappa(\eta^h)$, where here and in the sequel the indices $\kappa, \lambda, \mu, \nu, \cdots$ run over the range $\{1', 2', \cdots, n', (n+1)', (n+2)'\}$ and the indices $h, i, j, k, \cdots$, the range $\{1, 2, \cdots, n\}$. Since the submanifold M^n is invariant, we have

$$F_\lambda{}^\kappa B_i{}^\lambda = F_i{}^h B_h{}^\kappa, \tag{4.12}$$

where $F_i{}^h$ is the complex structure of M^n and satisfies

$$F_j{}^t F_i{}^s g_{ts} = g_{ji}, \tag{4.13}$$

that is, (F, g) gives a Hermitian structure of M^n.

We choose 2 mutually orthogonal unit normals C^κ and $D^\kappa = F_\lambda{}^\kappa C^\lambda$ to M^n. Then the equations of Gauss for M^n are

$$\nabla_j B_i{}^\kappa = H_{ji} C^\kappa + K_{ji} D^\kappa, \tag{4.14}$$

H_{ji} and K_{ji} being the second fundamental tensors. Differentiating (4.12) covariantly along M^n and using $\nabla_\mu F_\lambda{}^\kappa = 0$ and (4.14), we have

$$\nabla_j F_i{}^h = 0, \tag{4.15}$$

which shows that M^n is a Kaehlerian manifold and

$$H_{ji} = K_{jt} F_i{}^t, \qquad K_{ji} = -H_{jt} F_i{}^t. \tag{4.16}$$

If we put

$$H = (H_i{}^h), \qquad K = (K_i{}^h), \tag{4.17}$$

where $H_i^h = H_{it} g^{th}$ and $K_i^h = K_{it} g^{th}$, we have, from (4.16),

$$H = KF, \qquad K = -HF, \tag{4.18}$$

where $F = (F_i{}^h)$, from which we see that

$$H^{(1)} = 0, \qquad K^{(1)} = 0, \tag{4.19}$$

that is, M^n is minimal. From (4.16) and the symmetry of H_{ji} and K_{ji} in j and i, we have

$$HF = -FH, \qquad KF = -FK, \tag{4.20}$$

and consequently, using (4.18),

$$H^2 = K^2, \tag{4.21}$$

$$HK = -KH. \tag{4.22}$$

Now starting from the equations of Gauss

$$(4.23)\qquad R_{kji}^{h} = B_{kji\kappa}^{\nu\mu\lambda h} R_{\nu\mu\lambda}^{\kappa} + H_k^h H_{ji} - H_j^h H_{ki} + K_k^h K_{ji} - K_j^h K_{ki},$$

we can prove the following equations of Gauss involving Bochner curvature tensors of ambient manifold and that of immersed manifold:

$$\begin{aligned}
B_{kji}{}^{h} = {} & B_{kji\kappa}^{\nu\mu\lambda h} B_{\nu\mu\lambda}{}^{\kappa} - \frac{1}{n+4}[\delta_k^h B_\kappa^\nu B_{ji}^{\mu\lambda} B_{\nu\mu\lambda}{}^{\kappa} - \delta_j^h B_\kappa^\nu B_{ki}^{\mu\lambda} B_{\nu\mu\lambda}{}^{\kappa} \\
& + B_\kappa^\nu B_{kt}^{\mu\lambda} B_{\nu\mu\lambda}{}^{\kappa} g^{th} g_{ji} - B_\kappa^\nu B_{jt}^{\mu\lambda} B_{\nu\mu\lambda}{}^{\kappa} g^{th} g_{ki} \\
& - F_k{}^h B_\kappa^\nu B_{jt}^{\mu\lambda} B_{\nu\mu\lambda}{}^{\kappa} F_i{}^t + F_j{}^h B_\kappa^\nu B_{kt}^{\mu\lambda} B_{\nu\mu\lambda}{}^{\kappa} F_i{}^t \\
& - B_\kappa^\nu B_{kt}^{\mu\lambda} B_{\nu\mu\lambda}{}^{\kappa} F^{ht} F_{ji} + B_\kappa^\nu B_{jt}^{\mu\lambda} B_{\nu\mu\lambda}{}^{\kappa} F^{ht} F_{ki} \\
(4.24)\qquad & + 2(B_k^\nu B_{kt}^{\mu\lambda} B_{\nu\mu\lambda}{}^{\kappa} F_j{}^t F_i{}^h + F_{kj} B_\kappa^\nu B_{it}^{\mu\lambda} B_{\nu\mu\lambda}{}^{\kappa} F^{ht})] \\
& + \frac{1}{(n+2)(n+4)} B_\kappa^\nu B^{\mu\lambda} B_{\nu\mu\lambda}{}^{\kappa} (\delta_k^h g_{ji} - \delta_j^h g_{ki} + F_k{}^h F_{ji} - F_j{}^h F_{ki} - 2F_{kj} F_i{}^h) \\
& + H_k{}^h H_{ji} - H_j{}^h H_{ki} + K_k{}^h K_{ji} - K_j{}^h K_{ki} \\
& + \frac{1}{n+4}[\delta_k^h H_j{}^t H_{ti} - \delta_j^h H_k{}^t H_{ti} + H_k{}^t H_t^h g_{ji} - H_j{}^t H_t{}^h g_{ki} \\
& + F_k{}^h H_j{}^t K_{ti} - F_j{}^h H_k{}^t K_{ti} + H_k{}^t K_t{}^h F_{ji} - H_j{}^t K_t{}^h F_{ki} \\
& - 2(H_k{}^t K_{tj} F_i{}^h + F_{kj} H_i{}^t K_t{}^h)] \\
& - \frac{2}{(n+2)(n+4)} H_s{}^t H_t{}^s (\delta_k^h g_{ji} - \delta_j^h g_{ki} + F_k{}^h F_{ji} - F_j{}^h F_{ki} - 2F_{kj} F_i{}^h).
\end{aligned}$$

Assuming that the Bochner curvature tensor $B_{\nu\mu\lambda}^{\kappa}$ of the ambient manifold vanishes, we have

$$(4.25)\qquad H_h{}^k H^{ji} B_{kji}{}^h = \frac{1}{(n+2)(n+4)}[(n^2+6n+16)(H^{(2)})^2 - 8(n+2)H^{(4)}].$$

Assuming that the Bochner curvature tensor $B_{kji}{}^h$ of the submanifold vanishes also, we can prove that, for $n \geqq 4$, $H = 0$, $K = 0$. Thus we have

THEOREM 4.4 (YAMAGUCHI AND SATO [17]). *A Kaehler hypersurface of a Kaehler manifold with vanishing Bochner curvature tensor has also vanishing Bochner curvature tensor if and only if it is totally geodesic.*

BIBLIOGRAPHY

1. S. Bochner, *Curvature and Betti numbers*, Ann. of Math. (2) **49** (1948), 379–390. MR **9**, 618.

2. ———, *Curvature and Betti numbers*. II, Ann. of Math. (2) **50** (1949), 77–93, MR **10**, 571.

3. Bang-yen Chen and K.Yano, *Conformally flat spaces of codimension* 2 *in a Euclidean space*, Canad. J. Math. **25** (1973), 1170–1173.

4. ———, *Some results on conformally flat submanifolds*, Tamkang J. Math. **4** 1973), 167–174.

5. ———, *Manifolds with vanishing Weyl or Bochner curvature tensor*, J. Math. Soc. Japan (to appear).

6. S.I. Goldberg, *On conformally flat spaces with definite Ricci curvature*, Kōdai Math. Sem. Rep. **21** (1969), 226–232. MR **40** #6450.

7. R. S. Kulkarni, *Curvature structures and conformal transformations*, J. Differential Geometry **4** (1970), 425–451. MR **44** #2173.

8. S. Nishikawa and Y. Maeda, *Conformally flat hypersurfaces in a conformally flat Riemannian manifold*, Tôhoku Math. J. **26** (1974), 159–168.

9. P.J. Ryan, *A note on conformally flat spaces with constant scalar curvature*, Proceedings of the Thirteenth Biennial Seminar of the Canadian Mathematical Congress on Differential Topology, Differential Geometry and Applications, vol. 2 (1972), 119–124.

10. J.A. Schouten, *Über die konforme Abbildung n-dimensionaler Mannigfaltigkeiten mit quadratischer Massbestimmung auf eine Mannigfaltigkeit mit euklidischer Massbestimmung*, Math. Z. **11** (1921), 58–88.

11. ———, *Ricci-calculus, An introduction to tensor analysis and its geometrical applications*, 2nd ed., Die Grundlehren der math. Wissenschaften, in Einzeldarstellungen mit besonderer Berücksichtigung der Anwendungsgebiete, Band X, Springer-Verlag, Berlin, 1954. MR **16**, 521.

12. S. Tachibana, *On the Bochner curvature tensor*, Natur. Sci. Rep. Ochanomizu Univ. **18** (1967), 15–19. MR **36** #7089.

13. S. Tachibana and R.C. Liu, *Notes on Kählerian metrics with vanishing Bochner curvature tensor*, Kōdai Math. Sem. Rep. **22** (1970), 313–321. MR **42** #1030.

14. H. Takagi and Y. Watanabe, *On the holonomy groups of Kaehlerian manifolds with vanishing Bochner curvature tensor*, Tôhoku Math. J. **25** (1973), 177–184.

15. M. Tani, *On a conformally flat Riemannian space with positive Ricci curvature*, Tôhoku Math. J. (2) **19** (1967), 227–231. MR **36** #3279.

16. H. Weyl, *Zur Infinitesimalgeometrie: Einordnung der projektiven und der konformen Auffassung*, Göttingen Nachr. **1921**, 99–112.

17. S. Yamaguchi, and S. Sato, *On complex hypersurfaces with vanishing Bochner curvature tensor in Kählerian manifolds*, Tensor **22** (1971), 77–81. MR **43** #2649.

18. K. Yano, *Sur les équations de Gauss dans la géométrie conforme des espaces de Riemann*, Proc. Imp. Acad. Tokyo **15** (1939), 247–252. MR **1**, 175.

19. ———, *Differential geometry on complex and almost complex spaces*, Internat. Ser. of Monographs in Pure and Appl. Math., vol. 49, Pergamon Press, New York, 1965. MR **32** #4635.

20. ———, *On complex conformal connections*, Kōdai Math. Sem. Rep. (to appear).

21. K. Yano and S. Bochner, *Curvature and Betti numbers*, Ann. of Math. Studies, no. 32, Princeton Univ. Press, Princeton, N.J., 1953. MR **15**, 989.

22. K. Yano and S. Ishihara, *Kaehlerian manifolds with constant scalar curvature whose Bochner curvature tensor vanishes*, Hokkaido Math. J. **3** (1974), 297–304.

TOKYO INSTITUTE OF TECHNOLOGY

FOLIATIONS

Proceedings of Symposia in Pure Mathematics
Volume 27, 1975

ON THE DE RHAM COMPLEX OF $B\hat{\Gamma}$

B. CENKL*

A classifying space $B\Gamma''$ for any topological groupoid Γ'' was defined by Haefliger [5]. If Γ'' is a category with objects being the units of the groupoid Γ'' and the elements of Γ'' being the morphisms then the classifying space $B\Gamma''$, defined by Segal [9], is homotopy equivalent with Stasheff's $\mathscr{B}\Gamma''$ [11]. In particular, let Γ be the groupoid of germs given by the pseudogroup Π of local C^∞-diffeomorphisms of R^q, and $\hat{\Gamma}$ the category with objects P^1 (1-jets of Π with target $0 \in R^q$) and morphisms being the germs of local diffeomorphisms of P^1 commuting with the action of the structure group G^1 of P^1. Then $B\hat{\Gamma} \to B\Gamma$ is the universal normal bundle of the universal Γ-structure on $B\Gamma$.

On the bundle $P^\infty \to R^q$ of ∞-jets of Π with a fixed target, the groupoid Γ^∞ of ∞-jets of Γ acts naturally. P^∞ has a structure of a topological group. The cohomology of the topologized algebra of Γ^∞-invariant differential forms on P^∞ is isomorphic with the cohomology of the algebra L of formal vector fields on R^q.

The category $\tilde{\Gamma}^\infty$ with objects P^∞ and morphisms $P^\infty \times P^\infty$ gives the classifying space $B\tilde{\Gamma}^\infty$. If P^∞ is the category with one object and P^∞ is the set of morphisms there is a homotopy equivalence $\varepsilon : B\Gamma^\infty \to BP^\infty$. Furthermore we have a homotopy equivalence $\nu : BO(q) \to BP^\infty$. If $B_\nu \tilde{\Gamma}^\infty$ is the pull back, by ν of $B\tilde{\Gamma}^\infty$ we can describe the situation by the following diagram:

$$\begin{array}{ccccc}
B\hat{\Gamma} & \longrightarrow & B\tilde{\Gamma}^\infty & & B_\nu\tilde{\Gamma}^\infty \\
\downarrow & & \downarrow & & | \\
B\Gamma & \longrightarrow & B\Gamma^\infty & & | \\
\downarrow & & \varepsilon\downarrow & & \downarrow \\
BO(q) & \xrightarrow{\nu} & BP^\infty & \longleftarrow & BO(q)
\end{array}$$

AMS (*MOS*) *subject classifications* (1970). Primary 57D30.

*This research was supported by NSF grant GP-16354X1.

We can think of $B\tilde{\Gamma}^\infty \to BP^\infty$ as the universal bundle for the topological group P^∞.

The de Rham complex $F\tilde{\Gamma}^\infty$ is the double complex of Γ^∞-invariant vertical forms.

The projection $B_\nu \tilde{\Gamma}^\infty \to BO(q)$ comes from the transformation of functors $N_\nu\tilde{\Gamma}^\infty \to NO(q)$. The de Rham complex $F\Lambda N_\nu\tilde{\Gamma}^\infty$ is the double complex of Γ^∞-invariant forms on $N_\nu\tilde{\Gamma}^\infty$ which are vertical with respect to the projection $N_\nu\tilde{\Gamma}^\infty \to NO(q)$. The spectral sequence argument together with the results from the continuous cohomology of topological groups [7] show that the cohomology of the single complex $F\tilde{\Gamma}^\infty$ associated with $F\Lambda N_\nu \tilde{\Gamma}^\infty$ is equal to the cohomology $H(L, R)$ of the topologized Lie algebra L of formal vector fields at 0 in R^q. The computation of the latter can be found in [3]. The homotopy equivalence of $B\tilde{\Gamma}^\infty$ with $B_\nu \tilde{\Gamma}^\infty$ together with the jet functor $\hat{\Gamma} \to \tilde{\Gamma}^\infty$ give the natural map of the cohomology of this de Rham complex $H(F\tilde{\Gamma}^\infty)$ into the singular cohomology $H(B\hat{\Gamma}, R)$.

Because in the analytic case $H(B\hat{\Gamma}, R)$ is equal to the cohomology of the de Rham bicomplex $FA^*N\hat{\Gamma}[p]$, $p = 0, 1, \cdots$, of all forms on the nonhausdorff manifolds, and because there are good reasons to believe that the complexes $F\Lambda N\hat{\Gamma}$ and $F\Lambda N_\nu \tilde{\Gamma}^\infty$ have the same cohomology even in the C^∞ case we call the complex $F\Lambda N_\nu \Gamma^\infty$ the de Rham complex.

The work in this paper has been inspired by the investigations, done by Bott and Hochschild [1], of the relations between the continuous cohomology of Lie groups and ordinary characteristic classes.

1. Let Π be a pseudogroup of all local C^∞-diffeomorphisms of R^q [4]. The set of germs of the maps from Π has a structure of a groupoid. This groupoid Γ together with the natural topology is a topological groupoid [8]. Let $\alpha: \Gamma \to R^q$ and $\beta: \Gamma \to R^q$ be the source and the target projections or projections on the right and left units respectively. The groupoid Γ will be identified with the topological category with points of R^q as objects and elements of Γ as morphisms.

With the pseudogroup Π can be associated the groupoid Γ^k of k-jets of diffeomorphisms of Π. This groupoid has a structure of a differential groupoid and in fact a structure of a Lie groupoid [8]. The inverse limit $\Gamma^\infty = \text{proj lim } \Gamma^k$ together with the limit topology is the groupoid of infinite jets of diffeomorphisms from Π. The source and the target projections are again denoted by α and β respectively. And Γ^∞ is identified with the category whose objects are R^q and morphisms are Γ^∞.

The mapping which to each germ $a \in \Gamma$ at $x \in R^q$ assigns the infinite jet of a at x, $j_x^\infty a \in \Gamma^\infty$ gives the covariant functor (jet functor) $j: \Gamma \to \Gamma^\infty$.

Let $P^\infty = \{a \in \Gamma^\infty | \beta(a) = 0 \in R^q\}$ be the principal bundle over R^q with the projection α and structure group G^∞. If G^k is the Lie group of k-jets of all $\phi \in \Pi$, $\phi(0) = 0$, at the point 0, then $G^\infty = \text{proj lim } G^k$. P^∞ is the inverse limit of the system $\pi_{k-1}^k: P^k \to P^{k-1}$, $P^k = \{a \in \Gamma^k | \beta(a) = 0 \in R^q\}$. $\pi_k: P^\infty \to P^k$, $k \geqq 0$, is the projection; $\pi_0 = \alpha$, $P^0 = R^q$. The G^∞-bundle $\alpha: P^\infty \to R^q$ can be given a structure of a topological group. Namely, each element $v \in P^\infty$ can be identified with the pair (v, t_v) where t_v is the infinite jet of the translation of R^q which maps $\alpha(v)$ into 0. With this identification we can define the composition $P^\infty \times P^\infty \to P^\infty, (v, w) \to v \cdot w$, as a composition of maps $v \cdot w = v \cdot t_v^{-1} \cdot w \cdot t_v$. Because $t_{v \cdot w} = t_v \cdot t_w = t_w \cdot t_v$ we have

$$(v \cdot w) \cdot u = (v \cdot w) \cdot t_{v \cdot w}^{-1} \cdot u \cdot t_{v \cdot w}$$
$$= v \cdot t_v^{-1} \cdot (w \cdot v) \cdot t_v = v \cdot (w \cdot u).$$

And if e is the identity then $v \cdot e = v \cdot t_v^{-1} \cdot e \cdot t_v = v$ and $e \cdot v = e \cdot t_e^{-1} \cdot v \cdot t_e = v$. For each $v \in P^\infty$ there is an inverse $v' = t_v \cdot v^{-1} \cdot t_v^{-1}$. Obviously

$$v \cdot v' = v \cdot t_v^{-1} \cdot t_v \cdot v^{-1} \cdot t_v^{-1} \cdot t_v = e,$$

and because $t_{t_v} = t_v$, $t_{t_v^{-1}} = t_v^{-1}$;

$$t_{v'} = t_v \cdot t_v^{-1} \cdot t_v^{-1} = t_v^{-1}.$$

Therefore $v' \cdot v = v' \cdot t_v \cdot v \cdot t_v^{-1} = t_v \cdot v^{-1} \cdot t_v^{-1} \cdot t_v \cdot v \cdot t_v^{-1} = e$. And because P^∞ is the inverse limit the continuity of the operation $\cdot$ follows.

The tangent bundle $T(P^\infty)$ is the limit proj lim $T(P^k)$; i.e. a tangent vector $X \in T_a(P^\infty)$ is a system of vectors

$$X^k \in T_{\pi_k(a)}(P^k), \quad k \geqq 0, \qquad (\pi_l^k)_* X_k = X_l, \quad k > l.$$

The vector bundle $\pi\colon V^\infty \to R^q$ of infinite jets of smooth vector fields on R^q is the bundle of Lie algebras. The bracket [,] on the module of smooth vector fields on R^q gives the Lie algebra structure to $V^\infty = \pi^{-1}(x)$, $x \in R^q$. For any vector fields X, Y on R^q, $j_x^\infty X$ and $j_x^\infty Y$ belong to V_x^∞ and $[j_x^\infty X, j_x^\infty Y] = j_x^\infty [X, Y]$. The bracket operation on V_x^∞ is linear over the reals. V_x^∞ together with its limit topology is isomorphic with the topological Lie algebra L of derivations in the ring of formal power series in q variables $R[[x^1, \cdots, x^q]]$ [3]. Furthermore there is an isomorphism of $T_a(P^\infty)$ with $V_{\alpha(a)}^\infty$ [4]. Therefore the tangent bundle $T(P^\infty)$ can be looked at as a bundle of topological Lie algebras. Let $T_a^*(P^\infty)$ be the topological dual to $T_a(P^\infty)$. The smooth sections of the bundle

$$\wedge T^*(P^\infty) = \bigcup \wedge T_a^*(P^\infty)$$

are called differential forms on P^∞. There is a well-defined differential

$$d\colon \wedge^r T_a^*(P^\infty) \to \wedge^{r+1} T_a^*(P^\infty) \quad \text{for any } a \in P^\infty,$$

which gives a differential d on the exterior algebra of differential forms on P^∞ [3].

The groupoid Γ^∞ acts transitively on P^∞ on the right by the composition of maps in Γ^∞, whenever defined. (Note that G^∞ acts on P^∞ on the left as P^∞ are jets of elements with target 0.) The differential of this action maps $T(P^\infty)$ into itself. And the algebra of right invariant vector fields and differential forms Λ_I on P^∞ are isomorphic with $T_a(P^\infty)$ and $T_a^*(P^\infty)$, for any $a \in P^\infty$, respectively. Because $d(\Lambda_I) \subset \Lambda_I$ we have a differential complex $\{\Lambda_I, d\}$ of right Γ^∞-invariant differential forms on P^∞. The cohomology of this complex is isomorphic with the cohomology of the complex $\{T_a^*(P^\infty), d\}$. We have therefore

PROPOSITION 1. *The cohomology of the complex $\{\Lambda_I, d\}$ of right Γ^∞-invariant* *differential forms on P^∞ is isomorphic with the cohomology $H(L,R)$ of the topological Lie algebra of derivations L of $R[[x^1,\cdots,x^q]]$.*

The Lie algebra $T_*(P^\infty)$, $* =$ the ∞-jet of the identity at $0 \in R^q$, associated with the topological group P^∞ is an analogue of the Lie algebra of a Lie group.

2. Let M be a C^∞-differentiable manifold with a Γ-foliation of class C^∞ [5]. Any Γ-foliation can be given by a cocycle $\{\gamma_{ij}\}$ on an open countable numerable covering $U = \{U_i\}_{i=0,1,\cdots}$ of M such that $\gamma_i = \gamma_{ii}\colon U_i \to R^q$ are submersions of class C^∞.

PROPOSITION 2. *To a Γ-foliation of class C^∞ on M is canonically associated a bundle $\pi\colon B^\infty \to M$ whose fibre is isomorphic with P^∞.*

PROOF. Suppose that Φ is the transitive pseudogroup of local diffeomorphisms of M preserving the Γ-foliation.

For any $m \in M$ let $0 \leqq i_0 < i_1 < \cdots < i_n$ be a sequence of integers such that for the partition of unity $\{\lambda_i\}$ subordinate to $U = \{U_i\}$, $\lambda_{i_0}(m) \neq 0, \cdots, \lambda_{i_n}(m) \neq 0$ and $\lambda_j(m) = 0$ if j is not in the sequence. Then $m \in U_{i_0} \cap \cdots \cap U_{i_n}$. Let

$$V_{i_0} = \gamma_{i_0}(U_{i_0}) \subset R^q, \qquad x = \gamma_{i_0}(m).$$

If $m' \in M$ is any other point then, because Φ is transitive, there is $\phi \in \Phi$ such that $\phi(m') = m$ and a neighborhood 0 of m', $\phi(0) \subset U_{i_0}$. Let $\gamma\colon 0 \to Q \subset R^q$, $Q = \gamma(0)$, be a submersion given by the Γ-foliation. $x' = \gamma(m')$. As Π is transitive there is $g \in \Gamma^\infty$ such that $g \cdot j^\infty_{m'}(\gamma_{i_0} \cdot \phi) = j^\infty_{m'}(\gamma)$. Then define $B^\infty|_m = \{g |$for any choice of $m' \in M$, ϕ and $\gamma\}$. Then $B^\infty = \bigcup B^\infty_m$. If we choose some other countable numerable covering we get an equivalent bundle.

From this observation and Proposition 2 follows by a standard spectral sequence argument

PROPOSITION 3. *For any C^∞-smooth Γ-foliation on M there is a canonical homomorphism $H(L, R) \to H(P, R)$, where $P \to M$ is the principal $GL(q, R)$-bundle normal to the foliation.*

REMARK. Another construction of this homomorphism was given by Bott, Bernstein and Rozenfeld. The canonical connection on the jet bundles gives another way of seeing this map [2].

3. Let Γ and Γ^∞ be the categories defined above. Recall the definition of the classifying space $B\Gamma$ as defined by Segal [9]. The category Ord has as objects the finite sequences of integers $[n] = (0, 1, \cdots, n)$, $n \geqq 0$; and morphisms are the monotonic maps $\mu\colon[n] \to [m]$ ($i < j$ implies $\mu(i) < \mu(j)$). A semisimplicial groupoid $N\Gamma$ is a contravariant functor from the category Ord into the category of sets, defined as follows: if (n) is the category whose objects are the integers 0, 1, $\cdots$, n with exactly one morphism for each pair of objects then define $N\Gamma[n]$ to be the set of covariant functors from (n) into Γ. We can think of $N\Gamma[n]$ as of the set of n-tuples of composable elements from Π. And we shall write $\Gamma * \cdots * \Gamma$ (n times) for $N\Gamma[n]$, $N\Gamma[0] = R^q$. This definition of a semisimplicial set makes sense even if Γ is replaced by another category, for example Γ^∞ or the category $O(q)$ with a single object and morphisms being the orthogonal group $O(q)$.

The semisimplicial set $N\Gamma$ is graded by the nonnegative integers $(\cdots, N\Gamma[n-1], N\Gamma[n], N\Gamma[n+1], \cdots)$ and there are given the face and degeneracy operators

$$\begin{aligned} &\partial_i : N\Gamma[n] \to N\Gamma[n-1], \qquad 0 \leqq i \leqq n; \\ &s_i : N\Gamma[n] \to N\Gamma[n+1], \qquad 0 \leqq i \leqq n; \\ &\partial_i(x_0, \cdots, x_n) = (x_0, \cdots, x_{i-1}, x_{i+1}, \cdots, x_n), \\ &s_i(x_0, \cdots, x_n) = (x_0, \cdots, x_i, x_i, \cdots, x_n) \quad \text{for any } (x_0, \cdots, x_n) \in N\Gamma[n]. \end{aligned}$$

Let $\Delta_n = \{t_0, t_1, \cdots, t_n \mid 0 \leqq t_0 < t_1 < \cdots < t_n \leqq 1, \sum_{i=1}^n t_i = 1\}$ be the standard n-simplex and $\delta_i : \Delta_{n-1} \to \Delta_n$, $\sigma_i : \Delta_{n+1} \to \Delta_n$ the maps $\delta_i(t_0, \cdots, t_{n-1}) = (t_0, \cdots, t_{i-1}, 0, t_i, \cdots t_{n-1})$, $\sigma_i(t_0, \cdots, t_{n+1}) = (t_0, \cdots, t_i + t_{i+1}, \cdots t_{n+1})$. Let Δ_n be given the usual topology, $N\Gamma[n]$ the discrete topology and let $(\partial_i b_n, a_{n-1}) \sim (b_n, \delta_i a_{n+1})$, $(s_i b_n, a_{n+1}) \sim (b_n, \sigma_i a_{n+1})$, $a_{n\pm 1} \in \Delta_{n\pm 1}$, $b_n \in N\Gamma[n]$ be the equivalence relation on the topological sum $\amalg (\Delta_n \times N\Gamma[n])$. Then the CW complex

$$B\Gamma = \amalg (\Delta_n \times N\Gamma[n]) / \sim,$$

called the geometric realization of the semisimplicial set $N\Gamma$, is the classifying space for the category Γ. Similarly we have $B\Gamma^\infty$ and $BO(q)$. The space $B\Gamma$ is the classifying space for Γ-structures on topological spaces [5]. $BO(q)$, which is homotopy equivalent to $B\Gamma^\infty$ [9], is the classifying space for $O(q)$-bundles.

Now, let us consider the category $\tilde{\Gamma}^\infty$ whose objects are elements of the groupoid Γ^∞ with target 0, i.e. $P^\infty = \{a \in \Gamma^\infty \mid \beta(a) = 0\}$. And morphisms of $\tilde{\Gamma}^\infty$ are the pairs $P^\infty \times P^\infty$. There is a covariant functor τ_0 from $\tilde{\Gamma}^\infty$ into Γ^∞. To an object a associate $\alpha(a) \in R^q$ and to a morphism $(a, b) \in \tilde{\Gamma}^\infty \times \tilde{\Gamma}^\infty$, $a^{-1} \cdot b \in \Gamma^\infty$. Hence for any integer $n \geqq 0$ we have the projection

$$\pi : N\tilde{\Gamma}^\infty[n] \to N\Gamma^\infty[n],$$

$\pi(a_0, a_1, \cdots, a_n) = (a_0^{-1} a_1, a_1^{-1} a_2, \cdots, a_{n-1}^{-1} a_n)$, and the induced map for the classifying spaces.

PROPOSITION 4. *The fibre of the fibration* $\pi : B\tilde{\Gamma}^\infty \to B\Gamma^\infty$ *is isomorphic with* G^∞.

PROOF. The topological group G^∞ acts on $N\tilde{\Gamma}^\infty[n]$, $n \geqq 0$, on the left by a $\cdot$-action, $G^\infty \times N\tilde{\Gamma}^\infty[n] \xrightarrow{\cdot} N\tilde{\Gamma}^\infty[n]$, $v \cdot (a_0, a_1, \cdots, a_n) = (v \cdot a_0, v \cdot a_1, \cdots, v \cdot a_n)$ where for any $a \in \tilde{\Gamma}^\infty$, $v \in G^\infty$, $v \cdot a$ is simply the composition $v \cdot a = v \cdot a$. For any $(a, b) \in \tilde{\Gamma}^\infty \times \tilde{\Gamma}^\infty$, $v \in G^\infty$, $(v \cdot a, v \cdot b) \in \tilde{\Gamma}^\infty \times \tilde{\Gamma}^\infty$, and $(v \cdot a)^{-1} \cdot (v \cdot b) = a^{-1} \cdot b$. Therefore the orbits of the $\cdot$-action by G^∞ on $N\tilde{\Gamma}^\infty[n]$ are subspaces of the fibres of the π-projection. But P^∞ acts on the fibres transitively and effectively, therefore it is isomorphic with the fibres by an argument similar to that for the Lie groups.

Let P^∞ be the category with a single object and the group P^∞ as a set of morphisms. There is a covariant functor τ from $\tilde{\Gamma}^\infty$ to P^∞ which to a morphism $(a, b) \in P^\infty \times P^\infty$ in the category $\tilde{\Gamma}^\infty$ associates the morphism $\cdot\, t_a \cdot a^{-1} \cdot b \cdot (t_a)^{-1}$ (t_a is the ∞-jet of the translation $\alpha(a) \to 0\cdot)$ in the category P^∞.

PROPOSITION 5. *The functor* $\kappa : \tilde{\Gamma}^\infty \to P^\infty$ *gives the fibration*

$$\pi : B\tilde{\Gamma}^\infty \longrightarrow BP^\infty$$

with fibre being the topological group P^∞.

PROOF. Follows by an argument similar to the one used in Proposition 4.

Or, we can observe that κ is a composition of the functor τ_0 with the functor $\tau_1 : \Gamma^\infty \to P^\infty$ which to a morphism v of Γ^∞ associates the morphism $t_v - 1\cdot v\cdot (t_v - 1)^{-1}$. And the fibre of the fibration $\pi : B\Gamma^\infty \to BP^\infty$ is R^q.

Because BP^∞ and $BO(q)$ are homotopy equivalent there is a continuous map $\nu : BO(q) \to BP^\infty$ such that composed with the projection $\tau : BP^\infty \to BO(q)$, $\tau\cdot\nu$ and $\nu\cdot\tau$ are homotopic to the identity.

A map $P^\infty \to O(q)$ defined as a composition of the jet projection $P^\infty \to P^1 = R^q \times GL(q, R)$, projection on $GL(q, R)$ and homotopy $GL(q, R) \to O(q)$ induces the transformation of functors

$$\tau : NP^\infty[n] \longrightarrow NO(q)[n]$$

and ultimately the projection of bundles. And ν is induced by continuous maps $\nu : NO(q)[n] \to NP^\infty[n]$. Let $B_\nu\tilde{\Gamma}^\infty = \nu^*(B\tilde{\Gamma}^\infty)$ be the pull back of $B\tilde{\Gamma}^\infty$ by ν. Then $\pi : B_\nu\tilde{\Gamma}^\infty \to BO(q)$ is a bundle with fibre P^∞.

PROPOSITION 6. *Let* $\pi : B^\infty \to M$ *be the bundle associated with a* C^∞*-differentiable* Γ*-foliation on* M. *Then, up to a homotopy there is a map* η *and a bundle map* ζ *such that*

$$\begin{array}{ccc} B^\infty & \xrightarrow{\ \zeta\ } & B_\nu\tilde{\Gamma}^\infty \\ \downarrow & & \downarrow \\ M & \xrightarrow{\ \eta\ } & BO(q) \end{array}$$

commutes.

PROOF. The existence of η and ζ follows from the standard classification theorems for bundles.

A nice simple description of this map, using the transition functions, for bundles and even Γ-structures was given for example in [11].

4. For any $n \geqq 0$ we have the bundle map

$$\begin{array}{ccc} N\tilde{\Gamma}^\infty[n] & \longrightarrow & N_\nu\tilde{\Gamma}^\infty[n] \\ \pi\downarrow & & \downarrow\pi \\ NP^\infty[n] & \xrightarrow{\ \tau\ } & NO(q)[n] \end{array}$$

where π are bundle projections, $N_\nu\tilde{\Gamma}^\infty[n]$ is induced by ν; $\nu^*(N\tilde{\Gamma}^\infty[n]) = N_\nu\tilde{\Gamma}^\infty[n]$. $B_\nu\tilde{\Gamma}^\infty$ is the geometric realization of $N_\nu\tilde{\Gamma}^\infty$. So far we have been treating $N\tilde{\Gamma}^\infty[n], \cdots$ as sets. But if the topology of P^∞, $O(q)$, is taken into account we get for

$n \geqq 0$ the topological spaces $N\tilde{\Gamma}^{\infty}[n]$, $N_{\nu}\tilde{\Gamma}^{\infty}[n]$, $NP^{\infty}[n]$ and a manifold $NO(q)[n]$. $N_{\nu}\tilde{\Gamma}^{\infty}$ is a semisimplicial space with the left $*$-action by P^{∞}.

And now we define the de Rham complex associated with $B\hat{\Gamma}$ to be the complex $\Lambda N_{\nu}\tilde{\Gamma}^{\infty}[n]$ of differential forms along the fibres of $\pi: N_{\nu}\tilde{\Gamma}^{\infty}[n] \to NO(q)[n]$ invariant with respect to the right action by Γ^{∞} and which depend differentiably on the base. Hence Λ is a contravariant functor from the category of semisimplicial spaces and maps into the category of semisimplicial modules and morphisms. The covariant functor F from the category of sets into the category of abelian groups A which to each set S associates the free group FS generated by S composed with the covariant functor $\Lambda N_{\nu}\tilde{\Gamma}^{\infty}$ defines the covariant functor

$$F\Lambda N_{\nu}\tilde{\Gamma}^{\infty}: \mathrm{Ord} \longrightarrow A.$$

This functor may be given a structure of a single complex

$$\Big\{F^k\tilde{\Gamma}^{\infty} = \bigoplus_{p+q=k} F\Lambda^q N_{\nu}\tilde{\Gamma}^{\infty}[p], D\Big\}, \qquad p > 0,\ q \geqq 0,$$

where $D = d_F \pm \delta$, d_F: $F\Lambda^q N_{\nu}\tilde{\Gamma}^{\infty}[p] \to F\Lambda^{q+1}N_{\nu}\tilde{\Gamma}^{\infty}[p]$ is the ordinary differential of forms along the fibres multiplied by $(-1)^p$, and $\delta = \sum_{i=0}^{p}(-1)^i \partial_i$, $\delta: F\Lambda^q N_{\nu}\tilde{\Gamma}^{\infty}[p] \to F\Lambda^q N_{\nu}\tilde{\Gamma}^{\infty}[p+1]$ is the semisimplicial differential. Similarly we can define the covariant functor $F\Lambda NO(q): \mathrm{Ord} \to A$.

DEFINITION. The de Rham complex associated with $B\hat{\Gamma}$ is the complex $\{F\tilde{\Gamma}^{\infty}, D\}$.

THEOREM 1. *The cohomology of the de Rham complex of $B\hat{\Gamma}$ is*

$$H(F\tilde{\Gamma}^{\infty}) = H(L,R).$$

The proof consists of several steps.

PROPOSITION 7. *There is a spectral sequence associated with the complex $\{F\Lambda N_{\nu}\tilde{\Gamma}^{\infty}, D\}$ whose*

$$E_2^{p,q} = H_{\delta}^{p}(H_{dR}^{q}(N_{\nu}\tilde{\Gamma}^{\infty}))$$

where H_{δ} is the δ-cohomology of the semisimplicial module $H_{dR}^{q}(N_{\nu}\tilde{\Gamma}^{\infty})$ and H_{dR} stands for the de Rham cohomology.

PROOF. Let $A^{p,q}$ be the bigraded module $F\Lambda^q N_{\nu}\tilde{\Gamma}^{\infty}[p]$ and $A^{p,q} = 0$ for $p < 0$ or $q < 0$. The differential on $A^{p,q}$ is $D = d_F + \delta$. $F^p = \sum_{p \leqq v}\sum_m A^{v,m}$ gives the filtration of $A = \sum\sum A^{p,q}$. $F^0 \supset F^1 \supset \cdots \supset F^p \supset F^{p+1} \supset \cdots$. If $F^{p,q} = F^p \cap \sum_{j\geqq 0} A^{q-j,p+j}$ then $F^{p,q} \supset F^{p+1,q-1}$ and we define

$$E_0^{p,q} = F^{p,q}/F^{p+1,q-1} = A^{p,q} \quad \text{and} \quad E_0^{p} = \sum_{q\geqq 0} E_0^{p,q}.$$

As d_0 is defined by d_F we get from the commutative square

$$\begin{array}{ccc} E_0^{p,q} & \xrightarrow{\ d_0\ } & E_0^{p,q+1} \\ \downarrow & & \downarrow \\ \sum_{i\geqq 0} A^{p,i} & \xrightarrow{\ d_F\ } & \sum_{i\geqq 0} A^{p,i+1} \end{array}$$

the isomorphism

$$E_1^{p,q} = H_{dR}^q(N_\nu \tilde{\Gamma}^\infty[p])$$

where on the right is the cohomology of the complex $\{F\Lambda^* N_\nu \tilde{\Gamma}^\infty[p], d_F\}$.

Let $x \in F^{p,q}$ be a cocycle representing $\tilde{x} \in \tilde{E}_1^{p,q}$, i.e. $dx = 0$. Then $Dx = (-1)^q \delta x \in F^{p+1,q}$ is a d_F-cocycle. Hence $(Dx)^\sim \in E_1^{p+1,q}$. The homomorphism $\tilde{x} \to (Dx)^\sim$ is the differential $d_1 \colon E_1^{p,q} \to E_1^{p+1,q}$. An element $\tilde{x} \in E_1^{p,q}$ is a d_1-cocycle if and only if its representative d_F-cocycle is a δ-cocycle. Therefore

$$E_2^{p,q} = H_\delta^p(H_{dR}^q(N_\nu \tilde{\Gamma}^\infty[p])).$$

In order to study the convergence of this spectral sequence we have to investigate the E_2-term more closely.

PROPOSITION 8. *There is a spectral sequence which converges to $H_{dR}(N_\nu \tilde{\Gamma}^\infty)$ and whose $\tilde{E}_2$-term is $\tilde{E}_2^{0,r} = H^r(L, R) \otimes \Lambda^0 NO(q)$, $\tilde{E}_2^{j,r} = 0$, $j > 0$, $r \geqq 0$.*

PROOF. We shall use the standard spectral sequence for the fibre bundle π: $N_\nu \tilde{\Gamma}^\infty[p] \to NO(q)[p]$. Let us denote simply by $\wedge^r F^*$ the bundle $F\Lambda^r N_\nu \tilde{\Gamma}^\infty[p]$ over $N_\nu \tilde{\Gamma}^\infty[p]$. Therefore the sections of $\wedge^r F^*$ are the global r-forms right invariant with respect to the Γ^∞-action and which are π-vertical. Denote by $r_x z$ the restriction of the section $z \in \Gamma(\wedge^r F^*)$ to $\pi^{-1}(x)$, $x \in NO(q)[p]$. The differential d_F : $\Gamma(\wedge^r F^*) \to \Gamma(\wedge^{r+1} F^*)$ is characterized by $r_x(d_F z) = d(r_x z)$, where d is the ordinary differential. The differential d_F induces the morphism of sheaves $d_F \colon \mathscr{F}^r \to \mathscr{F}^{r+1}$, where $\mathscr{F}^r$ is the sheaf over $NO(q)[p]$ of germs of sections of $\wedge^r F^*$.

Then we get the exact sequence of sheaves over $NO(q)[p]$:

$$0 \to \mathscr{Z}^r \to \mathscr{F}^r \to d_F(\mathscr{F}^r) \to 0,$$

$\mathscr{Z}^r = \operatorname{Ker} d_F$, $r \geqq 0$. If we denote by $\mathscr{A}^0$ the sheaf of germs of C^∞-functions on $NO(q)[p]$ we get another exact sequence of sheaves:

$$0 \to \mathscr{Z}^r \otimes \mathscr{A}^0 \to \mathscr{F}^r \otimes \mathscr{A}^0 \to d_F(\mathscr{F}^r) \otimes \mathscr{A}^0 \to 0.$$

And because $\mathscr{Z}^r \otimes \mathscr{A}^0$ is a fine sheaf, the Čech cohomology $H^i(NO(q)[p], \mathscr{Z}^r \otimes \mathscr{A}^0) = 0$, $i > 0$. Therefore

$$0 \to \Gamma(\mathscr{Z}^r \otimes \mathscr{A}^0) \to \Gamma(\mathscr{F}^r \otimes \mathscr{A}^0) \to \Gamma(d_F(\mathscr{F}^r) \otimes \mathscr{A}^0) \to 0$$

is the exact sequence of $\mathscr{A}^0$-modules.

Let $H_x[p]$ be the d_F-cohomology of the complex $\{\wedge F^* | \pi^{-1}(x), d_F\}$, $x \in NO(q)[p]$. The bundle $H[p] = \bigcup H_x[p]$ is trivial because it is the bundle of cohomologies of Γ^∞-invariant forms along the fibres. Therefore $H^*[p] = \bigoplus H^r[p]$, $H^r[p] = NO(q)[p] \times H^r(L, R)$ by Proposition 1. Denote by $\mathscr{H}^r$ the sheaf of C^∞-sections of $H^r[p]$ over $NO(q)[p]$. We get the exact sequence of sheaves

$$0 \to d_F \mathscr{F}^{r-1} \to \mathscr{Z}^r \to \mathscr{H}^r \to 0, \qquad r \geqq 0.$$

The same argument as above guarantees the exactness of the sequence

$$0 \to \Gamma(d_F \mathscr{F}^{r-1} \otimes \mathscr{A}^0) \to \Gamma(\mathscr{Z}^r \otimes \mathscr{A}^0) \to \Gamma(\mathscr{H}^r \otimes \mathscr{A}^0) \to 0.$$

If $\mathfrak{H}^r[p]$ is the sheaf of C^∞ sections of $H^r\,[p]$ over $NO(q)\,[p]$ we can write $\Gamma(\mathfrak{H}^r[p]) = \Gamma(\mathscr{A}^0 \otimes \mathfrak{H}(L))$, where $\mathfrak{H}(L)$ is the trivial sheaf of sections of $H^*(L, R)$. Therefore we get the isomorphisms

$$\Gamma(\mathscr{H}^r \otimes \mathscr{A}^0) \cong \Gamma(\mathfrak{H}^r\,[p]) = \Gamma(\mathscr{A}^0 \otimes \mathfrak{H}^r(L)) \cong A^0(NO(q)\,[p]) \otimes H^r(L,R),$$

where A^0 stands for the ring of global functions.

The filtration of the complex $F\Lambda N_\nu \tilde{\Gamma}^\infty[p]$ by the subspaces L^k of forms with horizontal degree $k \geqq 0$ is trivial. $L^k = 0,\ k > 0,\ L^0 = L^{0,0} \oplus L^{0,1} \oplus \cdots$, where $L^{0,j}$ are forms of vertical degree $j \cdot d_F(L^0) \subset L_0$. The spectral sequence corresponding to this filtration $\{K_r^{k,j},\ d_r\}$ is also trivial. Namely $K_0^{k,j} = 0,\ k > 0,\ K_0^{0,j} = L_0^{0,j}$ and $d_0 = d_F$. There is the obvious isomorphism

$$k_0^j : K_0^{0,j} \to \Gamma(\mathscr{F}^j \otimes \mathscr{A}^0)$$

and the canonical isomorphism

$$k_1^j : K_1^{0,j} \to \Gamma(\mathscr{H}^j \otimes \mathscr{A}^0).$$

Because $d_r = 0,\ r \geqq 1$, we get

$$H^j_{dR}(N_\nu \tilde{\Gamma}^\infty\,[p]) = K_\infty^{0,j} = K_1^{0,j}.$$

Therefore we can conclude that

$$H^r_{dR}\,(N_\nu\,\tilde{\Gamma}^\infty\,[p]) = A^0(NO(q)[p]) \otimes H^r(L,R).$$

Proof of the theorem. The E_1-term of the spectral sequence $\{E_r,\ d_r\}$ from Proposition 7 is $E_1^{p,r} = A^0(N_\nu\ \tilde{\Gamma}^\infty\ [p]) \otimes H^r\ (L,\ R)$. By the Eilenberg-Zilber theorem

$$E_2^{p,r} = \bigoplus_{u+v=p} H^u_\delta(A^0NO(q)) \otimes H^v_\delta(H^r(L,R)).$$

But $H^v_\delta(H^r(L,\ R)) = 0$ for $v > 0$ and $H^0_\delta(H^r(L,R)) = H^r(L,\ R)$. And by the definition $H^u_\delta(A^0\ NO(q)) = H^u_c(O(q),\ R)$, where H^u_c stands for the continuous cohomology of the group $O(q)$ [7]. The theorem of van Est [7] gives the isomorphism $H^u_c(O(q),\ R) \cong H^u(\mathfrak{O}, \mathfrak{K}; R)$, where $\mathfrak{O}$ is the Lie algebra of $O(q)$, $\mathfrak{K}$ the Lie algebra of a maximum compact subgroup of $O(q)$. As $\mathfrak{K} = \mathfrak{O}$, $H^u_c(O(q),\ R) = 0$, $u > 0$, $H^0_c(O(q),\ R) = R$. Hence $E_2^{p,r} = 0,\ \ p > 0,\ \ E_2^{0,r} = H^r(L,\ R)$. Therefore $d_r = 0$ for $r \geqq 2$, $E_\infty^{0,r} = H^r(L,\ R)$, $E_\infty^{p,r} = 0,\ p > 0$; and $H^r_{dR}(B\Gamma) = E_\infty^{0,r}$.

5. The universal normal bundle $B\hat{\Gamma}$ associated with the universal Γ-structure on $B\Gamma$ is the classifying space for the category $\hat{\Gamma}$ whose objects are points of P^1 and whose morphisms are germs of local diffeomorphisms of P^1 into itself commuting with the left action by G^1 (structure group of P^1). Let $\tilde{\Gamma}^1$ be the category with objects P^1 and morphisms $P^1 \times P^1$. The jet projection gives the homotopy equivalence $B\tilde{\Gamma}^1 \to B\tilde{\Gamma}^\infty$. Then we get the commutative diagram, where the horizontal arrows are defined by the jet functor:

$$\begin{array}{ccc} B\hat{\Gamma} & \longrightarrow & B\hat{\Gamma}^{\infty} \\ \downarrow & & \downarrow \\ B\Gamma & \longrightarrow & B\Gamma^{\infty} \end{array}$$

Because $N\Gamma[p]$ is locally euclidean, the de Rham complex of all forms $FAN\hat{\Gamma}[p]$ is well defined, and so is the complex $FAN\hat{\Gamma}^{\infty}[p]$ of all forms on $N\hat{\Gamma}^{\infty}[p]$. Finally let us denote by $CN\hat{\Gamma}[p]$ the complex of singular real cochains. Then we have the morphisms of double complexes

$$FAN_{\nu}\hat{\Gamma}^{\infty} \xrightarrow{\phi_1} FAN\hat{\Gamma}^{\infty} \xrightarrow{\phi_2} FAN\hat{\Gamma} \xrightarrow{\phi_3} CN\hat{\Gamma},$$

where ϕ_1 is the pull back by the homotopy equivalences $N\Gamma^{\infty}[p] \to NO(q)[p]$; ϕ_2 is the pull back induced by the jet functor and ϕ_3 is defined by the usual de Rham cochain map for each p. The composition $\phi_3 \cdot \phi_2 \cdot \phi_1 = \phi: FAN_{\nu}\hat{\Gamma}^{\infty} \to CN\hat{\Gamma}$ induces the map on the cohomology level. Hence we get

THEOREM 2. *There is a natural map* $\phi^*: H^*(FAN_{\nu}\hat{\Gamma}^{\infty}) \to H^*(B\hat{\Gamma}; R)$.

In the analytic category ϕ_3^* is an isomorphism on the cohomology level. One hopes to prove that $(\phi_2 \cdot \phi_1)^*$ is an isomorphism of cohomologies always, even in the C^{∞} category.

Similar theorems for somewhat different categories were proved by Bott and Haefliger **[6]**, and results closely related to this work were observed by Shulman and Stasheff **[10]**.

BIBLIOGRAPHY

1. R. Bott, *Lectures at Harvard University*, Cambridge, Mass., 1972/73.

2. B. Cenkl, *Secondary characteristic classes*, Journèes Exotiques, Lille, 1973.

3. I. M. Gelfand and D. B. Fuks, *The cohomology of the Lie algebra of formal vector fields*, Izv. Akad. Nauk SSSR Ser. Mat. **34** (1970), 322–337. Math. USSR Izv. **4** (1970), 327–342. MR **42** #1103.

4. V. Guillemin and S. Sternberg, *Deformation theory of pseudo-group structures*, Mem. Amer. Math. Soc. No. 64 (1966). MR **35** #2302.

5. A. Haefliger, *Homotopy and integrability*, Manifolds (Proc. Nuffic Summer School, Amsterdam, 1970), Springer, Berlin, 1971, pp. 133–163. MR **44** #2251.

6. ———, *Lectures at the Institute for Advanced Study*, Princeton, N.J., 1972/73.

7. G. Hochschild and G. D. Mostow, *Cohomology of Lie groups*, Illinois J. Math. **6** (1962), 367–401. MR **26** #5092.

8. A. Kumpera and D. C. Spencer, *Lie equations*. Vol. I: *General theory*, Princeton Univ. Press, Princeton, N.J., 1972.

9. G. Segal, *Classifying spaces and spectral sequences*, Inst. Hautes Etudes Sci. Publ. Math. No. 34 (1968), 105–112. MR **38** #718.

10. H. Shulman and J. Stasheff, *de Rham theory for classifying spaces* (to appear).

11. J. Stasheff, *Construction of BC*; *Classification via* $\mathscr{B}\Gamma^q$, Lectures on Algebraic and Differential Topology, Lecture Notes in Math., vol. 279, Springer-Verlag, Berlin and New York, 1972, pp. 81–94.

NORTHEASTERN UNIVERSITY

Proceedings of Symposia in Pure Mathematics
Volume 27, 1975

LOCALLY FREE LIE TRANSFORMATION GROUPS OF CODIMENSION TWO

LAWRENCE CONLON

The background of the work to be announced here is formed by a number of known results on the rank and file of manifolds. Recall that the *file* of M is the largest integer k for which $\boldsymbol{R}^k$ has a smooth locally free action on M, while the *rank* of M is the largest k for which M admits a global k-frame of commuting vector fields. Clearly $\operatorname{rank}(M) \geqq \operatorname{file}(M)$ and equality holds if M is compact.

We will present a theorem generalizing the following results.

(a) $\operatorname{rank}(S^3) = 1$ (Lima [1]).

(b) $\operatorname{rank}(S^1 \times S^2) = 1$ (Rosenberg [3]).

(c) $\operatorname{rank}(T^{n-2} \times S^2) = n - 2$ (Novikov [2]).

(d) M^n compact, $\pi_1(M)$ finite $\Rightarrow \operatorname{rank}(M^n) \leqq n - 2$ (Sacksteder [6]; cf. also Rosenberg [4]).

(e) M a 3-manifold, $\pi_2(M) \neq 0 \Rightarrow \operatorname{file}(M) = 1$ (Rosenberg [5]).

If $\mathfrak{F}$ is a foliation of M, a normal vector field to $\mathfrak{F}$ is said to be *parallel* along $\mathfrak{F}$ if it is invariant under the linear holonomy of each leaf. We define $\rho(\mathfrak{F})$ to be the largest integer k for which there exists a global normal frame field $(X_1, \cdots, X_k)$ with each X_i a complete vector field parallel along $\mathfrak{F}$. Thus, e.g., $\rho(\mathfrak{F}) = 0$ means that $\mathfrak{F}$ is not invariant under any nonsingular flow transverse to the leaves.

If $\boldsymbol{R}^{k+1}$ has a smooth locally free action on M, then the foliation $\mathfrak{F}$ by orbits of the subgroup $\boldsymbol{R}^k \subset \boldsymbol{R}^{k+1}$ has $\rho(\mathfrak{F}) \geqq 1$. Thus, if we prove that every $\mathfrak{F}$ produced on M by a locally free action of $\boldsymbol{R}^k$ has $\rho(\mathfrak{F}) = 0$, we conclude that $\operatorname{file}(M) \leqq k$ (and $\operatorname{rank}(M) \leqq k$ for M compact). Of course, $\rho(\mathfrak{F}) = 0$ is a stronger statement.

AMS (MOS) subject classifications (1970). Primary 57D30, 57E99.

By these remarks, the following theorem generalizes the above results on rank and file. The proof will appear elsewhere.

THEOREM. *Let M be a smooth connected n-manifold, G a connected locally free Lie transformation group of M of dimension $n - 2$.* Let $\mathfrak{F}$ *be the foliation of M by G-orbits. Then* $\rho(\mathfrak{F}) \geqq 1$ *implies*

(1) $\pi_2(M) = 0$.

(2) *If M is compact then* $\pi_1(M)$ *is infinite and, if abelian,* $\pi_1(M)$ *has* rank $\geqq 2$.

(3) *If M is compact,* $\pi_1(M)$ *abelian, and G contractible, then* rank $(\pi_1(M)) \geqq n - 1$.

REFERENCES

1. E. Lima, *Commuting vector fields on* S^3, Ann. of Math. (2) **81** (1965), 70–81. MR **30** #1517.

2. S. P. Novikov, *Topology of foliations*, Trudy Moskov. Mat. Obšč. **14** (1965), 248–278. Trans. Moscow Math. Soc. **1965**, 268–304. MR **34** #824.

3. H. Rosenberg, *The rank of* $S^2 \times S^1$, Amer. J. Math. **87** (1965), 11–24. MR **31** #764.

4. ———, *Actions of* R^n *on manifolds*, Comment, Math. Helv. **41** (1966–67), 170–178. MR **34** #6794.

5. ———, *Singularities of* R^2 *actions*, Topology **7** (1968), 143–145. MR **37** #3596.

6. R. Sacksteder, *Foliations and pseudogroups*, Amer. J. Math. **87** (1965), 79–102. MR **30** #4268.

WASHINGTON UNIVERSITY

Proceedings of Symposia in Pure Mathematics
Volume 27, 1975

ON COMPACT FOLIATIONS

K. DECESARE AND T. NAGANO*

0. Introduction. A compact foliation is one in which every leaf is compact. When attempting to determine the structure of such objects, one is almost immediately confronted with the following question: Is the natural projection onto the space of leaves (in the quotient topology) a closed map? In his thesis [5], Reeb proved that this was the case for codimension one compact foliations and gave an example of a compact codimension 2-foliation of a noncompact manifold in which the projection was not closed. However, the question remains open when the ambient manifold is assumed to be compact, an equivalent statement being: Is the space of leaves a Hausdorff space? (It should be mentioned that Epstein [2] has answered the question affirmatively for compact 3-manifolds with boundary.)

In this paper, a sufficient condition is given in order that the space of leaves be Hausdorff, and some consequences of the Hausdorff property are proved. Specifically, let M be a compact manifold and $\mathscr{F}$ a compact codimension q-foliation of M. Let $M/\mathscr{F}$ denote the space of leaves in the quotient topology.

THEOREM 1. *If the volume of the leaves is bounded, or if the "spinning function" (defined in* § 1) *takes only finite values, then* $M/\mathscr{F}$ *is Hausdorff.*

THEOREM 2. *If* $M/\mathscr{F}$ *is Hausdorff, then* M *is a* $\mathscr{V}$*-bundle over the* $\mathscr{V}$*-manifold* $M/\mathscr{F}$ (*Satake* [6]).

The authors wish to thank K. Millett for a very timely and valuable conversation during the conference.

AMS (MOS) subject classifications (1970). Primary 57D30, 57E30.
Key words and phrases. Bundle-like metrics, exotic classes, foliations, leaf spaces.
*Partially supported by NSF grant GP 29662.

1. The proof of Theorem 1. A neighborhood U of a point in M is called *flat* if the induced foliation on U is isomorphic with the trivial one on the cartesian space. If L is a leaf of $\mathscr{F}$ (abbreviated $L \in \mathscr{F}$), a nonempty connected component of $U \cap L$ is called a *plaque* of U. The *spinning function* s for $\mathscr{F}$, $s : M \to N \cup \{\infty\}$ is defined by

$$s(x) = \inf_U \sup_{L \in M/\mathscr{F}} \#\{\text{plaques of } L \text{ contained in } U\},$$

where U runs over the flat neighborhoods of x. s is constant along each leaf.

The force of Theorem 1 is that the open problem mentioned in the Introduction is restated as: Can s take the value ∞?

For the proof of Theorem 1, first note that a sequence (L_ν) of leaves converges to a leaf L in $M/\mathscr{F}$ if and only if there is a sequence (x_ν) of points $x_\nu \in L_\nu$ which converges to a point x in L (in the topology of M). The proof is preceded by a lemma.

LEMMA. *Every sequence (L_ν) in $M/\mathscr{F}$ converges to at most a finite number of leaves (if any).*

PROOF. Suppose not; say (L_ν) converges to each of (Λ_n). By compactness, we may assume that (Λ_n) converges to some leaf L. Take $x \in L$ and a flat neighborhood U of x. Then there is a k such that $\Lambda_n \cap U$ is nonempty for $n \geqq k$. Pick a positive integer m. Then $\Lambda_k, \cdots, \Lambda_{k+m}$ all have plaques in U, say ρ_j is a plaque of Λ_{k+j}, $j = 0, \cdots, m$. Now within U we can pick disjoint flat neighborhoods $U_0, \cdots, U_m$ saturated for the projection of U, so that $\rho_j \subseteq U_j$, $j = 0, \cdots, m$. But since (L_ν) converges to each of $\Lambda_k, \cdots, \Lambda_{k+m}$, there must be a ν_0 so large that L_ν has at least one plaque in U_j, $j = 0, \cdots, m$, if $\nu \geqq \nu_0$. Thus, L_ν intersects U at least $m + 1$ times if $\nu \geqq \nu_0$. Since m was arbitrary it follows both that $s(x) = \infty$ and that the volume of the leaves is unbounded contrary to the assumption.

REMARK. That $s(x)$ is always finite is a purely topological condition so this lemma and, in fact, the theorem, is valid in the C^0 category. Evidently, one can interpret "volume of the leaves" in either the measure-theoretic or the Riemannian sense.

To prove the theorem, it is enough to show that any convergent sequence L_ν converges to exactly one leaf. Thus, suppose $L_\nu \to L$ in the topology of $M/\mathscr{F}$. By Lemma 1.1, there are only finitely many leaves $L_1, \cdots, L_m$ such that $L_\nu \to L_i$, $i = 0, \cdots, m$, $L = L_0$. Pick disjoint open sets $U_0, \cdots, U_m$ about $L_0, \cdots, L_m$ respectively. It is clear that if ν is large enough, $L_\nu \cap U_i$ is nonempty for all $i \in \{0, \cdots, m\}$. Thus, if it is shown that infinitely many L_ν's are contained in $\bigcup U_i$, it follows from the connectedness of any L_ν that there is only one U_i, finishing the proof.

Suppose to the contrary that there are only finitely many L_ν's contained in $\bigcup U_i$. Then there exists a sequence (x_ν) with $x_\nu \in L_\nu$, such that some subsequence converges to a point x^1 which is necessarily not in $\bigcup U_i$. Call this subsequence (L_ν^1). It is clear that (L_ν^1) converges to each of $L_0, \cdots, L_m$ and at least one more leaf, namely, the leaf through x^1. Let U_{m+1} be open, disjoint from the other U_i's, and such that the leaf through x^1 is contained in U_{m+1}.

If infinitely many L_ν^1's are contained in $\bigcup U_i$, $i = 1, \cdots, m + 1$, then again there can be only one U_i, again finishing the proof. If not, repeat the process to get x^2, (L_ν^2) and U_{m+2}.

Now either this process stops after finitely many steps, in which case the proof is finished, or there is an infinite sequence of subsequences $(L_\nu) \supseteq (L_\nu^1) \supseteq \cdots \supseteq (L_\nu^k) \supseteq \cdots$ and infinitely many distinct U_i's. However, choosing the subsequence (L_k^ω) defined by $L_k^\omega \in (L_\nu^{k-1}) \backslash (L_\nu^k)$, one easily checks that (L_k^ω) converges to every U_i, contradinting the lemma and finishing the proof.

2. The Hausdorff case. Throughout this section M will be a compact, C^2 manifold and $\mathscr{F}$ a compact, C^2, codimension q-foliation of M such that $M/\mathscr{F}$ is Hausdorff. This condition implies that each leaf is stable in the sense that it has a neighborhood basis (in M) consisting of saturated open sets. Since $\mathscr{F}$ is differentiable, the normal bundle $Q(L)$ to each leaf L possesses a natural parallel translation induced by $\mathscr{F}$ that gives a codimension q-foliation of $Q(L)$ transverse to the fibers.

PROPOSITION 2.1. *In the above situation, each leaf L has finite holonomy.*

PROOF. Let $x \in L \in \mathscr{F}$ and $L \xrightarrow{i} T \xrightarrow{\pi} L$ be a tubular neighborhood of L in M, chosen so that the fibers of π are transverse to $\mathscr{F}$. Let $N \subseteq T$ be a saturated neighborhood of L. Then N inherits the projection of T.

Set D equal to the connected component about x of $\pi^{-1}(x) \cap N$. Since π is C^2, it is a covering space of finite degree if it is restricted to any leaf $L' \subseteq N$. Thus, by lifting paths, we have a homomorphism $\Phi : \pi_1(L, x) \to \text{Diff}(D)$. Set $G = \Phi(\pi_1(L, x))$. Then the proof will be finished in two steps:

(1) *G is finite.* Certainly G is countable. Moreover, it acts on D as a transformation group. By the Baire Category Theorem, the isotropy group of some $d \in D$ must be trivial. Thus, the cardinality of G is the cardinality of the orbit of d, which in turn is the degree of the covering of L by the leaf through y.

(2) *G is isomorphic to the holonomy group of L at x.* To prove this, first note that the holonomy group of L at x is isomorphic to the group of germs at x of the diffeomorphisms in G. Thus, it is enough to show that the assignment $\phi \to$ (germ of ϕ at x), $\phi \in G$, is injective. Accordingly, suppose that ϕ is the identity on a neighborhood of x. Since all covering degrees are finite, ϕ is a pointwise periodic diffeomorphism of D. By a theorem of Montgomery (cf. [3, p. 224]) ϕ must be periodic. But a periodic diffeomorphism whose fixed point set has nonempty interior must be the identity (cf. [3] again), finishing the proof.

REMARK. This proof is valid in the C^1 case. The C^2 assumption will now be used to show that the holonomy and the linear holonomy are isomorphic. Thus, assume $\phi \in G$ and $(d\phi)_x$ is the identity map. Pick a metric in which ϕ is an isometry. In the C^2 case, x has a neighborhood in which any point can be joined to x by a unique geodesic contained in the neighborhood. Certainly, ϕ is the identity on this neighborhood.

PROPOSITION 2.2. *In the above situation, each leaf L has a neighborhood basis consisting of saturated open sets on which the foliation $\mathscr{F}$ is conjugate to the foliation of Q (L) (i. e., there is a diffeomorphism from Q (L) onto any one of these neigh-*

borhoods that carries the foliation on $Q(L)$ to the foliation on the neighborhood).

PROOF. Pick $x \in L$ and let G, D and T, be as in the proof of Proposition 2.1. Since G is a finite group of diffeomorphisms of D with fixed point x, we can choose a neighborhood D' of x that is invariant under the action of G and diffeomorphic to R^q (e.g., take a G-invariant Riemannian metric and exponentiate a small open q-disc in TD_x). If $y \in L$, pick a path from x to y and lift the path for every $d \in D'$. This gives a diffeomorphism from D' into the fiber of T over y. The G-invariance of D' says precisely that the image of this diffeomorphism is independent of the path chosen from x to y. Doing this for every y, we obtain the saturation of D', since this is precisely the set of lifts of all paths from x to y, $y \in L$. The saturation of D', call it B, is certainly an open subbundle of T with the foliation transverse to its fibers. Since this foliation is determined up to conjugacy by the homomorphism $\Phi : \pi_1(L, x) \to G$ and since G is naturally isomorphic to the linear holonomy, the result follows.

3. Further remarks. It follows from Theorem 2.2 and Lemma 1.2 that $M/\mathscr{F}$ is a compact $\mathscr{V}$-manifold and $M \to M/\mathscr{F}$ is a $\mathscr{V}$-bundle (Satake **[6]**). Thus, picking a metric (in the $\mathscr{V}$-manifold sense) on $M/\mathscr{F}$, the resulting metric on M is a bundle-like metric for the foliation, i.e., the foliation is Riemannian. In particular, the Bott vanishing theorem can be strengthened (Pasternack **[4]**) and the exotic classes (Bott **[1]**) all vanish.

By a closer examination of the neighborhoods given in Theorem 2.2, one can easily show that the nonholonomic leaves form an open, dense, connected subset of $M/\mathscr{F}$ that is a q-dimensional manifold, and, further, that there is an integer k such that all holonomy groups have order $\leqq k$. It turns out that there is a very nice stratification of $M/\mathscr{F}$. (Of course, this also follows from $M/\mathscr{F}$ being a $\mathscr{V}$-manifold.)

Since the order of the holonomy groups is bounded, the volume of the leaves is bounded also. Thus, the sufficient condition of Theorem 1.1 is also necessary in the C^2 case.

The simplest way of generating examples (in the Hausdorff case) is to take a compact $\mathscr{V}$-manifold with some metric and use for M its (oriented) frame bundle. Of course, one may not be able to recognize the resulting manifold very easily. For example, if $M/\mathscr{F}$ is the 2-disc, one obtains the frame bundle by suspending the diffeomorphism of $S^2 \subseteq R^3$ given by reflection through a hyperplane (in R^3). Along these lines it should be mentioned also that any compact foliation coming from a locally constant bundle has a Hausdorff leaf space.

ADDED IN PROOF. Millett has also proved the results of §2 in the C^0 case, using the work of Kirby and Siebenmann on topological tubular neighborhoods.

REFERENCES

1. R. Bott, *Lectures on characteristic classes and foliations*, Lecture Notes in Math., vol. 279, Springer-Verlag, Berlin and New York, 1972, pp. 1–80.

2. D. B. A. Epstein, *Periodic flows on three-manifolds*, Ann. of Math. (2) **95** (1972), 66–82. MR **44** #5981.

3. D. Montgomery and L. Zippin, *Topological transformation groups*, Interscience, New York, 1955. MR 17, 383.

4. J. Pasternack, *Foliations and compact Lie group actions*, Comment. Math. Helv. **46** (1971), 467–477. MR **45** #9353.

5. G. Reeb, *Sur certaines propriétés topologiques des variétés feuilletées*, Actualités Sci. Indust., no. 1183, Hermann, Paris, 1952. MR **14**, 1113.

6. I. Satake, *On a generalization of the notion of manifold*, Proc. Nat. Acad. Sci. U.S.A. **42** (1956), 359–363. MR **18**, 144.

UNIVERSITY OF NOTRE DAME

Proceedings of Symposia in Pure Mathematics
Volume 27, 1975

SEMISIMPLICIAL WEIL ALGEBRAS AND CHARACTERISTIC CLASSES FOR FOLIATED BUNDLES IN ČECH COHOMOLOGY

FRANZ W. KAMBER AND PHILIPPE TONDEUR

0. Introduction. The authors have generalized the Chern-Weil construction for the characteristic classes of a principal bundle $G \to P \to M$ in [12] to [17]. In view of applications to complex analytic manifolds and algebraic varieties the requirement of the existence of a global connection in P is dropped. What is used instead is a family of local connections on P restricted to the sets of a cohomologically trivial open covering of M. Even in the smooth case connections are often given in this way, and a direct construction of the characteristic classes of P via these data seems desirable, regardless of the fact that they can be constructed by using a global connection in P. If one wants to work with these data directly, one is then lead automatically to semisimplicial methods, as the resulting invariants are defined via Čech cohomology. In the papers indicated above the phenomenon of secondary characteristic classes for foliated bundles is shown to be a consequence of a filtration preserving property of the Weil homomorphism. An account of the developments leading to this point of view with appropriate references to the literature can be found in the introduction to [17], to which we also refer for the basic *concepts of* foliated bundle, $\mathfrak{g}$-DG-algebra, etc.

The basic idea for the construction of the generalized characteristic homomorphism in [16], [17] is as follows. Let $\mathfrak{U} = (U_j)$ be a sufficiently fine open covering of the base space of the foliated G-bundle $P \xrightarrow{\pi} M$. Then there exists a family $\omega = (\omega_j)$ of local connections ω_j on $P \mid U_j$ representing the flat connection mod Ω, where Ω

AMS (MOS) subject classifications (1970). Primary 57D20; Secondary 57D30.

$\subset \Omega_M^1$ is the given foliation on M. ω defines then a formal connection in the $\mathfrak{g}$-DG-algebra of Čech cochains $\check{C}^{\cdot}(\mathfrak{U}, \pi_*\Omega_P^{\cdot})$ of $\mathfrak{U}$ with coefficients in the direct image of the de Rham complex $\Omega_P^{\cdot}$ of P. A semisimplicial model $W_1(\mathfrak{g})$ of the Weil algebra $W(\mathfrak{g})$ is constructed, on which ω defines a $\mathfrak{g}$-DG-algebra homomorphism

$$k_1(\omega): W_1^{\cdot}(\mathfrak{g}) \to \check{C}^{\cdot}(\mathfrak{U}, \pi_*\Omega_P^{\cdot}).$$

This generalized Weil homomorphism restricts then on H-basic elements for a closed subgroup H to the map giving rise to the characteristic homomorphism Δ_* of P [**17**, Theorem 3.7]. Details of this construction are given as follows.

In § 1 we define a sequence of semisimplicial models $W_s(\mathfrak{g})$ for the Weil algebra $W(\mathfrak{g})$. The construction of $W_1(\mathfrak{g})$ is similar to the Amitsur complex of $W(\mathfrak{g})$. $W_s(\mathfrak{g})$ for $s > 1$ is obtained by an iterative process and $W_0(\mathfrak{g}) = W(\mathfrak{g})$. The essential feature of these algebras is that they behave cohomologically exactly like the Weil algebra. More precisely, there are even filtrations $F_s(\mathfrak{g})$ on $W_s(\mathfrak{g})$ for $s \geqq 0$ and canonical filtration preserving homomorphisms $\rho_s : W_s(\mathfrak{g}) \to W_{s-1}(\mathfrak{g})$ for $s > 0$ inducing cohomology-isomorphisms of the associated graded algebras and hence also of the truncated algebras $W_s(\mathfrak{g})_k = W_s/F_s^{2(k+1)}$, $s \geqq 0, 0 \leqq k \leqq \infty$. The same is true for $\mathfrak{h}$-basic subalgebras provided $\mathfrak{h} \subset \mathfrak{g}$ satisfies certain conditions (Theorem 1.10). The algebra $W_1(\mathfrak{g})$ is noncommutative except in trivial cases. Using this it is possible to show that the projection $\rho_1: W_1(\mathfrak{g}) \to W(\mathfrak{g})$ does not admit a splitting by a $\mathfrak{g}$-DG-algebra-homomorphism. However a splitting $\lambda: W(\mathfrak{g}) \to W_1(\mathfrak{g})$ of ρ_1 by a $\mathfrak{g}$-DG-module map λ does exist (Theorem 1.12).

In § 2 we construct first the homomorphism $k_1(\omega)$ mentioned above. It is filtration preserving, where the filtration on the target complex $\check{C}^{\cdot}(\mathfrak{U}, \pi_*\Omega_P^{\cdot})$ is given by the ideals generated by the powers of Ω in $\pi_*\Omega_P^{\cdot}$. For an H-structure of P defined by a cross-section $s: M \to P/H$ the generalized characteristic homomorphism Δ_* is then defined in Theorem 2.8. Note that we have in fact more generally a homomorphism of filtered $\mathfrak{g}$-DG-algebras $\Delta(\omega): W_1^{\cdot}(\mathfrak{g}, H)_q \to \check{C}^{\cdot}(\mathfrak{U}, \Omega_M^{\cdot})$. This map induces a homomorphism of the associated multiplicative spectral sequences. The spectral sequence converging to $H(W_1(\mathfrak{g}, H)_q)$ is given in (1.11). The spectral sequence for $H_{DR}^{\cdot}(M)$ associated to the Ω-filtration has been announced ([**12**], [**16**, III]) and computed in [**17**]. The spectral sequence is of the general form

$$E_1^{s,t}(\Omega) = \operatorname{Ext}^t_{\mathscr{U}(\mathscr{L})}(M; \mathscr{O}_M, \Lambda^s\Omega) \Rightarrow H_{DR}^{s+t}(M)$$

where $\mathscr{L} = (\Omega_M^1/\Omega)^* \subset \mathscr{T}_M$ is the annihilator sheaf of Ω in $\mathscr{T}_M$ and $\mathscr{U}(\mathscr{L})$ is the universal envelope of the sheaf of Lie algebras $\mathscr{L}$. This spectral sequence is the Leray spectral sequence for de Rham cohomology in the case where the foliation Ω is defined by a global submersion, and it may therefore be considered as a proper generalization of the Leray spectral sequence to foliations. It takes a more familiar form in other cases as well. The filtration preserving homomorphism $\Delta(\omega)$ induces then a multiplicative morphism of the spectral sequences as indicated in the following diagram:

$$\begin{array}{ccc} H^{2s+t}(W_1(\mathfrak{g}, H)_q) & \xrightarrow{\Delta_*} & H^{2s+t}_{DR}(M) \\ \Uparrow & & \Uparrow \\ E_2^{2s,t} = H^t(\mathfrak{g}, H) \otimes I(G)_q^{2s} & \xrightarrow{\Delta_1^{s,t}} & E_1^{s,s+t}(\Omega) \end{array}$$

The homomorphism Δ_1 is called the first derived characteristic homomorphism of the foliated bundle in question. On $I(G)_q = E_2^{\cdot,0}$ the map Δ_1 gives a generalization of the characteristic classes of a complex-analytic principal bundle in Hodge cohomology defined by Atiyah in **[1]**, while the invariants defined by Δ_1 on $E_2^{0,\cdot} = H^\cdot(\mathfrak{g}, H)$ are invariants associated to the partial flat structure in P given by the foliation. All of this has been explained in detail in **[17]** and will not be included in these notes.

At the end of § 2 we describe explicitly some invariants associated to a foliated holomorphic vector bundle whose first integral Chern class is zero.

1. Semisimplicial Weil algebras. We use the notion of a $\mathfrak{g}$-DG-algebra as defined in **[17]** (compare also **[7]**, **[16, I, II]**). Algebras are over a fixed ground field K of characteristic zero. From the introduction we see that we have to consider two types of semisimplicial $\mathfrak{g}$-DG-algebras, namely the algebras $W_s(\mathfrak{g})$ (to be defined below) and $\check{C}^\cdot(\mathfrak{U}, \pi_*\Omega_P^\cdot)$, and also homomorphisms between them. The common concept is the algebra $\check{C}^\cdot(S, \boldsymbol{A}^\cdot)$ of cochains on a semisimplicial set with coefficients in a local system $\boldsymbol{A}$ of $\mathfrak{g}$-DG-algebras as S.

Recall that a semisimplicial set S is given by p-simplices S_p for $p \geqq 0$, and face operators $\varepsilon_i^p: S_{p+1} \to S_p$, degeneracy operators $\mu_i^p: S_p \to S_{p+1}$, subject to the usual conditions under composition (see e.g. Godement, *Théorie des faisceaux*, p. 271). A local system $\boldsymbol{A}^\cdot$ of $\mathfrak{g}$-DG-algebras on S is the assignment of a graded $\mathfrak{g}$-DG-algebra $A_\sigma^\cdot$ for every $\sigma \in S_p$ and $\mathfrak{g}$-DG-homomorphisms

$$\bar{\varepsilon}_i^p(\sigma): A^\cdot_{\varepsilon_i^p(\sigma)} \to A^\cdot_\sigma, \qquad \bar{\mu}_i^p(\sigma): A^\cdot_{\mu_i^p(\sigma)} \to A^\cdot_\sigma$$

with obvious composition rules. Local systems pull back canonically under maps of semisimplicial sets. For S a semisimplicial set and $\boldsymbol{A}$ a local system of $\mathfrak{g}$-DG-algebras, the $\boldsymbol{A}$-valued cochains on S

$$(1.1) \qquad C^\cdot(S, \boldsymbol{A}^\cdot) = \bigoplus_p C^p(S, \boldsymbol{A}^\cdot), \qquad C^p(S, \boldsymbol{A}^\cdot) = \prod_{\sigma \in S_p} \boldsymbol{A}^\cdot_\sigma$$

together with the maps $\varepsilon_i^p: C^p \to C^{p+1}$, $\mu_i^p: C^{p+1} \to C^p$ defined by

$$(\varepsilon_i^p c)(\sigma) = \bar{\varepsilon}_i^p\, c(\varepsilon_i^p \sigma), \qquad (\mu_i^p c)(\sigma) = \bar{\mu}_i^p c(\mu_i^p \sigma)$$

are a (co-) semisimplicial object in the category of $\mathfrak{g}$-DG-algebras. $C^\cdot(S, \boldsymbol{A}^\cdot)$ is functorial, covariant for maps of local systems $\boldsymbol{A}^\cdot \to \boldsymbol{A}'^\cdot$ and contravariant for maps of semisimplicial sets $S \to S'$.

The Alexander-Whitney multiplication in $C^\cdot(S, \boldsymbol{A}^\cdot)$ is defined as follows. Denote $C^{p,q} = C^p(S, \boldsymbol{A}^q)$. Then

(1.2) $$m_C\colon C^{p,q}\otimes C^{p',q'}\to C^{p+p',q+q'}$$

is the composition $m_C = (-1)^{p'q}\mu_p^{p+p'}\circ m_A\circ\eta\otimes\zeta$, where

$$\eta = \varepsilon_{p+p'+1}^{p+p'}\circ\cdots\circ\varepsilon_{p+1}^{p},\qquad \zeta = \varepsilon_p^{p+p'}\circ\cdots\circ\varepsilon_0^{p'}$$

and m_A is induced by the (associative) multiplication in $\boldsymbol{A}$. This turns $C^\cdot(S, \boldsymbol{A}^\cdot)$ into an associative graded algebra. It is noncommutative, even if $\boldsymbol{A}$ is a local system of commutative $\mathfrak{g}$-*DG*-algebras. The differential d_A in $\boldsymbol{A}$ and the semisimplicial boundary operator $\delta = \sum_{i=0}^{p+1}(-1)^i\varepsilon_i^p$ turn $C^\cdot(S, \boldsymbol{A}^\cdot)$ into a double complex with total differential $d_C = \delta + (-1)^p d_A$ on $C^{p,q}$. Then $(C^\cdot(S, \boldsymbol{A}^\cdot), d_C)$ is a $\mathfrak{g}$-*DG*-algebra, where the operators $\theta(x)$, $i(x)$ are defined for each simplex, i.e.

$$(i(x)c)(\sigma) = (-1)^p i(x)(c(\sigma)),\qquad (\theta(x)c)(\sigma) = \theta(x)(c(\sigma)),\qquad c\in C^{p,q},\ x\in\mathfrak{g}.$$

The example of interest in our geometric context is the case $S = N(\mathfrak{U})$ for the nerve of a covering $\mathfrak{U} = (U_j)$ of M, and $\boldsymbol{A} = \pi_*\Omega_P^\cdot$ the local system defined by the direct image of the de Rham complex of P under $P\xrightarrow{\pi}M$. This is the algebra $\check{C}^\cdot(\mathfrak{U}, \pi_*\Omega_P^\cdot)$ discussed in the introduction. A family $\omega = (\omega_j)$ of local connections in $P|U_j$ representing the given flat connection in P mod Ω **[17]** defines then a formal connection in $\Lambda\mathfrak{g}^*\xrightarrow{\omega}\check{C}^0(\mathfrak{U}, \pi_*\Omega_P^1)\subset\check{C}^\cdot(\mathfrak{U}, \pi_*\Omega_P^\cdot)$. But since this is a noncommutative $\mathfrak{g}$-*DG*-algebra, we cannot use the universal property of $W(\mathfrak{g})$ to extend ω to an algebra homomorphism on $W(\mathfrak{g})$. The remedy consists in applying the construction above once more to obtain a semisimplicial algebra $W_1(\mathfrak{g})$ replacing $W(\mathfrak{g})$.

To explain the construction of $W_1(\mathfrak{g})$, we consider first a semisimplicial object in the category of Lie algebras defined by $\mathfrak{g}$ as follows. Let $\mathfrak{g}^{l+1}$ denote for $l\geqq 0$ the $(l+1)$-fold product of $\mathfrak{g}$ with itself. Define for $0\leqq i\leqq l+1$, $0\leqq j\leqq l$,

$$\varepsilon_i^l\colon \mathfrak{g}^{l+2}\to\mathfrak{g}^{l+1},\qquad \varepsilon_i^l(x_0,\cdots,x_{l+1}) = (x_0,\cdots,x_{i-1},x_{i+1},\cdots,x_{l+1}),$$
$$\mu_j^l\colon \mathfrak{g}^{l+1}\to\mathfrak{g}^{l+2},\qquad \mu_j^l(x_0,\cdots,x_l) = (x_0,\cdots,x_j,x_j,x_{j+1},\cdots,x_l).$$

Then ε and μ are the face and degeneracy maps for the semisimplicial object in question and satisfy the usual relations.

Next consider the Weil algebra as a contravariant functor from Lie algebras to $\mathfrak{g}$-*DG*-algebras and apply it to the semisimplicial object discussed. This gives rise to a cosemisimplicial object $W_1(\mathfrak{g})$ in the category of $\mathfrak{g}$-*DG*-algebras. Note that

$$W_1^l(\mathfrak{g}) = W(\mathfrak{g}^{l+1}) = W(\mathfrak{g})^{\otimes l+1}$$

and the face and degeneracy maps $\varepsilon_i^l = W(\varepsilon_i^l)\colon W_1^l\to W_1^{l+1}$, $\mu_i^l = W(\eta_i^l)\colon W_1^{l+1}\to W_1^l$ are given by the inclusions omitting the ith factors and multiplication of the ith and $(i+1)$th factors. Thus $W_1(\mathfrak{g})$ is the Amitsur complex of $W(\mathfrak{g})$.

Let now *Pt* be the semisimplicial point (terminal object) with exactly one simplex σ_l in each dimension $l\geqq 0$ and canonical face and degeneracy maps. As a local system on *Pt* is precisely given by a cosemisimplicial object of $\mathfrak{g}$-*DG*-algebras, we may consider $W_1^\cdot(\mathfrak{g})$ as the cochain complex on *Pt* with coefficients in the local system $\boldsymbol{W}: \sigma_l \to W(\mathfrak{g}^{l+1}),\ l \geqq 0 : W_1^{\cdot}(\mathfrak{g}) = C^{\cdot}(Pt, \boldsymbol{W})$. As such $W_1(\mathfrak{g})$ is equipped with the Alexander-Whitney multiplication and thus carries canonically the structure of a $\mathfrak{g}$-DG-algebra. Observe that the $\mathfrak{g}$-actions on $W(\mathfrak{g}^{l+1})$ are induced by the diagonal $\Delta: \mathfrak{g} \to \mathfrak{g}^{l+1}$. The construction performed with the functor W can now obviously be repeated with the functor W_1. By iteration we thus obtain a sequence of (co-)semisimplicial $\mathfrak{g}$-DG-algebras $W_s(\mathfrak{g}),\ s > 0$, which will turn out to be proper substitutes for the commutative Weil algebra $W(\mathfrak{g}) = W_0(\mathfrak{g})$. Note that by construction

(1.3) $$W_s(\mathfrak{g}) = \bigoplus_{l \geqq 0} W_s^l(\mathfrak{g}) = \bigoplus_{l \geqq 0} W_{s-1}(\mathfrak{g}^{l+1}).$$

The canonical projections

(1.4) $$\rho_s: W_s(\mathfrak{g}) \to W_s^0(\mathfrak{g}) = W_{s-1}(\mathfrak{g}), \qquad s > 0$$

are $\mathfrak{g}$-DG-algebra homomorphisms.

We proceed now to define inductively even filtrations $F_s^{\cdot}(\mathfrak{g})$ with respect to $\mathfrak{g}$ on $W_s(\mathfrak{g}^m)$ $(s \geqq 0,\ m \geqq 1)$ such that $F_0^{\cdot}(\mathfrak{g})$ on $W_0(\mathfrak{g}) = W(\mathfrak{g})$ is given by the canonical filtration

$$F^{2p}W(\mathfrak{g}) = S^p(\mathfrak{g}^*)\cdot W(\mathfrak{g}), \qquad F^{2p-1} = F^{2p}.$$

Let

(1.5) $$\begin{aligned} F_0^{2p}(\mathfrak{g})W(\mathfrak{g}^m) &= \mathrm{id}\{(W^+(\mathfrak{g}^m)^{i(\mathfrak{g})})\}^p, \\ F_s^{\cdot}(\mathfrak{g})W_s(\mathfrak{g}^m) &= \bigoplus_{l \geqq 0} F_s^{\cdot}(\mathfrak{g})W_s^l(\mathfrak{g}^m) = \bigoplus_{l \geqq 0} F_{s-1}^{\cdot}(\mathfrak{g})W_{s-1}\{(\mathfrak{g}^m)^{l+1}\}, \qquad s \geqq 1. \end{aligned}$$

The odd filtrations are defined by $F_s^{2p-1} = F_s^{2p}$. The face and degeneracy operators of W_s are filtration-preserving. The filtration F_s is functorial for maps $W_s(\mathfrak{g}) \to W_s(\mathfrak{g}')$ induced by Lie homomorphisms $\mathfrak{g}' \to \mathfrak{g}$. One verifies that $F_s^{\cdot}\, W_s^{\cdot}$ is an even, bihomogeneous and multiplicative filtration by $\mathfrak{g}$-DG-ideals.

The split exact sequence

$$0 \to \mathfrak{g} \xrightarrow{\ \Delta\ } \mathfrak{g}^{l+1} \to V_l \to 0$$

defines the $\mathfrak{g}$-module V_l, whose dual is given by $V_l^* = \ker \Delta^* = \{(\alpha_0, \cdots, \alpha_l) | \sum_{i=0}^{l} \alpha_i = 0\}$. The filtration $F_1^{2p}\, W^l(\mathfrak{g}) = F_0^{2p}\, W(\mathfrak{g}^{l+1})$ is then given by

(1.6) $$F_1^{2p}\, W_1^l(\mathfrak{g}) \cong \bigoplus_{|r| \geqq p} \Lambda^{\cdot}\mathfrak{g}^* \otimes \{\Lambda^{\cdot} V_l^* \otimes S^{\cdot}(\mathfrak{g}^{*l+1})\}^{|r|}$$

where the weight $|r|$ is determined by $\deg \Lambda^1 V_l^* = \deg S^1(\mathfrak{g}^{*l+1}) = 1$. For the graded object we have therefore

$$G_1^{2p}\, W_1^l(\mathfrak{g}) = \Lambda\mathfrak{g}^* \otimes \{\Lambda V_l^* \otimes S(\mathfrak{g}^{*l+1})\}^{|p|}.$$

For every subalgebra $\mathfrak{h} \subset \mathfrak{g}$ the filtrations $F_s^{\cdot}$ induce filtrations on the relative algebras

(1.7) $$W_s(\mathfrak{g}, \mathfrak{h}) = \{W_s(\mathfrak{g})\}_{\mathfrak{h}}, \qquad s \geqq 0.$$

It is immediate that the canonical projections $\rho_s\colon W_s \to W_{s-1}$ are filtration-preserving. Define for $s \geqq 0$

$$(1.8) \qquad W_s(\mathfrak{g}, \mathfrak{h})_k = W_s(\mathfrak{g}, \mathfrak{h})/F_s^{2(k+1)} W_s(\mathfrak{g}, \mathfrak{h}), \qquad k \geqq 0.$$

For $k = \infty$ we set $F^\infty = \bigcap_{p\geqq 0} F^{2p} = 0$, so that

$$(1.9) \qquad W_s(\mathfrak{g}, \mathfrak{h})_\infty = W_s(\mathfrak{g}, \mathfrak{h}).$$

The main result concerning the relationship between the W_s is then as follows. The proof will appear in a forthcoming paper.

1.10. THEOREM. *Let $(\mathfrak{g}, \mathfrak{h})$ be a reductive pair of Lie algebras. The homomorphisms of spectral sequences induced by the filtration-preserving canonical projections ρ_s: $W_s(\mathfrak{g}, \mathfrak{h}) \to W_{s-1}(\mathfrak{g}, \mathfrak{h})$ induce isomorphisms on the E_2-level and hence isomorphisms for every $0 \leqq k \leqq \infty$,*

$$H(\rho_s)\colon H\{W_s(\mathfrak{g}, \mathfrak{h})_k\} \xrightarrow{\cong} H\{W_{s-1}(\mathfrak{g}, \mathfrak{h})_k\}, \qquad s > 0.$$

The E_2-term of the spectral sequence for $s = 0$ has been determined **[15]**. Hence there exists an even multiplicative spectral sequence

$$(1.11) \qquad E_2^{2p,q} = H^q(\mathfrak{g}, \mathfrak{h}) \otimes I(\mathfrak{g})_k^{2p} \Rightarrow H^{2p+q}(W_s(\mathfrak{g}, \mathfrak{h})_k)$$

for $s \geqq 0, 0 \leqq k \leqq \infty$.

It is useful for explicit calculations to have a cohomology-inverse of the isomorphism $H(\rho_1)$ in (1.10) at disposal. While the noncommutativity of $W_1(\mathfrak{g})$ implies that the projection $\rho_1\colon W_1(\mathfrak{g}) \to W(\mathfrak{g})$ does not admit a splitting by a $\mathfrak{g}$-DG-algebra homomorphism, there exist linear maps which split ρ_1 and preserve all other structures.

1.12. THEOREM. *There exists a canonical linear map $\lambda = (\lambda^l)_{l\geqq 0}\colon W^\cdot(\mathfrak{g}) \to W_1^\cdot(\mathfrak{g})$ of degree 0 satisfying*

(i) *λ is a $\mathfrak{g}$-DG-module map;*
(ii) $\rho_1 \circ \lambda = \mathrm{id}_W$;
(iii) *$\lambda^l(w) = 0$, if $w \in W^{q,2p}(\mathfrak{g})$, $l > p$;*
(iv) *λ preserves filtrations on $W(\mathfrak{g})$, $W_1(\mathfrak{g})$.*

Properties (i), (ii), (iv) and (1.10) imply

1.13. COROLLARY. *λ induces a mapping of complexes*

$$\lambda\colon W(\mathfrak{g}, \mathfrak{h})_k \to W_1(\mathfrak{g}, \mathfrak{h})_k$$

for any subalgebra $\mathfrak{h} \subset \mathfrak{g}$, $0 \leqq k \leqq \infty$ and under the conditions of (1.10) *we have $H(\lambda) = H(\rho_1)^{-1}$. Thus $H(\lambda)$ is multiplicative.*

As indicated λ is given by linear maps $\lambda^l\colon W^\cdot(\mathfrak{g}) \to W(\mathfrak{g}^{l+1})^{\cdot - l}$ of degree $-l$, preserving filtrations and commuting (up to sign) with the operators $i(x)$, $\theta(x)$, $x \in \mathfrak{g}$. The commutativity with the differentials reads explicitly as

$$\lambda^l \circ d_W + (-1)^{l+1} d_{W_1} \circ \lambda^l = \delta \circ \lambda^{l-1} = \sum_{j=0}^{l} (-1)^j \varepsilon_j^{l-1} \circ \lambda^{l-1}. \tag{1.14}$$

Thus for $l = 1$, λ^1 is a "universal" homotopy operator between the two "universal" connections ε_i: $W(\mathfrak{g}) \to W(\mathfrak{g} \times \mathfrak{g}) = W(\mathfrak{g}) \otimes W(\mathfrak{g})$:

$$\lambda^1 d + d\lambda^1 = \varepsilon_0 - \varepsilon_1. \tag{1.15}$$

Some explicit formulas are given below.

$$\begin{aligned} \lambda(\alpha) &= (\alpha, 0, \cdots), & \alpha \in \Lambda^1(\mathfrak{g}^*) &= W^{1,0}; \\ \lambda(\tilde{\alpha}) &= (\tilde{\alpha}, \delta\alpha, 0, \cdots), & \tilde{\alpha} \in S^1(\mathfrak{g}^*) &= W^{0,2}. \end{aligned} \tag{1.16}$$

For $\alpha \in \Lambda^1(\mathfrak{g}^*)^{\mathfrak{g}}$ and $\tilde{\alpha}$ the corresponding element in $S^1(\mathfrak{g}^*)^{\mathfrak{g}} = I^2(\mathfrak{g})$:

$$\lambda(\alpha\tilde{\alpha}) = (\alpha\tilde{\alpha}, -(\alpha \otimes 1)\, \delta\alpha, 0, \cdots). \tag{1.17}$$

Thus $\lambda(\alpha\tilde{\alpha}) = \lambda(\alpha) \cdot \lambda(\tilde{\alpha}) = \lambda(\tilde{\alpha}) \cdot \lambda(\alpha)$ in this case and hence also

$$\lambda(\alpha\tilde{\alpha}^q) = \lambda(\alpha) \cdot \lambda(\tilde{\alpha})^q, \qquad q \geqq 1. \tag{1.18}$$

2. The generalized characteristic homomorphism of a foliated bundle. We consider a foliated G-bundle $P \to M$, $\mathfrak{U} = (U_j)$ a sufficiently fine open covering of M and $\omega = (\omega_j)$ a family of connections ω_j in $P|U_j$ representing the flat connection mod Ω in P. These data define then a homomorphism

$$k_1(\omega)\colon W_1^{\cdot}(\mathfrak{g}) \to \check{C}^{\cdot}(\mathfrak{U}, \pi_* \Omega_P^{\cdot}) \tag{2.1}$$

as follows. For $l \geqq 0$, let $\sigma = (i_0, \cdots, i_l)$ be an l-simplex of the nerve $N(\mathfrak{U})$. Consider the compositions

$$\omega_{i_j}\colon \Lambda(\mathfrak{g}^*) \to \Gamma\{U_{i_j}, \pi_* \Omega_P^{\cdot}\} \to \Gamma(U_\sigma, \pi_* \Omega_P^{\cdot}) \quad \text{for } j = 0, \cdots, l.$$

This defines

$$k(\omega_\sigma)\colon W(\mathfrak{g}^{l+1}) \to \Gamma(U_\sigma, \pi_* \Omega_P^{\cdot}) \tag{2.2}$$

as the universal $\mathfrak{g}$-DG-algebra homomorphism extending

$$\Lambda(\omega_\sigma)\colon \Lambda(\mathfrak{g}^{*l+1}) \to \Gamma(U_\sigma, \pi_* \Omega_P^{\cdot}) \tag{2.3}$$

given on the factor j by ω_{i_j}. We get therefore a homomorphism $k_1(\omega)\colon W_1^l(\mathfrak{g}) \to \check{C}^l(\mathfrak{U}, \pi_* \Omega_P^{\cdot})$ by setting $k_1(\omega)_\sigma = k(\omega_\sigma)$. We observe that this homomorphism is the composition of the following two $\mathfrak{g}$-DG-homomorphisms. The unique map of semi-simplicial sets $N(\mathfrak{U}) \to Pt$ first pulls the canonical system $\boldsymbol{W}$ on Pt back to a local system $\boldsymbol{W}_{\mathfrak{U}}$ on $N(\mathfrak{U})$. Then the assignment $\sigma \mapsto k(\omega_\sigma)$ defines a map of local systems $\boldsymbol{W}_{\mathfrak{U}} \to \pi_* \Omega_P^{\cdot}$. The induced homomorphisms compose to $k_1(\omega)$. Thus (2.1) is a homomorphism of $\mathfrak{g}$-DG-algebras, the generalized Weil homomorphism of P. Together with (1.10) the following result is crucial for our construction.

2.4. PROPOSITION. $k_1(\omega)$ *is filtration-preserving in the sense that*

$$k_1(\omega)\colon F_1^{2p}W_1 \to F^p\check{C}(\mathfrak{U}, \pi_*\Omega_P^{\cdot}); \qquad p \geqq 0.$$

PROOF. The filtration on the image complex is defined by

$$(2.5) \qquad F^p\check{C}(\mathfrak{U}, \pi_*\Omega_P^{\cdot}) = \check{C}^{\cdot}(\mathfrak{U}, F^p\Omega_P^{\cdot}) = \check{C}^{\cdot}\{\mathfrak{U},(\Omega\cdot\pi_*\Omega_P^{\cdot})^p\}.$$

Similarly $C^{\cdot}(\mathfrak{U}, \Omega_M^{\cdot})$ is filtered by

$$(2.6) \qquad F^p\check{C}(\mathfrak{U}, \Omega_M^{\cdot}) = \check{C}(\mathfrak{U}, F^p\Omega_M^{\cdot}) = \check{C}\{\mathfrak{U},(\Omega\cdot\Omega_M)^p\}.$$

By the multiplicativity of the filtration and (1.6) it is sufficient to verify the relations

$$k(\omega_\sigma)u \in \Gamma(U_\sigma, F^1\pi_*\Omega^2), \quad \text{for } u \in S^1(\mathfrak{g}^{*l+1})$$

and

$$k(\omega_\sigma)v \in \Gamma(U_\sigma, F^1\pi_*\Omega_P^1), \quad \text{for } v \in \Lambda^1 V_l^*.$$

But $S^1(\mathfrak{g}^{*l+1})$ and $\Lambda^1 V_l^*$ are generated linearly by the elements $\tilde{\alpha}_j = (0,\cdots, \tilde{\alpha},\cdots,0)$, $\tilde{\alpha} \in S^1(\mathfrak{g}^*)$ and $\alpha_{jk} = (0,\cdots, -\alpha,\cdots, \alpha,.., 0)$ for $\alpha \in \Lambda^1\mathfrak{g}^*$ and it is sufficient to check the relations on these elements. Now for $\sigma = (i_0,\cdots, i_l)$ we get $k(\omega_\sigma)\tilde{\alpha}_j = k(\omega_{i_j})\tilde{\alpha} \in \Gamma(U_{i_j}, F^1\pi_*\Omega_P^2)$ and $k(\omega_\sigma)\alpha_{jk} = (\omega_{i_k} - \omega_{i_j})\alpha \in \Gamma(U_{i_j i_k}, F^1\pi_*\Omega_P^1)$ since the local connections ω_i in $P|U_i$ represent the given flat connection ω_0 mod Ω in P.

Let now q be the geometric codimension of the foliation $\Omega \subset \Omega_M^1$ as defined in [17, 1.3]. It is then clear that the filtrations (2.5), (2.6) are zero for $p > q$. Thus $k_1(\omega)$ induces by (2.4) a homomorphism

$$k_1(\omega)\colon W_1(\mathfrak{g})_q \to \check{C}(\mathfrak{U}, \pi_*\Omega_P).$$

More generally for a closed subgroup $H \subset G$ with finitely many connected components and Lie algebra $\mathfrak{h} \subset \mathfrak{g}$ we have an induced map between the H-basic algebras of (2.1). If $\hat{\pi}\colon P/H \to M$ denotes the projection induced from $\pi\colon P \to M$, then $(\pi_*\Omega_P^{\cdot})_H = \hat{\pi}_*\Omega_{P/H}^{\cdot}$ and hence

$$k_1(\omega)\colon W_1(\mathfrak{g}, H) \to \check{C}(\mathfrak{U}, \hat{\pi}_*\Omega_{P/H}^{\cdot}).$$

Since this map is still filtration-preserving, and the filtration on the *RHS* is zero for degrees exceeding q, we get

$$(2.7) \qquad k_1(\omega)\colon W_1(\mathfrak{g}, H) \to \check{C}(\mathfrak{U}, \hat{\pi}_*\Omega_{P/H}).$$

To define invariants in the base manifold M, we need an H-reduction of P given by a section $s\colon M \to P/H$ of $\hat{\pi}\colon P/H \to M$ as the pull-back $P' = s^*P$. It defines a homomorphism $s^*\colon \check{C}(\mathfrak{U}, \hat{\pi}_*\Omega_{P/H}^{\cdot}) \to \check{C}(\mathfrak{U}, \Omega_M^{\cdot})$. Note that the cohomology of this target complex maps canonically into the de Rham cohomology of M (viewed as hypercohomology) $j\colon H^{\cdot}(\check{C}(\mathfrak{U}, \Omega_M^{\cdot})) \to \boldsymbol{H}^{\cdot}(M; \Omega_M^{\cdot}) = H_{DR}^{\cdot}(M)$. Thus we can finally write down the definition of the generalized characteristic homomorphism

$$\Delta_*\colon H^{\cdot}(W(\mathfrak{g}, H)_q) \to H_{DR}^{\cdot}(M)$$

of a foliated bundle P equipped with an H-reduction s^*P as

(2.8) $$\Delta_* = j \circ s^* \circ (k_1)_* \circ \lambda_*.$$

This leads to the following result.

2.9. THEOREM. *Let $P \to M$ be a foliated G-bundle with an H-reduction, $H \subset G$ a closed subgroup with finitely many connected components. Assume that the pair* $(\mathfrak{g}, \mathfrak{h})$ *is a reductive pair of Lie algebras.*

(i) *There is a canonical multiplicative homomorphism*

$$\Delta_*: H(W(\mathfrak{g}, H)_q) \to H_{DR}(M)$$

where q is the geometric codimension of the given foliation on M. Δ_* is the generalized characteristic homomorphism of the bundle P equipped with the foliation ω_0 and the H-reduction s.

(ii) *Δ_* is functorial under pull-backs:*

$$f^*\Delta_*(P, \omega_0, s) = \Delta_*(f^*P, f^*\omega_0, f^*s).$$

(iii) *Δ_* is invariant under integrable homotopies.*

The reductivity assumption is unnecessary in the smooth case. We remark that Theorem 1.10 (formulated for Lie algebras only) remains true under the assumptions made on the pair (G, H) in 2.9. This construction contains beside the usual (primary) characteristic classes of P secondary invariants, which therefore are naturally explained in the framework of the Chern-Weil theory. For the particular foliated bundle obtained from the normal bundle of a nonsingular foliation or from a Haefliger Γ_q-cocycle, the relationship of these invariants with the invariants of Godbillon-Vey [**9**] and Bott-Haefliger ([**6**],[**10**]) has been explained in [**17**].

As an example consider a G-principal bundle $P \to M$ with its unique (up to homotopy) K-reduction for the maximal compact subgroup $K \subset G$. Thus we disregard the foliation on M and P $(q = \infty)$. The spectral sequence (1.11) is now of the form $(m = 2p + q)$:

(2.10) $$E_2^{2p,q} = H^q(\mathfrak{g}, K) \otimes I(G)^{2p} \Rightarrow H^m(W_1(\mathfrak{g}, K)) \cong I(K)^m$$

by (1.10) and [**15**, Proposition 1, (iii)]. The composition of the edge homomorphism $I(G)^{2p} \to H^{2p}(W_1(\mathfrak{g}, K))$ with Δ_* is precisely the usual Chern-Weil homomorphism of the principal bundle P constructed via Čech-cohomology, i.e. using local connection data only. For the universal bundle $T_G \to B_G$ the homomorphism Δ_* turns out to be an isomorphism since it is contained in the following diagram:

$$\begin{array}{ccc} H^\cdot(W_1(\mathfrak{g}, K)) & \xrightarrow{\Delta_*} & H^\cdot(B_G, \mathbf{R}) \\ \downarrow \cong & & \downarrow \cong \\ I(K) & \xrightarrow{\cong} & H^\cdot(B_K, \mathbf{R}) \end{array}$$

Hence we obtain a spectral sequence

$$(2.11)\qquad E_2^{2p,q} = H^q(\mathfrak{g}, K) \otimes I(G)^{2p} \Rightarrow H^m(W_1(\mathfrak{g}, K)) \cong H^m(B_G, \boldsymbol{R})$$

with edge homomorphism given by the universal Chern-Weil homomorphism $I(G) \to H(B_G, \boldsymbol{R})$. This spectral sequence and the semisimplicial realization of $H^\cdot(B_G, \boldsymbol{R})$ coincide with the results in **[5]**, **[18]** via the van Est theorem **[11]**.

As another application we consider the case of a foliated vector bundle in the complex-analytic category. Let $\Omega \subset \Omega_M^1$ be a nonsingular complex-analytic foliation of a complex manifold $(M, \mathcal{O})$ and let E^r be a holomorphic vector bundle of rank r such that the holomorphic frame bundle $F(E)$ carries an Ω-foliation ω_0. This means that E is equipped with a holomorphic partial flat connection along the Lie algebra subsheaf $\mathscr{L} = (\Omega_M^1/\Omega)^* \subset \mathscr{T}_M$ defined by Ω. If the holomorphic line bundle $\Lambda^r E$ admits a nonzero holomorphic section, i.e. a holomorphic section $s: M \to F(E)/SL(r, \boldsymbol{C}) = F(\Lambda^r E)$, we obtain by our general procedure a characteristic homomorphism

$$(2.12)\qquad \Delta_*: H^\cdot(W(\mathfrak{gl}(r, \boldsymbol{C}), \mathfrak{sl}(r, \boldsymbol{C}))_q) \to H^\cdot_{DR}(M)$$

where $q = \operatorname{rank}_{\mathcal{O}} \Omega$ is the complex codimension. By the computation in **[15]** we have

$$(2.13)\qquad \begin{aligned} H^i(W(\mathfrak{gl}(r,\boldsymbol{C}),\mathfrak{sl}(r,\boldsymbol{C}))_q) &\cong \boldsymbol{C}[c_2,\cdots,c_r]_q^i, & i &\leqq 2q,\\ &\cong \boldsymbol{C}\{\alpha\} \otimes \boldsymbol{C}[c_1,\cdots,c_r]^{2q}, & i &= 2q+1,\\ &\cong 0, & i &> 2q+1, \end{aligned}$$

where $c_j(A) = \operatorname{tr}(\Lambda^j((i/2\pi)A))$ represent the Chern polynomials in $I(\mathfrak{gl}(r, \boldsymbol{C}))$ and $\alpha = (i/2\pi)\operatorname{tr} \in (\Lambda^1\mathfrak{gl}^*)^{\mathfrak{gl}} \subset W^{1,0}$, $d\alpha = c_1$. The $\Delta_*(c_j) \in H^j_{DR}(M)$, $j \geqq 2$, are the Chern classes of E and for $c^\lambda = c_1^{\lambda_1} \cdots c_r^{\lambda_r}$, $\sum_{j=1}^r j\cdot\lambda_j = q$, the classes $\Delta_*(\alpha \otimes c^\lambda)$ are secondary invariants of the Ω-foliated holomorphic bundle E. The Čech cocycle in $\check{C} = \check{C}^\cdot(\mathfrak{U}, \Omega_M^\cdot)$ representing the class $\Delta_*(\alpha \otimes c_1^q)$ can be computed as follows. Let $\mathfrak{U} = (U_i)$ be a sufficiently fine open covering of M and choose local holomorphic frames $\bar{s}_i = (\bar{s}_i^1,\cdots, \bar{s}_i^r)$ of $E|U_i$. By assumption we may adjust the frames $\bar{s}_i$ so that the transition functions $\bar{g}_{ij} : U_{ij} \to GL(r, \boldsymbol{C})$ given by $\bar{s}_j = \bar{s}_i \cdot \bar{g}_{ij}$ satisfy $\det(\bar{g}_{ij}) = 1$. The local sections $s_i = \bar{s}_i^1 \wedge \cdots \wedge \bar{s}_i^r$ define then a global trivialization of $\Lambda^r E$. A family of local holomorphic connections (ω_i) representing the given foliation on $F(E)$ is then determined by the local $\mathfrak{gl}$-valued connection forms $\bar{\theta}_i = \bar{s}_i^* \omega_i$ on U_i. For $\theta_i = \operatorname{tr}(\bar{\theta}_i)$ we have the relations $d\theta_i \in \Gamma(U_i, \Omega\cdot\Omega_M^1)$ and $\theta_{ij} = \theta_j - \theta_i \in \Gamma(U_{ij}, \Omega)$.

For the chains $u = (i/2\pi)(\theta_j) \in \check{C}^{0,1}$ and $v = (i/2\pi)(d\theta_j, \theta_{jk}) \in \check{C}^{0,2} \otimes \check{C}^{1,1}$ satisfying $du = v$, we have by (1.16), (1.17) $\Delta(\omega)\alpha = u$, $\Delta(\omega)c_1 = v$ and hence, by (1.18),

$$(2.14)\qquad \Delta(\omega)(\alpha \otimes c_1^q) = u \cdot v^q.$$

Explicitly for $q = 1$:

$$(2.15)\qquad \Delta(\omega)(\alpha \otimes c_1) = -(2\pi)^{-2}(\theta_j \wedge d\theta_j, -\theta_j \wedge \theta_{jk}) \in F^1\check{C}^3.$$

We can actually do better than that. It is well known that the class $c_1(E) \in H^2(M, Z)$ can be realized as the coboundary of the exact sheaf sequence

$$0 \to \boldsymbol{Z} \to \mathcal{O}_M \xrightarrow{\exp} \mathcal{O}_M^* \to 1$$

where $\mathcal{O}^*$ is the sheaf of units in the structure sheaf $\mathcal{O}_M$ and $\exp = e^{2\pi i -}$. On a simple covering $\mathfrak{U}$ the class $c_1(M) \in H^2(M, Z)$ is then represented by the 2-cycle $c = (i/2\pi)(-\delta \log g) \in \check{C}^2(\mathfrak{U}, \boldsymbol{Z})$ where $g = (g_{ik}) = (\det \bar{g}_{ik}) \in \check{C}^1(\mathfrak{U}, \mathcal{O}^*)$. Hence if $c_1(E) = 0$ we may adjust the local sections $\bar{s}_i$ of $F(E)$ such that $\delta(\log g) = 0$ for a convenient choice of $\log g_{ij}$. If we define $u = (i/2\pi)(\theta_j, \log g_{jk}) \in \check{C}^1$ and $v = (i/2\pi) \cdot (d\theta_j, \theta_{jk}) \in F^1\check{C}^2$, where now $\theta_{jk} = \theta_k - \theta_j - d \log g_{jk}$, we have the general relation $d_c(u) = v - c$ and hence by our assumption $d_c(u) = v$. As $v^{q+1} \in F^{q+1}\check{C}$ is necessarily zero, it follows that $u \cdot v^q$ is closed and defines a cohomology class in $H^{2q+1}_{DR}(M) \cong H^{2q+1}(M, \boldsymbol{C})$. For $q = 1$ the cocycle $u \cdot v$ is explicitly given by

$$u \cdot v = \frac{-1}{4\pi^2}(\theta_j \wedge d\theta_j, -\theta_j \wedge \theta_{jk} + \log g_{jk} \cdot d\theta_k, \log g_{jk} \cdot \theta_{kl}) \in \check{C}^3. \tag{2.16}$$

Hence $u \cdot v$ is composed of chains of bidegree (0, 3), (1, 2) (2, 1) respectively.

More generally we have, for $q > 0$,

2.17. PROPOSITION. *If $c_1(E) \in H^2(M, Z)$ is zero, the characteristic homomorphism Δ_* in* (2.12) (2.13) *is still defined. The classes $\Delta_*(\alpha \otimes c^\lambda)$ satisfying $\lambda_1 > 0$ are independent of all choices made and therefore define invariants of the Ω-foliated holomorphic bundle E.*

2.18. COROLLARY. *If $\Lambda^r E$ admits a global nonzero holomorphic section s, then the characteristic homomorphism Δ_* is independent of s on the classes $\alpha \otimes c^\lambda$, $\lambda_1 > 0$.*

If we take for E the vector bundle associated to the locally free sheaf Ω equipped with its canonical Ω-foliation, the invariants defined by Δ_* are closely related to those defined in [**4**, Theorem 4] and in [**3**] at the end of § 10.

REFERENCES

1. M. F. Atiyah, *Complex analytic connections in fibre bundles,* Trans. Amer. Math. Soc. **85** (1957), 181–207. MR **19**, 172.

2. R. Bott, *On a topological obstruction to integrability*, Proc. Sympos. Pure Math., vol. 16, Amer. Math. Soc., Providence, R. I., 1970, pp. 127–131. MR **42** #1155.

3. ———, *Lectures on characteristic classes and foliations,* Lecture Notes in Math., vol. 279, Springer-Verlag, Berlin and New York, 1972.

4. ———, *On the Lefschetz formula and exotic characteristic classes,* Proc. Diff. Geom. Conf., Rome, 1971.

5. ———, *On the Chern-Weil homomorphism and the continuous cohomology of Lie groups,* Advances in Math. **11** (1973), 289–303.

6. R. Bott and A. Haefliger, *On characteristic classes of Γ foliations,* Bull. Amer. Math. Soc. **78** (1972), 1039–1044. MR **46** #6370.

7. H. Cartan, *Cohomologie réelle d'un espace fibré principal différentiable,* Séminaire Cartan, Exposés 19 et 20, 1949/50.

8. S. S. Chern and J. Simons, *Some chohomology classes in principal fiber bundles and their applications to riemannian geometry*, Proc. Nat. Acad. Sci. U.S.A. **68** (1971), 791–794. MR **43** #5453.

9. C. Godbillon and J. Vey, *Un invariant des feuilletages de codimension* 1, C. R. Acad. Sci. Paris Sér. A-B **273** (1971), A92–A95. MR **44** #1046.

10. A. Haefliger, *Sur les classes caractéristiques des feuilletages*, Séminaire Bourbaki (Juin 1972), Exposé 412.

11. G. Hochschild and G. Mostow, *Cohomology of Lie groups*, Illinois J. Math. **6** (1962), 367–401. MR **26** #5092.

12. F. Kamber and Ph. Tondeur, *Characteristic classes of modules over a sheaf of Lie algebras*, Notices Amer. Math. Soc. **19** (1972), A-401. Abstract #693-D5.

13. ———, *Characteristic invariants of foliated bundles*, University of Illinois, August 1972 (preprint).

14. ———, *Derived characteristic classes of foliated bundles*, University of Illinois, August 1972 (preprint).

15. ———, *Cohomologie des algèbres de Weil relatives tronquées*, C. R. Acad. Sci. Paris **276** (1973), 459–462.

16. ———, *Algèbres de Weil semi-simpliciales*, C. R. Acad. Sci. Paris **276** (1973), 1177–1179.

———, *Homomorphisme caractéristique d'un fibré principal feuilleté*, C. R. Acad. Sci. Paris **276** (1973), 1407–1410.

———, *Classes caractéristiques dérivées d'un fibré principal feuilleté*, C. R. Acad. Sci. Paris **276** (1973), 1449–1452.

17. ———, *Characteristic invariants of foliated bundles*, Manuscripta Math. **11** (1974), 51–89.

18. H. Shulman, *Characteristic classes of foliations*, Ph. D. Thesis, University of California, Berkeley, 1972.

UNIVERSITY OF ILLINOIS, URBANA

Proceedings of Symposia in Pure Mathematics
Volume 27, 1975

LINE FIELDS TRANSVERSAL TO FOLIATIONS

ULRICH KOSCHORKE*

In this paper we show that every line field on a closed smooth manifold is bordant to one which is transverse regular to a smooth foliation of codimension 1. If we assume that the original manifold or the line field is oriented, then this condition can be preserved throughout the bordism.

Similar results for higher codimensions have been obtained in many cases (see [**6**]) by applying Thurston's general classification theorem for foliations [**14**]. However, in codimension 1, this theorem is not available,[1] and we have to resort to a rather explicit construction of enough line fields transversal to foliations, to generate the bordism groups of line fields. Here we use the determination of these groups given in [**5**].

First we define appropriate bordism groups of manifolds with foliations, resp. plane fields. Our main result can then be phrased as a surjectivity statement for obvious forgetful homomorphisms.

Consider pairs $(M, \mathfrak{F})$, where M is a closed oriented n-manifold and $\mathfrak{F}$ is a smooth foliation on M with arbitrary (resp. oriented) q-dimensional normal bundle. Two such pairs $(M_0, \mathfrak{F}_0)$ and $(M_1, \mathfrak{F}_1)$ are called bordant if there exists an oriented bordism X from M_0 to M_1 with a (co-oriented) q-codimensional foliation $\tilde{\mathfrak{F}}$ on it such that $\tilde{\mathfrak{F}}$ meets ∂X transversally (i.e. $TX|\partial X$ is spanned by $T(\partial X)$ and $T\tilde{\mathfrak{F}}|\partial X$), and $\tilde{\mathfrak{F}}$ induces the foliations $\mathfrak{F}_0$ and $\mathfrak{F}_1$ (with their original co-orientation) on $\partial X = M_1 \cup -M_0$. This defines an equivalence relation between foliated manifolds; transitivity follows from the fact that, in a suitable neighborhood U of ∂X, we can find a vector field normal to ∂X and tangent to $\tilde{\mathfrak{F}}$, and hence there is a collar represen-

AMS (*MOS*) *subject classifications* (1970). Primary 57D27, 57D30, 57D19, 58A30.

*Research partially supported by SFB Reine Mathematik, Universität Bonn, Germany.

[1]*Added in proof.* Thurston [**15**] has now extended many of his results to codimension 1. See also the remark at the end of this paper.

tation $U \cong \partial X \times [0, 1)$ which makes $\tilde{\mathfrak{F}}|U$ correspond to the foliation $p^*(\mathfrak{F}_0 \cup \mathfrak{F}_1)$ extended from ∂X to the collar by means of the obvious projection.

DEFINITION 1. The bordism group of oriented closed n-manifolds with arbitrary (resp. co-oriented) q-codimensional foliations is denoted by $\mathfrak{F}\,\Omega_n(q)$ (resp. $\mathfrak{F}\,\Omega_n^{\mathrm{or}}(q)$). The corresponding group based on unoriented manifolds is denoted by $\mathfrak{F}\mathfrak{N}_n(q)$ (resp. $\mathfrak{F}\mathfrak{N}_n^{\mathrm{or}}(q)$).

Observe that

$$\mathfrak{F}\,\Omega_*^{(\mathrm{or})}(*) = \sum_{n,q;n\geqq q} \mathfrak{F}\Omega_n^{(\mathrm{or})}(q)$$

also has a natural structure as a bigraded algebra over $\Omega_* \subset \mathfrak{F}\Omega_*^{(\mathrm{or})}(0)$; and, for fixed q, $\mathfrak{F}\Omega_*^{(\mathrm{or})}(q) = \sum_{n\geqq q} \mathfrak{F}\Omega_n^{(\mathrm{or})}(q)$ is a (sub) Ω_*-module. Corresponding statements hold for the $\mathfrak{N}_*$-algebra $\mathfrak{F}\mathfrak{N}_*^{(\mathrm{or})}(*)$.

Similarly we define $\Omega_n(q)$, $\Omega_n^{\mathrm{or}}(q)$, $\mathfrak{N}_n(q)$ and $\mathfrak{N}_n^{\mathrm{or}}(q)$ to be the bordism groups of closed n-manifolds equipped with a q-plane field, with the indicated orientedness assumptions.

If we associate to a bordism class $[M, \mathfrak{F}]$ the class of M together with a q-plane field normal to $\mathfrak{F}$, we obtain the Ω_*-linear homomorphism

$$\nu: \mathfrak{F}\,\Omega_*^{(\mathrm{or})}(q) \to \Omega_*^{(\mathrm{or})}(q),$$

and the $\mathfrak{N}_*$-linear homomorphism

$$\nu: \mathfrak{F}\mathfrak{N}_*^{(\mathrm{or})}(q) \to \mathfrak{N}_*^{(\mathrm{or})}(q).$$

We want to study these homomorphisms in the case $q = 1$. If $n = 2$, results of Thurston [13] imply that ν is an isomorphism under all four orientedness conditions:

$$\mathfrak{F}\,\Omega_2(1) = \Omega_2(1) = 0; \qquad \mathfrak{F}\,\Omega_2^{\mathrm{or}}(1) = \Omega_2^{\mathrm{or}}(1) = 0;$$
$$\mathfrak{F}\,\mathfrak{N}_2(1) = \mathfrak{N}_2(1) = \boldsymbol{Z}_2; \qquad \mathfrak{F}\mathfrak{N}_2^{\mathrm{or}}(1) = \mathfrak{N}_2^{\mathrm{or}}(1) = 0.$$

However, if $n = 3$, the kernel of ν need not even be countably generated. Indeed, $\Omega_3^{(\mathrm{or})}(1) = 0$, but $\Omega_3^{(\mathrm{or})}(1) = \operatorname{Ker} \nu$ maps homomorphically onto $\boldsymbol{R}$. (See [12].) For general n the following construction will permit us to generate most of the image of ν.

PROPOSITION 2. *Let M be a manifold admitting a closed, nowhere vanishing* 1-*form β (e.g. such a β arises if M fibers over S^1). Let η be a vector bundle over M, and $\xi \subset \eta$ a line-(sub)-bundle.*

Then there exists a foliation on the projective space bundle $RP(\eta)$ whose normal bundle is isomorphic to $\pi^*(\xi) \otimes \lambda$ (where $\pi: RP(\eta) \to M$ is the projection, and λ is the canonical line bundle over $RP(\eta)$).

PROOF. Fix a smooth function $\psi: [0, 1] \to [0, 1]$ which takes the value 1 around 0, and 0 around 1.

Also choose a euclidean structure on η. Let $\xi^{\perp}$ be the orthogonal complement of ξ in η, and let $\rho(x) = \langle x, x\rangle$ for $x \in \xi^{\perp}$. Then we define a nowhere vanishing 1-

form on the disk bundle $D(\xi^\perp)$ of $\xi^\perp$ by

$$\tilde{\beta} = \psi(\rho)\cdot\pi^*(\beta) + (1 - \psi(\rho))\,d\rho,$$

where $\pi\colon D(\xi^\perp) \to M$ is the projection. Clearly $d\tilde{\beta} = \psi'(\rho)\,d\rho \wedge \pi^*(\beta)$, and hence $\tilde{\beta} \wedge d\tilde{\beta} = 0$. Use the projection along ξ to lift $\tilde{\beta}$ to the sphere bundle $S(\eta)$ in η. The resulting foliation is invariant under the antipodal map and hence gives rise to a foliation $\mathfrak{F}$ on $RP(\eta)$.

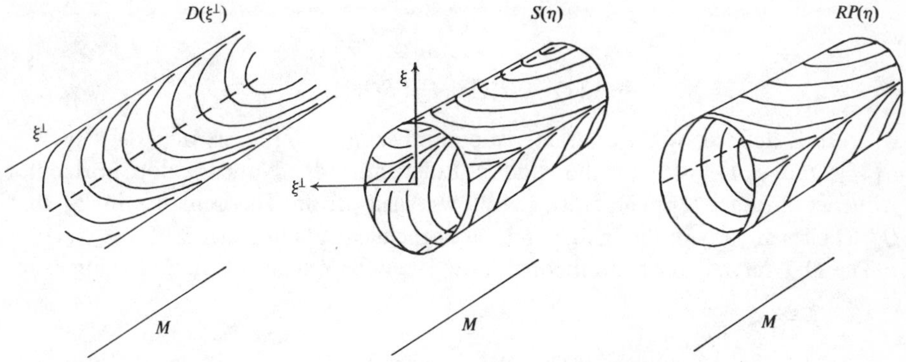

CONSTRUCTION OF THE FOLIATION $\mathfrak{F}$ ON $RP(\eta)$

To determine the normal bundle $\nu(\mathfrak{F})$, observe that according to the theorem of Leray-Hirsch we can write

$$w_1(\nu(\mathfrak{F})) = \pi^*(\alpha_1) + \pi^*(\alpha_0)\,w_1(\lambda), \qquad \alpha_i \in H^i(M;\mathbf{Z}_2).$$

Clearly $\alpha_0 = 1$, since $\nu(\mathfrak{F})$ restricts to the nontrivial line bundle on each fiber of $RP(\eta)$. Also if $s\colon M \to RP(\eta)$ is the section given by $\xi \subset \eta$, then the foliation $\mathfrak{F}$ is induced by $\pi^*(\beta)$ near $s(M)$, and hence has trivial normal bundle there. Thus

$$0 = s^*(w_1(\nu(\mathfrak{F}))) = \alpha_1 + s^*(w_1(\lambda)) = \alpha_1 + w_1(\xi).$$

Therefore $\alpha_1 = w_1(\xi)$, and $w_1(\nu(\mathfrak{F})) = w_1(\pi^*(\xi) \otimes \lambda)$.

THEOREM 3. *The homomorphisms* $\nu\colon \mathfrak{F}\Omega_n(1) \to \Omega_n(1)$ *and* $\nu\colon\ \mathfrak{F}\Omega_n^{\mathrm{or}}(1) \to \Omega_n^{\mathrm{or}}(1)$ *are surjective for all positive n. Furthermore the first* ν *has a right inverse at least for even n.*

PROOF. Let $\mathfrak{N}_{n-1}^{\mathrm{fib}}(S^1)$ denote the bordism group of pairs (N, p), where N is a closed $(n-1)$-manifold and $p\colon N \to S^1$ is a submersion. Via the mapping torus construction, this group corresponds to the bordism group of $(n-2)$-manifolds equipped with a smooth automorphism.

The submersion p induces the nowhere vanishing closed 1-form $p^*(d\phi)$ on N, where ϕ is the angle function on S^1. Hence, if we apply Proposition 2 to $\eta = \xi_N \oplus \boldsymbol{R}$ and $\xi = \boldsymbol{R}$, we obtain a foliation $\mathfrak{F}$ on $RP(\xi_N \oplus \boldsymbol{R})$ with normal bundle λ.[2] This construction is compatible with bordism and therefore gives rise to a

[2] ξ_N denotes the orientation line bundle of N.

homomorphism

$$\omega_{\mathfrak{F}}: \mathfrak{N}^{\text{fib}}_{n-1}(S^1) \to \mathfrak{F}\Omega_n(1).$$

According to Remark 3.4 in [5], $\omega_{\mathfrak{F}}$ maps already into the kernel $\mathfrak{F}\tilde{\Omega}_n(1)$ of the forgetful map $\mathfrak{F}\Omega_n(1) \to \Omega_n$; and furthermore, the following diagram is commutative:

$$\begin{array}{ccccc} \mathfrak{N}^{\text{fib}}_{n-1}(S^1) & \longrightarrow & \mathfrak{N}_{n-1} & \xrightarrow{\chi_2} & \mathbf{Z}_2 \\ \downarrow{\scriptstyle \omega_{\mathfrak{F}}} & & {\scriptstyle \omega}\downarrow\cong & \nearrow{\scriptstyle \theta} & \\ \mathfrak{F}\tilde{\Omega}_n(1) & \xrightarrow{\nu} \tilde{\Omega}_n(1) \xrightarrow{f} & \tilde{\Omega}_n(BO(1)) & & \end{array}$$

where θ is defined by the characteristic number $\theta([M, f]) = w(M)(1 + f^*(w_1))^{-1} \cdot [M]$, and χ_2 is given by the mod 2 Euler number. Now the top horizontal sequence is exact [2]. Therefore it follows easily from Theorem 6.2 in [5] that $\tilde{\Omega}^{(\text{or})}_n(1)$ lies already in the image under ν of the elements of order 2 in $\mathfrak{F}\tilde{\Omega}^{(\text{or})}_n(1)$.

The first statement of the theorem now is a consequence of the following exact sequence

$$\mathfrak{F}\Omega^{(\text{or})}_n(1) \longrightarrow \Omega_n \xrightarrow{\chi_2} \mathbf{Z}_2.$$

Here we use the fact [9], [3] that every $x \in \Omega_n$ with vanishing signature can be represented by a manifold which fibers over S^1; in addition we exploit the existence of a 4-manifold whose signature equals 2 and which carries a transversally orientable codimension one foliation (see Mizutani [8]).

To obtain a right inverse of $\nu: \mathfrak{F}\Omega_n(1) \to \Omega_n(1)$, the only difficulty now is to find, for x in the 2-torsion of Ω_n, an element $[M, \mathfrak{F}]$ in the 2-torsion of $\mathfrak{F}\Omega_n(1)$ with $[M] = x$. But according to Proposition 5 in [1], each such x can be represented by a union of manifolds of the form $RP(\xi_N + \mathbf{R}^{2k+1})$, where N is an $(n - 2k - 1)$-dimensional manifold. If n is even, then $\chi(N)_2 = 0$, and we may assume that N fibers over S^1. From Proposition 2 we now obtain a foliation $\mathfrak{F}$ on $RP(\xi_N + \mathbf{R}^{2k+1})$ which is compatible with an orientation reversing diffeomorphism (reflect along a 1-codimensional subbundle of $\xi^\perp = \xi^\perp_N$). Thus $[RP(\xi_N + \mathbf{R}^{2k+1}), \mathfrak{F}]$ lies in the two-torsion of $\mathfrak{F}\Omega_n(1)$.

Another possible way to approach this problem is to check (e.g. in [3]) whether x can be represented by a manifold M which fibers over S^1 and which admits an orientation reversing diffeomorphism compatible with the fibering and hence with the foliation it defines. This approach might still work in the case of $\Omega^{\text{or}}_n(1)$, provided we consider also the liftings to $M \times I$ of nontrivial foliations on $S^1 \times I$ (e.g., the foliation transverse to the Reeb foliation) when trying to construct a bordism for $2(M, \mathfrak{F})$.

THEOREM 4. *The homomorphisms* $\nu: \mathfrak{F}\mathfrak{N}_n(1) \to \mathfrak{N}_n(1)$ *and* $\nu: \mathfrak{F}\mathfrak{N}^{\text{or}}_n(1) \to \mathfrak{N}^{\text{or}}_n(1)$ *are surjective for all positive n.*

PROOF. According to Theorem 6. 2 in [5] we can identify $\mathfrak{N}_n(1)$ with the kernel of $\theta: \mathfrak{N}_n(BO(1)) \to \boldsymbol{Z}_2$ where $\theta[M, f] = w(M)(1 + f^*(w_1))^{-1}[M]$. Now Stong **[10]** has shown that

$$x_0 = [\text{point}, \boldsymbol{R}],\ x_1 = [S^1, \gamma], \text{ and } x_i = [RP(\gamma + \boldsymbol{R}^{i-1}), \lambda],\ i = 2,3,\cdots,$$

form a basis of the $\mathfrak{N}_*$-module $\mathfrak{N}_*(BO(1))$; here γ denotes the nontrivial line bundle over S^1. Thus, if $x \in \mathfrak{N}_n(1)$, we can write

$$x = [N^n]\cdot x_0 + [N^{n-1}]\cdot x_1 + \sum_{i\geqq 2} [N^{n-i}]\cdot x_i.$$

Applying Proposition 2 to $M = S^1$, $\eta = \gamma \oplus \boldsymbol{R}^{i-1}$ and $\xi = \boldsymbol{R}$, we see that, for $i \geqq 2$, x_i lies in the image of ν, and so does the right-hand sum above. Also $\theta(x_i) = 0$, but $\theta(x_0) = \theta(x_1) = 1$. Hence

$$0 = \theta(x) = \chi_2[N^n]\cdot 1 + \chi_2[N^{n-1}]\cdot 1 \quad (\chi_2 = \text{mod } 2 \text{ Euler number}),$$

and for dimension reasons $\chi_2[N^n] = \chi_2[N^{n-1}] = 0$. (Compare **[10]**.) According to **[2]** we may assume that N^n and N^{n-1} are fibered over S^1 by maps p_n and p_{n-1}. The resulting foliation on N^n has trivial normal bundle, hence $[N^n]\cdot x_0$ lies in the image of ν. Also, $[N^{n-1}]\cdot x_1$ can be represented by $N^{n-1} \times S_1$, together with the Möbius strip γ pulled back by the projection

$$N^{n-1} \times S^1 \xrightarrow{p_{n-1} \times \text{Id}} S^1 \times S^1 \xrightarrow{\pi_2} S^1.$$

But γ, pulled back by π_2, occurs as the normal bundle of a foliation on the torus $S^1 \times S^1$. Therefore $[N^{n-1}]\cdot x_1$ also lies in the image of ν, and so does x.

The rest of Theorem 4 follows also from Theorem 6.2 in **[5]** and from the fact **[2]** that every bordism class in the kernel of $\chi_2: \mathfrak{N}_n \to \boldsymbol{Z}_2$ can be represented by a manifold which is fibered over S_1.

REMARK (ADDED IN PROOF). The results of this paper were essentially obtained in 1972 (cf. **[4]**). Since then W. Thurston has developed an extremely far-reaching theory of codimension one foliations **[15]**, which allows in particular to construct Ω_*-linear (resp. $\mathfrak{N}_*$-linear) homomorphisms

$$\phi: \Omega_*^{(\mathrm{or})}(1) \longrightarrow \mathfrak{F}\Omega_*^{(\mathrm{or})}(1) \quad (\text{resp. } \phi: \mathfrak{N}_*^{(\mathrm{or})}(1) \longrightarrow \mathfrak{F}\mathfrak{N}_*^{(\mathrm{or})}(1))$$

such that under all four orientedness assumptions $\nu\cdot\phi = \text{Id}$.

To define ϕ, observe that a line field ζ on a closed manifold N gives rise to a codimension one *Haefliger structure* $\mathfrak{H}_0$ on N (via the "horizontal" foliation on the $O(1)$-bundle ζ). According to **[15]** we can deform $\mathfrak{H}_0$ into a foliation $\mathfrak{F}_0$ with normal line field homotopic to ζ. Since the bordism class of $(N, \mathfrak{F}_0)$ depends only on the bordism class of (N, ζ), this construction induces the desired homomorphisms ϕ.

Interestingly, this approach also leads to the foliation $\mathfrak{F}$ defined in the proof of Proposition 2, at least if a decreasing function ψ is used there. Indeed, note that the basic model for the construction of $\mathfrak{F}$ is (half of) a *Reeb foliation* $\mathscr{R}_2$ on the strip $I \times \boldsymbol{R}$ in the (x, y)-plane (where x stands for $2\pi^{-1}\cdot\arcsin(\rho^{1/2})$, suitably smoothed around $\rho = 0$; and y stands for any locally defined function with $dy = \beta$). In order to deform $\mathscr{R}_2$, we first extend it in (x, y, z)-space to a (suitably adapted three-dimensional Reeb) foliation $\mathscr{R}_3$ with the following properties:

(i) $\mathscr{R}_3$ is defined on the set $\{(x, y, z) \in \boldsymbol{R}^3 \mid 0 \leqq x \leqq 1, -\varepsilon < z < f(x) + \varepsilon\}$, where f is a nonnegative function such that $f(x) = 1 - x$ for x close to 1. Moreover for all constants c, $|c| < \varepsilon$, the graph of $f + c$ (as a function in x and y) is a leaf of $\mathscr{R}_3$.

(ii) $\mathscr{R}_3$ is transversal to $I \times \boldsymbol{R} \times \{0\}$ and induces $\mathscr{R}_2$ there.

(iii) $\mathscr{R}_3$ is transversal to all lines parallel to the z-axis.

(iv) $\mathscr{R}_3$ is invariant under translation in the y-direction.

Now consider the pullback of $\mathscr{R}_3$ under the vertical deformation map

$$d : (I \times \boldsymbol{R} \times (-\varepsilon, \varepsilon)) \times I \longrightarrow \boldsymbol{R}^3,$$

defined by $d(x, y, z; t) = (x, y, (1 - t)f(x) + z)$. This pullback provides the needed *model deformation* $\mathscr{D}$ from the trivial Haefliger structure on $I \times \boldsymbol{R}$ (given by the horizontal foliation on $I \times \boldsymbol{R} \times (-\varepsilon, \varepsilon)$) to the foliation $\mathscr{R}_2$. Next, using local orientations of the line bundle ξ (of Proposition 2), parametrize the fibers of ξ by z, and pull $\mathscr{D}$ (locally) back to $\pi^*(\xi) \mid S^+(\eta)$, where $S^+(\eta) \subset S(\eta)$ is the halfsphere bundle defined by $z \geqq 0$. This fits together to give a global smooth deformation over all of $S(\eta)$ which is compatible with the antipodal map. So we obtain finally, over $RP(\eta)$, an explicit deformation from the horizontal Haefliger structure with normal bundle $\pi^*(\xi) \otimes \lambda$ to the foliation $\mathfrak{F}$ of Proposition 2. Note that a foliation given by a fibration over the circle is a special case of such an $\mathfrak{F}$ (take $\eta = \xi = \boldsymbol{R}$).

It is now clear that all the bordism classes of foliations used in the proof of Theorems 3 and 4 (with the possible exception of those derived from Mizutani's example) coincide with the ones given by the homomorphisms ϕ. For trivial Haefliger structures, therefore, our constructions add a high degree of visibility to Theorem 1(c) of **[15]** in a situation which is most of the time representative up to bordism.

I would like to thank William Thurston for a conversation which helped to reassess the relevance of this paper in the light of his new theory.

REFERENCES

1. P. G. Anderson, *Cobordism classes of squares of orientable manifolds*, Ann. of Math. (2) **83** (1966), 47–53. MR **32** #4700.

2. P. E. Conner and E. E.Floyd, *Fibering within a cobordism class*, Michigan Math. J. **12** (1965), 33–47. MR **31** #4038.

3. U. Karras, M. Kreck, W. O. Neumann and E. Ossa, *Cutting and pasting of manifolds, SK-groups*, Publish & Perish, Waltham, Mass. (to appear).

4. U. Koschorke, *Bordism of manifolds with linefields*, Notices Amer. Math. Soc. **20** (1973), A289–A290. Abstract #73T–G31.

5. ———, *Concordance and bordism of line fields*, Invent. Math. **24** (1974), 241–268.

6. ———, *Singularities and bordism of q-planefields and of foliations*, Bull. Amer. Math. Soc. **80** (1974), 760–765.

7. B. Lawson, *Foliations*, Bull. Amer. Math. Soc. **80** (1974), 369–418.

8. T. Mizutani, *A codimension-*1 *foliation on a* 4*-manifold of signature* 2, Topology (to appear).

9. W. Neumann, *Fibering over the circle within a bordism class*, Math. Ann. **192** (1971), 191–192. MR **44** #4758.

10. R. Stong, *Subbundles of the tangent bundle* (preprint).

11. I. Tamura, *Lectures at the foliation seminar*, Princeton University, Princeton, N.J., 1973.

12. W. Thurston, *Noncobordant foliations of S^3*, Bull. Amer. Math. Soc. 78 (1972), 511–514. MR **45** #7741.

13. ———, *Bordism of foliated surfaces* (to appear); see also [**15**].

14. ———, *A theory of foliations of codimension greater than one*, Comment. Math. Helv. **49** (1974), 214–231.

15. ———, *Existence of codimension one foliations* (preprint).

QUEENS COLLEGE OF THE CITY UNIVERSITY OF NEW YORK

Proceedings of Symposia in Pure Mathematics
Volume 27, 1975

CLASSIFYING SPACES FOR RIEMANNIAN FOLIATIONS

JOEL PASTERNACK

Introduction. This paper is about Riemannian foliations and Haefliger structures, and their classifying spaces. A Riemannian foliation on a manifold is a foliation whose normal bundle is equipped with a Riemannian metric which is preserved under the natural parallel translation of normal vectors along the leaves. The theory of Riemannian foliations is in some ways analogous to the theory of holomorphic foliations developed by Bott [**2**] and Baum and Bott [**1**]. In fact, one has in the Riemannian case, as in the holomorphic case, a strong vanishing theorem for the characteristic ring of the normal bundle and a similar construction of surjections from homotopy groups of the classifying spaces onto uncountable groups. This latter development will be described elsewhere.

This paper presents the definitions and basic results on Riemannian Haefliger structures in §§ 1–3. In § 4 it is shown that the classifying space $BSR\Gamma_1$ has the homotopy type of an Eilenberg-Mac Lane space $K(R, 1)$.

1. $R\Gamma_q$-structures. An $R\Gamma_q$-structure is a special case of a Γ_q-structure (compare [**4**, p. 184]).

DEFINITION. A Riemannian Haefliger cocycle, of codimension q, on a topological space X is given by

(1) an open covering $\{U_\alpha\}_{\alpha\in I}$ of X for I some indexing set,

(2) for each α a Riemannian metric $\langle\ ,\ \rangle_\alpha$ on R^q and a continuous map $f_\alpha : U_\alpha \to R^q$,

(3) for all pairs, α, β and $x \in U_\alpha \cap U_\beta$ an isometry $\gamma^x_{\beta\alpha}$ of a neighborhood of

---

AMS (MOS) subject classifications (1970). Primary 57D30; Secondary 53C10, 57D20.

$f_\alpha(x)$ contained in $(R^q, \langle\ ,\ \rangle_\alpha)$ onto a neighborhood of $f_\beta(x)$ in $(R^q, \langle\ ,\ \rangle_\beta)$ satisfying

$$f_\beta(y) = \gamma^x_{\beta\alpha} \circ f_\alpha(y)$$

for all y in a neighborhood of x,

(4) the assignment $x \to \gamma^x_{\beta\alpha}$ is continuous into the Etale space of germs of isometries, and, for $x \in U_\alpha \cap U_\beta \cap U_\delta$,

$$\gamma^x_{\alpha\delta} = \gamma^x_{\alpha\beta} \cdot \gamma^x_{\beta\delta}.$$

If M is a smooth (i.e. C^∞) manifold, a *smooth* $R\Gamma_q$-cocycle is one in which the f_α are smooth maps for all $\alpha \in I$.

DEFINITION. Two $R\Gamma_q$-cocycles $A = \{U_\alpha, f_\alpha, \gamma_{\alpha\beta}\}_{\alpha,\beta\in I}$ and $A' = \{U_\alpha, f_\alpha, \gamma_{\alpha\beta}\}_{\alpha,\beta\in I'}$ are equivalent if there exists an $R\Gamma_q$-cocycle corresponding to the covering $\{U_\alpha\}_{\alpha\in I\cup I'}$ which restricts to A and A' in the obvious sense on the coverings $\{U_\alpha\}_{\alpha\in I}$ and $\{U_\alpha\}_{\alpha\in I'}$ respectively. An equivalence class of $R\Gamma_q$-cocycles is called an $R\Gamma_q$-structure.

The normal bundle η of an $R\Gamma_q$-cocycle $\{U_\alpha, f_\alpha, \gamma_{\alpha\beta}, \langle\ ,\ \rangle_\alpha\}$ is the normal bundle of the underlying Γ_q-cocycle. That is,

(1.1)
(i) $\eta | U_\alpha \cong f_\alpha^{-1}(T(R^q))$ where $T(R^q)$ is a tangent bundle of R^q,
(ii) for $x \in U_\alpha \cap U_\beta$, $(x, h) \sim (x, d\gamma^x_{\beta\alpha}(h))$ for all $h \in T(R^q)|_{f_\alpha(x)}$.

Further η has induced Riemannian metric $\langle\ ,\ \rangle$ given by

$$\langle\ ,\ \rangle | U_\alpha = f_\alpha^{-1}(\langle\ ,\ \rangle_\alpha). \tag{1.2}$$

Equivalent $R\Gamma_q$-cocycles have isometric normal bundles.

Given topological spaces X and Y, a continuous map $f: X \to Y$, and a $R\Gamma_q$ structure H on Y, then $f^*(H)$ denotes the $R\Gamma_q$-structure on X pulled back in the obvious way.

A smooth $R\Gamma_q$-foliation is a smooth $R\Gamma_q$-structure for which the $df_\alpha|_x$ have rank q for all $x \in U_\alpha$ and $\alpha \in I$. This type of foliation was first studied by Reinhart [8]. The principal easy-to-find examples of $R\Gamma_q$-foliations on a manifold M are the following:

EXAMPLE 1. If there is a submersion $f: M \to N$ where N is a manifold, then any Riemannian metric on N will define an $R\Gamma_q$-foliation on M.

EXAMPLE 2. Suppose that $(\ ,\ )$ is a Riemannian metric on M and a Lie group acts by isometries on $(M, (\ ,\ ))$. If all the orbits of the action are of the same dimension, then M is foliated by these orbits and the metric on the normal vectors to the orbits yields an $R\Gamma_q$-foliation. (Cf. [7] or [8].)

REMARK ON EXAMPLE 2. In case there are "exceptional" orbits of less than maximal dimension, the $R\Gamma_q$-foliation on the complement of these "exceptional" orbits cannot in general be extended to an $R\Gamma_q$-structure on all of M; the type of singularity in the foliation at the exceptional orbits is in general not of the type arising in a Riemannian Haefliger structure.

2. The classifying spaces: $BR\Gamma_q$, $BSR\Gamma_q$.

DEFINITION. Let H_0 and H_1 be $R\Gamma_q$-structures on a space X. H_0 and H_1 are homotopic if and only if there exists an $R\Gamma_q$-structure H on $X \times [0, 1]$ satisfying:

$$i_0^*(H) = H_0, \qquad i_1^*(H) = H_1,$$

where $i_t : X \to X \times [0, 1]$ is given by $i_t(x) = (x, t)$.

The construction given by Haefliger [5] yields a space $BR\Gamma_q$ equipped with an $R\Gamma_q$-structure H; the space $BR\Gamma_q$ is a classifying space for $R\Gamma_q$-structures in the sense that given a paracompact topological space X there is a one-to-one correspondence between homotopy classes of maps of X into $BR\Gamma_q$ and homotopy classes of $R\Gamma_q$-structures on X. The correspondence associates to a map $f, f: X \to BR\Gamma_q$, the $R\Gamma_q$-structure $f^*(H)$. The $R\Gamma_q$-structure H on $BR\Gamma_q$ is called the universal $R\Gamma_q$-structure. Similarly, one has classifying spaces $B\Gamma_q$, respectively $BSR\Gamma_q$, equipped with a universal Γ_q-structure, respectively $SR\Gamma_q$-structure.

The space $BR\Gamma_q$ has associated with it the following maps:

(1) $\nu_q : BR\Gamma_q \to BO_q$, ν_q classifies the universal normal bundle,

(2) $\omega_q : BR\Gamma_q \to B\Gamma_q$, ω_q classifies the universal $R\Gamma_q$-structure.

Similarly there are maps $\mu_q : BSR\Gamma_q \to BSO_q$ and $\delta_q : BSR\Gamma_q \to BS\Gamma_q$.

These classifying spaces are central in developing an obstruction theory for determining whether a particular manifold admits an $R\Gamma_q$-foliation or whether a particular foliation of a manifold admits or may be deformed into one admitting an $R\Gamma_q$-structure. In this regard one would like to be able to compute the homotopy groups of the fiber of the maps ν_q and ω_q. Also, one would like to know the cohomology of $BR\Gamma_q$ and $BSR\Gamma_q$ in order to develop a theory of characteristic classes for $R\Gamma_q$-foliations. The remainder of this paper discusses this latter problem. The main geometric objects employed will be the basic Riemannian connection on the normal bundle of a smooth $R\Gamma_q$-structure and the volume form of a smooth $SR\Gamma_q$-structure.

3. The basic Riemannian connection. In this section we define the canonical Riemannian connection on the normal bundle of a smooth $R\Gamma_q$-structure and discuss some applications.

3.1. *Definition and basic applications.* Let H be a smooth $R\Gamma_q$-structure on a manifold M; let $\{U_\alpha, f_\alpha, \gamma_{\alpha\beta}, \langle\ ,\ \rangle_\alpha\}$ be a representative cocycle for H and let η be the normal bundle.

DEFINITION. The canonical Riemannian connection ∇ on η the normal bundle of H is defined as follows: Let D_α be the torsion-free Riemannian connection on $(R^q, \langle\ ,\ \rangle_\alpha)$ and let

$$\nabla \mid U_\alpha = f_\alpha^{-1}(D_\alpha). \tag{3.1}$$

The definitions of ∇ on $U_\alpha \cap U_\beta$ agree because the $\gamma_{\alpha\beta}$ are local isometries and the torsion-free Riemannian connection is unique. This connection preserves the natural Riemannian metric (1.2) on η. In case the $R\Gamma_q$-structure is a foliation, then ∇ is also a *basic* connection in the sense used by Bott [2, p. 33].

The curvature of ∇, denoted $K(\nabla)$, is locally a matrix of 2-forms pulled up from R^q and is skew-symmetric in local orthonormal frames. An Ad-invariant polynomial on the Lie algebra of skew-symmetric matrices can be applied to the curvature to yield globally defined differential forms on the manifold M (cf. [6]). Recall that the Chern-Weil theory gives that these differential forms are closed and represent elements in the Pontryagin ring of η. Now observe that

$$\varphi(K(\nabla)) \equiv 0 \quad \text{on } M \tag{3.2}$$

if φ is an Ad-invariant homogeneous polynomial of degree greater than $q/2$. Applying this fact one can prove the following results about the normal bundle η of a smooth $R\Gamma_q$-structure.

THEOREM [7]. $\mathrm{Pont}^{(r)}(\eta; Q) = 0$ *for* $r > q$, *where* $\mathrm{Pont}^{(r)}(\eta; Q)$ *is the Pontryagin ring generated by the rational Pontryagin classes of the normal bundle* η. *Moreover, in case* η *is orientable the theorem remains true with the Euler class of* η *adjoined.*

COROLLARY. *The maps* $\nu_q^{(r)}$ *and* $\mu_q^{(r)}$, *given by*

$$\nu_q^{(r)} : H^{(r)}(BO_q; Q) \longrightarrow H^{(r)}(BR\Gamma_q; Q),$$
$$\mu_q^{(r)} : H^{(r)}(BSO_q; Q) \longrightarrow H^{(r)}(BSR\Gamma_q; Q),$$

are the zero maps for $r > q$.

THEOREM (SHULMAN [2] OR [9]). *If* $a^{(s)}, b^{(t)}, c^{(u)}$ *are elements of* $\mathrm{Pont}(\eta; Q)$ *with* $s + t > q$ *and* $t + u > q$, *then the Massey triple product* $\langle a^{(s)}, b^{(t)}, c^{(u)}\rangle$ *is defined and* $\langle a^{(s)}, b^{(t)}, c^{(u)}\rangle = 0$.

Open question. Suppose $s + t + u - 1 > q$ and $a^{(s)} b^{(t)} = 0 = b^{(t)} c^{(u)}$, then does it follow that $\langle a^{(s)}, b^{(t)}, c^{(u)} \rangle = 0$?

Using the connection ∇ and equation (3.2) one can apply the ideas of Baum and Bott [1] to show that certain homotopy groups of the spaces $BR\Gamma_q$ are "huge." For example, a surjection can be constructed from $\pi_3(BR\Gamma_3)$ onto the real numbers modulo the integers. Details will be given elsewhere.

3.2. *Characteristic classes for foliations.* It is of interest to develop a theory of characteristic classes for foliations analogous to the classical theory of characteristic classes for vector bundles. As in the classical case, the Λ-characteristic classes for foliations (for Λ a given ring) can be thought of as coming from the elements in the cohomology of the classifying space; that is $H^*(B\Gamma_q; \Lambda)$, $H^*(BR\Gamma_q; \Lambda)$ and $H^*(BSR\Gamma_q; \Lambda)$ are the set of universal Λ-characteristic classes for foliations, $R\Gamma_q$-foliations and $SR\Gamma_q$-foliations respectively. Given a smooth manifold M with foliation E and classifying map $f_E : M \to B\Gamma_q$, then the Λ-characteristic classes of E are the elements of $f_E^*(H^*(B\Gamma_q; \Lambda))$.

In [2] certain "exotic" C-characteristic classes for foliations are defined and it is shown how these classes can be viewed as elements in the cohomology of the classifying space $B\Gamma_q$. More specifically a cochain complex WO_q is constructed and a map λ^* is defined

$$\lambda^* : H^*(WO_q) \longrightarrow H^*(B\Gamma_q; C).$$

The image of λ^* includes the characteristic ring of the normal bundle of the universal Γ_q-structure and in addition contains other elements of $H^*(B\Gamma_q; C)$ which are referred to as "exotic" characteristic classes. These classes include the Godbillon-Vey classes [3]. Given a manifold with a smooth foliation, the construction given in [2] of the differential forms which represent in the de Rham cohomology the "exotic" classes of the foliation involves the comparison of a basic connection and a Riemannian connection on the normal bundle of the foliation. For an $R\Gamma_q$-foliation one has the canonical connection ∇ which is both basic and Riemannian and it follows directly that the exotic classes are zero on $R\Gamma_q$-foliations. Thus as in [2] we have

THEOREM. *The map* $\omega_q^* : H^*(B\Gamma_q; C) \to H^*(BR\Gamma_q; C)$ *is the zero map on the exotic classes.*

To begin a theory of characteristic classes for Riemannian foliations one needs to produce nonzero cohomology elements in $H^*(BR\Gamma_q)$ or $H^*(BSR\Gamma_q)$ in addition to the elements in the characteristic ring of the universal normal bundle. The volume form introduced in the next section is such a nonzero element in $H^q(BSR_q; R)$.

4. The volume form for $SR\Gamma_q$ -structures. In this section we define the volume form for a smooth $SR\Gamma_q$-structure and use this form to show that $BSR\Gamma_1$ has the homotopy type of an Eilenberg-Mac Lane space $K(R, 1)$.

4.1. *Definition and basic properties.*

DEFINITION. Let H be a smooth $SR\Gamma_q$-structure on a manifold M with representative cocycle $\{U_\alpha, f_\alpha, \gamma_{\alpha\beta}, \langle\ ,\ \rangle_\alpha\}$. Denote by ω_α the volume form associated to the Riemannian metric $\langle\ ,\ \rangle_\alpha$ on R^q. Define a q-form $\theta(H)$ first locally over U_α by

$$\theta(H) \,|\, U_\alpha = f_\alpha^{(q)}(\omega_\alpha).$$

Then observe that $\theta(H)$ is a well-defined q-form on M because for $x \in U_\alpha \cap U_\beta$ we have $f_\beta = \gamma_{\beta\alpha}^x \circ f_\alpha$ in a neighborhood of x, with $\gamma_{\beta\alpha}^x$ a local isometry and since H is orientable we can choose $\gamma_{\beta\alpha}^x$ so that

(4.2) $$(\gamma_{\beta\alpha}^x)^{(q)}(\omega_\beta) = \omega_\alpha.$$

Also observe that $\theta(H)$ does not depend on the representative cocycle. This q-form $\theta(H)$ is called the volume form of H.

It can easily be seen that the volume form $\theta(\cdot)$ satisfies the following properties:

(4.3)
(i) For any smooth $SR\Gamma_q$-structure H, $\theta(H)$ is a closed q-form.
(ii) If H_1 and H_2 are homotopic $SR\Gamma_q$-structures on M then $\theta(H_1)$ is cohomologous to $\theta(H_2)$.
(iii) If $f : N \to M$ is a smooth map of manifolds, then

$$\theta(f^*(H)) = f^{(q)}(\theta(H)).$$

As in [2, p. 70], $\theta(\cdot)$ determines a cohomology class in $H^q(BSR\Gamma_q; R)$.

REMARK. In case the $SR\Gamma_q$-structure H is in fact an $SR\Gamma_q$-foliation, then $\theta(H)$ is nowhere zero on M and is the defining q-form for the subbundle of tangents to the leaves. Since $\theta(H)$ is closed it follows immediately that the Godbillon-Vey classes of $SR\Gamma_q$-foliations are zero.

4.4. $BSR\Gamma_1 \cong K(R, 1)$. In case $q = 1$ we can use the volume form to prove:

THEOREM 4.1. *The classifying space $BSR\Gamma_1$ has the homotopy type of the Eilenberg-Mac Lane space $K(R, 1)$.*

The theorem follows from the following three propositions:

Let M be a smooth manifold.

PROPOSITION 1. *If ω is a closed 1-form on M, then there exists an $SR\Gamma_1$-structure on M for which $\theta(H) = \omega$.*

PROPOSITION 2. *Let H_1 and H_2 be $SR\Gamma_q$-structures on M. Then $\theta(H_1) = \theta(H_2)$ implies that H_1 equals H_2.*

PROPOSITION 3. *Let H_1 and H_2 be $SR\Gamma_q$-structures on M. Then H_1 is homotopic to H_2 if and only if $\theta(H_1)$ is cohomologous to $\theta(H_2)$.*

PROOF OF 1. By the Poincaré lemma, M can be covered by open sets U_α with map $f_\alpha : U_\alpha \to R$ satisfying

$$df_\alpha \mid U_\alpha = \omega \mid U_\alpha. \tag{4.5}$$

Then take $\langle \partial/\partial s, \partial/\partial s \rangle_\alpha \equiv 1$ and

$$\gamma_{\alpha\beta}(s) = s + (f_\alpha(x) - f_\beta(x)). \tag{4.6}$$

PROOF OF 2. Suppose H_1 is represented by $\{U_\alpha, f_\alpha, \gamma^x_{\alpha\beta}, \langle\ ,\ \rangle_\alpha\}$ and H_2 is represented by $\{U_i, f_i, \gamma_{ij}, \langle\ ,\ \rangle_i\}$ and let g_α (similarly g_i) be a smooth positive real valued function on R defined by

$$g_\alpha(s) = \langle \partial/\partial s, \partial/\partial s \rangle_\alpha \big|_s .$$

By definition H_1 equals H_2 if and only if there exists for each $x \in U_\alpha \cap U_i$ a local isometry $\gamma^x_{i\alpha}$ such that the following diagram commutes in a neighborhood of x:

$$\begin{array}{ccc} & U_\alpha \cap U_i & \\ {\scriptstyle f_\alpha} \swarrow & & \searrow {\scriptstyle f_i} \\ (R, \langle\ ,\ \rangle_\alpha) & \xrightarrow{\ \gamma^x_{i\alpha}\ } & (R, \langle\ ,\ \rangle_i) \end{array} \tag{4.7}$$

Now simply define $\gamma^x_{i\alpha}$ in a neighborhood of $f_\alpha(x)$ by

$$\text{(i)}\quad \frac{d\gamma^x_{i\alpha}}{ds}\bigg|_s = \frac{(g_\alpha(s))^{1/2}}{(g_i(s))^{1/2}}, \qquad \text{(ii)}\quad \gamma^x_{i\alpha}(f_\alpha(x)) = f_i(x). \tag{4.8}$$

Equation (4.8) (i) means that

$$(\gamma^x_{i\alpha})^{(1)}((g_i(s))^{1/2}\,ds) = (g_\alpha(s))^{1/2}ds \tag{4.9}$$

and indeed $\gamma^x_{i\alpha}$ is a local isometry.

If $\theta(H_1) = \theta(H_2)$ then, over $U_\alpha \cap U_i$,

$$f_\alpha{}^{(1)}((g_2)^{1/2}\,ds) = f_i^{(1)}((g_i)^{1/2}\,ds) \tag{4.10}$$

since $(g_\alpha)^{1/2}\,ds$ and $(g_i)^{1/2}\,ds$ are respectively the volume forms for $(R, \langle\ ,\ \rangle_\alpha)$ and $(R, \langle\ ,\ \rangle_i)$.

Equations (4.9) and (4.10) and the fact that $(g_i)^{1/2}$ is nowhere zero imply that $\gamma^x_{i\alpha} \circ f_\alpha$ and f_i have the same differential in a neighborhood of x. Thus by (4.8)(ii) $f_i = \gamma^x_{i\alpha} \circ f_\alpha$ in a neighborhood of x and the diagram (4.7) is commutative as required.

PROOF OF 3. Suppose $\theta(H_1)$ and $\theta(H_2)$ are cohomologous and let $\psi : M \to R$ be a smooth function satisfying

$$\theta(H_2) - \theta(H_1) = d\psi.$$

Let ω be a 1-form on $M \times R$ defined by

$$\omega = \pi^*(\theta(H_1)) + t(\pi^*(\theta(H_2) - \theta(H_1))) + \psi\,dt \tag{4.11}$$

where $\pi : M \times R \to M$ is the projection $\pi(m, t) = m$. The form ω is closed, and by Proposition 1 there exists an $SR\Gamma_1$-structure H on $M \times R$ with $\theta(H) = \omega$. Now

$$\begin{aligned} &(\text{a})\quad i_0^{(1)}(\omega) = \theta(H_1), \\ &(\text{b})\quad i_1^{(1)}(\omega) = \theta(H_2), \end{aligned} \tag{4.12}$$

where $i_t(m) = (m, t)$. Therefore by property (4.3) (iii) and Proposition 2 it follows that $i_0^*(H) = H_1$ and $i_1^*(H) = H_2$ and thus H_1 and H_2 are proven to be homotopic.

The only if part of Proposition 3 is (4.3) (ii). Q.E.D.

$R\Gamma_1$-structures are particularly simple because up to isometry there is only one Riemannian metric on the real line. However, even in this simple case we see the phenomenon of "huge" homotopy groups for the classifying spaces.

REFERENCES

1. P. Baum and R. Bott, *Singularities of holomorphic foliations* (preprint).

2. R. Bott, *Lectures on characteristic classes and foliations*, Lecture Notes in Math., vol. 279, Springer-Verlag, Berlin and New York, 1972, pp. 1–80.

3. C. Godbillon and J. Vey, *Un invariant des feuilletages de codimension* 1, C. R. Acad. Sci. Paris Sér A-B **273** (1971), A92–A95. MR **44** #1046.

4. A. Haefliger, *Feuilletages sur les ouverts*, Topology **9** (1970), 183–194. MR **41** #7709.

5. ———, *Homotopy and integrability*, Manifolds (Proc. Nuffic Summer School, Amsterdam, 1970), Lecture Notes in Math., vol. 197, Springer-Verlag, Berlin, 1971, pp. 133–163. MR **44** #2251.

6. S. Kobayashi and K. Nomizu, *Foundations of differential geometry*. Vol. II, Interscience Tracts in Pure and Appl. Math., no. 15, vol. II, Interscience, New York, 1969, MR **38** #6501.

7. J. Pasternack, *Foliations and compact Lie group actions*, Comment. Math. Helv. **46** (1971), 467–477. MR **45** #9353.

8. B. Reinhart, *Foliated manifolds with bundle-like metrics*, Ann. of Math. (2) **69** (1959), 119–132. MR **21** #6004.

9. H. Shulman, *Characteristic classes and foliations*, Ph.D. Thesis, University of California, Berkeley, 1972.

UNIVERSITY OF ROCHESTER

Proceedings of Symposia in Pure Mathematics
Volume 27, 1975

CODIMENSION ONE PLANE FIELDS AND FOLIATIONS

PAUL A. SCHWEITZER, S. J.*

The results reported here, whose proofs will appear in [3], treat foliations on a compact m-dimensional C^∞ manifold M. All plane fields and foliations will be assumed to be of codimension one.

The author would like to thank William Thurston who suggested the following theorem and also supplied an essential part of the proof.

THEOREM I. *If M has a transversally orientable (t.o.) C^∞ foliation, then every t.o. plane field on M is homotopic to a plane field tangent to a C^∞ foliation.*

Phillips [1] has shown that all codimension one plane fields are homotopic to foliations on *open* manifolds. Phillips, Gromov, Haefliger, and Thurston [4] have reduced the analogous question in higher codimension to a homotopy question.

LEMMA 1. *Let $\mathcal{F}_0$ be a C^∞ t.o. foliation of M with $m \geqq 3$. Then there exist*
(i) *a C^∞ t.o. foliation $\mathcal{F}_1$ of M, and*
(ii) *disjoint subspaces $T_1, T_2, \cdots, T_k$ of M, each diffeomorphic to $S^1 \times D^{m-1}$, such that each closed leaf of $\mathcal{F}_1$ is contained in the interior of some T_i $(1 \leqq i \leqq k)$.*

The first step in the proof of Lemma 1 is to modify $\mathcal{F}_0$ so that all closed leaves are isolated. Then each isolated closed leaf l contains a simple closed curve α with nontrivial holonomy on both sides of l. This holonomy can be exploited to modify the foliation inside a tubular neighborhood $T \cong S^1 \times D^{m-1}$ of α so that in the

---

AMS (MOS) subject classifications (1970). Primary 57D30.

*Supported by NSF grant GP-31359-X at Harvard University.

resulting foliation the leaf containing $l - T$ is not closed, and all new closed leaves are contained in T.

It is not difficult to show that the foliation $\mathscr{F}_1$ of Lemma 1 satisfies the hypotheses of the following theorem of John Wood. Theorem I then follows.

THEOREM [5, THEOREM 9.3]. *If M has a C^∞ t.o. foliation $\mathscr{F}_1$ and S is a nonempty disjoint union of circles in M transverse to $\mathscr{F}_1$ such that $H_1(S) \to H_1(M)$ is onto, then every t.o. plane field on M is homotopic to a C^∞ foliation.*

THEOREM II. *If $m \geqq 5$ and M has a C^k foliation, where $1 \leqq k \leqq \infty$, then M has a C^0 foliation with no closed leaf.*

In fact, the foliation is constructed so that it has a C^0 tangent bundle and every leaf is a C^k submanifold of M.

In the proof of this theorem the construction used in Lemma 1 is adapted using B. Raymond's example of a C^∞ foliation of S^3 containing an invariant compact set $\bar{L}_\gamma$ that separates S^3 but contains no closed leaves [2]. The foliation constructed in Theorem II contains an invariant set C homeomorphic to $S^1 \times S^1 \times S^{m-5} \times \bar{L}_\gamma$. The leaves on opposite sides of C spiral asymptotically in different directions, so that C plays a role analogous to the role of the closed leaf of the Reeb foliation of S^3.

ADDED IN PROOF. Thurston (*Existence of codimension one foliations,* preprint) has recently shown that *every* codimension one plane field on a closed C^∞ manifold is homotopic to a plane field tangent to a C^∞ foliation, thus strengthening Theorem I.

REFERENCES

1. A. Phillips, *Foliations on open manifolds.* I, Comment. Math. Helv. **43** (1968), 204–211. MR **37** #4829.

2. B. Raymond, *Cantor sets and foliations* (to appear).

3. P.A. Schweitzer, *Codimension one foliations without compact leaves* (to appear).

4. W. Thurston, *The theory of foliations of codimension greater than one*, Comment. Math. Helv. (to appear).

5. J. W. Wood, *Foliations on 3-manifolds*, Ann. of Math. (2) **89** (1969), 336–358. MR **40** #2123.

PONTIFÍCIA UNIVERSIDADE CATÓLICA, RIO DE JANEIRO

Proceedings of Symposia in Pure Mathematics
Volume 27, 1975

THE DOUBLE COMPLEX OF Γ_k

HERBERT SHULMAN*

This is joint work with R. Bott and J. Stasheff. S. Halperin provided useful suggestions concerning $A^*_\infty(\Gamma^{(*)})$.

Let $\Gamma = \Gamma_k =$ the space of germs of C^∞ diffeomorphisms of R^k. We give it the sheaf topology using the source map $s : \Gamma \to R^k$. Denote by $\Gamma^{(p)}$ the p-tuples of composable elements of Γ. For any cochain theory (C^*, δ), there is a naturally defined differential $\delta' : C^*(\Gamma^{(p)}) \to C^*(\Gamma^{(p+1)})$ commuting with δ so that $(C^*(\Gamma^{(*)}), \delta + (-1)^* \delta')$ is a bicomplex, denoted $C(\Gamma)$.

THEOREM 1. $H^*(C(\Gamma)) = H^*(B\Gamma)$. *(See* [1] *for details.)*

It can be seen that $\Gamma^{(p)}$ is a smooth k-dimensional manifold, though non-Hausdorff. Let $A^*(\Gamma^{(p)})$ denote smooth differential forms on $\Gamma^{(p)}$, and $A(\Gamma)$ the associated double complex. There is a natural map $H^*(A(\Gamma)) \to H^*(B\Gamma; \boldsymbol{R})$ which is not an isomorphism (unless we restrict to germs of analytic diffeomorphisms) because of the failure of the de Rham isomorphism for Γ.

For $G = \mathrm{Gl}(k; \boldsymbol{R})$ and $\nu : \Gamma \to G$, the map taking each germ to its differential, we have a commutative diagram

$$\begin{array}{ccc} H^*(A^*(G^*)) & \xrightarrow{\nu^*} & H^*(A^*(\Gamma^{(*)})) \\ \downarrow \approx & & \downarrow \\ H^*(BG; \boldsymbol{R}) & \xrightarrow{B\nu^*} & H^*(B\Gamma; \boldsymbol{R}) \end{array}$$

AMS (MOS) subject classifications (1970). Primary 57D30, 55F40; Secondary 58A10, 58A20, 58D99, 55J10.

* *Current address.* University of Pennsylvania, Philadelphia.

Now for $q > k$, $A^q(\Gamma^{(p)}) = 0$ so if $F_k = \bigoplus_{q>k} A^q(G^p)$ we get a factoring

$$\begin{array}{ccc} & A^*(G^*)/F_k & \\ \nearrow & & \searrow \bar{\nu}^* \\ A^*(G^*) & \xrightarrow{\quad \nu \quad} & A^*(\Gamma^{(*)}) \end{array}$$

Using a result of Bott (and Hochschild), it is an easy calculation to show $H^*(A^*(G^*)/F_k) = H^*(WO_k)$ where WO_k is the complex defined by Bott and Gelfand and Fuks.

On the other hand, the image of ν^* is contained in a subcomplex $A^*_\infty(\Gamma^{(*)}) \subset A^*(\Gamma^{(*)})$ defined as follows:

Let J^r denote (the smooth manifold of) r-jets of C^∞ diffeomorphisms of R^k, and $A^*(J^\infty) = \text{inv lim } A^*(J^r)$, the map $J^r : \Gamma \to J^r$ taking each germ to its r-jet induces $A^*(J^r) \xrightarrow{(j^r)} A^*(\Gamma)$. In a similar way we get $A^*(J^{\infty(p)}) \xrightarrow{(j^\infty)^*} A^*(\Gamma^{(p)})$. We denote image $((j^\infty)^*)$ by $A^*_\infty(\Gamma^{(*)})$. This is a subcomplex of $A^*(\Gamma^{(*)})$ containing image (ν^*). Bott and Haefliger have shown that $H^*(A^*_\infty(\Gamma^{(*)}))$ is isomorphic to the cohomology of formal vector fields on R^k (the Gelfand-Fuks cohomology).

Conjecture. $\bar{\nu}^* : H^*(A^*(G^*)/F_k) \to H^*(A^*_\infty(\Gamma^{(*)}))$ is an isomorphism.

If true, this would provide an alternative calculation for the Gelfand-Fuks cohomology and perhaps new insight into it.

Reference

1. R. Bott, H. Shulman and J. Stasheff, *de Rham theory for classifying spaces*, Advances in Math. (to appear).

Yale University

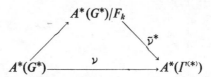

Proceedings of Symposia in Pure Mathematics
Volume 27, 1975

A LOCAL CONSTRUCTION OF FOLIATIONS FOR THREE-MANIFOLDS

WILLIAM P. THURSTON*

This paper presents a method for constructing foliations of 3-manifolds, which resolves the last nagging questions about existence of codimension one foliation on 3-manifolds: What 2-plane fields are homotopic to foliations? What 3-manifolds have foliations with boundary components as leaves? And, what codimension one foliations on the boundary of a 3-manifold extend over the 3-manifold? The answers are "all of them" except for some well-known restrictions for the second question.

The construction makes use of the theorem of Moser [5] and Herman and Sergeraert [2] that $\mathrm{Diff}_+^\infty(S^1)$ is perfect, where $\mathrm{Diff}_+^r(S^1)$ denotes the group of C^r orientation-preserving diffeomorphisms of the circle. This implies, by a theorem of Epstein [1], that $\mathrm{Diff}_+^\infty(S^1)$ is a simple group. Mather ([4], [5]) has recently extended this to the group $\mathrm{Diff}_+^{r+}(S^1)$ $[2 \leqq r \leqq \infty]$. Cf. Thurston [8] and Mather [4] for similar results on other manifolds.

The background concerning existence of foliations on 3-manifolds began with Reeb [7] who first constructed a foliation of S^3. Then Lickorish [3] and Novikov and Zieschang [6] showed that every closed oriented 3-manifold has a foliation. These constructions begin with Reeb's foliation of S^3, and make use of the fact that any closed orientable 3-manifold may be obtained from S^3 by cutting out the tubular neighborhood of a link—which can be deformed to a braid—and gluing it back in a different way. Finally, John Wood in his thesis [10] found a related con-

---

AMS (MOS) subject classifications (1970). Primary 57D30.
* *Current address*. Princeton University.

struction for foliations of closed, unoriented 3-manifolds, and proved that any transversely oriented plane field on a closed 3-manifold is homotopic to a foliation. The proof of the latter result made use of the classification of homotopy classes of nonsingular vector fields on a manifold.

For foliations in higher codimensions, I have recently shown that a plane field is homotopic to a foliation iff its normal bundle admits a Haefliger structure [9]. If this theorem were true in codimension one,[1] it would imply the present results; however, at least the relative version of this theorem is false in codimension one, because of the Reeb stability theorem.

I would like to thank Daniel Asimov for help with the idea of using counter-orientations in this construction.

There is a special class of foliations of the 2-torus T^2 defined by orientation-preserving diffeomorphisms of the circle. Begin with the trivial foliation of $S^1 \times I$ by leaves $S^1 \times t$. Then, given an orientation-preserving, C^r, diffeomorphism φ of the circle S^1, glue $S^1 \times 1$ to $S^1 \times 0$ by φ. This produces a foliation $\mathscr{S}\varphi$ transverse to the fibers of a circle bundle over the circle, which is T^2 since φ is isotopic to the identity, e. If we choose an isotopy $\{\varphi_t\}$ to e, this determines an element $\tilde{\varphi}$ of the universal covering group $(\mathrm{Diff}^r_+(\tilde{S}^1))^\sim$ of $\mathrm{Diff}^r_+(S^1)$, covering φ; and it also determines an isomorphism of the underlying manifold of $\mathscr{S}\varphi$ with $S^1 \times S^1$. Thus we can speak of $\mathscr{S}\tilde{\varphi}$ as a foliation of $S^1 \times S^1$ with no ambiguity.

LEMMA 1. *$\mathscr{S}\tilde{\varphi}$ bounds a codimension-one foliation $\mathscr{F}$ of $S^1 \times D^2$, such that $T\mathscr{F}$ is homotopic*, rel $S^1 \times S^1$, *to a plane field transverse to the factors $S^1 \times x$.*

PROOF OF LEMMA 1. Suppose first that φ is a rotation ρ_α of the circle, where $\alpha \in S^1$. Then $\mathscr{S}\tilde{\varphi}$ is a foliation determined by a closed form ω, and it bounds in a very standard way, by spinning the leaves around a torus leaf, then putting a Reeb component inside. More precisely, there is a foliation $\mathscr{F}$ of $S^1 \times D^2$ determined by an integrable form

$$\bar{\omega} = \lambda_1(r)\omega + \lambda_{1/2}(r)dr + \lambda_0(r)d\theta$$

where we parametrize $S^1 \times D^2$ by coordinates $(\theta; r, \varphi)$, with $\theta, \varphi \in S^1$, $r \in I$, and where $\{\lambda_0, \lambda_{1/2}, \lambda_1\}$ is a C^∞ partition of unity on I. Also, we extend ω naturally to $S^1 \times D^2 - S^1 \times 0$.

The homotopy of $T\mathscr{F}$ is produced by the nonintegrable nonsingular one-form

$$\bar{\omega}_t = \lambda_1\omega + \lambda_{1/2}(td\theta + (1-t)dr) + \lambda_1(d\theta).$$

To prove the lemma in the general case, it suffices to observe that the set of $\tilde{\varphi}$ for which the lemma holds is a normal subgroup of $(\mathrm{Diff}^r_+(S^1))^\sim$. First, the set of such $\tilde{\varphi}$ is closed under conjugacy: the foliation $\mathscr{S}_{\tilde{\alpha}\tilde{\varphi}\tilde{\alpha}^{-1}}$ is conjugate to $\mathscr{S}\tilde{\varphi}$ by the diffeomorphism $\alpha^{-1} \times e$ of $S^1 \times S^1$, isotopic to the identity, so if one foliation extends so does the other. Second, the set of such $\tilde{\varphi}$ is closed under composition: if $\mathscr{S}\tilde{\varphi}$ and $\mathscr{S}\tilde{\psi}$ extend to $S^1 \times D^2$, we can glue the two extended foliations together

[1] *Added in proof.* It is; cf. *Existence of codimension one foliations,* to appear.

along a neighborhood of $S^1 \times x$ where $x \in \partial D^2$. The resulting 3-manifold is again $S^1 \times D^2$, and the foliation on the boundary is $\mathscr{S}\tilde{\varphi}\circ\tilde{\psi}$. And, finally, the set of such $\tilde{\varphi}$ is clearly closed under formation of inverses. This subgroup contains all covering transformations and projects to a nontrivial normal subgroup of $\mathrm{Diff}^r_+(S^1)$, by the previous case. Hence it is all of $(\mathrm{Diff}^r_+(S^1))^\sim$, by Mather's result **[4]**.

This concludes the proof of Lemma 1.

We define a *counter-orientation* of a triangulated manifold to be an assignment of an orientation to each simplex, such that the orientations assigned to two simplices with a common hyperface are opposite.

LEMMA 2. *The barycentric subdivision of any triangulation of any manifold is counter-orientable.*

PROOF. The vertices of an n-simplex in the barycentric subdivision have a natural order (by dimension of the face of the simplex). This ordering imparts a counter-orientation.

We are now ready for the construction.

THEOREM 1. (a) *Let M^3 be a closed* 3-*manifold, and τ^2 a* 2-*plane field on M^3. Then is homotopic to an integrable C^∞* 2-*plane field.*

(b) *Let M^3 be a compact* 3-*manifold with a C^{r+} plane field τ^2 transverse to ∂M^3 and integrable in a neighborhood of ∂M^3. Then τ^2 is homotopic*, rel ∂M^3, *to an integrable C^{r+} plane field* $[2 \leqq r \leqq \infty]$.

(c) *Let M^3 be a compact* 3-*manifold such that each boundary component $(\partial M^3)_i$ has $H^1(\partial M^3_i; \mathbf{Z}) \neq 0$ (equivalently, each boundary component has Euler characteristic $\leqq 0$) and let τ^2 be a* 2-*plane field on M^3 tangent to ∂M^3. Then τ^2 is homotopic* rel ∂M^3 *to an integrable C^∞ plane field.*

(d) *If the condition in* (b) *is met on some boundary components, and the condition in* (c) *is met on others, then the conclusion is still valid.*

COROLLARY 1. *M^3 has foliation tangent to ∂M^3 iff*

(a) *no component of ∂M^3 is P^2 or S^2, and M^3 has a line field transverse to ∂M^3, or*

(b) *M^3 is $S^2 \times I$, $P^2 \times I$, or $S^2 \times_{z_2} I$ where z_2 acts on S^2 and I by the antipodal map.*

PROOF OF COROLLARY 1. By the Reeb stability theorem [7], a 3-manifold of which a boundary component is P^2 or S^2 has a foliation tangent to the boundary iff one of the three alternatives listed in (b) occurs. The rest is immediate from Theorem 1(c).

COROLLARY 2. *Foliated bordism in dimension* 2 *is the same as bordism of* 2-*plane fields, in any of the cases (involving oriented or nonoriented manifolds, transversely oriented or nontransversely oriented line fields, oriented or nonoriented line fields).*

REMARK. I had previously a proof of Corollary 2, based on Lemma 1 together with knowledge of the qualitative nature of line fields on surfaces, which used a simple geometrical construction to get rid of unnecessary Reeb components.

PROOF OF THEOREM 1. See [9] for an elaboration of the first paragraph of this proof.

PROOF OF THEOREM 1(a). Take a fine triangulation of M^3 with simplices not too misshapen, and then take its first barycentric subdivision α, so that the triangulation α has a counter-orientation. Homotope τ^2 so it is transverse to each simplex of α. Now, a slight perturbation of τ^2 makes it integrable in a neighborhood of the 2-skeleton, and the foliation around the boundary of each 3-simplex α_i is then determined by a diffeomorphism φ_i of the unit interval I, with support (φ_i) contained in the interior of I, and φ_i determined only up to conjugacy by arbitrary diffeomorphisms (possibly orientation-reversing) of I. By the Reeb stability theorem, such a foliation extends over α_i iff φ_i is the identity. The plan is to disguise each 3-cell as a solid torus, giving us more freedom to extend the foliations. The next step is to modify the foliation of a neighborhood of the two-skeleton so that φ_i has as few fixed points as possible: φ_i will have fixed points only near the top and bottom of I. This step makes use of the counter-orientation. If φ_i has no fixed points in the middle of I the foliation near $\partial\alpha_i$ has leaves which form helices. A helix determines an orientation. Two helical staircases (actually they are called spiral staircases) which are adjacent and have a few steps in common must have helices going in opposite senses. Thus, if we use the counter-orientation for each 3-simplex to determine the sense of the helix, then the direction we wish to twist the leaves of the foliation on each 2-simplex is well defined. (A counter-orientation defines an orientation for each 2-simplex; the direction of twist is in fact given by the orientation of the 2-simplex.)

The induced foliation on $\partial\alpha_i$ has two singularities, and both are like the singularity of level surfaces of a function near a maximum or minimum. If we pick a local transverse orientation for τ^2, one of the singularities is at the top of α_i and the other at the bottom. The next step is to construct for each α_i an arc γ_i contained in the neighborhood of the two-skeleton of M^3 transverse to the foliation, and going from the top of α_i to the bottom. In fact, given two points x, y on $\partial\alpha_j$, for any j, such that x and y are not near the top or bottom of $\partial\alpha_j$ (where φ_j has fixed points) there is a closed transverse curve contained in a neighborhood of $\partial\alpha_j$ and containing x and y, since the leaf of x spirals above and below y. Choose a 1-simplex going upward from the top of α_i. There is a chain of closed transversals leading from the midpoint of this 1-simplex to the midpoint of a 1-simplex going down from the bottom of α_i (because each connected component of M^3 is connected). Using this chain, the requisite transverse arc γ_i is constructed.

For each i, push α_i inward a little, obtaining α_i', round it, and extend the transverse arc γ_i to meet $\partial\alpha_i'$. All the γ_i's can be perturbed so they are simple and disjoint. Hollow out a neighborhood N_i of γ_i so that $N_i \cup \alpha_i'$ forms a solid torus, with boundary $= S^1 \times S^1$ contained in a neighborhood of the two-skeleton and transverse to the foliation there. The induced foliation is transverse to each factor $x \times S^1$. By Lemma 1 it follows that the foliation extends over $N_i \cup \alpha_i'$. The foliation $\mathscr{F}$ of M^3 so constructed has a tangent plane field $T\mathscr{F}$ homotopic to τ^2.

PROOF OF THEOREM 1(b). Suppose now that M^3 has a boundary, and τ^2 is transverse to ∂M^3 and integrable in a neighborhood of ∂M^3. The foliation near ∂M^3 is then a product foliation, induced from projection to ∂M^3 in a collar neighborhood. Triangulate M^3 minus the collar neighborhood, transverse to τ^2, with a counter-oriented foliation, and complete the construction as before. φ_i can still be made so it has fixed points only near ∂I, so there are enough transverse curves.

PROOF OF THEOREM 1(c). By hypothesis, for each boundary component ∂M_i^3, $H^1(\partial M_i^3; \mathbf{Z})$ is nontrivial. S^1 is the classifying space for $H^1(\ ;\mathbf{Z})$: let $f: \partial M_i^3 \to S^1$ be a nontrivial map. f can be homotoped so it is in general position, and there are no singularities of the maximum or minimum type. f induces a closed 1-form ω on ∂M_i^3, from the fundamental form of the circle. Homotope τ^2 in a collar neighborhood of ∂M_i^3 so it is given by the integrable 1-form $dr + \lambda_1(r)\omega$ where r is the normal parameter of the collar neighborhood, and λ_1 is a C^∞ function of r, 0 at ∂M_i^3 where $r = 0$, positive for $r > 0$ and 1 where $r = 1$. Now triangulate M^3 minus a neighborhood of ∂M^3 counter-orientably and transverse to τ^2. Because ω has no singularities of the maximum or minimum type, every leaf of the foliation defined by ω on ∂M_i^3 has a closed transversal passing through any other leaf. Thus, if vertices or edges are near ∂M_i^3, there are transverse arcs joining them. Using the counter-orientation, enough transverse arcs can be guaranteed, so that each 3-cell can be disguised as a solid torus and the construction goes through.

PROOF OF THEOREM 1(d). This is simply a combination of 1(b) and 1(c).

ADDENDUM. In relation to Theorem 1(c), I have since discovered the following generalization of Reeb's stability theorem: If $\mathscr{F}$ is a codimension one, transversely oriented, C^1 foliation of a compact manifold M^n with one compact leaf L such that $H^1(L; \mathbf{Z}) = 0$, then all leaves of $\mathscr{F}$ are diffeomorphic to L. This is false in the C^0 case.

REFERENCES

1. D. B. A. Epstein, *The simplicity of certain groups of homeomorphisms*, Compositio Math. **22** (1970), 165–173. MR **42** #2491.

2. M. Herman et F. Sergeraert, *Sur un théorème d'Arnold et Kolmogorov*, C. R. Acad. Sci. Paris Sér. A-B **273** (1971), A409–A411. MR **44** #7586.

3. W. B. R. Lickorish, *A foliation for* 3-*manifolds*, Ann. of Math. (2) **82** (1965), 414–420. MR **32** #6488.

4. J. Mather, *Commutators of diffeomorphisms*. I, Comment. Math. Helv. (to appear).

5. ———, *Commutators of diffeomorphisms*. II (to appear).

6. S. P. Novikov, *Topology of foliations*, Trudy Moskov. Mat. Obšč. **14** (1965), 248–278=Trans. Moscow Math. Soc. **1965**, 268–304. MR **34** #824.

7. G. Reeb, *Sur certaines propriétés topologiques des variétés feuilletées*, Actualités Sci. Indust., no. 1183, Hermann, Paris, 1952. MR **14**, 1113.

8. W. Thurston, *Foliations and groups of diffeomorphisms*, Bull. Amer. Math. Soc. **81** (1975) (to appear).

9. ———, *The theory of foliations in codimension greater than one*, Comment. Math. Helv. **49** (1974), 214–231.

10. J. Wood, *Foliations on* 3-*manifolds*, Ann. of Math. (2) **89** (1969), 336–358. MR **40** #2123.

MASSACHUSETTS INSTITUTE OF TECHNOLOGY

Proceedings of Symposia in Pure Mathematics
Volume 27, 1975

THE THEORY OF FOLIATIONS OF CODIMENSION GREATER THAN ONE

WILLIAM P. THURSTON*

A method is developed to circumvent the Gromov-Phillips theorem and to extend Haefliger's classifying theorem for foliations [1] to the case of foliations of codimension greater than one on compact manifolds. In particular, a plane-field τ^{n-k} on M^n, $k \geqq 2$, is homotopic to a foliation iff TM^n/τ^{n-k} admits a Haefliger structure. Every plane field of codimension greater than one is homotopic to a C^0 foliation with C^∞ leaves. If S^n has a k-plane field, $k \leqq n/2$, then S^n has a C^∞, codimension k foliation. Every 2-dimensional plane-field is homotopic to a C^∞ foliation. (Cf. [3] for the codimension one case.)

ADDED IN PROOF. These results have now been extended to codimension 1.

REFERENCES

1. A. Haefliger, *Feuilletages sur les variétés ouvertes*, Topology **9** (1970), 183–194. MR **41** #7709.
2. W. Thurston, *The theory of foliations of codimension greater than one*, Comment. Math. Helv. **49** (1974), 214–231.
3. ———, *A local construction of foliations for three-manifolds*, these PROCEEDINGS.

MASSACHUSETTS INSTITUTE OF TECHNOLOGY

AMS (MOS) subject classifications (1970). Primary 57D30.
**Current address*. Princeton University.

Proceedings of Symposia in Pure Mathematics
Volume 27, 1975

LOCALLY FREE ACTIONS AND STIEFEL-WHITNEY NUMBERS

H. B. WINKELNKEMPER

0. Introduction. If ξ is a vector bundle over a closed manifold X, let $RP(\xi, X)$ denote the associated projective space bundle over X. Let M^n be a closed smooth n-manifold and $j\colon M \to M$ an involution, then one knows [**1**, Theorem 24.2] that the fixed point set F of j is a union of closed m-manifolds F^m and in the unoriented cobordism ring $\mathcal{N}^*$:

(0.1) $$\sum_m [RP(\nu, F^m)] = 0,$$

(0.2) $$\sum_m [RP(\nu \oplus R, F^m)] = [M].$$

Here ν is the normal bundle to each F^m and R denotes the trivial 1-bundle. Let G be a compact Lie group which acts on M; consider the involution $i\colon M \times G \to M \times G$ defined by $i(x, g) = (gx, g^{-1})$, $x \in M$, $g \in G$; if G has only a finite number of elements g_l $(l = 1,2,\cdots)$ of order 2 (if $G = S^1$, S^3, or their products for example) then the fixed point set of i is the union of the sets (F_{g_l}, g_l) where F_{g_l} denotes the fixed point set of the involution on M defined by g_l; hence, applying the above result to i, we have

PROPOSITION 0.3. *If G is a k-dimensional compact Lie group, with only a finite number of elements of order 2, acting on a closed smooth manifold M, then the subset F of M, consisting of points whose isotropy groups contain a nontrivial element of order 2, is a union of not necessarily disjoint closed m-manifolds F^m and in the unoriented cobordism ring $\mathcal{N}^*$:*

AMS (MOS) subject classifications (1970). Primary 57E15; Secondary 57D30.

$$(0.4) \qquad \sum_m [RP(\nu \oplus R^k, F^m)] = [M \times RP^{k-1}],$$

$$(0.5) \qquad \sum_m [RP(\nu \oplus R^{k+1}, F^m)] = [M \times RP^k].$$

Here RP^k denotes real projective space and R^k is the trivial k-bundle. For (0.5) we have used the fact that a compact Lie group, being parallelizable, is always cobordant to 0.

An action of a k-dimensional Lie group G on an n-manifold M^n is called locally free if all isotropy groups are discrete subgroups of G. Such an action defines k linearly independent vector fields on M^n [**4**, Proposition 5.2] and hence for $i > n - k$ the Stiefel-Whitney classes w_i vanish; furthermore, in certain cases this property is inherited by the submanifolds of M mentioned in Proposition 0.3.

In this note, we are interested in using these facts and Proposition 0.3 to obtain certain Stiefel-Whitney numbers as obstructions to the existence of locally free group actions:

Let $G = S^1$, then it is easy to see directly that the Euler characteristic of M, $\chi = 0$ and, if M is orientable, that the Pontriagin numbers of M have to vanish. This is the simplest case of a theorem of Pasternack [**4**, Corollary 1], proved by using the differential geometric methods of Bott. Conversely, it follows from ([**2**], [**6**]) that any cobordism class of $\mathcal{N}^n$ (or Ω^n) such that $w_n = 0$ (or the Pontriagin numbers vanish) contains a manifold on which S^1 acts locally freely; hence, if $G = S^1$, there are no more characteristic number obstructions to the existence of a locally free S^1-action, except the above and these can be considered to lie in the *real* cohomology $H^*(M, R)$.

In § 2, as a first application of Proposition 0.3 we begin to obtain nontrivial necessary conditions, different from the above, on the Stiefel-Whitney numbers of an n-manifold M^n, on which $S^1 \times S^1$ acts locally freely:

PROPOSITION 2.4. *If $S^1 \times S^1$ acts locally freely on a closed* 4 *or* 6*-manifold, M is cobordant to* 0 *in $\mathcal{N}^*$.*

In these dimensions there exist manifolds, admitting 2 linearly independent vector fields, which are not cobordant to 0; furthermore, we can choose these manifolds so that all real Pontriagin classes vanish and hence the above-mentioned criteria, arrived at via differential geometrical methods, do not give any information.

Similarly one can prove:

PROPOSITION 2.13. *If $S^1 \times S^1$ acts locally freely on an* 8*-manifold, M^8, then* $w_2 w_6[M^8] = 0$.

1. Some elementary consequences of Wu's formulas. Let M^n be a closed, smooth n-manifold with Stiefel-Whitney classes $w_1, \cdots, w_n$. The Wu classes, $u_p \in H^p(M,Z_2)$, are characterized by the equations

$$(1.1) \qquad u_p \cdot x_{n-p} = Sq^p x_{n-p} \quad \text{for all} \quad x_{n-p} \in H^{n-p}(M, Z_2).$$

Wu [7] proved the formulas

(1.2) $$w_i = \sum_{p=0}^{i} Sq^{i-p} u_p, \qquad i = 0, \cdots, n,$$

(1.3) $$Sq^r w = \sum_{t=0}^{r} \binom{i-r+t-1}{t} w_{r-t} w_{i+t}.$$

For the binomial coefficients $\binom{a}{b}$ taken mod 2, one convenes that $\binom{a}{b} = 1$ for $b = 0$, $\binom{a}{b} = 0$ for $a < b$, $b \neq 0$.

One also has the Cartan formula

(1.4) $$Sq^p(x \cdot y) = \sum_{l=0}^{p} Sq^l x \cdot Sq^{p-l} y.$$

Since (1.2) can also be written as a recursion formula

(1.5) $$u_i = w_i + \sum_{p=0}^{i-1} Sq^{i-p} u_p,$$

the Wu classes and the Stiefel-Whitney classes determine each other; thus, for example, using (1.3), (1.4), (1.5),

(1.6) $$u_1 = w_1, \quad u_2 = w_2 + w_1^2, \quad u_3 = w_1 w_2, \cdots.$$

This leads to the so-called Wu relations among Stiefel-Whitney numbers: If $P_q(w_i)$ is a homogeneous polynomial in the w_i of weight q (the weight of a monomial $\Pi_{i=0}^n w_i^{q_i}$ is $\Sigma_{i=0}^n iq_i$), then from (1.1):

(1.7) $$u_{n-q} \cdot P_q(w_i) = Sq^{n-q} P_q(w_i).$$

Using equations (1.3), (1.4) and (1.5) the Wu relations are obtained as consequences of writing the left-hand side as well as the right-hand side of (1.7) in functions of only Stiefel-Whitney classes.

We will need the following explicit Wu relations, which hold on any closed smooth n-manifold M^n:

(1.8) $$w_1 w_2 w_{n-3} + w_3 w_{n-3} + n w_2 w_{n-2} = 0,$$

(1.9) $$(n+1) w_3 w_{n-3} + w_1 w_3 w_{n-4} = 0,$$

(1.10) $$n w_1^3 w_{n-3} + w_1 w_3 w_{n-4} + (n+1) w_1 w_2 w_{n-3} = 0,$$

which follow from $w_1 w_2 w_{n-3} = Sq^1 w_2 w_{n-3}$, $w_1 w_2 w_1 w_{n-4} = Sq^3 w_1 w_{n-4}$ and $w_1 w_3 w_{n-4} = Sq^1 w_3 w_{n-4}$, respectively.

If n is odd, from (1.9) and (1.10):

(1.11) $$w_1^3 w_{n-3} = 0;$$

if $w_{n-1} = w_n = 0$, from $(w_2 + w_1^2) w_{n-2} = Sq^2 w_{n-2}$:

(1.12) $$w_1^2 w_{n-2} = 0.$$

Let ξ be a k-dimensional vector bundle over M^n; denote by $v_1, v_2, \cdots, v_k$ its Stiefel-Whitney classes and by $S(\xi, M)$, $RP(\xi, M)$ the associated sphere and projective space bundles over M; let $c \in H^1(RP(\xi, M), Z_2)$ denote the first Stiefel-

Whitney class of the associated line bundle to the 2-fold covering p: $S(\xi, M) \to RP(\xi, M)$, defined by the antipodal map in each fiber, then [**1**, p. 61]:

(1.13) $$w_m RP(\xi, M) = \sum_{p+q+r=m} \binom{k-p}{q} p^*(w_r v_p) c^q;$$

furthermore $p^*: H^*(M, Z_2) \to H^*(RP(\xi, M), Z_2)$ is a monomorphism and the cohomology ring structure of $H^*(RP(\xi, M), Z_2)$ is entirely determined by the single relation

(1.14) $$c^k = \sum_{j=1}^{k} p^*(v_j) c^{k-j}.$$

REMARK. Since p^* is a monomorphism in the following, for the sake of notation we will denote $p^*(w_r)$ and $p^*(v_p)$ again by w_r, v_p respectively. Thus we write (1.13) and (1.14) as:

(1.15) $$w_m RP(\xi, M) = \sum_{p+q+r=m} \binom{k-p}{q} w_r v_p c^q,$$

(1.16) $$c^k = \sum_{j=1}^{k} v_j c^{k-j},$$

and one computes:

(1.17) $$\begin{aligned} c^{k+1} &= (v_1^2 + v_2)c^{k-1} + \text{terms of lower order in } c, \\ c^{k+2} &= (v_1^3 + v_3)c^{k-1} + \text{terms of lower order in } c, \\ c^{k+3} &= (v_1^4 + v_1^2 v_2 + v_2^2 + v_4)c^{k-1} + \text{terms of lower order in } c. \end{aligned}$$

These facts and the Wu relations above applied to closed manifolds such as $RP(\xi, M)$ lead to the

LEMMA. *Let ξ be a k-dimensional vector bundle over a closed smooth n-manifold M^n, with Stiefel-Whitney classes $v_1, \cdots, v_k$. If $w_{n-1}M = w_n M = 0$, then the following relations always hold:*

(1.18) *if n is odd,* $v_2 w_{n-2} + v_1 v_2 w_{n-3} + v_3 w_{n-3} = 0$,

(1.19) *if n is even,* $v_1 v_2 w_{n-3} + v_3 w_{n-3} = 0$,

(1.20) *for all n,* $v_1^2 w_{n-2} = 0$.

PROOF OF (1.18). Let n be odd and $k \equiv 1 \bmod 4$, since dim $RP(\xi, M) = n + k - 1 \equiv 1 \bmod 2$ from (1.11) we have

(1.21) $$w_1^3 w_{n+k-4} RP(\xi, M) = 0;$$

we now express this equation in a function of v_p, w_r using (1.15), (1.16), as well as $w_{n-1} = w_n = 0$:

$$\begin{aligned} w_1 RP(\xi, M) &= v_1 + w_1 + c, \\ w_{n+k-4} RP(\xi, M) &= v_2 w_{n-2} c^{k-4} + v_2 w_{n-3} c^{k-3} + w_{n-3} c^{k-1}; \end{aligned}$$

using (1.17), one computes:

$$w_1^3 w_{n+k-4}\, RP(\xi, M) = (v_2 w_{n-2} + v_1 v_2 w_{n-3} + v_3 w_{n-3})c^{k-1}$$

and (1.18) follows from (1.21) and the fact that (1.16) is the only relation on the cohomology ring of $RP(\xi, M)$. For other values of k apply the case just proved to the bundle $\xi \oplus R^q$ over M where $q + k \equiv 1 \bmod 4$.

PROOF OF (1.19). Let n be even and k odd, then $\dim RP(\xi, M) = n + k - 1 \equiv 0 \bmod 2$ and from (1.8)

$$(w_1 w_2 w_{n+k-4} + w_3 w_{n+k-4})\, RP(\xi, M) = 0;$$

as in the proof of (1.18), (1.19) follows if we write this identity in functions of w_r, v_p.

PROOF OF (1.20). For all $x \in H^1(M^n, Z_2)$, the Wu relation $w_1 x w_{n-2} = Sq^1 x w_{n-2}$ gives $x^2 w_{n-2} = 0$.

REMARK. Let $S^1 \times S^1$ act locally freely on M and let F denote the fixed point set of the nontrivial element of order 2 of $S^1 \times R$. Since $S^1 \times R$ is connected and since the quotient of $S^1 \times R$ by any closed discrete subgroup is again $S^1 \times R$, we can suppose $S^1 \times R$ acts locally freely and effectively on each m-dimensional component F^m of F and hence $w_{m-1} F^m = w_m F^m = 0$. We compute using (1.15), (1.16), (1.17) and the lemma:

$$\text{for all } n: \quad w_1^2 w_{n-2}[M^n] = \sum_m w_1^2 w_{m-2}[F^m];$$

$$\text{for even } n: \quad (w_1^3 w_{n-3} + w_2 w_{n-2})[M^n] = \sum_{\text{even } m} (w_1^3 w_{m-3} + w_2 w_{m-2})[F^m]$$

and obtain, by induction, the *trivial* statements:

(1.21) If $S^1 \times R$ acts locally freely on M^n, $w_1^2 w_{n-2}[M] = 0$.

(1.22) If $S^1 \times R$ acts locally freely on M^n and n is even, $(w_1^3 w_{n-3} + w_2 w_{n-2}) \cdot [M^n] = 0$.

(1.21) is trivial because by (1.12) it already follows from the Wu relations and $w_{n-1} = w_n = 0$; one can also prove (1.22) is trivial.

QUESTION. Are all relations obtained like (1.21) and (1.22) trivial?

2. Stiefel-Whitney numbers as obstructions to locally free actions of $S^1 \times S^1$. Suppose $S^1 \times S^1$ acts locally freely on M^n; let F_i, $i = 1,2,3$, denote the fixed point sets of the involutions of M corresponding to the 3 nontrivial elements of order 2 of $S^1 \times S^1$; then we know from (0.1), (0.3) that $F_i = \bigcup_m F_i^m$, where F_i^m denotes the union of the m-dimensional components of F_i, $m = 0,1,\cdots,n$; the F_i^m are closed m-manifolds and

$$3[M] = [M] = \sum_i \sum_m [RP(\nu \oplus R, F_i^m)], \tag{2.1}$$

$$[M \times RP^2] = \sum_i \sum_m [RP(\nu \oplus R^3, F_i^m)]; \tag{2.2}$$

hence

$$\sum_i \sum_m [RP(\nu \oplus R^3, F_i^m) + RP^2 \times RP(\nu \oplus R, F_i^m)] = 0. \tag{2.3}$$

REMARK. As remarked at the end of §1, $S^1 \times S^1$ acts locally freely on each component of F_i^m.

We use (2.3) to prove

PROPOSITION 2.4. (a) *Let $S^1 \times S^1$ act locally freely on the closed, smooth 4-manifold M^4; then $w_1^4[M^4] = 0$ and hence M^4 is cobordant to 0 in $\mathcal{N}^4$.*

(b) *Let $S^1 \times S^1$ act locally freely on the closed, smooth 6-manifold M^6; then $w_2 w_4 [M^6] = w_1^6 [M^6] = 0$ and hence M^6 is cobordant to zero in $\mathcal{N}^6$.*

REMARK. $M^4 = RP(L_1 \oplus L_2 \oplus R, S^1 \times S^1)$, where L_1, L_2 are the pullbacks of the nontrivial bundle L over S^1 under the obvious projections, has 2 linearly independent vector fields, but $w_1^4[M^4] \neq 0$; $M^6 = RP^2 \times M^4$ has 2 such vector fields, but $w_2 w_4 [M^6] \neq 0$; furthermore, their real Pontriagin classes vanish.

PROOF OF (a). With the notation from (1.15), we have, if ξ is a k-dimensional vector bundle over X:

$$(2.4)\quad \begin{aligned} w_3\, RP(\xi, X) &= w_3 + v_1 w_2 + v_2 w_1 + v_3 + v_1 w_1 c + v_1 c^2 && \text{if } k \equiv 0 \bmod 4, \\ &= w_3 + v_1 w_2 + v_2 w_1 + v_3 + (w_2 + v_2) c && \text{if } k \equiv 1 \bmod 4. \end{aligned}$$

$$(2.5)\quad \begin{aligned} w_2\, RP(\xi, X) &= w_2 + v_1 w_1 + v_2 + v_1 c + c^2 && \text{if } k \equiv 2 \bmod 4, \\ &= w_2 + v_1 w_1 + v_2 + w_1 c + c^2 && \text{if } k \equiv 3 \bmod 4. \end{aligned}$$

If M is a 4-manifold, $w_3^2 [M \times RP^2] = w_2^2[M]$. Since $S^1 \times S^1$ is an effective, locally free action, we know that $\dim F_i^m = 2$ or 3. From (2.3),

$$(2.6)\quad \begin{aligned} &\Sigma\, (w_3^2 [RP(\nu \oplus R^3, F^2)] + w_3^2 [RP^2 \times RP(\nu \oplus R, F^2)]) \\ &\quad + \Sigma\, (w_3^2 [RP(\nu \oplus R^3, F^3)] + w_3^2 [RP^2 \times RP(\nu \oplus R, F^3)]) = 0; \end{aligned}$$

the first (second) summation is over all 2-dimensional (3-dimensional) components of F.

By (2.4), (2.5), (2.6), (1.17), and (1.20), for each 2-dimensional component F^2:

$$\begin{aligned} w_3^2\, RP(\nu \oplus R^3, F^2) &= w_3^2 + v_1^2 w_2^2 + v_2^2 w_1^2 + v_3^2 + (w_2^2 + v_2^2)\, c^2 = 0, \\ w_2^3 [RP^2 \times RP(\nu \oplus R, F^2)] &= w_2^2 [RP(\nu \oplus R, F^2)] \\ &= (w_2^2 + v_1^2 w_1^2 + v_2^2 + w_1^2 c^2 + c^4)\, [RP(\nu \oplus R\ F^2)] \\ &= (w_1^2 + v_1^2 + v_2)\, [F^2] = v_2[F^2]. \end{aligned}$$

Similarly, for each 3-dimensional component F^3:

$$\begin{aligned} w_2^3 [RP(\nu \oplus R^3, F^3)] &= v_1^3 [F^3], \\ w_3^2 [RP^2 \times RP(\nu \oplus R, F^3)] &= w_2^2 [RP(\nu \oplus R, F^3)] = v_3 [F^3] = 0, \end{aligned}$$

since $\dim \nu = 1$; summarizing we have proved $\Sigma v_2 [F^2] + \Sigma v_1^3 [F^3] = 0$; but since $S^1 \times S^1$ acts locally freely on each F^3, it follows easily from a table of Mostert (completed by W. D. Neumann [3]) and [5], that $v_1^3 [F^3] = 0$, i. e.,

$$(2.7)\qquad \Sigma v_2 [F^2] = 0;$$

but from (0.2) and computing as before:

$$w_1^4[M^4] = \sum w_1^4[RP(\nu \oplus R, F^2)] + \sum w_1^4[RP(\nu \oplus R, F^3)]$$
$$= \sum v_2[F^2]$$

and Proposition 2.4 (a) follows from 2.7.

REMARK. Proposition 2.4 (a) (i.e. (2.7)) is nontrivial in the following sense: if M^2 denotes the Moebius band it is easy to construct a locally free $S^1 \times S^1$-action on the closed 4-manifold $M^4 = \partial(M^2 \times M^2 \times I)$, $I = [0, 1]$, such that for some component, F_0^2, of F^2, $v_2[F_0^2] \neq 0$.

In general, if n is even and $S^1 \times S^1$ acts locally freely on M one proves, using § 1, the nontrivial relation

$$\sum_{\text{even } m} v_2 w_{m-2}[F^m] = \sum_{\text{all } m} v_1^3 w_{m-3}[F^m].$$

PROOF OF (b). Let N^4 be any smooth, closed 4-manifold, then for any $x \in H^1(N^4, Z_2)$, the Wu relation $w_1 x^3 = Sq^1 x^3$ gives

(2.8) $$x^4 + x^3 w_1 = 0;$$

substituting $x + w_1$ for x gives

(2.9) $$x^2 w_1^2 + x w_1^3 = 0.$$

To prove (b) we do what we did with w_3^2 in (a) with the following Stiefel-Whitney numbers:

$$w_1^6 w_2, \quad w_2^4 + w_4^2, \quad w_1^4 w_2^2, \quad w_1^8, \quad w_1^2 w_3^2 + w_1^2 w_2^3, \quad w_1^5 w_3, \quad w_2 w_3^2$$

obtaining respectively after somewhat tedious computations:

$$\sum(v_1^4 + v_1^2 w_1^2)[F^4] = 0,$$
$$\sum(v_2 w_1^2 + v_1^2 w_1^2)[F^4] = 0,$$
$$\sum(v_1^4 + v_1^2 w_1^2 + v_2^2)[F^4] = 0,$$
$$\sum v_2[F^2] + \sum(v_1^4 + v_2^2 + v_1^2 v_2)[F^4] = 0,$$
$$\sum(v_1^2 v_2 + v_1^2 w_1^2 + v_1 w_1^3)[F^4] = 0,$$
$$\sum(v_1^4 + v_1^3 w_1 + v_2^2 + v_1^2 v_2 + v_2 w_2)[F^4] = 0,$$
$$\sum(v_1^4 + v_1^2 w_1^2 + v_2^2 + v_2 w_2 + v_2 w_1^2)[F^4] = 0.$$

Combining these relations among themselves and with (2.8), (2.9):

(2.10) $$\sum v_2[F^2] = 0,$$

(2.11) $$\sum v_2 w_2[F^4] = 0,$$

(2.12) $$\sum v_2 w_1^2[F^4] = 0.$$

From (0.2), computing, as usual,

$$w_2 w_4[M^6] = \sum(w_2 + v_2)[F^2] + \sum(w_2^2 + v_2 w_2)[F^4] = 0$$

by (2.10), (2.11) and (a);

$$w_1^6[M^6] = \sum v_2[F^2] + \sum(w_1^4 + v_1^4 + v_2^2 + v_1^2 w_1^2 + v_2 w_1^2)[F^4] = 0$$

by the above relations, (2.12) and (a).

COROLLARY. *Let M denote a closed riemannian manifold of dimenison 4 or 6 such that $[M] \neq 0$ in $\mathcal{N}^*$ and let V denote a nonsingular Killing vector field on M (such manifolds exist: see M^4 below); then V has a closed orbit.*

Indeed, it is not difficult to see, since the group of isometries $I(M)$ of M is a finite-dimensional compact Lie group, that if V had no closed orbit the closure of the subgroup of $I(M)$ defining V would contain $S^1 \times S^1$, which would act locally freely on M.

If dimension $M = 8$ and $S^1 \times S^1$ acts locally freely on M^8, then M^8 need not be cobordant to 0: Pasternack [**4**, §6] showed that S^1 acts locally freely on $M^4 = CP^2 \# 3RP^4$, hence $S^1 \times S^1$ acts locally freely on $M^4 \times M^4$, which is not cobordant to 0 in $\mathcal{N}^8$. However, by computing as above one can prove:

PROPOSITION 2.13. *If $S^1 \times S^1$ acts locally freely on M, $w^2 w^6[M^8] = 0$.*

REMARKS. Although Proposition 3.4 (a) already seems out of the range of known classification theorems for group actions, Propositions 2.4 (b) and 2.13 are definitely out of the range of such classification theorems.

For 5-manifolds on which $S^1 \times S^1$ acts locally freely, (0.5) gives no relations on the Stiefel-Whitney numbers.

REFERENCES

1. P. E. Conner and E. E. Floyd, *Differential periodic maps*, Ergebnisse der Mathematik und ihrer Grenzgebiete, N.F., Band 33, Academic Press, New York; Springer-Verlag, Berlin, 1964, MR **31** #750.

2. ———, *Fibering within a cobordism class*, Michigan Math. J. **12** (1965), 33–47. MR **31** #4038.

3. W. D. Neumann, *3-dimensional G-manifolds with 2-dimensional orbits*, Proc. Conference on Transformation Groups (New Orleans, La., 1967), Springer, New York, 1968, pp. 220–222. MR **39** #6355.

4. J. S. Pasternack, *Foliations and compact Lie group actions*, Comment. Math. Helv. **46** (1971), 467–477. MR **45** #9353.

5. H. Rosenberg, R. Roussarie and D. Weil, *A classification of closed orientable 3-manifolds of rank two*, Ann. of Math. (2) **91** (1970), 449–464. MR **42** #5280.

6. R. Stong, *Stationary point free group actions*, Proc. Amer. Math. Soc. **18** (1967), 1089–1092. MR **36** #895.

7. Wen-tsun, Wu *Les i-carrés de une variété grassmannienne*, C.R. Acad. Sci. Paris **230** (1950), 918–920. MR **12**, 42.

UNIVERSITY OF MARYLAND

ALGEBRAIC AND PIECEWISE LINEAR TOPOLOGY

Proceedings of Symposia in Pure Mathematics
Volume 27, 1975

STIEFEL-WHITNEY HOMOLOGY CLASSES AND SINGULARITIES OF PROJECTIONS FOR POLYHEDRAL MANIFOLDS

THOMAS F. BANCHOFF

Abstract. Stiefel-Whitney classes of simplicial manifolds are described in terms of projections into Euclidean spaces. An explicit chain associated with a projection is shown to determine a homology class which is independent of the projection, and this class is dual to a Stiefel-Whitney cohomology class of the manifold.

In this paper we shall describe the Stiefel-Whitney classes of simplicial manifolds by means of singularities of projections into Euclidean spaces. For a simplexwise linear map $f: M \to E$ of a q-dimensional triangulated manifold into Euclidean $(q-i+1)$-dimensional space which is one-to-one on each simplex of dimension $q-i+1$, we shall define an index of singularity $u(s, f)$ for each $(q-i)$-simplex s of M (an integer if i is odd and an integer modulo 2 if i is even). We then will show that the $(q-i)$-chain $\sum_{s \in M} u(s, f)s$ is a cycle, that its homology class is independent of the mapping f used to define it, and that this homology class is the Poincaré dual of the i-dimensional Stiefel-Whitney cohomology class w^i of the manifold M.

In the first section we shall consider the case $i = 1$, so that the singularity set of the mapping $f: M^q \to E^q$ is a fold chain of dimension $q-1$, and this will be shown to represent the zero homology class if and only if the manifold is orientable.

In the second section we consider the other extreme $i = q$, so that the singularity set of the mapping $f: M \to E^1$ is the collection of critical points of the mapping,

AMS (MOS) subject classifications (1970). Primary 53A05, 57C35; Secondary 55A25.
Key words and phrases. Stiefel-Whitney class, characteristic class, singularities, projections, polyhedral manifold, simplicial manifold.

and the homology class is determined by the Euler characteristic of the manifold.

In the third section, we combine the two approaches to define the singularity index of $f: M^q \to E^{q-i+1}$ and to prove that it has the required properties.

The author wishes to acknowledge the helpful collaboration of Richard Porter and Robert MacPherson in the seminar at Brown University during which these results were formulated. Very similar results have been obtained completely independently by Clint McCrory [7].

1. Mappings of surfaces into the plane. Most of the interesting phenomena concerning mappings from a manifold into a Euclidean space of the same dimension already appear in the case of mappings of surfaces into the plane, so we shall study this case in some detail, and then go on to generalize the results.

If $f: M \to E$ is a simplexwise linear mapping of a triangulated surface into the plane such that f is one-to-one on each triangle, then each edge of M is either a *fold* edge or an *ordinary* edge for f.

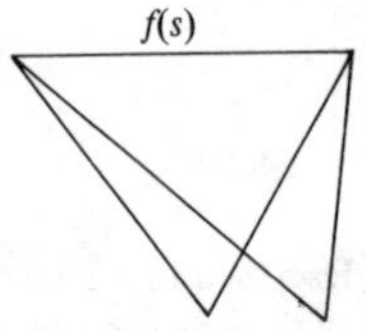

s IS A FOLD FOR f

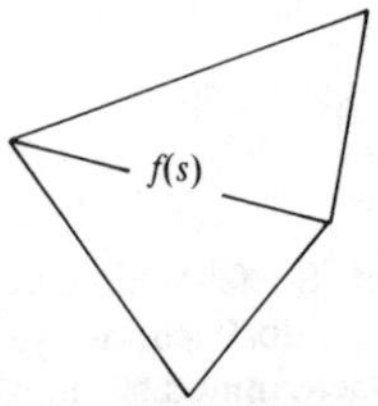

s IS ORDINARY FOR f

The edge s of M is said to be a *fold* edge for f if the images of both of the triangles adjacent to s in M lie to the same side of the line determined by the image $f(s)$ in E; otherwise s is ordinary for f. The collection of fold edges is called the *fold chain* of f, denoted by $u(f)$.

The boundary ds of an edge with vertices v and w is the set consisting of v and w. The boundary of a chain is the collection of vertices appearing in the edges of the chain, each taken with a multiplicity equal to the number of times the vertex appears as an endpoint of an edge in the chain. If each vertex in the boundary of a chain appears with an even multiplicity, then the chain is said to be a *mod* 2 *cycle*. We shall now show that any fold chain $u(f)$ is a mod 2 cycle.

In order to prove that $u(f)$ is a mod 2 cycle, we must show that any vertex of the surface M is an endpoint of an even number of fold edges. To see this, we may first number the edges appearing at a vertex v in cyclic order: $vw_1, vw_2, \cdots, vw_m$ so that each triple vw_iw_{i+1} determines one of the triangles at v (where we understand the vertex w_{m+1} to mean w_1). We say that a triangle $vw_i\, w_{i+1}$ is *positive* for f if the angle of rotation from the ray $f(v)f(w_i)$ to the ray $f(v)f(w_{i+1})$ is positive, and otherwise we say that the triangle is negative for f. Note that this concept is well defined since $f(v)$, $f(w_i)$, and $f(w_{i+1})$ are never collinear by our hypothesis on f so the angle of rotation is always uniquely determined.

The edge vw_i is a fold edge for f if and only if one of the adjacent triangles is positive and the other is negative. If we start at the triangle vw_1w_2 and proceed around the vertex v keeping track of the number of changes from positive to negative or back again, then by the time we return to the original triangle, we must have encountered an even number of such sign changes. From this we conclude that there must be an even number of fold edges at the vertex v, and this completes the proof that the fold chain $u(f)$ is a mod 2 cycle.

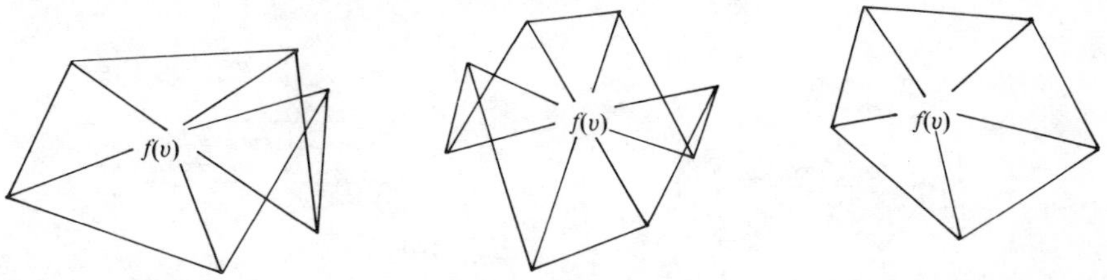

2 FOLD EDGES 4 FOLD EDGES 0 FOLD EDGES

We would now like to examine the way that the fold chain changes as we change the mapping f. Since a simplexwise linear mapping is completely determined by the images of the vertices of M, we want to see what happens to the fold chain of f when we move the vertices around in a continuous manner.

Observe first of all that if f is one-to-one on each triangle of M then every mapping sufficiently close to f will have the same property. It follows that if f' is a mapping sufficiently close to f, then we can deform f to f' by moving one vertex at a time so that at any intermediate stage the mapping f_t will also be one-to-one on each triangle of M. During such a deformation a triangle always keeps the same orientation, so a fold edge of f is a fold edge of each f_t and $u(f)$ is precisely the same collection of edges of M as is $u(f')$.

Even if f' is not close to f, it is still possible to deform f into f' through a family of mappings f_t moving one vertex at a time, but it may not be possible to find such a family for which f_t is always one-to-one on triangles. We can however find such a family f_t such that there are only a finite number of times that f_t is not one-to-one on all triangles, and that at each such exceptional time, there is precisely one triangle of M with vertices having images lying in a straight line. We may describe such a situation by saying that one vertex of a triangle passes through the opposite edge as we go from a position f_- to a nearby position f_+ so that exactly one triangle, say uvw, changes its orientation during this passing through. We want to compare $u(f_-)$ and $u(f_+)$, and to see which edges of M are in one but not the other.

Since uvw is the only triangle which changes its orientation, the only edges which can change from fold to ordinary or conversely during the passing through are the three edges uv, uw, and vw. But if the edge uv was a fold for f_- because both adjacent triangles uvw and uvw' had images on the same side of the image $f_-(uv)$, then these triangles will have images under f_+ which are on opposite sides of the image $f_+(uv)$, and conversely. Therefore if uv was a fold for f_-, then it is not a fold for f_+ and conversely. The same is true for the edges uw and vw, so the fold chains $u(f_-)$ and $u(f_+)$ differ only by the collection of edges in the boundary of the triangle uvw.

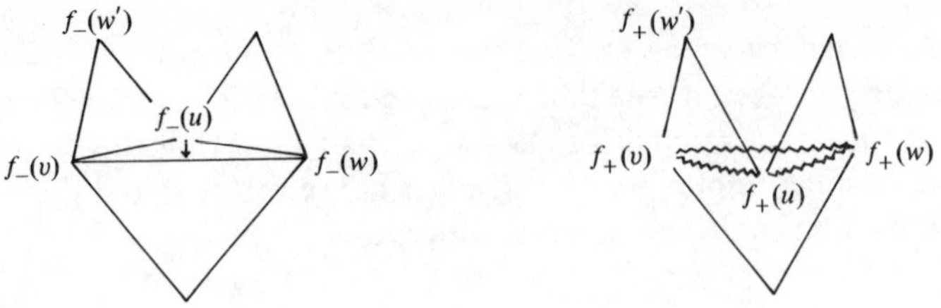

CASE 1

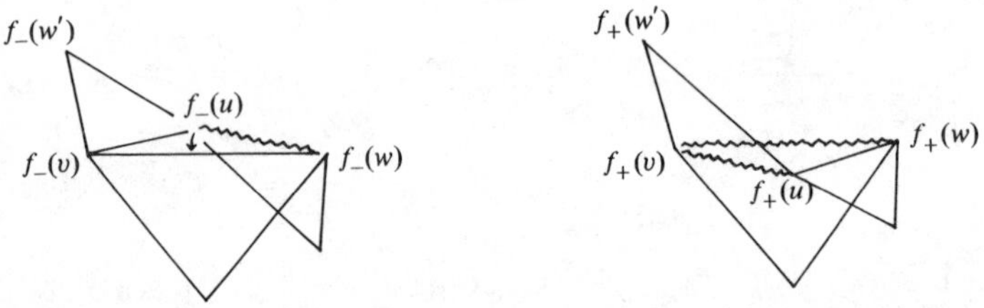

CASE 2

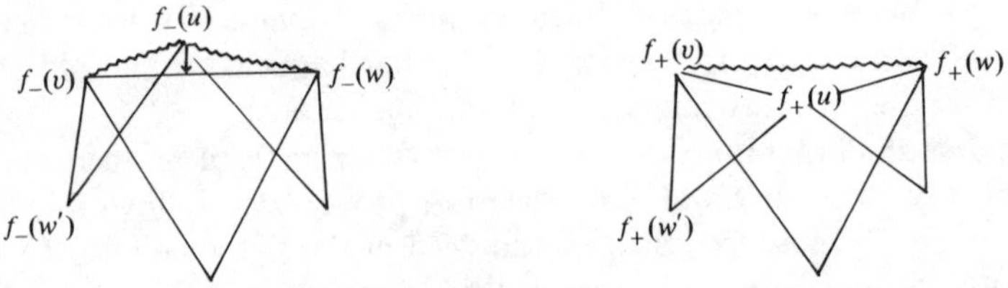

CASE 3

Each time we pass such an exceptional position in the family of mappings f_t, we change the fold chain by the boundary of a triangle, so it follows that the fold chains $u(f)$ and $u(f')$ themselves differ by the collection of boundary edges of a finite number of triangles, each edge taken with a multiplicity equal to the number of times it appears in the boundary of a triangle involved in a passing through. Two chains which differ only by such a collection of edges are said to be *homologous mod* 2, or to determine *the same homology class mod* 2. We may express the result of the foregoing paragraphs, then, by saying that any two simplexwise linear mappings f and f' which are one-to-one on each triangle determine fold chains $u(f)$ and $u(f')$ which are homologous mod 2, or that the homology class mod 2 determined by such a mapping f is independent of the mapping used to define it. This homology class is then a characteristic of the surface itself, and it is one of the simplest examples of a *characteristic homology class*.

We now give a specific way of describing this characteristic homology class for a triangulated surface. The simplest such description would be to find a mapping f such that every edge of the surface would be a fold, but this in general is impossible since we have already shown that there must be an even number of folds at each vertex. We can obtain a triangulation of a surface which does have this property however by the process of *barycentric subdivision*. To obtain this, we take a new vertex at the center of gravity of each edge and then form new edges by joining this point to the vertices of the edge. Then take the center of gravity of each triangle and join it to the vertices and centers of the edges. The triangles of the barycentric subdivision then are described by taking a vertex of the surface, the center of an edge at that vertex, and the center of a triangle at that edge. In this subdivision, there are six edges around the barycenter of every triangle, four edges around the barycenter of every edge, and twice as many around every vertex as there were in the original triangulation.

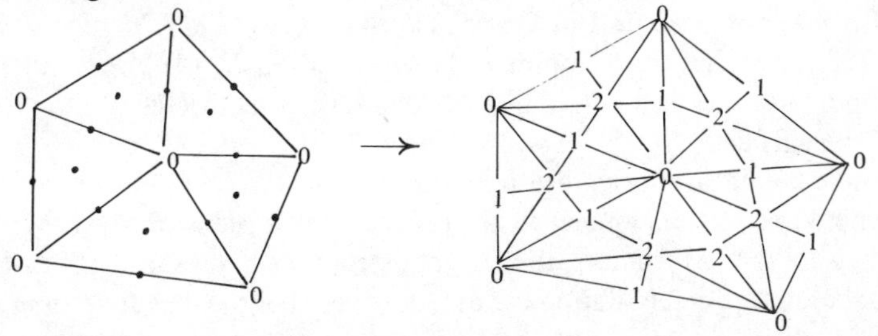

We may now describe a mapping on this first barycentric subdivison by defining the image of each vertex in the new triangulation. To do this we choose any three points in general position in the plane, say the points (0, 0), (1, 1), and (2, 4), and we define the image of the barycenter of each triangle to be (2, 4), the image of the barycenter of each edge to be (1, 1), and the image of each vertex of the original triangulation to be (0, 0). Since each triangle of the barycentric subdivision has vertices of the three distinct types, when we extend this mapping linearly, each triangle will be mapped in a one-to-one way. But each edge of the barycentric subdivision will be a fold edge for this mapping, so this gives an explicit representative of the characteristic class we have defined.

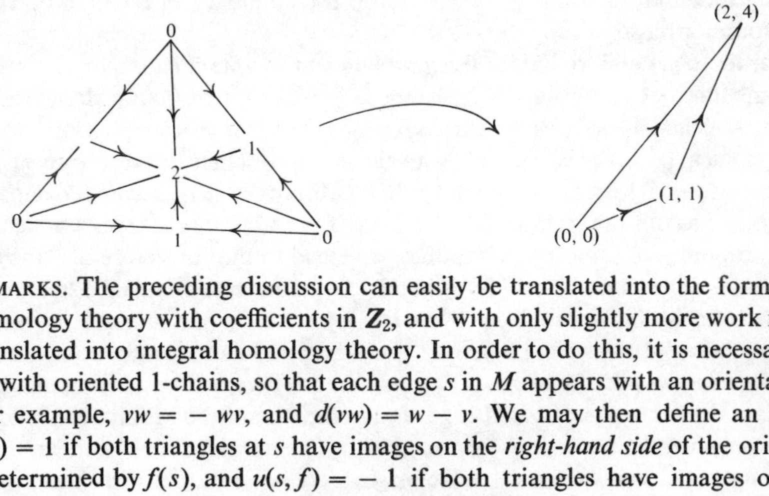

REMARKS. The preceding discussion can easily be translated into the formalism of homology theory with coefficients in $\mathbf{Z}_2$, and with only slightly more work it can be translated into integral homology theory. In order to do this, it is necessary to work with oriented 1-chains, so that each edge s in M appears with an orientation, so for example, $vw = -wv$, and $d(vw) = w - v$. We may then define an index $u(s, f) = 1$ if both triangles at s have images on the *right-hand side* of the oriented line determined by $f(s)$, and $u(s, f) = -1$ if both triangles have images on the left-hand side. Set $u(s, f) = 0$ otherwise, so in all cases we have $u(-s, f) = -u(s, f)$. This enables us to define a chain $u(f) = \sum_{s \text{ in } M} u(s, f)s$ which will be independent of the orientations chosen for the edges of M since $u(-s, f)(-s) = u(s, f)s$ for each s.

If we select the orientation vw_i, $i = 1, 2, \cdots, m$, for the edges at a vertex v in cyclic order, then the fold edges alternately have indices $+1$ and -1 as we proceed around the vertex so the contribution of the boundary $du(f)$ to the vertex c is 0. Thus the 1-chain $u(f)$ with coefficients in $\boldsymbol{Z}$ is also a cycle.

Moreover in a passing through such as described in the paragraphs on deformations of a mapping, we find in every case that $u(f_+) - u(f_-) = uw - uv - vw = d(uvw)$. It follows that the homology class of $u(f)$ in $H_1(M; \boldsymbol{Z})$ is independent of the mapping used to define it, and this defines the first Stiefel-Whitney homology class of the surface.

If we take the natural orientation on the edges of the first barycentric subdivision of M, where the vertices appear in the order of the dimensions of the simplexes of which they are the barycenters, then the construction given shows that a representative of the first Stiefel-Whitney class is given by the negative of $u(f)$ for this mapping f.

An Alternate Description of the Fold Chain of a Mapping $f: M^2 \to E^2$. Let $f: M^2 \to E^2$ be a simplexwise linear mapping on a triangulated surface which is 1-1 on each triangle. We say that an edge s of M is a *fold edge* of f if the two adjacent triangles to s are mapped to the same side of the line determined by the image $f(s)$ in E^2. If the adjacent triangles are mapped to opposite sides, then we say that the edge is *ordinary* for f.

If the mapping f is the orthogonal projection of a convex polyhedron into the plane, then the fold edges determine a simple closed polygon on the polyhedron which decomposes the surface into two regions with the collection of fold edges of f as the common boundary. We would like to know under what circumstances we can conclude that the fold edge collection is the boundary of a region on the surface, for an arbitrary map $f: M^2 \to E^2$.

In order to get an insight into this problem and to obtain an important lemma at the same time, let us consider what happens to a thin strip about a closed curve on the surface that does not pass through any vertex of the triangulation. We may think of such a strip as a rectangle with the curve as center line and with two sides pasted together. The resulting strip is either a Möbius band or a cylinder, depending on whether or not the rectangle is given a half twist before the sides are identified. A few examples of projections into the plane lead to the conjecture that the center line crosses an even number of folds in the case of a cylinder and on odd number in the case of a Möbius band. To prove this conjecture, we first think of the rectangle as being blue on one side and red on the other, and as we move along the center curve, we keep track of which color is facing up. There will be a color change each time that the curve passes a fold edge, and in the case of the Möbius band, one further change when the curve passes the seam where the sides are identified. Since in any complete circuit we return to the original color we started with, there must be an even number of color changes in any such circuit. Thus for the cylinder, where the only changes occur at folds, we have an even number of folds, and in the case of the Möbius band, just one change occurs at the seam so there must be an odd number which occur at the fold edges.

We can now answer the question completely. If the surface M is orientable, then we may color all the triangles uniformly red on one side and blue on the other and an edge s will be a fold edge for the mapping f if and only if one triangle at s has red side facing up and the other has blue side facing up. It follows that the fold edges do form the boundary of a collection of triangles, specifically those with red side facing up. Thus, in the case of an orientable surface, the collection of fold edges always bounds.

On the other hand, if the surface is nonorientable so that some curve in the surface has a neighborhood strip which is a Möbius band, then the fold edge collection cannot bound a collection of triangles, since in a complete circuit of the curve it would be necessary alternately to leave and to enter the collection, ending up in the same situation as at the beginning. This would mean that the curve would meet an even number of fold edges, but the center curve of a Möbius band meets an odd number of fold edges. Thus in the case of a nonorientable surface, the collection of fold edges never bounds.

Precisely the same proof shows that for a simplexwise linear mapping $f : M^q \to E^q$ which is 1-1 on each q-simplex of a q-manifold, the collection of $(q-1)$-simplexes which have both adjacent q-simplexes mapped to the same side of the hyperplane determined by the image of the $(q-1)$-simplex is the boundary of a collection of q-simplexes if and only if M is orientable.

Remarks on the Differentiable Case. The idea of describing the Stiefel-Whitney classes of a smooth manifold in terms of the singularities of mappings of the manifold into Eucildean spaces goes back to Pontrjagin [**8**, Theorem 4], where it was shown that a representative of the $(q-i)$-dimensional homology class dual to the ith Stiefel-Whitney cohomology class is carried by the set of points of M where the Jacobian of a mapping $f : M^q \to E^{q-i+1}$ has rank less than $q - i + 1$. This notion was developed considerably by Thom [**11**], who gave a much more precise description of the nature of the generic singularities of such smooth mappings. This approach has been extended by Porter [**9**]. In the case of mappings of the plane into the plane, these singularities were described fully by Whitney, who showed that any smooth mapping of a surface M^2 into the plane E^2 could be approximated by one whose only singularities were folds and cusps, and that the collection of such points in M was an embedded submanifold, i.e. a set of finitely many disjoint simple closed curves.

Note that the situation is quite different in the simplexwise linear case, even for mappings of surfaces into the plane, since small perturbations of the images of the vertices of a fixed triangulation cannot separate into simpler form the image of a vertex star with more than two fold edges, so the fold chain will not in general be an embedded 1-manifold. Also ramification behavior is stable in the simplexwise linear case, so there may be isolated points where the mapping is not one-to-one but where there are no fold edges for any nearby mapping. (Such behavior is examined in [**4**].)

REMARKS ON THE CASE OF MAPPINGS OF A q-MANIFOLD INTO q-SPACE. For a simplexwise linear mapping $f : M^q \to E^q$ of a triangulated q-dimensional manifold into Euclidean q-dimensional space which is one-to-one on each of the q-simplexes of M, we may define a fold index for each simplex s^{q-1} of dimension $q - 1$ by setting $u(s^{q-1}, f) = 1$ if both adjacent q-simplexes to s^{q-1} in M have image in the upper half-space of the oriented hyperplane determined by the image $f(s^{q-1})$ in E^q, and $u(s^{q-1}, f) = -1$ if both images lie in the lower half-space and $u(s^{q-1}, f) = 0$ otherwise. Then the chain $u(f) = \sum_{s^{q-1} \text{ in } M} u(s^{q-1}, f)s^{q-1}$ in the group $C_{q-1}(M; \mathbf{Z})$ of $(q - 1)$-chains with integral coefficients will have the property that $du(f) = 0$ by the same argumentation that applied in the case $q = 2$. The proof that the homology class of $u(f)$ is independent of the mapping f is somewhat more complicated than in the case $q = 2$ since there are more ways for a mapping f_t in a deformation to collapse a q-simplex into a hyperplane than simply by causing a vertex to pass through the opposite face. For example, in the case of a family of mappings $f_t : M^3 \to E^3$, an exceptional mapping which collapses one tetrahedron into a plane might have either three or four extreme vertices. It is possible to handle all the cases which arise, but it is not much harder to handle the general case of a mapping $f : M^q \to E^{q-i+1}$ so we will leave the details until then.

Again we can get a representative for this characteristic homology class by describing a specific mapping on the first barycentric subdivision of M. If we denote the barycenter of the r-simplex s^r by the symbol $\hat{s}^r$, then we may describe the k-simplexes of the barycentric subdivision $\hat{M}$ by sequences of simplexes $(s^{i_0}, s^{i_1}, \cdots, s^{i_k})$ of M where each simplex is in the boundary of the following simplex. A simplexwise linear mapping $g : \hat{M}^q \to E^q$ is then completely determined by the images of the vertices $\hat{s}^r$, and we define the mapping g by setting $g(\hat{s}^r) = (r, r^2 \cdots, r^q)$. Since the points $(r, r^2, \cdots, r^q)$ with $r = 0, 1, \cdots, q$ are linearly independent points of E^q, it follows that the mapping g is one-to-one on each q-simplex of M^q. Moreover the image of each $(q - 1)$-simplex of M is an extreme simplex of the image of M, so every $(q - 1)$-simplex has index $+1$ or -1. Specifically a $(q - 1)$-simplex of the barycentric subdivision is given by a sequence of q-distinct simplexes of M, each in the boundary of the next, so exactly one dimension is missing from the sequence. If there is no simplex of dimension r, then the index of the $(q - 1)$-simplex is given by $(-1)^r$. This chain agrees, up to a sign independent of the number r, with the chain described by Whitney in **[12]** and developed explicitly by Halperin and Toledo **[6]**. See also the original paper of Stiefel **[10]**. The mod 2 aspect of this problem was handled independently by Cheeger **[5]**. All of these papers use vector fields rather than projections in defining the characteristic classes.

2. Mappings of manifolds into the line. The case $i = q$ for a mapping $f : M^q \to E^{q-i+1}$ leads to the study of singularities of real-valued functions on a triangulated manifold, a generalization of portions of classical Morse theory. The reader is referred to the papers **[1]** and **[3]** of the author where the subject is treated in detail from a slightly different point of view. We shall give a very condensed review of these concepts in the context of Stiefel-Whitney homology classes.

If $f: M^q — E^1$ is a simplexwise linear mapping into an oriented line such that f is one-to-one on each edge of M, then we may define an index for each simplex s^r and each vertex v with tells whether or not v is the highest vertex of s^r with respect to the mapping f. We set $A(s^r, v, f) = 1$ if v is in the closure of s^r and if $f(v) \geqq f(x)$ for all x in s^r, and $A(s^r, v, f) = 0$ otherwise. We then index a point v by computing the Euler characteristic of the collection of simplexes of M which have v as their highest vertex for f, i.e. we set

$$a(v, f) = \sum_{r=0}^{q} (-1)^r \sum_{s^r \text{ in } M} A(s^r, v, f).$$

For example, if v is the absolute minimum of the function f, then $a(v, f) = 1$ since $A(s^r, v, f) = 1$ if and only if s^r equals v itself. If v is an absolute maximum, then v is the highest vertex for all of the simplexes in its star, and this is an open q-cell so $a(v, f) = (-1)^q$. In general for a vertex v of a q-manifold we have $a(v, -f) = (-1)^q a(v, f)$.

Note that if f resembles a smooth mapping with a nondegenerate critical point of Morse index r at the point v, then the collection of simplexes with v as highest vertex is an open r-cell and $a(v, f) = (-1)^r$.

We use the indices $a(v, f)$ to define a 0-chain $u(f) = \sum_{v \text{ in } M} a(v, f)v$, and we will show that this chain satisfies the conditions of the theorem.

Since the boundary of any 0-chain is 0, the chain $u(f)$ is automatically a cycle in this case.

To show that $u(f)$ is independent of the mapping used to define it, we consider what happens when one vertex passes through another during a deformation f_t. Assume that $f_-(v) < f_-(w)$ and that $f_+(v) > f_+(w)$, and that all other pairs of vertices have the same order both for f_- and f_+. Then

$$A(s^r, x, f_+) = A(s^r, x, f_-)$$

unless x is either v or w and then

$$A(s^r, v, f_+) + A(s^r, w, f_+) = A(s^r, v, f_-) + A(s^r, w, f_-).$$

It follows that

$$\begin{aligned} u(f_+) - u(f_-) &= \sum_{x \text{ in } M} (a(x, f_+) - a(x, f_-))x \\ &= \sum_{x \text{ in } M} \sum_{r=0}^{q} (-1)^r \sum_{s^r \text{ in } M} (A(s^r, x, f_+) - A(s^r, x, f_-))x \\ &= \sum_{r=0}^{q} (-1)^r \sum_{s^r \text{ in } M} (A(s^r, v, f_+) - A(s^r, v, f_-))(v - w) \\ &= (a(v, f_+) - a(v, f_-))\, d(wv). \end{aligned}$$

Therefore $u(f_+)$ differs from $u(f_-)$ by a boundary and the homology class of $u(f)$ is independent of the mapping used to define it.

In order to obtain an explicit representative of this 0-dimensional characteristic homology class, we take the mapping defined on the first barycentric subdivision by $g(\hat{s}^r) = r$ for each simplex $\hat{s}^r$ of $\hat{M}$. Then $\hat{s}^r$ is the highest vertex of a simplex $(s^{i_0}, s^{i_1}, \cdots, s^{i_k})$ for g if and only if s^{i_k} is precisely s^r. The collection of such simplexes of $\hat{M}$ is exactly the collection of simplexes of M that are contained in the simplex s^r of the original subdivision, so it follows that $a(\hat{s}^r, g)$ is the Euler characteristic of the open r-simplex s^r, i.e. $a(\hat{s}^r, g) = (-1)^r$.

If we use the further fact that in a connected manifold M, any 0-chain $\sum_{x \text{ in } M} a(x, f)x$ is homologous to the chain $[\sum_{x \text{ in } M} a(x, f)]v$ for a fixed vertex v, then we may identify the characteristic class $u(f)$ with the integer $\sum_{x \text{ in } M} a(x, f)$ for any mapping f. In particular if we use the mapping g defined on the first barycentric subdivision, then this integer is identified as

$$\sum_{s^r \text{ in } M} a(s^r, g) = \sum_{s^r \text{ in } M} (-1)^r = \sum_{r=0}^{q} (-1)^r \alpha_r,$$

where α_r is the number of r-simplexes in M. Thus the integer is the Euler characteristic of the manifold M, which corresponds to the description of the dual q-dimensional cohomology class as the Euler class of the manifold.

3. Mappings of q-manifolds to euclidean spaces of codimension $i - 1$. We now consider the general case, leading to the description of a homology class dual to the ith Stiefel-Whitney cohomology class of M.

Let $f : M^q \to E^{q-i+1}$ be a *general* simplexwise linear mapping, i.e. a mapping which is one-to-one on all simplexes of dimension $\leqq q - i + 1$. We wish to define a $q - i$ chain on M^q by giving an index $u(\sigma^{q-i}, f)$ to each oriented $q - i$ cell of M^q. To do this, we consider the dual cell $St(\hat{\sigma}^{q-i})$ in the first barycentric subdivision $\hat{M}^q$ of M^q, and the natural extension $\hat{f} : \hat{M}^q \to E^{q-i+1}$ of the mapping f to this subdivided complex. The mapping $\pi \circ f : St(\hat{\sigma}^{q-i}) \to E^1$ is a *general map* in the sense of CPC,[1] where E^1 is the line orthogonal to the oriented affine hyperplane determined by $f(\sigma^{q-i})$ in E^{q-i+1} and determining the orientation of $E^{q-i+1} = E^{q-i} \oplus E^1$, and $\pi : E^{q-i+1} \to E^1$.

We index the cell σ^{q-i} by taking the index $a(\sigma^{q-i}, f) = \pm\, a(\hat{\sigma}^{q-i}, \pi \circ f \,|\, St(\hat{\sigma}^{q-i}))$ as defined in CPC. A direct definition is given by defining an index

$$\begin{aligned} A(\sigma^r, \sigma^{q-i}, f) &= 1 \quad \text{if } \pi f(v) \leqq \pi f(w) \text{ for } \sigma^{q-i} \subset \sigma^r,\ v \in \sigma^{q-i},\ w \in \sigma^r, \\ &= 0 \quad \text{otherwise,} \end{aligned}$$

and setting $a(\sigma^{q-i}, f) = \sum_{r=0}^{q} (-1)^r \sum_{\sigma^r} A(\sigma^r, \sigma^{q-i}, f)$.

The chain $Z_{q-i}(f)$ is thus defined by $\sum_{\sigma^{q-i}} a(\sigma^{q-i}, f)\, \sigma^{q-i}$. Note that, if we change the orientation of the cell σ^{q-i}, then this changes the orientation of the line E^1 and we have $a(-\sigma^{q-i}, f) = a(\sigma^{q-i}, -f)$.

In order for the chain Z_{q-i} to be defined over the *integers* independent of the

[1]CPC refers to *Critical points and curvature for embedded polyhedra* [1].

orientation, this index must have the property that $a(\sigma^{q-i}, -f) = (-1)^i a(\sigma^{q-i}, f)$. This property is satisfied, for example, if M^q is a simplicial manifold, and if i is odd, since in general for a manifold $a(\sigma^{q-i}, -f) = (-1)^i a(\sigma^{q-i}, f)$. (See Appendix for a proof of this result.)

Notice that even in the case of a simplicial manifold, the indices $a(\sigma^{q-i}, f)$ are not restricted to be ± 1, as remarked in CPC (which was involved with the case $i = q$). Note that in any case $Z_{q-i}(f)$ is well defined as a chain with coefficients in the integers modulo 2.

We now wish to demonstrate that $Z_{q-i}(f)$ is a cycle, i.e., that

$$dZ_{q-i}(f) = \sum_{\sigma^{q-i}} a(\sigma^{q-i}, f)\, d\sigma^{q-i} = 0.$$

For every cell σ^{q-i-1}, we must then show that

$$0 = \sum_{\sigma^{q-i}} a(\sigma^{q-i}, f)\, I(\sigma^{q-i}, \sigma^{q-i-1})$$

where $I(\sigma^{q-i}, \sigma^{q-i-1}) = \pm 1$ if $\sigma^{q-i-1} \subset \sigma^{q-i}$ and the orientation of σ^{q-i-1} induced by that of σ^{q-i} is (± 1) times the orientation of σ^{q-i-1}. We define $I(\sigma^r, \sigma^{q-i-1})$ similarly. But

$$\begin{aligned}\sum_{\sigma^{q-i}} a(\sigma^{q-i}, f)\, I(\sigma^{q-i}, \sigma^{q-i-1}) &= \sum_{\sigma^{q-i}} \Big(\sum_{r=0}^{q} A(\sigma^r, \sigma^{q-i}, f) \Big) I(\sigma^{q-i}, \sigma^{q-i-1}) \\ &= \sum_{r=0}^{q} (-1)^r \sum_{\sigma^r} \sum_{\sigma^{q-i}} A(\sigma^r, \sigma^{q-i}, f)\, I(\sigma^{q-i}, \sigma^{q-i-1}).\end{aligned}$$

Since we may reorient $(q-i)$-cells without changing the cycle, we may simplify the computation by assuming that all indices $I(\sigma^{q-i}, \sigma^{q-i-1})$ are either 1 or 0, so we must examine the expression

$$\sum_{\sigma^{q-i}} A(\sigma^r, \sigma^{q-i}, f)\, I(\sigma^{q-i}, \sigma^{q-i-1}) = \sum_{\sigma^{q-i-1} \subset \sigma^{q-i}} A(\sigma^r, \sigma^{q-i}, f).$$

Consider first the case when $f(\sigma^{q-i-1})$ lies in the boundary of $f(\sigma^r)$. Then the image $f(\sigma^r)$ in E^{q-i+1} will be a cell of dimension $q - i + 1$ if $r \geqq q - i$ and there will only be two $(q-i)$-cells with images in the boundary of $f(\sigma^r)$ which contain σ^{q-i-1}. For one of these cells, we have $A(\sigma^q, \sigma^{q-i}, f) = 1$ and for the other, we get 0, so $\sum_{\sigma^{q-i-1} \subset \sigma^{q-i}} A(\sigma^r, \sigma^{q-i}, f) = 1$.

The proposition that $dZ_{q-i} = 0$ then follows if we know that the set of cells σ^r such that $\sigma^{q-i-1} \not\subset d\sigma^r$ and $f(\sigma^{q-i-1})$ lies in the point set boundary of $f(\sigma^r)$ has zero Euler characteristic if i is odd and even Euler characteristic if i is even. This result is contained in another paper of the author, but we include a proof in the Appendix of this paper for completeness.

It follows that $dZ_{q-i} = \sum_{r=0}^{q} (-1)^r \sum_{\sigma^r} U(\sigma^r, \sigma^{q-i}, f)$ where $U(\sigma^r, \sigma^{q-i}, f) = 1$ if and only if $\sigma^{p-i} \not\subset d\sigma^r$ and $f(\sigma^{q-i-1}) \subset df(\sigma^r)$, so Z_{q-i} is a cycle if i is odd and a cycle mod 2 if i is even.

Next we must show that the cycle $Z_{q-i}(f)$ is independent of the mapping f. We approach this by a deformation argument, i.e. we take a 1-parameter family $\{f_t\}$ of simplicial mappings so that all but a finite number are general mappings and so that for each of the exceptional mappings, there is precisely one $(q-i+1)$-simplex which is mapped into an oriented hyperplane. Let $f_0 : M^q \to E^{q-i+1}$ be an exceptional mapping with $f_0(\sigma_0^{q-i+1})$ lying in an oriented affine hyperplane H wth $f_+(\sigma_0^{q-i+1}) \subset H^+$ and $f_-(\sigma_0^{q-i+1}) \subset H^-$, where H^+ and H^- are closed half-spaces bounded by H.

We would like to show that

$$Z_{q-i}(f_+) - Z_{q-i}(f_-) = d\,C_{q-i+1}$$

for some $(q-i+1)$-chain in M^q.

Now

$$Z_{q-i}(f_+) - Z_{q-i}(f_-) = \sum_{\sigma^{q-i}} (a(\sigma^{q-i}, f_+) - a(\sigma^{q-i}, f_-))\sigma^{q-i}$$

$$= \sum_{\sigma^{q-i}} \sum_{r=0}^{q} (-1)^r \sum_{\sigma^r} (A(\sigma^r, \sigma^{q-i}, f_+) - A(\sigma^r, \sigma^{q-i}, f_-))\sigma^{q-i}.$$

Note first of all that the only cells σ^{q-i} entering into this sum with nonzero coefficients are the cells in the boundary of σ_0^{q-i+1}. Note also that $A(\sigma^r, \sigma^{q-i}, f_+) - A(\sigma^r, \sigma^{q-i}, f_-) = 0$ unless $\sigma^r \subset \sigma_0^{q-i+1}$, and $f(\sigma^r) \subset H^+$, and moreover under this condition

$$A(\sigma^r, \sigma^{q-i}, f_+) - A(\sigma^r, \sigma^{q-i}, f_-) = A(\sigma_0^{q-i+1}, \sigma^{q-i}, f_+) - A(\sigma_0^{q-i+1}, \sigma^{q-i}, f_-).$$

Let $J(\sigma^r) = 1$ if $\sigma^r \subset \sigma_0^{q-i+1}$ and $f(\sigma^r) \subset H^+$, and 0 otherwise.

We may assume, furthermore, that the $(q-i)$-cells σ^{q-i} in $d(\sigma_0^{q-i+1})$ are oriented so that

$$d(\sigma_0^{q-i+1}) = \sum_{\sigma^{q-i}} (A(\sigma_0^{q-i+1}, \sigma^{q-i}, f_+) - A(\sigma_0^{q-i+1}, \sigma^{q-i}, f_-))\sigma_{q-i}.$$

It then follows that

$$Z_{q-i}(f_+) - Z_{q-i}(f_-)$$

$$= \sum_{\sigma^{q-i}} \left[\sum_{r=0}^{q} (-1)^r \sum_{\sigma^r \subset \sigma_0^{q-i+1}}^{q} J(\sigma^r)\, A(\sigma_0^{q-i+1}, \sigma^{q-i}, f_+) - A(\sigma_0^{q-i+1}, \sigma^{q-i}, f_-) \right] \sigma^{q-i}$$

$$= \left(\sum_{r=0}^{q} (-1)^r \sum_{\sigma^r \subset \sigma_0^{q-i+1}} J(\sigma^r) \right) d\,\sigma_0^{q-i+1}$$

so the difference between the two cycles is a boundary and the homology class $W_{q-i}(M^q)$ is well defined, independently of the map f used to determine a representative.

We may note that the mapping $f: M \to E^{q-i+1}$ induces a mapping $f' : M' \to E^{q-i+1}$ on any subdivision M' of M and that the cycles $Z_{q-i}(f)$ and $Z_{q-i}(f')$ determine corresponding homology classes in the naturally isomorphic homology groups $H_{q-i}(M; Z(\text{or } Z_2))$ and $H_{q-i}(M'; Z(\text{or } Z_2))$.

In order to identify the homology class determined by mappings from a q-manifold to E^{q-i+1}, we consider a specific simplexwise linear map defined on the first barycentric subdivision $\hat{M}^q$.

The vertices of $\hat{M}$ are barycenters $\hat{\sigma}^r$ of faces of M, and we define a mapping $g : \hat{M} \to E^{q-i+1}$ by sending $g(\hat{\sigma}^r) = (r, r^2, r^3, \cdots, r^{q-i+1})$ and extending linearly over all simplexes of $\hat{M}$.

The curve $X(t) = (t, t^2, t^3, \cdots, t^{q-i+1})$ has the property that any $q - i + 1$ points $X(t_0), X(t_1), \cdots, X(t_{q-i+1}), X(t)$ with $t_0 < t_1 < \cdots < t_{q-i+1}$ are linearly independent if and only if $t \neq t_0, t_1, \cdots, t_{q-i+1}$, and moreover, the $(q - i + 1)$-simplex determined by the points will have positive orientation if and only if $t < t_0$, or $t_{2j-1} < t < t_{2j}$ for some index j, or $t_{q-i} < t$ if $q-i$ is odd.

To compute the index $a(\sigma^{i_0} \subset \sigma^{i_1} \subset \cdots \subset \sigma^{i_{q-i}}, g)$ of a $(q - 1)$-cell in $\hat{M}$ determined by a sequence of cells $\sigma^{i_0} \subset \sigma^{i_1} \subset \cdots \subset \sigma^{i_{q-i}}$ of increasing dimension in M^q, we must first identify the link of this cell in $\hat{M}$ and then see how the cells in the star are mapped relative to the hyperplane spanned by the points $X(i_j) = g(\hat{\sigma}^{i_j}), j = 0, \cdots, q - i$. The $q - i + 1$ cells in the star of this cell in $\hat{M}$ correspond to vertices $\hat{\sigma}^r$ in the link of the cell in $\hat{M}$ where σ^r is a cell that fits into the sequence, i.e. $\sigma^r \subset \sigma^{i_0}$ or $\sigma^{i_j} \subset \sigma^r \subset \sigma^{i_{j+1}}$ or $\sigma^{i_{q-i}} \subset \sigma^r$. The vertices $\hat{\sigma}^r$ with $\sigma^{i_j} \subset \sigma^r \subset \sigma^{i_{j+1}}$ determine the link of σ^{i_j} in $d\sigma^{i_{j+1}}$, and the link of a cell of dimension i_j in a sphere of dimension $i_{j+1} - 1$ is a sphere of dimension $i_{j+1} - i_j - 2$. The vertices $\hat{\sigma}^r$ with $\sigma^r \subset \sigma^{i_0}$ give a sphere of dimension $i_0 - 1$ and the vertices $\hat{\sigma}^r$ with $\sigma^{i_{q-i}} \subset \sigma^r$ determine the link of $\sigma^{i_{q-i}}$ in M^q, and this is a sphere of dimension $q - i_{q-i} - 1$. (Note: $S^{-1} = \emptyset$ and $A * \emptyset = A$.)

The link of the $(q - i)$-simplex is then given as the join of a set of spheres:

$$S^{i_0-1} * S^{i_1-i_0-2} * \cdots * S^{i_{q-i}-i_{q-i-1}-2} * S^{q-i_{q-i}-1}.$$

Since the join of a pair of spheres $S^a * S^b$ is a sphere of dimension $a + b + 1$, it follows that the link of the $(q - i)$-simplex has dimension $i_0 - 1 + i_1 - i_0 - 1 + \cdots + i_{q-i} - i_{q-i-1} - 1 + q - i_{q-i} = q - (q - i + 1) = i - 1$, as expected.

To compute the index $a(\sigma^{i_0} \subset \cdots \subset \sigma^{i_{q-i}} g)$ we must subtract from 1 the Euler characteristic of the collection of all simplexes in the link all vertices of which lie above the oriented hyperplane determined by $g(\sigma^{i_0} \subset \cdots \subset \sigma^{i_{q-i}})$. But this set of simplexes is precisely the join of spheres

$$S^{i_0-1} * S^{i_2-i_1-2} * \cdots * S^{i_{2m}-i_{2m-1}-2} \quad \text{if } q - i = 2m,$$

and the join

$$S^{i_0-1} * S^{i_2-i_1-2} * \cdots * \tilde{u}\, S^{i_{2m}-i_{2m-1}-2} * S^{q-i_{2m+1}-1} \quad \text{if } q - i = 2m + 1.$$

In the case $q - i = 2m$, we have a sphere of dimension $i_0 - 1 + i_2 - i_1 - 1 + \cdots + i_{2m} - i_{2m-1} - 1 = i_0 - i_1 + i_2 - \cdots - i_{2m-1} + i_{2m} - (m + 1)$, and in the case $q - i = 2m + 1$, we have a sphere of dimension $i_0 - i_1 + i_2 - \cdots - i_{2m-1} + i_{2m} - i_{2m+1} + q - m - 1$.

Since 1 minus the Euler characteristic of a sphere of dimension r is $1 - (1 + (-1)^r) = (-1)^{r+1}$, it follows that the index $a(\sigma^{i_0} \subset \cdots \subset \sigma^{i_{q-i}}, g)$ is given by

$$(-1)^{i_0-i_1+\cdots-i_{2m-1}+i_{2m}-m} \quad \text{if } q - i = 2m$$

and by

$$(-1)^{i_0 - i_1 + \cdots - i_{2m-1} + i_{2m} - i_{2m+1} + q - m} \quad \text{if } q - i = 2m + 1.$$

This is precisely the chain, up to a factor of -1 depending only on q and i, which describes the homology class dual to the ith Stiefel-Whitney cohomology class of a smoothly triangulated differentiable manifold, as described in [6] and this completes the main result of this paper.

APPENDIX. In this appendix we sketch the proofs of two results contained in the author's paper [2] which are needed in § 3.

Let $f : M^q \to E^{q-i+1}$ be a simplexwise linear mapping on a q-dimensional simplicial manifold M^q such that f is one-to-one on each simplex of dimension $\leqq q - i + 1$.

PROPOSITION 1. *$a(\sigma^{q-i}, -f) = (-1)^i a(\sigma^{q-i}, f)$ for any simplex σ^{q-i}.*

PROPOSITION 2. *For any simplex σ^{q-i-1}, let $U(\sigma^r, \sigma^{q-i-1}, f) = 1$ if and only if $\sigma^{q-i-1} \nsubseteq \sigma^r$ and $f(\sigma^{q-i-1})$ is contained in $df(\sigma^r)$, the point set boundary of $f(\sigma^r)$. Then*

$$\tilde{u}(\sigma^{q-i-1}, f) = \sum_{r=0}^{q} (-1)^r \sum_{\sigma^r} U(\sigma^r, \sigma^{q-i-1}, f)$$

equals 0 *if i is odd and is even if i is even.*

Both propositions rely on an observation that goes back at least to Seifert, namely that for an even-dimensional manifold N which is the boundary of an open manifold with boundary N^+, we have $\chi(N) + 2\chi(N^+) = 0$. This follows since two copies of N^+ together with N give an odd-dimensional manifold, with 0 Euler characteristic. Thus if N is even dimensional and $N = dN^+ = dN^-$, then $\chi(N^+) = \chi(N^-)$.

PROOF OF PROPOSITION 1. Each oriented σ^{q-i} determines an oriented hyperplane H and two open half-spaces H^+ and H^-. Let $L = \{p \in LK(\sigma^{q-i}) \mid f(p) \in H\}$, an $(i-2)$-dimensional cell submanifold of the $(i-1)$-sphere $LK(\sigma^{q-i})$ formed by the preimages of intersections of simplexes with H. Then $L^+ = \{p \in LK(\sigma^{q-i}) \mid f(p) \in H^+\}$ consists of all σ^t in $LK(\sigma^{q-i})$ with $A(\sigma^t * \sigma^{q-i}, \sigma^{q-i}, f) = 1$ as well as one cell of dimension $t+1$ for very t cell in L. Thus

$$\chi(L^+) = \sum_{t=0}^{i-1} \sum_{\sigma^t} A(\sigma^t * \sigma^{q-i}, \sigma^{q-i}, f) - \chi(L).$$

If follows that

$$\begin{aligned} a(\sigma^{q-i}, f) &= (-1)^{q-i} + \sum_{r=q-i+1}^{q} (-1)^r \sum_{\sigma^r} A(\sigma^r, \sigma^{q-i}, f) \\ &= (-1)^{q-i} + (-1)^{q-i-1} \sum_{t=0}^{i-1} (-1)^t \sum_{\sigma^t} A(\sigma^t * \sigma^{q-i}, \sigma^{q-i}, f) \\ &= (-1)^{q-i} + (-1)^{q-i-1}(\chi(L^+) - \chi(L)). \end{aligned}$$

Similarly

$$a(\sigma^{q-i}, -f) = (-1)^{q-i} + (-1)^{q-i-1}(\chi(L^-) - \chi(L)).$$

Now if i is even, then L is an even-dimensional manifold so $\chi(L^+) = \chi(L^-)$ and $a(\sigma^{q-i}, -f) = a(\sigma^{q-i}, f)$. If i is odd, then $\chi(L)$ is 0 and

$$\begin{aligned} a(\sigma^{q-i}, f) + a(\sigma^{q-i}, -f) &= 2(-1)^{q-i} + (-1)^{q-i-1}(\chi(L^+) + \chi(L^-) + \chi(L)) \\ &= 2(-1)^{q-i} + (-1)^{q-i-1}\chi(LK(\sigma^{q-i})) = 0. \end{aligned}$$

Therefore in either case $a(\sigma^{q-i}, -f) = (-1)^i a(\sigma^{q-i}, f)$.

PROOF OF PROPOSITION 2. For any σ^{q-i-1}, let C be the affine $(q-i-1)$-space spanned by $f(\sigma^{q-i-1})$ and let $R = \{p \in LK(\sigma^{q-i-1}) \mid f(p) \in C\}$. Then R is an $(i-2)$-dimensional submanifold of the i-dimensional sphere $LK(\sigma^{q-i-1})$ and $LK(\sigma^{q-i-1})$ contains a $(t+2)$-simplex for each t cell in R. Therefore

$$\begin{aligned} \chi(LK(\sigma^{q-i-1})) &= \chi(R) + \sum_{t=0}^{i} (-1)^t \sum_{\sigma^t} (\sigma^t * \sigma^{q-i-1}, \sigma^{q-i-1}, f) \\ &= \chi(R) + \tilde{u}(\sigma^{q-i-1}, f). \end{aligned}$$

If i is odd, then $\chi(LK(\sigma^{q-i-1})) = 0 = \chi(R)$, so $\tilde{u}(\sigma^{q-i-1}, f) = 0$.

If i is even, consider a $(q-i)$-dimensional affine half-space C^+ with C as boundary which is in general position with respect to the mapping f. Then $R^+ = \{p \in LK(\sigma^{q-i-1}) \mid f(p) \in C^+\}$ is an $(i-1)$-dimensional submanifold-with-boundary in $LK(\sigma^{q-i-1})$ and $dR^+ = R$. Then $\chi(R)$ is even and $\tilde{u}(\sigma^{q-i-1}, f) = \chi(LK(\sigma^{q-i-1})) \cdot \chi(R) = \chi(S^i) - \chi(R)$ is even.

The author wishes to acknowledge very helpful comments by Clint McCrory concerning the proof of Proposition 2 of this Appendix.

BIBLIOGRAPHY

1. T. Banchoff, *Critical points and curvature for embedded polyhedra*, J. Differential Geometry **1** (1967), 245–256.

2. ———, *Critical points and curvature for embedded polyhedra*. II (to appear).

3. ———, *Critical points and curvature for embedded polyhedral surfaces*, Amer. Math. Monthly **77** (1970), 475–485. MR **41** #4444.

4. ———, *Parity of cusps and local degrees for polyhedral manifolds* (to appear).

5. J. Cheeger, *A combinatorial formula for Stiefel-Witney classes*, Topology of Manifolds, Markham, New York, 1970.

6. S. Halperin and D. Toledo, *Stiefel-Whitney homology classes*, Ann. of Math. **9** (1972), 511–525.

7. C. McCrory, *Euler singularities and homology operations*, these PROCEEDINGS.

8. L. S. Pontrjagin, *Vector fields on manifolds*, Mat. Sb. **24 (66)** (1949), 129–162; English transl., Amer. Math. Soc. Transl. (1) **7** (1962), 220–278. MR **10**, 727.

9. R. Porter, *Characteristic classes and singularities of mappings*, these PROCEEDINGS.

10. E. Stiefel, *Richtungsfelder und Parallelismus in Mannigfaltigkeiten*, Comment. Math. Helv. **8** (1936).

11. R. Thom, *Les singularitiés des applications différentiable*, Ann. Inst. Fourier Grenoble **6** (1955/56), 43–87.

12. H. Whitney, *On the theory of sphere-bundles*, Proc. Nat. Acad. Sci. U.S.A. **26** (1940), 148–153. MR **1**, 220.

UNIVERSITY OF CALIFORNIA, LOS ANGELES

Proceedings of Symposia in Pure Mathematics
Volume 27, 1975

AXIOMS FOR CHARACTERISTIC CLASSES OF MANIFOLDS

JOHN D. BLANTON AND PAUL A. SCHWEITZER, S.J.*

In this paper we present axioms for the Stiefel-Whitney and Pontryagin classes of manifolds, the Chern classes of almost complex manifolds, and their homology duals. The axioms are closely related to the ordinary axiomatization of the corresponding characteristic classes of vector bundles ([**6**, pp. 58–59 and 73], [**8**]). It is hoped that these axioms will offer a direct method of proving that various geometric definitions of characteristic classes coincide, and perhaps assist in the discovery of other geometric interpretations.

As an application of the uniqueness Theorem I′, we shall show in a future note [**1**] that the homology classes defined by the combinatorial formula of Stiefel [**9**] satisfy the axioms (1′), (2′), and (4′). Since Halperin and Toledo [**3**] and Milnor [**7**] have shown that they satisfy (3′), Theorem I′ yields a new axiomatic proof of Stiefel's conjecture that the formula defines the Stiefel-Whitney homology classes. Halperin and Toledo recently published the first detailed proof of Stiefel's conjecture [**4**]. Earlier proofs were announced by Whitney [**10**] and outlined by Cheeger [**2**].

The uniqueness theorems for the various axiom systems are proved for smooth manifolds. It would be interesting to extend this result to other categories. For example, do our axioms or some modified axioms for the Stiefel-Whitney classes characterize them uniquely on PL manifolds, on topological manifolds, or on Euler spaces?

AMS (*MOS*) *subject classifications* (1970). Primary 55F40, 57D20; Secondary 02K15, 18F15.
*The authors would like to thank the Department of Mathematics of McGill University for its hospitality to the Clavius Group while this work was done.

1. Axioms for Stiefel classes and their duals. Consider the category $\mathscr{M}$ whose objects are C^∞ separable Hausdorff manifolds and whose morphisms are open embeddings, that is, $f:M \to N$ is a morphism of $\mathscr{M}$ if M and N are objects of $\mathscr{M}$ and f is a diffeomorphism of M onto an open subset of N. Cartesian product defines a product operation on $\mathscr{M}$.

The total Stiefel-Whitney class (also called the *Stiefel class*), $W(M) = W_0(M) + W_1(M) + \cdots + W_m(M) \in H^*(M; Z/2)$, where m is the dimension of M, satisfies the following axioms.

(1) For every $M \in \mathrm{Obj}(\mathscr{M})$ and every integer $i \geqq 0$ there is a Stiefel class $W_i(M) \in H^i(M; Z/2)$.

(2) If $f:M \to N$ is a morphism of $\mathscr{M}$, then $f^*W(N) = W(M)$. (Naturality)

(3) $W(M \times N) = W(M) \times W(N)$. (Whitney product formula)

(4) For every nonnegative integer i there exists a positive even integer $m \geqq i$ such that $W_i(P_m(R)) = \binom{m+1}{i} x^i$. (Normalization)

As usual, $Z/2$ is the ring of integers modulo two, $P_m(R)$ is the real projective space of dimension m, $x \in H^1(P_m(R); Z/2)$ generates its modulo two cohomology ring, and the coefficient is the usual binomial coefficient.

THEOREM I. *There exists a unique cohomology class* $W(M)$ *for each* $M \in \mathrm{Obj}(\mathscr{M})$ *such that the axioms* (1) *through* (4) *are satisfied.*

REMARK. In the axiomatization for vector bundles, the normalization axiom only normalizes W_1 of the canonical line bundle γ^1 over some real projective space, but the other W_i are implicitly normalized by requiring that $W_0(\gamma^1) = 1$ and $W_i(\gamma^1) = 0$ for $i > 1$. In the axiomatization for manifolds, it seems necessary to make the normalization (4) of each W_i explicit. For example, the class $W^*(M) = W(M)^3$ satisfies (1) through (3), and when $i = 0$ or $i = 1$ the normalization $W_i^*(P_m(R)) = \binom{m+1}{i} x^i$ holds, but $W_2^* \neq W_2$. Axiom (4) will be satisfied if, for arbitrarily large integers n, $W(P_{2n}(R)) = (1 + x)^{2n+1}$, or in particular $W(P_{2n}(R)) = 1 + x + x^{2n}$ for arbitrarily large n. Similar modifications can be made in the fourth axiom of each axiom system below.

Theorem I has an equivalent analogue in homology. Let $\bar{H}_*$ be the homology functor defined using infinite (but locally finite) chains, either singular or simplicial. (See [5, pp. 90–91]. On finite complexes, $\bar{H}_*$ coincides with ordinary homology.) Every manifold $M \in \mathrm{Obj}(\mathscr{M})$ has a fundamental class $[M] \in \bar{H}_m(M; Z/2)$ and the cap product defines an isomorphism between the simplicial cochain complex and the dual (infinite) chain complex, thus inducing the Poincaré duality isomorphism

$$[M] \cap \cdot : H^i(M; Z/2) \to \bar{H}_{m-i}(M; Z/2)$$

where m is the dimension of M, just as in the case of a closed manifold.

Observe that an open embedding $f: M \to N$ induces a restriction homomorphism $f^*: \bar{H}_*(N; Z/2) \to \bar{H}_*(M; Z/2)$, so that $\bar{H}_*(\cdot\,; Z/2)$ is a *contravariant* functor on the category $\mathscr{M}$. The homomorphism f^* can be defined by the fact that it is dual to the cohomology homomorphism f^* in the sense that the diagram

$$\begin{array}{ccc} H^i(N; Z/2) & \xrightarrow{f^*} & H^i(M; Z/2) \\ {\scriptstyle [N]\cap\cdot}\downarrow & & \downarrow{\scriptstyle [M]\cap\cdot} \\ \bar{H}_{m-i}(N; Z/2) & \xrightarrow{f^*} & \bar{H}_{m-i}(M; Z/2) \end{array}$$

commutes (so that f^* is the Umkehrungs homomorphism), or equivalently by a geometric process involving iterated subdivision of simplexes of N which overlap the boundary of $f(M)$ so that the restriction of a chain on N to M is defined.

Let $W'_i(M) = [M] \cap W_i(M) \in \bar{H}_{m-i}(M; Z/2)$, so that

$$W'(M) = W'_0(M) + W'_1(M) + \cdots + W'_m(M)$$

is the homology class Poincaré dual to $W(M)$.The axioms (1) through (4) for $W(M)$ are equivalent to the following properties.

(1′) For every $M \in \mathrm{Obj}(\mathscr{M})$ and every integer i, $0 \leqq i \leqq m$, there is a Stiefel homology class $W'_i(M) \in \bar{H}_{m-i}(M; Z/2)$.

(2′) If $f:M \to N$ is a morphism of $\mathscr{M}$, then $f^*W'(N) = W'(M)$.

(3′) $W'(M \times N) = W'(M) \times W'(N)$.

(4′) For every nonnegative integer i there exists a positive even integer $m \geqq i$ such that $W'_i(P_m(R)) = \binom{m+1}{i}x'^i$. Here $x' = [P_m(R)] \cap x$ and the multiplication is the intersection product, which is Poincaré dual to the cup product on cohomology, so that x'^i is the unique nonzero element in $\bar{H}_{m-i}(P_m(R); Z/2)$, $0 \leqq i \leqq m$.

It follows that Theorem I is equivalent to the following theorem.

THEOREM I′. *There exists a unique homology class $W'(M)$ for each $M \in$ Obj($\mathscr{M}$) such that the axioms* (1′) *through* (4′) *are satisfied.*

2. The proof of Theorem I. For the existence of $W(M)$, it suffices to observe that the total Stiefel-Whitney class $W(M) = W(\tau M)$ of the tangent bundle τM of M satisfies the axioms. An alternate existence proof is given by the fact that the homology Stiefel classes defined by Stiefel's combinatorial formula satisfy axioms (1′) through (4′), as mentioned in the introduction.

The proof of uniqueness depends on two lemmas, which allow us to mimic the uniqueness proof for Stiefel-Whitney classes of vector bundles. These lemmas assert, approximately, that up to homotopy every vector bundle is stably equivalent to the tangent bundle of a manifold and every stably tangential map (see below) is stably equivalent to an open embedding.

LEMMA 1. *Let X be a space of the homotopy type of a finite CW complex and let ξ^n be an n-plane bundle over X. Then there exist a manifold $M \in$ Obj($\mathscr{M}$) of dimension $m \geqq n$ and a homotopy equivalence $f : X \to M$ such that there is a vector bundle isomorphism $f^!(\tau M) \cong \xi^n \oplus (m - n)$, where $(m - n)$ is the trivial $(m - n)$-plane bundle over X.*

PROOF. Without loss of generality we may assume that X is an open set in R^k, by embedding it and replacing it by an open regular neighborhood if necessary.

Then $M = E(\xi^n)$ (the total space of the bundle ξ^n) and the zero section $f : X \to E(\xi^n)$ have the desired property.

LEMMA 2. *Suppose that $M, N \in \mathrm{Obj}(\mathcal{M})$ and $f : M \to N$ is a stably tangential map, that is, there exist nonnegative integers r and s such that $f^!(\tau N) \oplus (s) \cong \tau M \oplus (r)$. Then there exist a nonnegative integer k and an open embedding $g: M \times R^{r+k} \to N \times R^{s+k}$ such that the following diagram homotopy commutes, where P_1 is the projection on the first factor.*

$$\begin{array}{ccc} M \times R^{r+k} & \xrightarrow{\ g\ } & N \times R^{s+k} \\ {\scriptstyle P_1}\downarrow & & \downarrow{\scriptstyle P_1} \\ M & \xrightarrow{\ f\ } & N \end{array}$$

PROOF. Choose $k > \dim M - s$, a smooth embedding $h : M \to R^{s+k}$, and a smooth map $f_1 \simeq f : M \to N$. Then $g_1 = (f_1, h) : M = M \times 0 \to N \times R^{s+k}$ is a smooth embedding, and its normal bundle, which is in the stable range, is trivial. Thus g_1 extends to a smooth open embedding $g : M \times R^{r+k} \to N \times R^{s+k}$ such that the above diagram is homotopy commutative.

Before beginning the uniqueness proof we shall note some consequences of the axioms (1) through (4).

First observe that $W_0(M) = 1$ for every $M \in \mathrm{Obj}(\mathcal{M})$. By (4), $W_0(P_m(R)) = 1$ for some positive m. Since there exists an open embedding of R^m in $P_m(R)$, naturality (2) implies that $W_0(R^m) = 1$. Then by the product axiom (3) we conclude that this holds for $m = 1$ and consequently for all $m \geqq 0$. Since every component of $M \in \mathrm{Obj}(\mathcal{M})$ admits an open embedding of R^m, $W_0(M) = 1$ follows by (2).

Since $H^i(R^k; Z/2) = 0$ for $i > 0$, $W(R^k) = 1$. Consequently $W(M \times R^k) = W(M) \times 1 = p_1^* W(M)$ where $p_1 : M \times R^k \to M$ is the projection on the first factor.

LEMMA 3. *If $f : M \to N$ is a stably tangential map, then $f^* W(N) = W(M)$.*

PROOF. By Lemma 2, axiom (2), and the previous remark, $p_1^* f^* W(N) = p_1^* W(M)$. Since p_1 is a homotopy equivalence, p_1^* is an isomorphism and so the assertion follows.

Next we construct manifolds $N_n \in \mathrm{Obj}(\mathcal{M})$ and homotopy equivalences f_n: $P_n(R) \to N_n$ such that $f_n^!(\tau N_n) = \gamma_n^1 \oplus (k)$ (for some integer k depending on n), by applying Lemma 1 to the canonical line bundle γ_n^1 over $P_n(R)$ for every $n > 0$. We shall show that

$$W(N_n) = 1 + y_n \tag{2.1}$$

for all n, where $y_n = (f_n^*)^{-1}(x) \in H^1(N_n; Z/2)$.

From the vector bundle isomorphism $P_n(R) \oplus (1) \cong \gamma_n^1 \oplus \cdots \oplus \gamma_n^1 (n + 1$ terms) we obtain a bundle map

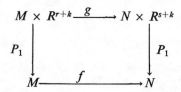

$$\begin{array}{ccc} E(\tau P_n(R) \oplus (1)) & \longrightarrow & E(\gamma_n^1 \times \cdots \times \gamma_n^1) \\ \downarrow & & \downarrow \\ P_n(R) & \xrightarrow{\quad d \quad} & (P_n(R))^{n+1} \end{array}$$

where d is the diagonal map. It follows that

$$d' \circ f_n = (f_n \times \cdots \times f_n) \circ d \colon P_n(R) \to (N_n)^{n+1},$$

where d' is the $(n+1)$-fold diagonal map of N_n, is a stably tangential map. Since $d'^*(W(N_n) \times \cdots \times W(N_n)) = W(N_n)^{n+1}$, Lemma 3 yields

$$f_n^* W(N_n)^{n+1} = W(P_n(R)). \tag{2.2}$$

From the inclusion map $i_{mn} \colon P_m(R) \to P_n(R)$ $(m < n)$ which is covered by a bundle map $\gamma_m^1 \to \gamma_n^1$, we obtain a stably tangential map $j_{mn} : N_m \to N_n$ such that the diagram

$$\begin{array}{ccc} P_m(R) & \xrightarrow{\quad i_{mn} \quad} & P_n(R) \\ {\scriptstyle f_m}\downarrow & & \downarrow{\scriptstyle f_n} \\ N_m & \xrightarrow{\quad j_{mn} \quad} & N_n \end{array}$$

homotopy commutes. Then $j_{mn}^*(y_n) = y_m$ and, by Lemma 3, $j_{mn}^* W(N_n) = W(N_m)$.

To prove (2.1), consider the assertion

A_i: For all $n > 0$, $W(N_n) = 1 + y_n +$ terms of degree $\geqq i$.

Since $W_0(N_n) = 1$, a_1 holds. Suppose a_i and let

$$W(N_n) = 1 + y_n + z_n + \text{terms of degree} > i$$

where $z_n \in H^i(N_n; Z/2)$. A short computation using (2.2) shows that

$$W(P_n(R)) = (1 + x)^{n+1} + (n+1) f_n^* z_n + \text{terms of degree} > i$$

so that $W_i(P_n(R)) = \binom{n+1}{i} x^i + (n+1) f_n^* z_n$. On the other hand, by axiom (4) there exists an even integer $m \geqq i$ such that $W_i(P_m(R)) = \binom{m+1}{i} x^i$, so that $f_m^* z_m$ and consequently z_m vanish. It is easy to check that $j_{mn}^*(z_n) = z_m$ and then conclude that $z_n = 0$ for all $n > 0$. By establishing A_{i+1}, this completes the inductive proof of (2.1).

To prove Theorem I, suppose that both $W(M)$ and $W^*(M)$ satisfy the axioms (1) through (4), and let $M \in \mathrm{Obj}(\mathcal{M})$ be any smooth manifold of dimension m. We must show that $W(M) = W^*(M)$.

Let γ^m be the restriction of the universal m-plane bun lle to the k-skeleton B of $BO(m)$, where k is some large integer. The m-plane bundles $\gamma_m^1 \times \cdots \times \gamma_m^1$ (m factors) and τM are classified by bundle maps, as in the following diagram.

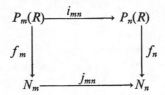

$$\begin{array}{ccccc} E(\gamma_m^1)^m & \longrightarrow & E(\gamma^m) & \longleftarrow & E(\tau M) \\ \downarrow & & \downarrow & & \downarrow \\ P_m(R)^m & \xrightarrow{\ f\ } & B & \xleftarrow{\ g\ } & M \end{array}$$

Lemma 1 yields a smooth manifold N whose tangent bundle is stably equivalent to γ_m, and then we get stably tangential maps homotopy equivalent to f and g:

$$(N_m)^m \xrightarrow{\ f'\ } N \xleftarrow{\ g'\ } M.$$

By (2.1), axiom (3), and Lemma 3, $f'^* W(N) = f'^* W^*(N)$. Since f' is homotopy equivalent to the classifying map of the bundle $\gamma_m^1 \times \cdots \times \gamma_m^1$ up to some large dimension, the usual computation shows that f'^* is injective on modulo two cohomology in dimensions up to m, so $W_i(N) = W_i^*(N)$ for $i \leqq m$. Then naturality with respect to g'^* shows that $W(M) = W^*(M)$.

3. Axioms for Pontryagin classes of manifolds. Corresponding to the axioms (1) through (4) given above for the Stiefel classes of manifolds, there are analogous axioms for the Pontryagin classes of manifolds in the category $\mathscr{M}$ and for the Chern classes on the corresponding category $\mathscr{M}_C$ of almost complex manifolds. These axioms and their duals are given without proof in this section and the next. The proofs are entirely analogous to those for Stiefel classes, but $P_n(R)$ is replaced by the n-dimensional complex projective space $P_n(C)$ and, in the proof for Chern classes, $BO(m)$ is replaced by $BU(m)$.

Let $\mathscr{M}$ be the same category as before and consider cohomology with coefficients in $Z[\frac{1}{2}]$, the integers localized away from 2.

THEOREM II. *There exists a unique total Pontryagin class*

$$p(M) = p_0(M) + p_1(M) + \cdots + p_i(M) + \cdots \in H^*(M; Z[\tfrac{1}{2}])$$

for each $M \in \mathrm{Obj}(\mathscr{M})$ satisfying the following axioms:

(1_p) *For every $M \in \mathrm{Obj}(\mathscr{M})$ and every integer $i \geqq 0$ there is a Pontryagin class $p_i(M) \in H^{4i}(M; Z[\frac{1}{2}])$.*

(2_p) *If $f: M \to N$ is a morphism of $\mathscr{M}$, then $f^* p(N) = p(M)$.*

(3_p) *$p(M \times N) = p(M) \times p(N)$.*

(4_p) *For every nonnegative integer i there exists a positive integer $m \geqq 2i$ such that $p_i(P_m(C)) = \binom{m+1}{i} u^{2i}$, where $u \in H^2(P_m(C); Z[\frac{1}{2}])$ is the positive generator of $H^*(P_m(C); Z[\frac{1}{2}])$.*

Poincaré duality yields the dual Pontryagin classes $p_i'(M) = [M] \cap p_i(M) \in \bar{H}_{m-4i}(M; Z[\frac{1}{2}] \otimes \mathscr{O}_M)$, where m is the dimension of M, $\bar{H}$ denotes infinite (locally finite) homology, and the local coefficients are twisted by the orientation sheaf $\mathscr{O}_M$ of M.

THEOREM II$'$. *There exists a unique homology class*

$$p'(M) = p_0'(M) + p_1'(M) + \cdots + p_i'(M) + \cdots \in \bar{H}_*(M; Z[\tfrac{1}{2}] \otimes \mathscr{O}_M)$$

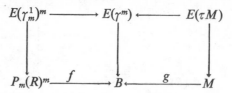

for each $M \in \text{Obj}(\mathscr{M})$ *satisfying the following axioms*:

($1_{p'}$) *For every* $M \in \text{Obj}(\mathscr{M})$ *and every integer* i, $0 \leqq i \leqq m/4$, *there is a Pontryagin homology class* $p'_i(M) \in \bar{H}_{m-4i}(M; Z[\frac{1}{2}] \otimes \mathcal{O}_M)$.

($2_{p'}$) *If* $f: M \to N$ *is a morphism of* $\mathscr{M}$, *then* $f^*p'(N) = p'(M)$.

($3_{p'}$) $p'(M \times N) = p'(M) \times p'(N)$.

($4_{p'}$) *For every nonnegative integer* i *there exists a positive integer* $m \geqq 2i$ *such that* $p'_i(P_m(C)) = \binom{m+1}{i} v^{2i}$, *where* $v = [P_{m-1}(C)] = [P_m(C)] \cap u$ *generates* $\bar{H}_{2m-2}(P_m(C); Z[\frac{1}{2}])$ *and the multiplication is the intersection product.*

Note. On the related category $\mathscr{M}_+$ of oriented manifolds and orientation preserving open embeddings, $p(M)$ and $p'(M)$ are still uniquely determined by the axioms. The proof is analogous but contains an additional step using (3_p) with N a point with negative orientation to show that $p(-M) = p(M)$. In this case, the coefficients in homology are not twisted.

4. Axioms for Chern classes of almost complex manifolds. Let $\mathscr{M}_C$ be the category whose objects are smooth separable Hausdorff manifolds which are almost complex that is, of even dimension $2n$ with a given $GL(n, C)$ reduction of the tangent bundle, and whose morphisms $f: M \to N$ are smooth open embeddings such that $f^!\tau N \cong \tau M$ as $GL(n, C)$ bundles.

THEOREM III. *There exists a unique total Chern class*

$$c(M) = c_0(M) + c_1(M) + \cdots + c_i(M) + \cdots \in H^{2i}(M; Z)$$

for each $M \in \text{Obj}(\mathscr{M}_C)$ *satisfying the following axioms*:

(1_c) *For every* $M \in \text{Obj}(\mathscr{M}_C)$ *and every integer* $i \geqq 0$ *there is a Chern class* $c_i(M) \in H^{2i}(M; Z)$.

(2_c) *If* $f: M \to N$ *is a morphism of* $\mathscr{M}_C$, *then* $f^*c(N) = c(M)$.

(3_c) $c(M \times N) = c(M) \times c(N)$.

(4_c) *For every nonnegative integer* i *there exists a positive integer* $m \geqq i$ *such that* $c_i(P_m(C)) = \binom{m+1}{i} u^i$ *where* u *is the positive generator of* $H^2(P_m(C); Z)$.

As in the cases of Stiefel and Pontryagin classes, the unique total Chern class of M coincides with the total Chern class of the tangent bundle of M.

THEOREM III$'$. *There exists a unique homology class*

$$c'(M) = c'_0(M) + c'_1(M) + \cdots + c'_i(M) + \cdots \in \bar{H}_*(M; Z)$$

for each $M \in \text{Obj}(\mathscr{M}_C)$ *satisfying the following axioms*:

($1_{c'}$) *For every* $M \in \text{Obj}(\mathscr{M}_C)$ *and every integer* i, $0 \leqq i \leqq m/2$, *there is a Chern homology class* $c'_i(M) \in H_{m-2i}(M; Z)$.

($2_{c'}$) *If* $f: M \to N$ *is a morphism of* $\mathscr{M}_C$, *then* $f^*c'(N) = c'(M)$.

($3_{c'}$) $c'(M \times N) = c'(M) \times c'(N)$.

($4_{c'}$) *For every nonnegative integer* i *there exists a positive integer* $m \geqq i$ *such that* $c'_i(P_m(C)) = \binom{m+1}{i} v^i$, *where* $v = [P_{m-1}(C)] = [P_m(C)] \cap u$ *generates* $\bar{H}_{2m-2}(P_m(C); Z)$.

REFERENCES

1. J. Blanton and P. Schweitzer, (in preparation).

2. J. Cheeger, *A combinatorial formula for Stiefel-Whitney classes*, Topology of Manifolds, Markham, Chicago, Ill., 1970, pp. 470–471.

3. S. Halperin and D. Toledo, *The product formula for Stiefel Whitney homology classes* (to appear).

4. ———, *Stiefel-Whitney homology classes*, Ann. of Math. (2) **96** (1972), 511–525.

5. P. J. Hilton and S. Wylie, *Homology theory: An introduction to algebraic topology*, Cambridge Univ. Press, New York, 1960. MR **22** #5963.

6. F. Hirzebruch, *Topological methods in algebraic geometry*, 3rd ed., Die Grundlehren der math. Wissenschaften, Band 131, Springer-Verlag, New York, 1966. MR **34** #2573.

7. J. Milnor, private communication.

8. J. Milnor and J. Stasheff, *Characteristic classes*, Ann. of Math. Study, no. 76, Princeton Univ. Press, Princeton, N.J., 1974.

9. E. Stiefel, *Richtungsfelder und Fernparallelismus in n-dimensionalen Mannigfaltigkeiten*, Comment. Math. Helv. **8** (1936), 305–353.

10. H. Whitney, *On the theory of sphere-bundles*, Proc. Nat. Acad. Sci. U.S.A. **26** (1940), 148–153. MR **1**, 220.

ST. JOHN FISHER COLLEGE

PONTIFICIA UNIVERSIDADE CATOLICA

Proceedings of Symposia in Pure Mathematics
Volume 27, 1975

ON THE GELFAND-FUKS COHOMOLOGY*

RAOUL BOTT**

Introduction. Roughly four years ago Gelfand and Fuks had the inspired idea of considering the *continuous* cohomology of the Lie-algebra of C^∞-vector fields on a manifold ([3], [4], [5]). They showed that the answer was finite dimensional for all compact manifolds and that for S^1 this cohomology is a tensor product of an exterior algebra in dimension 3 and a polynomial ring with generator in dimension 2:

$$H_c(\mathscr{L}S^1) = E(\omega) \otimes \boldsymbol{R}[y].$$

In this report I would like to present a new point of view towards their work which arose, on the one hand, out of the new insights into this cohomology due to Guillemin and Trauber, and on the other out of conversations between G. Segal, A. Haefliger and myself during the past year. I will finally formulate a conjecture which would completely settle the homotopy theoretic nature of this cohomology. Although one finds evidence for its validity on every side, its proof has so far eluded us. I should also mention that when I recently showed this conjecture to Gelfand he told me that it had also, quite independently, been formulated by Fuks.

To set the stage, recall that if L is any Lie algebra, then its cohomology can be defined as the cohomology of the complex

$$AL = \sum A^q L$$

of alternating forms on L, under the differential operator:

AMS (MOS) subject classifications (1970). Primary 17B65, 18H25, 17B55; Secondary 53C10, 55B25, 57D35, 58H05.

*General lecture given at the Institute.

**This research was partially supported by NSF grant GP 31359X-1.

$$d\varphi(x_1, \cdots, x_{q+1}) = \Sigma(-1)^{i+j}\varphi([x_i\, x_j], x_1 \cdots \hat{x}_i \cdots \hat{x}_j \cdots x_q).$$

Now if L carries a topology, the continuous alternating forms clearly form a subcomplex A_cL of AL and one may then call its cohomology the continuous cohomology of L:

$$H_c^*(L) \equiv H^*(A_cL).$$

With this understood, the Gelfand-Fuks cohomology of M is simply the continuous cohomology of its Lie algebra of vector fields, $\mathscr{L}M$, considered with the C^∞-topology:

$$\text{Gelfand-Fuks}\,(M) = H_c^*(\mathscr{L}M).$$

In the next section I will outline a new proof that $H_c^k(\mathscr{L}M)$ is finite dimensional for every k, provided that M admits a covering $\mathscr{U} = \{U_\alpha\}_{\alpha \in J}$ by a *finite* collection of open sets such that all nonvacuous intersections of the U_α are diffeomorphic to $\boldsymbol{R}^n$.

1. The case of $\boldsymbol{R}^n$. Let $\mathscr{L} = \mathscr{L}\boldsymbol{R}^n$, $n > 0$, and $A = A_c\mathscr{L}$. Consider the subcomplex $A_{pt} = A_{pt}\mathscr{L}$ consisting of all forms in A whose support is at the origin of $\boldsymbol{R}^n$. The aim of this section is to sketch in a proof of the following.

THEOREM 1. *The inclusion* $\iota: A_{pt} \to A$ *induces an isomorphism in cohomology*

$$\iota_*: H^*(A_{pt}) \cong H^*(\mathscr{L}\boldsymbol{R}^n). \tag{1.1}$$

Now as we will see in a moment, the complex $A_{pt}\mathscr{L}$ is much more tractable and it is not hard to show that it is finite dimensional. Hence Theorem 1 implies finite-dimensionality of the Gelfand-Fuks theory on $\boldsymbol{R}^n$. Note also that the cohomology $H^*(A_{pt}\mathscr{L})$ figures in the Gelfand-Fuks approach, as the *continuous* cohomology of the Lie algebra $\mathfrak{a}_n$ of formal vector fields on $\boldsymbol{R}^n$:

$$H^*(A_{pt}) \cong H_c^*(\mathfrak{A}_n),$$

but we will proceed independently of this assumption here.

To prove (1. 1), consider the action of $\boldsymbol{R}^*$ on $\boldsymbol{R}^n$ given by

$$f_t \cdot x = tx, \qquad x \in \boldsymbol{R},\ t \in \boldsymbol{R}^*.$$

Now each f_t is a diffeomorphism, and therefore acts on the vector fields and therefore acts also on A. Note that if $X = \Sigma\, a_i\,(\partial/\partial x^i)$ is a vector field, then

$$(f_t^*X)_x = t\,\Sigma\, a_i(x/t)\,\frac{\partial}{\partial x_i}. \tag{1.2}$$

Hence f_t^* acting on A shrinks supports towards 0 as $t \to \infty$. In particular, if $\delta_p^k \in A^1$, $k = 1, \cdots, n$, is given by

$$\delta_p^k\Big(\Sigma\, a_i \frac{\partial}{\partial x_i}\Big) = a_k(p), \qquad p \in \boldsymbol{R}^n,$$

then (1.2) implies that

(1.3) $$f_t^* \delta_p^k = t\delta_{(p/t)}^k.$$

More generally, if $D^\alpha = \partial^{\alpha_1}/\partial x_1^{\alpha_1} \cdots \partial^{\alpha_n}/\partial x_n^{\alpha_n}$ is a differential operator monomial, then one checks that

(1.4) $$f_t^*(D^\alpha \delta_p^k) = t^{1-|\alpha|} D^\alpha \delta_{(p/t)}^k, \qquad |\alpha| = \sum \alpha_i.$$

Now because every $\omega \in A^q$ can be represented as a finite sum of elements of the type

$$\omega = \int \mu(x_1, \cdots, x_q)\, D^{\alpha_1} \delta_{x_1}^{i_1} \wedge D^{\alpha_2} \delta_{x_2}^{i_2} \wedge \cdots \wedge D^{\alpha_q} \delta_{x_q}^{i_q}\, dx_1 \cdots dx_q$$

where $\mu(x_1, \cdots, x_q)$ has compact support and is continuous, the formula (1.4) explains how f_t^* acts on all of A. In particular we see that the action of $\boldsymbol{R}^*$ on A_{pt} is semisimple and decomposes A_{pt} according to its "weights":

(1.5) $$A_{pt} = A_{pt}^{(n)} + A_{pt}^{(n-1)} + \cdots$$

with $f_t^* | A_{pt}^{(k)} = t^k$, $k \leqq n$.

The crucial lemma for the proof (1.1) is the following:

Lemma. *For every $\omega \in A$, $\lim_{t\to\infty} (1/t^n)\omega$ exists and is a well-defined element $l_n(\omega)$ of $A_{pt}^{(n)}$. Furthermore l_n defines a surjection of A onto $A_{pt}^{(n)}$:*

(1.6) $$A \xrightarrow{\quad l_n \quad} A_{pt}^{(n)} \xrightarrow{\qquad} 0.$$

Explanation. By (1.4) all that has to be explained is why f_t^* never blows up to order greater than n. For instance what about

$$\omega = \int \mu(x_1, \cdots, x_q)\, \delta_{x_1}^1 \wedge \delta_{x_2}^1 \wedge \cdots \wedge \delta_{x_q}^1\, dx_1 \cdots dx_q$$

with $q > n$? Of course here it is the *alternation* that saves the day, so that the "worst" term is of the form $\delta_0^1 \wedge \delta_0^2 \wedge \cdots \wedge \delta_0^n$ in $A_{pt}^{(n)}$.

Granting this lemma, one extends it as follows:

Lemma. *Let $F^n A \supset F^{n-1} A \supset \cdots$ be the filtration on A defined by*

$$\omega \in F^k A \Leftrightarrow \lim_{t\to\infty} \frac{1}{t^k}\, \omega$$

exists, and let $l_k\, \omega$ be that limit. Then one has the exact sequence

(1.7) $$0 \to F^{k-1} A \to F^k A \xrightarrow{\quad l_k \quad} A_{pt}^{(k)} \to 0,$$

which is in fact split by the inclusion $A_{pt}^{(k)} \subset F^k A$.

Corollary. *If A_{pt} is graded by "weight" then the graded group GA associated to the filtration $\{F^k A\}$ is isomorphic to A_{pt}:*

(1.8) $$GA \cong A_{pt}.$$

To complete the proof of the theorem one still needs a homotopy formula. For this purpose let R be the radical vector field generating the action of $\boldsymbol{R}^n$ on $\boldsymbol{R}^n$: $R = \sum x^i(\partial/\partial x^i)$. Then the crucial homotopy formula in A is

$$\frac{d}{dt} f_t^* = \frac{1}{t}\{i(R)d + d\,i(R)\} f_t^*. \tag{1.9}$$

Applied with $t = 1$ to $A_{pt}^{(k)}$ we see that in $A_{pt}^{(k)}$ multiplication by k is homotopic to 0. Hence

COROLLARY. $H(A_{pt}^{(k)}) = 0$ *for* $k \neq 0$.

Next integrate (1.9) from 1 to T to obtain

$$f_T^* - 1 = K_T d + dK_T \tag{1.10}$$

with

$$K_T\omega = \int_1^T i(R) f_t^*\,\omega \frac{dt}{t}. \tag{1.11}$$

In general, the limit at $T \to \infty$ of K_T will not exist, but for F^kA with $k < 0$, it will and this limit will also lie in F^kA. Hence, in particular, $H(F^{-1}A) = 0$. This granted, we are finished for then $H(A/F^{-1}A) \cong H(A)$. But (1.8) implies that

$$G(A/F^{-1}A) \cong A_{pt}/F^{-1}A_{pt}, \tag{1.12}$$

whence by the corollary (just above) $H(A/F^{-1}A) \cong H\{A_{pt}^{(0)}\}$. Q.E.D.

It remains to understand why the limit K_T as $T \to \infty$ exists on and preserves $F^{-1}A$. For this purpose write

$$K_T = \int_1^T i(R)\, t f_t^* \frac{dt}{t^2}.$$

Now changing variables to $\lambda = 1/t$, we get $K_T = \int_{1/T}^1 \{i(R)(1/\lambda) f_{1/\lambda}^*\}\, d\lambda$ so that on $F^{-1}A$ this integral converges to

$$K_\infty = \int_0^1 \left\{i(R)\frac{1}{\lambda} f_{1/\lambda}^*\right\} d\lambda.$$

To see that $K_\infty F^{-1}A \subset F^{-1}A$, note that $i(R)$ commutes with f_t^* and furthermore $f_t^* f_u^* = f_{tu}^*$. Hence

$$f_t^* K_\infty = \int_0^1 \left\{i(R)\frac{1}{\lambda} f_{t/\lambda}^*\right\} d\lambda = \int_0^{1/t} \left\{i(R)\frac{1}{\lambda} f_{1/\lambda}^*\right\} d\lambda$$

which tends to 0 as $t \to \infty$.

2. The Künneth formula. If L_1 and L_2 are two finite-dimensional Lie algebras and $L_1 \oplus L_2$ denotes their direct sum then it is of course well known that

$$H^*(L_1 \oplus L_2) \cong H^*(L_1) \otimes H^*(L_2), \tag{2.1}$$

and this relation follows directly from the fact that on the level of complexes

$$A(L_1 \oplus L_2) \cong A(L_1) \otimes A(L_2). \tag{2.2}$$

These considerations immediately lead one to suspect that (2.1) is also valid in the Gelfand-Fuks theory, but as in this context (2.2) has to be replaced by a completed tensor-product,

$$A_c(L_1 \oplus L_2) \cong A_c(L_1) \hat{\otimes} A_c(L_2),$$

it is difficult to prove directly the Künneth relation. However, the techniques of the last section can be extended without difficulty to prove the Künneth formula for the disjoint unions of $\boldsymbol{R}^n$'s: Thus one has the formula

$$H^*\Big(\bigoplus_{\alpha=1}^{k} \mathscr{L}\boldsymbol{R}^n_\alpha\Big) = \bigotimes_{1}^{k} H^*(\mathscr{L}\boldsymbol{R}^n_\alpha). \tag{2.3}$$

At this stage one is ready to carry out a program suggested by G. Segal to get at the Gelfand-Fuks theory of a general manifold M.

For this purpose, let $\mathscr{U} = \{U_\alpha\}_{\alpha \in J}$ be a finite ordered open covering of M with the property that all intersections $U_{\alpha_1} \cap \cdots \cap U_{\alpha_k}$ are either empty or diffeomorphic to $\boldsymbol{R}^n$. (Such coverings always exist on compact manifolds.) Corresponding to $\mathscr{U}$, consider the *semisimplicial* space

$$\mathscr{U}M: \mathscr{U}_0 M \leftleftarrows \mathscr{U}_1 M \,\overset{\leftarrow}{\underset{\leftarrow}{\leftarrow}}\, \mathscr{U}_2 M \,\overset{\leftarrow}{\underset{\leftarrow}{\leftleftarrows}}\, \mathscr{U}_3 M \tag{2.4}$$

where $\mathscr{U}_k M$ is the disjoint union of the k-fold intersections of the $\{U_\alpha\}$, and the arrows indicate the inclusions

$$\mathscr{U}_{\alpha_0} \cap \cdots \cap \mathscr{U}_{\alpha_k} \subset \mathscr{U}_{\alpha_1} \cap \cdots \cap \hat{\mathscr{U}}_{\alpha_i} \cap \cdots \cap \mathscr{U}_{\alpha_k}.$$

Now then the functor $\mathscr{L}$ clearly behaves contravariantly vis-à-vis these maps (one can restrict vector fields to open subsets) so that applying $\mathscr{L}$ to $\mathscr{U}M$ gives rise to a "cosemisimplicial topological Lie algebra":

$$\mathscr{L}\circ\mathscr{U}M : \mathscr{L}\mathscr{U}_0 M \rightrightarrows \mathscr{L}\mathscr{U}_1 M \,\overset{\rightarrow}{\underset{\rightarrow}{\rightarrow}}.$$

As usual, the alternating sum of the arrows now defines a differential operator δ, and the classical proof of the fact that the Čech cohomology of a fine sheaf vanishes here leads to the result that

$$H_\delta\{\mathscr{L}\mathscr{U}M\} \cong \mathscr{L}M. \tag{2.5}$$

Now this same argument extends further to yield a corresponding formula for the simplicial *module* $A_c \circ \mathscr{L} \circ \mathscr{U}M$:

$$H_\delta\{A_c \circ \mathscr{L} \circ \mathscr{U}M\} = A_c \mathscr{L}M. \tag{2.6}$$

The module $A_c \circ \mathscr{L} \circ \mathscr{U}M$ of course also carries the differential operator d of the

Gelfand-Fuks theory, which combined with δ makes $A_c \circ \mathscr{L} \circ \mathscr{U} M$ into a double complex, whose total cohomology can, by (2.6), be identified with $H(A_c \mathscr{L} M)$.

On the other hand, filtering this double complex "the other" way produces a spectral sequence with E_2 terms:

$$E_2 = H_\delta \{H_d(A_c \circ \mathscr{L} \circ M)\} = H_\delta \{H^*(\mathscr{L} \mathscr{U} M)\}$$

and converging to the same total cohomology. In view of our Künneth formula however, the semisimplicial module $H^*(\mathscr{L}\mathscr{U}M)$ takes the form:

$$\overset{\alpha}{\bigotimes} H^*(\mathfrak{a}_n) \rightrightarrows \overset{\alpha\beta}{\bigotimes} H^*(\mathfrak{a}_n) \overset{\rightarrow}{\rightrightarrows} \tag{2.7}$$

whose δ-cohomology is now seen to be finite dimensional by virtue of known facts about $H^*(\mathfrak{a}_n)$. Actually with a little care (and the Eilenberg-Zilber theorem) one can push (2.7) to obtain the original spectral sequence of Gelfand-Fuks. Indeed it is not hard to evaluate our E_2 term to be

$$\tilde{H}_\delta\{H^*(\mathscr{L}\mathscr{U}M)\} = \sum_k H_*^a\left(M^{(k)}; M_{k-1}^{(k)}; \overset{k}{\bigotimes} \tilde{H}^*(\mathfrak{a}_n)\right) \tag{2.8}$$

where on the right M^k denotes $M \times \cdots \times M$ k times, and M_{k-1}^k denotes the subset consisting of points for which two or more of the coordinates are equal, H_* denotes homology (!) and the a indicates that we consider the homology which behaves alternatingly relative to the permutation group on k letters. Finally the ~ denotes augmented cohomology.

But rather than indicate how this is done, I will show in the next section how the recent work of Anderson [1] leads to a similar sequence in homology theory pure and simple.

3. The conjecture. We have already remarked on a likely Künneth formula:

$$H^*(\mathscr{L} M \oplus \mathscr{L} N) \cong H^*(\mathscr{L} M) \oplus H^*(\mathscr{L} N) \tag{3.1}$$

for disjoint unions, and using the spectral sequence just developed it is not hard to actually prove this result for the sort of manifolds under consideration. Now in topology this type of functional behavior is exhibited by the homology of *function spaces*. That is, if we define a functor $F(X)$ on spaces by setting

$$F(X) = H^*\{\text{Maps } (X; Y)\}$$

(real coefficients, say) then, under some finiteness conditions

$$F(X_1 \cup X_2) = F(X_1) \otimes F(X_2). \tag{3.2}$$

This suggests that the Gelfand-Fuks theory of an n-manifold M should be the standard cohomology of a function space: Maps $(M; Y_n)$. Furthermore, if such a formula held, then

$$H^*(Y_n) = H^*\{\text{Maps}(\mathbf{R}^n; Y_n)\}$$

would have to be equal to $H^*(\mathscr{L}\mathbf{R}^n) = H^*(\mathfrak{a}_n)$.

Now already in the original work of Gelfand-Fuks there occurs a space with this property and it is defined as follows:

Let $E_n \to^{\pi} BGL_n$ be the universal bundle over $BGL_n(\boldsymbol{C})$. Then the $2n$-skeleton of $BGL_n(\boldsymbol{C})$ is a well-defined space $BGL_n(2n)$, and Y_n is simply $\pi^{-1}\, BGL_n(2n)$. Note in particular that $GL(n, \boldsymbol{C})$ acts naturally on Y_n. In any case, this is then a natural candidate for our *universal space* vis à vis the Gelfand-Fuks theory. However, this hypothesis is not compatible with the work of Guillemin [6] and its extension by Trauber [8].

To explain these results one considers the *diagonal subcomplex*

$$A_\Delta \mathscr{L} M \subset A_c \mathscr{L} M$$

whose q-forms are precisely those with support on the diagonal in M^q. Now in [6] Guillemin evaluated $H(A_\Delta \mathscr{L} M)$ as follows:

Let PM be the principal $GL(n, \boldsymbol{R})$ bundle of M, and let Y_nM be the associated bundle

$$Y_nM = PM \times_{GL(n,\boldsymbol{R})} Y_n,[1]$$

so that $Y_nM \to M$ is a bundle with fiber Y_n.

Then Guillemin shows that

$$H^r(A_\Delta \mathscr{L} M) \cong H^{r+n}(Y_nM). \tag{3.3}$$

It follows therefore that in any case our sought-after homotopy functor for Gelfand-Fuks must admit a natural map from $H^*(Y_nM)$ decreasing dimension by n.

Now the construction which most naturally satisfies these conditions is the space $\Gamma(Y_nM)$ of continuous sections of Y_nM over M. Indeed if PM is trivial, then

$$\Gamma(Y_nM) = \mathrm{Map}(M; Y_n),$$

while, in all cases, there is the natural evaluation map

$$\Gamma(Y_nM) \times M \xrightarrow{\ e\ } Y_nM$$

which evaluates (s, m) at m to yield $s(m)$. In cohomology, this map induces a homomorphism

$$e^*: H^*(Y_nM) \to H^*(\Gamma\, Y_nM) \otimes H^*(M)$$

which after integration over the fiber produces the desired homomorphism of

$$H^*(Y_nM) \to H^*(\Gamma\, Y_nM).$$

Evaluation at several points similarly fits the results of [9]. All this leads us to the main

(3.4) *Conjecture.* $H^*(\mathscr{L} M) \cong H^*(\Gamma\, Y_n\, M)$.

Now apart from the heuristic arguments leading up to (3.4) there is considerable

[1] $GL(n, \boldsymbol{R}) \subset GL(n, \boldsymbol{C})$ acts on Y_n.

independent evidence. First of all, the Anderson spectral sequence [**1**] is applicable to the right-hand functor and *provides the same E_2-term as we found for the left-hand one*. Secondly, as Andre Haefliger has pointed out, (3.4) also fits the recent work of Mather and Thurston. Where they show an isomorphism of cohomology for two topological categories C_1 and C_2, our Conjecture (3.4) can be interpreted as an isomorphism between the corresponding *continuous* cohomologies.

But this connection would take me too far afield to explain here (see [**7**], [**2**] for these concepts).

Naturally the question now arises why such a reasonable conjecture once made is not immediately demonstrable. In view of the functional similarities of the two functors one clearly only needs a map between them which induces an isomorphism on $\boldsymbol{R}^n$. But unfortunately such a map seems hard to come by.

Let me conclude by observing that for spheres (3.4) rather easily leads to the result that

$$H^*(\mathscr{L} S^n) = H^*(\mathrm{Map}\,(S^n; Y_n))$$

which for $n = 1$ yields the desired $E(\omega) \otimes \boldsymbol{R}[y]$ but for $n \geqq 2$ gives a rather larger answer than indicated by Gelfand-Fuks. Furthermore, these results can be checked independently by yet another method, due to A. Haefliger.

BIBLIOGRAPHY

1. D. W. Anderson, *A generalization of the Eilenberg-Moore spectral sequence*, Bull. Amer. Math. Soc. **78** (1972), 784–786. MR **46** #9987.

2. R. Bott, *Some remarks on continuous cohomology*, Proc. Internat. Conference on Topology and Related Topics, University of Tokyo, Japan, April 1973.

3. I. M. Gel'fand and D. B. Fuks, *Cohomologies of the Lie algebra of tangent vector fields on a smooth manifold*. I, Funkcional. Anal. i Priložen **3** (1969), no. 3, 32–52. (Russian) MR **41** #1067.

4. ———, *Cohomologies of the Lie algebra of tangent vector fields on a smooth manifold*. II, Funkcional. Anal. i Priložen. **4** (1970), no. 2, 23–31 = Functional Anal. Appl. **4** (1970), 23–32. MR **44** #2248.

5. ———, *Cohomology of the Lie algebra of formal vector fields*, Izv. Akad. Nauk SSSR Ser. Mat. **34** (1970), 322–337 = Math. USSR Izv. **4** (1970), 327–342. MR **42** #1103.

6. V. Guillemin, *Cohomology of vector fields on a manifold*, Advances in Math. **10** (1973), no. 2, 192–220.

7. J. Mather, *Loops and foliations*, Proc. Internat. Conference on Topology and Related Topics, University of Tokyo, Japan, April 1973.

8. P. Trauber, *Gelfand-Fuks cohomology*, Ph. D. Thesis, Princeton University, Princeton, N. J., 1973.

HARVARD UNIVERSITY

Proceedings of Symposia in Pure Mathematics
Volume 27, 1975

ON $H^*(\tilde{G}_{s,q};R/Z)$, RANGE GROUP OF SOME CHERN-SIMONS INVARIANTS

JOHN S. MAFFEI*

Introduction. Let $\tilde{G}_{s,q}$ denote the real grassmannian of oriented s-planes in R^{s+q} through the origin. Let R denote the real numbers, Z the integers. This paper computes $H^*(\tilde{G}_{s,q}; R/Z)$.

§ I gives the notation and many of the definitions of this section are to be found in Pontrjagin. § II gives the basic definitions and derives some results about them. § III uses these results for the computations. § IV is a technical proof of Lemma 3.

I am happy to acknowledge helpful conversations with S.S. Chern, who proposed the problem to me.

I. If S is a set then $\langle S\rangle$ denotes the free abelian group generated by S.

Let $G_{s,q}$ denote the grassmannian of unoriented s-planes through the origin in R^{s+q}. Then $\tilde{G}_{s,q}$ is its oriented double cover. Let $T: \tilde{G}_{s,q} \to \tilde{G}_{s,q}$ be the nontrivial deck transformation. If $0 < s \in Z$, let $In(s) = \{1,2, \cdots, s\}$ and $Nn(s) = \{0\} \cup In(s)$. All congruences are mod 2. As $G_{s,q} \cong G_{q,s}$, if $s \equiv q + 1$, nothing is lost in assuming $s \equiv 0$.

Let $B = \{e_1, \cdots, e_{s+q}\}$ be a basis of R^{s+q}. Let K be a Schubert decomposition of $G_{s,q}$ relative to B. Elements of K will be denoted ω, σ and $[a_1 \cdots a_s]$ as usual. $\omega = [a_1 \cdots a_s]$ iff $\omega(i) = a_i$ for all $i \in In(s)$. Let $\tilde{K}$ be the lift of K. Orient the cells of $\tilde{K}$ as follows: If $\omega \in K$, let $p_i = e_{\omega(i)+i}$ for $i \in In(s)$. Define $R(\omega) = \langle p_1, \cdots, p_s\rangle$ and let this basis determine an orientation for $R(\omega)$. Let $\pi: R^{s+q} \to R(\omega)$ be the projection. Define $U(\omega) = \{P \in \tilde{G}_{s,q} | \pi(P) = R(\omega)$ and π preserves orientation$\}$. Let $\hat{U}(\omega) \equiv TU(\omega)$.

AMS (MOS) subject classifications (1970). Primary 55B99; Secondary 57D20.

*This research was supported by NSF GP-34785X.

For all $\omega \in K$, define $\omega(0) = (0)$, $\omega(s+1) = q$.

DEFINITION. For $\omega \in K$, $h \in In(s)$ let:

$$j(\omega) = \{j \in In(s) | \omega(j) > \omega(j-1)\}, \qquad I(\omega) = \{j \in In(s) | \omega(j) < \omega(j+1)\},$$

$$s(\omega, h) = s + h + \omega(h), \qquad t(\omega, h) = \sum_{i=1}^{h} \omega(i) + (s-h)\,\omega(h),$$

$$\lambda(\omega, h) = (-1)^{s(\omega,h)}, \qquad \mu(\omega, h) = (-1)^{t(\omega,h)}.$$

If $j \in J(\omega)$, let $\omega^j = [\omega(1) \cdots (\omega(j) - 1) \cdots \omega(s)]$. If $i \in I(\omega)$, let $\omega_i = [\omega(1) \cdots (\omega(i)+1) \cdots \omega(s)]$.

THEOREM (PONTRJAGIN). *The boundary operator on $C(\tilde{K}, Z)$is defined by*:

$$\partial U(\omega) = \sum_{j \in J(\omega)} (U(\omega^j) + \lambda(\omega, j)\,\hat{U}(\omega^j))\,\mu(\omega, j),$$
$$\partial \hat{U}(\omega) = \sum_{j \in J(\omega)} (\hat{U}(\omega^j) + \lambda(\omega, j)\,U(\omega^j))\,\mu(\omega, j).$$

Identify chains and cochains by taking the dual relative to the basis $\{U(\omega), \hat{U}(\omega) | \omega \in K\}$. Then

COROLLARY. *The coboundary operator on $C(\tilde{K}, Z)$ is defined by*:

$$\delta U(\omega) = -\sum_{i \in I(\omega)} (-1)^{n-i}(U(\omega_i) - \lambda(\omega, i)\,\hat{U}(\omega_i))\,\mu(\omega, i),$$
$$\delta \hat{U}(\omega) = -\sum_{i \in I(\omega)} (-1)^{n-i}(\hat{U}(\omega_i) + \lambda(\omega, i)\,U(\omega_i))\,\mu(\omega, i).$$

II.

DEFINITION. If $\omega \in K$, let

$$P(\omega) = U(\omega) + \hat{U}(\omega) \in C(\tilde{K}, Z), \qquad N(\omega) = U(\omega) - \hat{U}(\omega) \in C(\tilde{K}, Z),$$
$$P = \langle P(\omega) | \omega \in K\rangle, \qquad N = \langle N(\omega) | \omega \in K\rangle.$$

LEMMA 1. *P and N are subcomplexes of C with*

$$\partial P(\omega) = \sum_{j \in J(\omega)} (1 + \lambda(\omega, j))\,\mu(\omega, j)\,P(\omega^j),$$
$$\partial N(\omega) = \sum_{j \in J(\omega)} (1 - \lambda(\omega, j))\,\mu(\omega, j)\,N(\omega^j).$$

PROOF.

$$\begin{aligned}\partial P(\omega) &= \partial U(\omega) + \partial \hat{U}(\omega)\\ &= \sum_{j \in J(\omega)} \mu(\omega, j)(U(\omega^j) + \hat{U}(\omega^j) + \lambda(\omega, j)(U(\omega^j) + \hat{U}(\omega^j)))\\ &= \sum_{j \in J(\omega)} \mu(\omega, j)(1 + \lambda(\omega, j))\,P(\omega^j),\\ \partial N(\omega) &= \partial U(\omega) - \partial \hat{U}(\omega)\\ &= \sum_{j \in J(\omega)} \mu(\omega, j)(U(\omega^j) - \hat{U}(\omega^j) - \lambda(\omega, j)(U(\omega^j) - \hat{U}(\omega^j)))\\ &= \sum_{J \in (\omega)} \mu(\omega, j)(1 - \lambda(\omega, j))\,N(\omega^j). \quad \text{Q.E.D.}\end{aligned}$$

CONVENTION. $Q = N$ or P.

DEFINITION. If $\omega \in K$, and $\delta Q(\omega) \neq 0$, denote by $BQ(\omega)$ the cochain of Q such that $2BQ(\omega) = \delta Q(\omega)$.

DEFINITION.

$$\begin{aligned} IQ &= \{Q(\omega) \mid \delta Q(\omega) = 0 \,\&\, 2Q(\omega) \notin \delta Q\}, \\ IIQ &= \{Q(\omega) \mid \delta Q(\omega) = 0 \,\&\, 2Q(\omega) \in \delta Q\}, \\ IIIQ &= \{\gamma \in Q \mid \exists\, \omega \in K \text{ such that } \gamma = BQ(\omega) \,\&\, \gamma \notin IIQ\}. \end{aligned}$$

DEFINITION. If $s \equiv 0$, let

$$A_{s,q} = \{\omega \in K \mid \exists b_1, \cdots, b_e \in Z \text{ such that } \omega = [2b_1\, 2b_1 \cdots 2b_e 2b_e]\}.$$

If $s \not\equiv 0$, let $A_{s,q} = \varnothing$.
If $s \equiv 1 \equiv q$, let

$$\begin{aligned} B_{s,q} &= \{\omega \in K \mid \exists b_1, \cdots, b_e \in Z \text{ such that } \omega = [02b_1 2b_1 \cdots 2b_e 2b_e]\}, \\ C_{s,q} &= \{\omega \in K \mid \exists b_1, \cdots, b_e \in Z \text{ such that} \\ &\qquad \omega = [(q-2b_e)(q-2b_e)\cdots(q-2b_1)(q-2b_1)(q)]\}. \end{aligned}$$

Otherwise, let $C_{s,q} = \varnothing = B_{s,q}$.
Let $D_{s,q} = A_{s,q} \cup B_{s,q} \cup C_{s,q}$ and $\mathscr{D}_{s,q} = \langle P(\omega) \mid \omega \in D_{s,q} \rangle$.
DEFINITION. If $s \equiv 0$, let

$$\begin{aligned} W_{s,q} = \{\omega \in K \mid \exists b_1, \cdots, b_e \in Z \text{ such that} \\ \omega = [(2b_1+1)(2b_1+1)\cdots(2b_e+1)(2b_e+1)]\}. \end{aligned}$$

If $s \equiv 0 \equiv q$, let

$$\begin{aligned} U_{s,q} = \{\omega \in K \mid \exists b_1, \cdots, b_t \in Z \text{ such that} \\ \omega = [0 \cdots 02b_1 2b_1 \cdots 2b_t 2b_t q]\}. \end{aligned}$$

Otherwise, let $W_{s,q} = \varnothing = U_{s,q}$.
Let $Y_{s,q} = W_{s,q} \cup U_{s,q}$ and $\mathscr{Y}_{s,q} = \langle N(\omega) \mid \omega \in Y_{s,q} \rangle$.

LEMMA 2. *If $\alpha \in Q$ and $\delta\alpha = 0$, then α is cohomologous to a linear combination of cocycles in* $IQ \cup IIQ \cup IIIQ$. *If $\alpha \in \langle IIQ, IIIQ \rangle$, then $2[\alpha] = 0 \in H^*(Q; Z)$. Finally,* $H^*(Q; Z[\frac{1}{2}]) \cong \langle IQ \rangle \otimes Z[\frac{1}{2}]$.

PROOF. *Case* I. $Q = P$. Let $\pi\colon \tilde{G}_{s,q} \to G_{s,q}$ be the projection. Let each $\omega \in K$ be oriented so that $\pi(U(\omega)) = \omega$. Then $\phi\colon C(K) \to P$ defined by $\phi(\omega) = P(\omega)$ is a chain isomorphism. So the result is proved as in Chern's Theorem 1. Note that the orientations of the ω's given here are in general different from those given in Chern. This does not affect the proof.

Case II. $Q = N$. The proof is essentially the same as Chern's if one reads $N(\omega)$, $BN(\omega)$, etc. instead of ω, $B(\omega)$, etc. The comment "where $n + a_h + h$ is even" must be changed to "where $n + a_h + h$ is odd". Q.E.D.

LEMMA 3. $IP = \{P(\omega) \mid \omega \in D_{s,q}\}$, $In = \{N(\omega) \mid \omega \in Y_{s,q}\}$.

Since the proof of Lemma 3 will not clarify the following sections, it is given as the final section.

One can easily check that $C(\tilde{K})/P \oplus N \cong P \otimes Z_2$ as chain complexes. Consider these identified and let $V = P \otimes Z_2$.

From the exact sequence

$$0 \to P \oplus N \to C(\tilde{K}) \to V \to 0 \tag{1}$$

one has

$$0 \to \mathrm{Hom}(V, Z[\tfrac{1}{2}]) \to \mathrm{Hom}(C(\tilde{K}), Z[\tfrac{1}{2}]) \to \mathrm{Hom}(P \oplus N, Z[\tfrac{1}{2}]) \to \mathrm{Ext}(V, Z[\tfrac{1}{2}]) \to 0.$$

It is easy to show $\mathrm{Hom}(V, Z[\frac{1}{2}]) = 0$ and $\mathrm{Ext}(V, Z[\frac{1}{2}]) = 0$. Hence,

LEMMA 4. $\mathrm{Hom}(C(\tilde{K}), Z[\frac{1}{2}]) \cong \mathrm{Hom}(P \oplus N, Z[\frac{1}{2}])$.

THEOREM 5. $H^*(\tilde{G}_{s,q};\ Z[\frac{1}{2}]) \cong H^*(P;\ Z[\frac{1}{2}]) \oplus H^*(N;\ Z[\frac{1}{2}]) \cong (\mathscr{D}_{s,q} \otimes Z[\frac{1}{2}]) \oplus (\mathscr{Y}_{s,q} \otimes Z[\frac{1}{2}])$.

Let $\delta Q = QB$; then one can show that, for all $n \in Z$,

$$\mathrm{Tor}(H^{n+1}(Q;\ Z), R/Z) \cong \langle [\tfrac{1}{2}]\gamma \mid \gamma \in QB^{n+1} \rangle \cong QB^{n+1} \otimes Z_2.$$

Thus

THEOREM 6. *For all* $n \in Z$, $H^n(Q;\ R/Z) \cong (\langle IQ^n \rangle \otimes R/Z) \oplus (QB^{n+1} \otimes Z_2)$.

III. The exact sequence (1) implies

$$(2) \qquad 0 \to \mathrm{Hom}(V, R/Z) \to \mathrm{Hom}(C(\tilde{K}), R/Z) \to \mathrm{Hom}(P \oplus N, R/Z) \to \mathrm{Ext}(V, R/Z) \to 0.$$

The arguments that $\mathrm{Ext}(V, R/Z) = 0$, $\mathrm{Hom}(V, R/Z) \cong V$, and $\delta\,\mathrm{Hom}(V, R/Z) = 0$ are easy. Let $W = \mathrm{Hom}(V, R/Z)$. The long exact sequence of (2), using Theorem 6, is

$$\cdots \to W^{n-1} \to H^{n-1}(\tilde{G}_{s,q};\ R/Z) \to (\langle IP^{n-1} \rangle \otimes R/Z) \oplus (\langle IN^{n-1} \rangle \otimes R/Z) \oplus (PB^n \otimes Z_2) \oplus (NB^n \otimes Z_2) \xrightarrow{\delta^*} W^n \to H^n(\tilde{G}_{s,q};\ R/Z) \to \cdots.$$

Suppose $f \otimes [\frac{1}{2}] \in \mathrm{Hom}(P_n \oplus N_n, Z) \otimes R/Z$ such that $\delta(f \otimes [\frac{1}{2}]) = 0$ and $\delta f \neq 0$. It is easy to show that $(fBP(\sigma) + fBN(\sigma) \equiv 0 \equiv fBP(\sigma) - fBN(\sigma) \pmod 4)$ $(\forall \sigma \in K_{n+1}) \Leftrightarrow \delta^*(f \otimes [\frac{1}{2}]) = 0$, once one observes that $\partial U(\sigma) = BP(\sigma) + BN(\sigma)$, $\partial \hat{U}(\sigma) = BP(\sigma) - BN(\sigma)$. Let $\bar{D}_{s,q} = \{\omega \mid P(\omega) \in IP \cup IIP\}$. $\bar{Y}_{s,q} = \{\omega \mid N\omega \in IN \cup IIN\}$. Let $\Omega = K \sim (\bar{D}_{s,q} \cup \bar{Y}_{s,q})$. Then if $\delta^*(f \otimes [\frac{1}{2}]) = 0$, $(\forall \sigma \in K)$ $(fBP\sigma \equiv fBN\sigma \equiv \gamma(\sigma) \pmod 4)$ when $\gamma(\sigma) = 0$ or 2. In particular, if $\sigma \in D_{s,q} \cup Y_{s,q}$ then $fBQ\sigma \equiv 0 \pmod 4$. If $\sigma \in \Omega$, then $fBN\sigma \equiv fBP\sigma \equiv \gamma(\sigma) \pmod 4$. Then $\mathrm{Im}\ \delta_n^* \cong \langle K \rangle_n \otimes Z_2$. But then, by cardinality, $\mathrm{Im}\ \delta^* = W$.

THEOREM 7. $H^n(\tilde{G}_{s,q};\ R/Z) \cong (\langle IP_n, IN_n \rangle \otimes R/Z) \oplus (\langle \Omega_{n+1} \rangle \otimes Z_2)$.

THEOREM 8.

$$\begin{aligned} H^n(\tilde{G}_{1,q};\ R/Z) &\cong R/Z, \quad n = q, \\ &\cong R/Z, \quad n = 0, \\ &\cong 0, \quad \textit{otherwise}. \end{aligned}$$

IV.

PROOF OF LEMMA 3. The proof is by cases. The order of labeling will be: capital

Latin letters, capital Roman numerals, small Greek letters, small Roman numerals, small Latin letters.

(A) P. It is easy to check that $\delta\mathscr{D}_{s,q} = 0$ and $(\forall\tau \in K)(\forall\sigma \in D_{s,q})([\tau:\sigma] = 0)$. Therefore $\{P(\omega)|\omega \in D_{s,q}\} \subseteq IP$. The rest of (A) is to prove that $IP \subseteq \{P(\omega)| \omega \in D_{s,q}\}$.

Suppose $\omega \notin D_{s,q}$ and $\delta P(\omega) = 0$. Then $j \in I(\omega) \Rightarrow s(\omega, j) \equiv 0$.

Case I. Suppose $s \equiv 0 \equiv q$. Then $D_{s,q} = A_{s,q}$.

(α) $(\exists k \in In(s))(\omega(k) \equiv 1)$.

(i) $\omega(s) \equiv 1$. Then $s(\omega, s) \equiv 1$ and $\omega(s) < q$, so $\delta P(\omega) \neq 0$.

(ii) $\omega(s) \equiv 0$. Let $l \in In(s)$ be the greatest elements such that $l \leqq c \leqq s \Rightarrow \omega(c) = 0$. [Then $l > 1$. Since $\delta P(\omega) = 0$, $s(\omega, l-1) \equiv 0$. Therefore $l \equiv 0$. Define

$$\begin{aligned} \delta(i) &= \omega(i), & i &\neq l, \\ &= \omega(i) - 1, & i &= l. \end{aligned}$$

Then $\delta\,\frac{1}{2}P\sigma = \pm P(\omega)$.] Henceforth, Ψ will denote the statement in square brackets.

(β) $(\exists j \in J(\omega))(j \equiv 0)$. Then $j > 1$ and $s(\omega, j-1) \equiv 0$. Therefore $\omega(j-1) \equiv 1$. Hence (β) $\Rightarrow$ (α).

Case II. $s \equiv 0 \equiv q + 1$.

(α) $(\exists k \in In(s))(\omega(k) \equiv 1)$.

(i) $(\forall k \in In(s))(\omega(k) \equiv 1)$. Then $\omega(1) \geqq 1$. Let $l = 1$. If σ is defined as in Case I (α) (ii), $\delta\,\frac{1}{2}P(\sigma) = \pm P(\omega)$.

(ii) $(\exists\tilde{k} \in In(s))(\omega(\tilde{k}) \equiv 0)$.

(a) $\omega(s) \equiv 0$. Proceed as in Case I(α) (ii).

(b) $\omega(s) \equiv 1$. Let l be the least integer such that $l \leqq c \leqq s \Rightarrow \omega(c) \equiv 1$. {Then $l > 1$. Since $\delta P(\sigma) = 0$, $s(\omega, l-1) \equiv 0$. Therefore $l \equiv 1$. Define σ as in Case II(α) (ii). Then $\delta\,\frac{1}{2}P(\sigma) = \pm P(\omega)$.} Henceforth the statement in brackets will be denoted by Π.

(β) There exists $j \in J(\omega)$ such that $j \equiv 0$. As Case I(β).

Case III. $s \equiv q \equiv 1$. Then $D_{s,q} = B_{s,q} \cup C_{s,q}$.

(α) There exists $i \in In(s)$ such that $\omega(i) \equiv 0$. Let l be the least element of $In(s)$ such that $l \leqq c \leqq s \Rightarrow \omega(c) \equiv 0$.

(i) $l = 1$.

(a) $\omega(1) \neq 0$. Then $\omega(1) \geqq 2$. Let σ be defined as in Case II(α)(ii). Then $\delta\,\frac{1}{2}P(\sigma) = \pm P(\omega)$.

(b) $\omega(1) = 0$. Then there exists $j \in J(\omega)$ such that $j \equiv 1$. Then $j > 1$ and $s(\omega, j-1) \equiv 1$. Therefore $\delta P(\omega) \neq 0$.

(ii) $l > 1$. Then $l \in J(\omega)$ and $s(\omega, l-1) \equiv 0$. Π.

(β) $i \in In(s) \Rightarrow \omega(i) \equiv 1$.

(i) $\omega(s) < q$. Then $s \in I(\omega)$ and $s(\omega, s) \equiv 1$. Therefore $\delta P(\omega)$.

(ii) There exists $j \in J(\omega)$ such that $j \equiv 0$. Hence $j \geqq 2$ and $s(\omega, j-1) \equiv 1$. Therefore $\delta P(\omega) \neq 0$.

(B) $N_{s,q}$. It is not hard to show $\{N(\omega)|\omega \in Y_{s,q}\} \subseteq IN$. Then it must be shown that $\{N(\omega)|\omega \in Y_{s,q}\} \supseteq IN$.

Suppose $\omega \notin Y_{s,q}$ and $\delta N(\omega) = 0$. Then $j \in I(\omega) \Rightarrow s(\omega, j) \equiv 1$.

Case I. $s \equiv q \equiv 0$.

(α) $\omega(1) \neq 0$.

(i) There exists $i \in In(s)$ such that $\omega(i) \equiv 0$. Let l be the least such i.

(a) $l = 1$. (Let

$$\begin{aligned} \sigma(i) &= \sigma(i) - 1, \quad i = l, \\ &= \sigma(i), \qquad\quad i \neq l. \end{aligned}$$

Then $s(\sigma, l) \equiv 0$ and $\delta \frac{1}{2} N(\sigma) = \pm N(\omega)$.) Henceforth, let the sentence in parentheses be denoted Ω.

(b) $l > 1$. Then $s(\omega, l-1) \equiv 1 \Rightarrow l \equiv 1$. Ω.

(ii) There exists $j \in J(\omega)$ such that $j \equiv 0$. Then $j > 1$ and $s(\omega, j-1) \equiv 1$. Therefore $\omega(j-1) \equiv 0$. Therefore (ii) $\Rightarrow$ (i).

(β) $\omega(1) = 0$.

(i) There exists $i \in In(s)$ such that $\omega(i) \equiv 1$. Let l be the least integer such that $\omega(l) \equiv 1$. Then $l > 1$ and $s(\omega, l-1) \equiv 0$ imply $l \equiv 0$. Ω.

(ii) There exists $j \in J(\omega)$ such that $j \equiv 1$. Thus $j > 1$ and $s(\omega, j-1) \equiv 1 \Rightarrow \omega(l-1) \equiv 1$. Thus (ii) $\Rightarrow$ (i).

(iii) $\omega(s) \neq q$. Then $s(\omega, s) \equiv 1$. Therefore $\omega(s) \equiv 1$. Therefore (iii) $\Rightarrow$ (i).

Case II. $s \equiv q \equiv 1$. Then $\delta N(\omega) = 0 \Rightarrow \omega(s) = q$ or $s(\omega, s) \equiv 1$. Therefore $\omega(s) \equiv 1$. Let l be the least integer such that $\omega(l) \equiv 1$. Let σ be defined as in Ω. Then $s(\sigma, l) \equiv l + 1$.

(α) $l \equiv 0$. Then $l \geqq 2$ and $s(\omega, l-1) \equiv \omega(l-1) \equiv 0$. $\rightarrow\leftarrow$

(β) $l \equiv 1$. Then $s(\sigma, l) \equiv 0$ and $\delta \frac{1}{2} N(\sigma) = \pm N(\omega)$.

Case III. $s \equiv 0 \equiv q + 1$.

(α) There exists i such that $\omega(i) \equiv 0$.

(i) $\omega(s) \equiv 0$. Then $\omega(s) < q$ and $s(\omega, q) \equiv 0$. Therefore $\delta N(\omega) \neq 0$. $\rightarrow\leftarrow$

(ii) Let l be the least integer such that $\omega(l) \equiv 1$ and $l \leqq i \leqq s \Rightarrow \omega(l) \equiv 1$. Then $l > 1$ so $s(\omega, l-1) \equiv 1$. Therefore $l \equiv 0$. Ω.

(β) There exists $j \in J(\omega)$ such that $j \equiv 0$. Then $j \geqq 2$. Therefore $s(\omega, j-1) \equiv 1$. Therefore $\omega(j-1) \equiv 0$. Therefore (β) $\Rightarrow$ (α). Q.E.D.

BIBLIOGRAPHY

S. S. Chern, *On the multiplication in the characteristic ring of a sphere bundle*, Ann. of Math. (2) **49** (1948), 362–372. MR **9**, 456.

Charles Ehresmann, *Sur la topologie de certaines variétés algébriques réelles*, J. Math. Pures Appl. **16** (1937), 69.

L. S. Pontrjagin, *Characteristic cycles on differentiable manifolds*, Mat. Sb. **21 (63)** (1947), 233–284; English transl., Amer. Math. Soc. Transl. (1) **7** (1962), 149–219. MR **9**, 243.

Wu Wen-tsun, *Sur les espaces fibrés*, Hermann, Paris, 1952.

———, *On Pontrjagin classes*. I, Acta Math. Sinica **3** (1953), 291–315; English transl., Amer. Math. Soc. Transl. (2)**92**(1970), 49–62. MR **17**, 521; **42** #3.

UNIVERSITY OF MICHIGAN

Proceedings of Symposia in Pure Mathematics
Volume 27, 1975

EULER SINGULARITIES AND HOMOLOGY OPERATIONS

CLINT MCCRORY*

In any real analytic space V, there are canonical mod 2 homology classes which are Poincaré dual to the Stiefel-Whitney cohomology classes when V is nonsingular. These homology classes were defined by Sullivan [10], who observed that the long forgotten combinatorial Stiefel chains on smooth manifolds (which were rediscovered by J. Cheeger [14]) are still cycles in an analytic space. This is because analytic singularities are not topologically arbitrary—V must have odd local Euler number at each point; that is, V is an "Euler space".

Inspired by Thom's theory of singularities of smooth maps, we show that for any simplicial Euler space V of dimension n, the Stiefel-Whitney homology classes of V are represented by "Euler singularities" of maps to Euclidean spaces of dimension less than n (§ 2). This has been done independently by Banchoff [2]. (Although he works only with PL manifolds, our definitions of the Euler singularity are the same, stemming from his earlier work on PL Morse theory.)

The singularities of maps of V^n into Euclidean spaces of dimension greater than n represent "normal" Stiefel-Whitney homology classes (§ 3). In fact, our definition of normal classes uses only that V has a mod 2 fundamental class. Thus we obtain a simple geometric definition of a family of mod 2 homology operations, whose relation to the Steenrod cohomology operations is discussed in §4.

ADDED IN PROOF. I have verified the conjectures in §§ 3 and 4, using PL immersion theory.

I wish to thank the Leverhulme Foundation and the University of Warwick for

AMS (MOS) subject classifications (1970). Primary 57D20, 55G05; Secondary 57C99, 32C05.
*Supported in part by NSF grant GP-36418X1.

their support during the course of this research. I also thank Domingo Toledo for his interest in my work, and Thomas Banchoff for describing his own approach to me.

1. Linear maps of simplicial complexes. A simplical complex is the prototype of a smooth stratified set. Though simplicial methods have been largely replaced by other geometric techniques for manifolds, they are still a good tool for studying spaces with singularities. Many classical simplicial constructions generalize directly to smooth stratifications. This paper can be viewed as an attempt to set up a simplicial model for the theory of characteristic homology classes of varieties, a theory which should ultimately be described using general stratifications.

First we establish some notation. All simplicial complexes will be locally finite. The same letter will be used for a complex and its geometric realization. $\sigma < K$ means σ is a simplex of K, while $x \in K$ means that x is a point of the realization of K. If $\sigma, \tau < K, \sigma < \tau$ means σ is a face of τ (including $\sigma = \tau$). σ° denotes the interior of σ. $\mathrm{Star}(\sigma) = \mathrm{star}(\sigma, K) = \{\tau < K, \tau > \sigma\}$. $N(\sigma) = \{\omega < K, \omega < \tau$ for some $\tau > \sigma\}$. $\mathrm{Link}(\sigma) = \{\omega < K, \omega < N(\sigma)$ but $\omega \not> \sigma\}$. If the vertices of ω and σ span a simplex τ of K, then $\tau = \omega * \tau$, the *join* of ω and σ. Thus link(σ) comprises all simplexes of K which can be joined to σ, and star(σ) comprises the joins of simplexes in link(σ) with σ. Superscripts are used to denote dimension; σ^i stands for an i-simplex of K, and K^i is the i-skeleton of K.

By a *polyhedron*, we mean a locally compact polyhedron in a Euclidean space. A "subpolyhedron" of a simplicial complex K is just a subcomplex of some subdivision of K. A homeomorphism of complexes will always mean a piecewise-linear homeomorphism. A triangulation of a polyhedron means a PL triangulation.

Let K be a simplicial complex. A *linear map* $f: K \to R^m$ is a continuous map which is linear on each simplex of K. f is *general* if it puts the vertices of K in general position ln R^m. f is *full* if it is locally general, i.e., for each vertex v of K, $f|N(v)$ is general. f is an *immersion* if $f|N(v)$ is an embedding for each vertex v of K. If $m \leqq \dim(K)$, the linear map $f: K \to R^m$ is a *profile* if it collapses no simplexes of dimension $\leqq m$. (This is slightly weaker than f being full.)

Any linear map $g: K \to R^m$ can be closely approximated by a full map, in the following sense. Given $\varepsilon > 0$, there exists a full map $f: K \to R^m$ such that $|g(v) - f(v)| < \varepsilon$ for all vertices v of K.

LEMMA 1. *Let $g: K \to R^m$ be a linear map, and let $y \in R^m \backslash g(K^{m-1})$. Then g is locally trivial at y.*

In other words, there is a neighborhood U of y and a (PL) homeomorphism $g^{-1}(U) \cong g^{-1}(y) \times U$ such that the following diagram commutes.

$$\begin{array}{ccc} g^{-1}(U) & \cong & g^{-1}(y) \times U \\ & {\scriptstyle g|}\searrow \quad \swarrow{\scriptstyle pr_U} & \\ & U & \end{array}$$

Lemma 1 follows from the classical lemma that a simplicial map is a trivial projection over the interior of each simplex of the target (see [**6**, p. 236]). (This corresponds to Thom's "first isotopy lemma" for stratified sets [**5**, p. 217].)

In this paper, all homology groups will be ordinary mod 2 homology (based on locally finite, but possibly infinite, chains). If K is a simplicial complex, a simplicial mod 2 i-chain C in K will be identified with the subcomplex of K consisting of all faces of i-simplexes occurring with nonzero coefficient in K. A (mod 2) *n-circuit* is a simplicial complex Z^n which is a mod 2 cycle. In other words, Z^n is the union of its n-simplexes (or Z is *purely n-dimensional*), and the sum of all the n-simplexes of Z is a mod 2 cycle (i.e., every $(n-1)$-simplex is a face of an even number of n-simplexes). For example, any purely n-dimensional real analytic space can be triangulated by an n-circuit. If Z is an n-circuit, its *fundamental class* $[Z] \in H_n(Z)$ is the class represented by the sum of all the n-simplexes of Z. An *n-circuit with boundary* is a simplicial pair $(W, \partial W)$ such that W is purely n-dimensional, ∂W is collared in W, and the sum of all the n-simplexes in W is a mod 2 chain whose boundary is the sum of all the $(n-1)$-simplexes in ∂W. $[W] \in H_n(W, \partial W)$ denotes the homology class of this chain. (Note that ∂W is an $(n-1)$-circuit.)

A polyhedron X is purely n-dimensional if it has a purely n-dimensional triangulation K. The mod 2 boundary ∂X is then the subpolyhedron of X triangulated by the $(n-1)$-simplexes of K which are the face of an odd number of simplexes of K. (∂X may not be collared in X.) This definition is independent of the triangulation.

If $f : X \to Y$ is any map, let $D(f)$ be the closure of $\{y$, there exist $x_1 \neq x_2, f(x_1) = y = f(x_2)\}$, the set of double points of f.

LEMMA 2. *Let Z be an n-circuit, and let $f : Z \to R^k$ be a general linear immersion. Then $\partial D(f) = \emptyset$.*

The proof is left to the reader.

2. Euler singularities. A (mod 2) *Euler complex* is a simplicial complex V such that the link of each simplex has even Euler number χ. This is equivalent to the condition that V has odd local Euler number at each point. For example, a closed combinatorial manifold is an Euler complex. Furthermore, any real analytic space can be triangulated by an Euler complex, i.e., it is an *Euler space* [**10**]. Note that a purely n-dimensional Euler complex is an n-circuit. An *Euler complex with boundary* is a simplicial pair $(V, \partial V)$ such that ∂V is collared in V, and the link of each simplex in $V \backslash \partial V$ has even Euler number. It follows that ∂V is an Euler complex, and if $\sigma < \partial V$, $\chi(\text{link}(\sigma, V))$ is odd.

If V is an Euler complex and $\sigma < V$, then link(σ) is an Euler complex. If V is finite, and $\chi(V)$ is even, then the cone on V is also an Euler complex. Conversely, if V bounds the finite Euler complex W, then $\chi(V)$ is even. (This can be seen easily by expressing the barycentric subdivision of W as the union of the open "dual cones" of the simplexes of W. For another proof, see [**1**].)

Let V be an n-dimensional Euler complex (without boundary), and let $f : V^n \to R^k$, $n \geq k$, be a profile. The *Euler singularity* $S(f)$ is a mod 2 $(k-1)$-chain of V defined as follows.

Let $\sigma^{k-1} < V$. Let P_0 be the hyperplane of R^k containing $f(\sigma)$, and let P_1 and P_2 be the closures of the two components of $R^k \backslash P_0$. Let $L = \text{link}(\sigma)$ and $L_i = L \cap f^{-1}(P_i)$, $i = 0, 1, 2$. We let $\sigma < S(f)$ if and only if $\chi(L_1)$ and $\chi(L_2)$ are both even.

In fact, $\chi(L_1)$ and $\chi(L_2)$ have the same parity. L_0, L_1, and L_2 are subpolyhedra of L, $L = L_1 \cup L_2$, and $L_0 = L_1 \cap L_2$. Thus $\chi(L) = \chi(L_1) + \chi(L_2) - \chi(L_0)$. But $\chi(L)$ is even since V is an Euler complex, so we must show that $\chi(L_0)$ is even.

LEMMA 3. *$\chi(L_0)$ is even.*

PROOF. Let $g : L \to R^1$ be the composition of $f\, L$ with the projection $\pi : R^k \to R^1$ where R^1 is identified with the line perpendicular to P_0. Let $y = \pi(P_0)$, so $L_0 = g^{-1}(y)$. Now $y \notin g(L^0)$, or else f would collapse some k-simplex of star (σ). Thus L_0 is bicollared in L, by Lemma 1. It follows that L_0 is an Euler space, since L is. In fact L_0 is the boundary of a compact Euler space, so $\chi(L_0)$ is indeed even.

To see that L_0 bounds, let $G = \pi \circ (f|D)$, where D is the join of the barycenter b of σ with $L = \text{link}(\sigma)$. $G(b) = y$, but any point y' near y lies off $G(D^0)$. Now $g^{-1}(y') \cong L_0$ for y' near y, and $g^{-1}(y')$ bounds $G^{-1}(y')$, which is an Euler space with boundary, since G is locally trivial at y' and D is an Euler complex with boundary. This completes the proof.

A simple example of the Euler singularity of a map is the "apparent contour" of a triangulated surface M^2 in R^3. Figure 1 shows a "profile" of a polyhedral-torus.

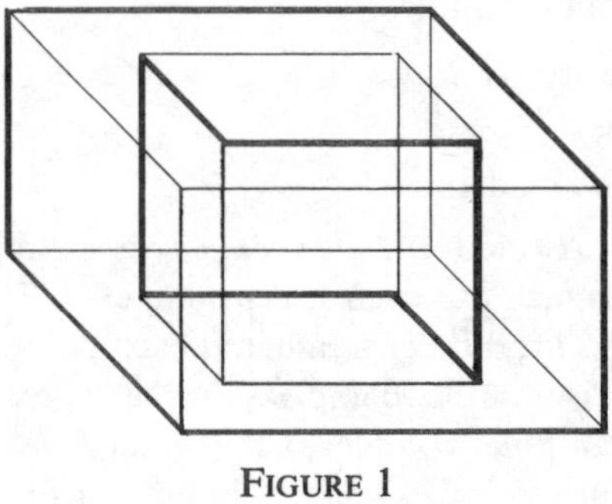

FIGURE 1

THEOREM 1. *Let $f : V^n \to R^k$, $n \geqq k$, be a profile of the Euler complex V. $S(f)$ is a* mod 2 *cycle, and the homology class of $S(f)$ is independent of f and subdivision of V.*

PROOF. Let $\tau^{k-2} < V$, and let $L = \text{link}(\tau)$. Let $g : L \to R^2$ be the composition of $f|L$ with the projection $\pi : R^k \to R^2$, where R^2 is identified with the plane perpendicular to the $(k-2)$-plane P spanned by $f(\tau)$. Let $y = \pi(P)$. Now if $\sigma_i^{k-1} > \tau$, let v_i be the vertex of L with $\sigma_i^{k-1} = v_i * \tau$. Let $L_i = \text{link}(\sigma_i, V) = \text{link}(v_i, L)$, and let $H_i \subset R^k$ be the closed half-plane to the right of the ray from y to $g(v_i)$. Finally, let $J_i = L_i \cap g^{-1}(H_i)$.

Now $\sigma_i^{k-1} < S(f)$ if and only if $\chi(J_i)$ is even, i.e., the number of simplexes

$\omega < \text{star}(v_i, L)$ such that $g(\omega) \subset H_i$ is odd. ($\omega < \text{star}(v_i, L)$ if $\omega = v_i * \rho$, $\rho \in L_i$, or $\omega = v_i$.) But if $\omega < L$, and $y \notin g(\omega)$, then $g(\omega) \subset H_i$ for precisely one vertex v_i of ω. However, if $y \in g(\omega)$, $g(\omega) \not\subset H_i$ for all vertices v_i of L. Therefore,

$$\# \{\sigma_i^{k-1} > \tau, \sigma_i^{k-1} < S(f)\} \equiv \# \{\omega < L, y \notin g(\omega)\} \pmod 2.$$

Now it follows from Lemma 1, by the same method used to prove Lemma 3, that $g^{-1}(y)$ is an Euler space which bounds, so $\chi(g^{-1}(y))$ is *even*. Furthermore, the intersection of $g^{-1}(y)$ with the simplexes of L induces a cell structure (in fact, a triangulation) of $g^{-1}(y)$, so $\chi(g^{-1}(y)) \equiv \# \{\omega < L, y \in g(\omega)\} \pmod 2$.

It follows that the number of simplexes $\sigma_i^{k-1} > \tau$ which lie in $S(f)$ has the same parity as $\chi(L)$, which is even since V is an Euler complex. Thus $S(f)$ is a mod 2 cycle.

Now suppose that V' is a subdivision of B. Then $f : V' \to R^k$ is still linear, but it may not be a profile. However, given $\varepsilon > 0$, there is a profile $f' : V' \to R^k$ which is an ε-approximation to f, and it is clear from the definition of $S(f)$ that if ε is small enough, $S(f')$ will be just the subdivision of $S(f)$.

Finally, $S(f)$ is independent of f because any two linear maps $f_0, f_1 : V \to R^k$ are piecewise-linearly homotopic, so there is a linear map $H : (J, V_0, V_1) \to (R^k \times I, R^k \times 0, R^k \times 1)$, where (J, V_0, V_1) is a simplicial subdivision of $(V \times I, V \times 0, V \times 1)$. Furthermore, we can choose H so that for $i = 0, 1$, if $t \in I$ is near i, then $H(x, t) = (f_i(x), t)$ for all $x \in V$. We now approximate H closely by a "relative profile" $\tilde{H} : (J, \partial J) \to (R^k \times I, R^k \times \partial I)$, for which the singularity $S(\tilde{H})$ will be a k-chain in J such that $\partial S(\tilde{H})$ is the sum of the subdivisions of $S(f_0)$ and $S(f_1)$. The details of this argument are left to the reader. This completes the proof of Theorem 1.

If V^n is an Euler complex, let $\mathscr{S}_i(V) \in H_i(V; \mathbf{Z}/2)$, $0 \leqq i < n$, be the homology class of $S(f)$, for f an $(i + 1)$-profile of V.

THEOREM 2. *If M is a smooth n-manifold, and $t : V \to M$ is a smooth triangulation, then $t_*(\mathscr{S}_i(V))$ is Poincaré dual to $w^{n-i}(M)$, the $(n - i)$th Stiefel-Whitney class of the tangent bundle of M.*

If $f : M^n \to R^k$, $n \geqq k$, is a smooth map, the smooth *singularity* $S(f)$ is $\{x \in M, \text{rank } (df_x) < k\}$. Pontryagin [7] and Thom [11] observed that for "generic" f, $S(f)$ is a mod 2 $(k - 1)$-cycle whose homology class is dual to $w^{n-k+1}(M)$. It should be possible to prove Theorem 2 by a direct argument, using Cairns' technique of triangulating a smooth manifold in Euclidean space by a linear complex close to it [3]. The crux of the proof is that the singularity of a generic smooth map can be detected using the local mod 2 Euler number. For according to Thom's stratification theory, away from "higher order" singularities (which have lower dimension), $S(f)$ is just the "suspension" of a classical Morse singularity [12, p. 78].

Instead of carrying through this delicate argument, we can derive Theorem 2 as a corollary of the following observation, due independently to Banchoff. If V is an Euler complex, and V' is its barycentric subdivision, the linear map $f : V' \to R^k$ which sends the barycenter of an i-simplex to $(i, i^2, \cdots, i^k)$ is a profile, and $S(f)$ is the sum of all the $(k - 1)$-simplexes in V', the classical Stiefel cycle. Now Theorem

2 follows from Whitney's theorem that the homology class of this cycle is Poincaré dual to w^{n-k+1} on a smooth manifold (see [4]).

REMARK. An *integral Euler complex* of dimension n is a complex V which has local Euler number $(-1)^n$ at each point. Our techniques specialize to integral Euler spaces, and we get that $\mathcal{S}_i(V^n)$ is an *integral* cycle when $n-i$ is odd (cf. [2], [4]).

It is not necessary to use a piecewise-linear structure to define Euler singularities. It is also possible to work with smooth stratifications—but some underlying geometric structure on a space V is needed to insure that any map $f: V \to R^k$ can be approximated by a "nice" stratified map. Suppose that $V \subset R^m$. Then we will call $f: V \to R^k$ a *profile* if there is a Whitney stratification of V such that f has maximal rank on each stratum, and f satisfies "condition A" for maps (cf. [5]). In principle, our techniques work for such a map, yielding $S(f)$ as a sum of $(k-1)$-strata which is a mod 2 cycle. In particular, suppose that V is an n-dimensional real analytic variety in R^m. If $n \geqq k$, is there an open dense set $\mathcal{P}$ in the Grassmann manifold of k-planes in R^m such that if $P \in \mathcal{P}$, then the projection $V \to P$ is a profile? (The fact that analytic varieties can be triangulated—indeed, the method used to triangulate them—indicates that this is true.)

3. Normal classes. Let Z be a (mod 2) n-circuit, and let $f: Z^n \to R^k$, $n < k$, be a full map. The *singularity* $\bar{S}(f)$ is a mod 2 $(2n-k-1)$-chain of Z defined as follows.

Let $\sigma^{2n-k-1} < Z$. Let $f_\sigma = f|N(\sigma)$. Then $D(f_\sigma)$, the closure of the double points of f_σ, is a purely $(2n-k)$-dimensional polyhedron in R^k (if it is not empty), since f is full and Z is purely n-dimensional. Now since f_σ is linear, it is the join of $f|\sigma$ with $f|\text{link}(\sigma)$. It follows that if $D(f_\sigma)$ is nonempty, $f(\sigma) \subset D(f_\sigma)$. We let $\sigma < \bar{S}(f)$ if and only if $f(\sigma) \subset \partial D(f_\sigma)$. (In fact, $f(\sigma) \subset \partial D(f_\sigma)$ means there is an odd number of pairs $\{\omega_1^n, \omega_2^n\}$ of n-simplexes in star(σ) such that $f((\omega_1^n)^\circ) \cap f((\omega_2^n)^\circ) \neq \varnothing$.)

$\bar{S}(f)$ is the "primary obstruction" to F being an immersion—by definition, if f is an immersion, $\bar{S}(f) = 0$. If $f: M^2 \to R^3$ is a full map of a triangulated surface into 3-space, f fails to be an immersion only at vertices of M. Figure 2 illustrates such a singularity, the apex of the cone on a polyhedral "figure 8".

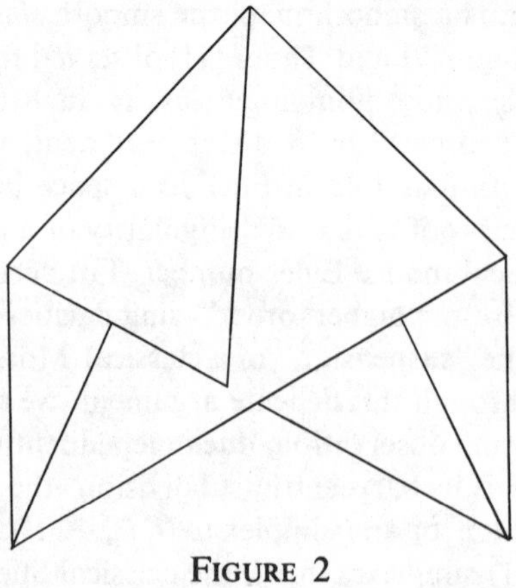

FIGURE 2

THEOREM 3. *Let Z be an n-circuit, and let $f: Z^n \to R^k$, $n < k$, be a full map. $\bar{S}(f)$ is a* mod 2 *cycle, and the homology class of $\bar{S}(f)$ is independent of f and subdivision of Z.*

Intuitively, $\bar{S}(f)$ is a cycle because it is locally a boundary. It is locally the pull-back of the boundary of $D(f)$, and $D(f)$ pulls back *locally* to two sheets in Z, either of which has as its boundary a piece of $\bar{S}(f)$.

LEMMA 3. *Let $\tau^{2n-k-2} < Z$.*

(i) *If $\sigma^{2n-k-1} > \tau$, then $f(\sigma^\circ)$ contains no double points of $f_\tau = f|N(\tau)$.*

(ii) *If $x \in$ star(τ), $f(x) \in \partial D(f_\tau)$ if and only if $x \in \bar{S}(f)$.*

PROOF. Since f is full, if $\omega^i > \tau$, $\dim(f(\omega^\circ) \cap f(\sigma^\circ)) \leqq i + (2n - k - 1) - k$, which is less than $2n - k - 1$ since $i \leqq n$ and $n < k$. But ω and σ have τ^{2n-k-2} as a common face, so $f(\omega)$ and $f(\sigma)$ intersect at most along $f(\tau)$. This implies (i).

(ii) holds because Z is a circuit. $f_\tau : N(\tau) \to R^k$ is a general linear map, and $N(\tau)$ is an n-circuit with boundary. (ii) says that the singularities of f_τ arise solely from *local* self-intersections, which follows from Lemma 2. We first note that if $\dim(\sigma) > 2n - k - 1$, f_σ is an embedding. For if $\omega_1^i, \omega_2^j > \sigma$, $\dim(f(\omega_1^\circ) \cap f(\omega_2^\circ)) \leqq i + j - k \leqq 2n - k \leqq \dim(\sigma)$. Now if $x \in$ star(τ), but x lies off the $(2n - k - 1)$-skeleton, a neighborhood of x does not intersect the $(2n - k - 1)$-skeleton under f_τ, by (i). But we have just seen that f_τ is locally an embedding off the $(2n - k - 1)$-skeleton. Therefore, $D(f_\tau)$ is a cycle near $f(x)$, by Lemma 2, so $f(x) \notin \partial D(f_\tau)$. On the other hand, if x lies in a $(2n - k - 1)$-simplex σ, x has a neighborhood U such that $y_1 \in U$ and $f_\tau(y_1) = y_2$ implies $y_2 \in U$, again by (i). Thus $f(x) \in \partial D(f_\tau)$ if and only if $f(x) \in \partial D(f_\sigma)$, i.e., $x \in \sigma < \bar{S}(f)$. This proves (ii).

We now prove Theorem 3. Let $\tau^{2n-k-2} < Z$. Let $\varphi : J \to K$ be a triangulation of f_τ such that

(a) φ collapses no simplexes;

(b) τ is a simplex of J;

(c) each $\sigma^{2n-k-1} > \tau$ is a simplex of J.

We can guarantee (a) because f embeds each simplex, and (c) by Lemma 3(i).

Let G be the set of simplexes $\omega <$ link(τ, J) such that $\varphi(\omega * \tau) \subset D(f_\tau)$. ($D(f_\tau)$ is clearly triangulated by a subcomplex L of K containing $f(\tau)$.) Since $D(f_\tau)$ is $(2n- k)$-dimensional, G is a 1-dimensional complex, or graph. Let $H = \varphi(G)$, a subcomplex of K with $f(\tau) * H = N(f(\tau), L)$. Now if v is a vertex of G, $v * \tau$ is a $(2n - k - 1)$-simplex $\sigma <$ star(τ, J), and $\varphi(\sigma) \subset \partial D(f_\tau)$ if and only if $\varphi(v)$ has odd valence in H (i.e., $\varphi(v)$ is incident to an odd number of edges of H). Lemma 3(ii) (and property (c) of φ) implies that $\varphi(\sigma) \subset \partial D(f_\tau)$ if and only if $\sigma < \bar{S}(f)$. Therefore, $v \mapsto \sigma = v * \tau$ induces a bijection between the vertices v of G such that $\varphi(v)$ has odd valence, and the $(2n - k - 1)$-simplexes $\sigma < \bar{S}(f)$ incident to τ.

Now Lemma 3(i) implies that if $v * \tau = \sigma < \bar{S}(f)$, there is no vertex v' of G with $\varphi(v') = \varphi(v)$. Thus the vertices v of G such that $v * \tau < \bar{S}(f)$ *inject* onto the vertices of odd valence of H. But any finite graph has an *even* number of vertices of odd valence, so there are an even number of $\sigma^{2n-k-1} < \bar{S}(f)$ incident to τ. Therefore, $\bar{S}(f)$ is a mod 2 cycle.

Once we know that $\bar{S}(f)$ is a cycle, the proof that the homology class of $\bar{S}(f)$ is independent of f and subdivision of Z is essentially the same as the proof for $S(f)$, $n \geqq k$ (Theorem 1). We again leave the details to the reader. This completes the proof of Theorem 3.

If Z is an n-circuit, let $\bar{\mathscr{S}}_i(Z) \in H_i(Z; \mathbf{Z}/2)$ be the homology class of $\bar{S}(f)$, for $f: Z^n \to R^{2n-i-1}$ a full map. Since we have assumed $n < k = 2n - i - 1$, $0 \leqq i < n - 1$.

We can also define $\bar{\mathscr{S}}_{n-1}(Z)$, as follows. Let $f: Z^n \to R^n$ be a linear map which embeds each simplex. The singularity $\bar{S}(f)$ will be a mod 2 $(n-1)$-chain of Z. Let $\sigma^{n-1} < Z$, and let P_0 be the hyperplane of R^n containing $f(\sigma)$. Let P_1 and P_2 be the closures of the two components of $R^n \backslash P_0$. Let $l = \#\{\omega^n > \sigma, f(\omega) \subset P_1\} - \#\{\omega^n > \sigma, f(\omega) \subset P_2\}$. l is defined only up to sign, and l is even, since the number of n-simplexes incident to σ is even. (Z is a mod 2 n-circuit.) Let the coefficient of σ in $\bar{S}(f)$ be $l/2$ mod 2. It is easy to see that the homology class of $\bar{S}(f)$ is none other than $\beta[Z]$, where $\beta: H_n(Z; \mathbf{Z}/2) \to H_{n-1}(Z; \mathbf{Z}/2)$ is the Bockstein homomorphism of the coefficient sequence $0 \to \mathbf{Z}/2 \to \mathbf{Z}/4 \to \mathbf{Z}/2 \to 0$.

CONJECTURE A. If M is a smooth n-manifold, and $t: Z \to M$ is a smooth triangulation, then $t_*(\bar{\mathscr{S}}_i(Z))$ is Poincaré dual to $\bar{w}^{n-i}(M)$, the $(n-i)$th Stiefel-Whitney class of the stable normal bundle of M.

If $f: M^n \to R^k$, $n \leqq k$, is a generic smooth map, the smooth singularity $S(f)$ is a mod 2 $(2n - k - 1)$-cycle whose homology class is dual to $\bar{w}^{k-n+1}(M)$ [**11**, p. 80]. As in the case of Theorem 2, it should be possible to prove this conjecture by a direct triangulation argument. In fact, $S(f) = f^{-1}(\partial D(f))$ for such a generic smooth map f. This is because there is a family of singularities of the "cone on a figure 8" type (obtained by generalizing the figure 8 to higher dimensions), such that $S(f)$ is always the "suspension" of a singularity in this family (away from "higher order" singularities).

Of course, our conjecture is supported by the fact that $\bar{w}^{k-n+1}(M)$ is the primary obstruction to immersing M^n in R^k, since the normal bundle of an immersion of M in R^k would have dimension $n - k$.

4. Homology operations. Let H be a homology theory. A (stable) *homology operation* on H is a natural transformation $\mathscr{O}: H \to H$ which commutes with the boundary map of H. If H is ordinary (singular) mod 2 homology, Thom [**12**] has defined stable operations $\mathscr{O}_i$ for each $i \geqq 0$, which the lower degree by i, roughly as follows. Any nice space X (such as a polyhedron) can be embedded in R^N for N large, and there is a Lefschetz duality isomorphism

$$\mathscr{D}: H^*(R^N, R^N \backslash X) \to \tilde{H}_{N-*}(X).$$

If $x \in H_*(X)$, $\mathscr{O}_i(x) = \mathscr{D}\mathrm{Sq}^i \mathscr{D}^{-1}(x)$, where Sq^i is the ith Steenrod cohomology operation. The definition of $\mathscr{O}_i$ is independent of the embedding because Sq^i is stable.

According to Steenrod [**9**, p. 32], if c is the conjugation of the Steenrod algebra, the total Thom operation $\mathcal{O} = \sum_i \mathcal{O}_i$ satisfies

$$\langle \mathcal{O}(x), y \rangle = \langle x, c(\mathrm{Sq})(y) \rangle.$$

($c(\mathrm{Sq})$ is the inverse Steenrod operation.)

If M is a smooth submanifold of R^N, the isomorphism $\mathscr{D}$ followed by Poincaré duality in M can be identified with the inverse Thom isomorphism for the normal bundle of M in R^N. Thus Thom's definition of characteristic classes ($w(\eta) = \varphi^{-1} \mathrm{Sq}\varphi(1)$, where φ is the Thom isomorphism for the bundle η) implies that $\mathcal{O}_i[M]$ is Poincaré dual to $\bar{w}^i(M)$.

Now singular mod 2 homology can be viewed as the bordism theory based on mod 2 circuits. The function $Z^n \mapsto \bar{\mathscr{S}}_{n-i}(Z^n)$ (§3) defines a *geometric homology operation* θ_i in the following sense. θ_i assigns to each n-circuit K with boundary an element $\theta_i(K) \in H_{n-i}(K, \partial K)$ such that

(i) $\partial_* \theta_i(K) = \theta_i(\partial K)$, where $\partial_* : H_{n-i}(K, \partial K) \to H_{n-i-1}(\partial K)$.

(ii) If L is an "n-subcircuit" of K (i.e., ∂L is bicollared in K and $L \cap \partial K$ is a component of ∂K), then $r_* \, \theta_i(K) = \theta_i(L)$, where r_* is restriction to L:

$$\begin{array}{ccc} H_{n-i}(K, \partial K) & \to & H_{n-i}(K, K \backslash \mathrm{int}(L)) \\ & \searrow^{r_*} & \uparrow \approx \\ & & H_{n-i}(L, \partial L) \end{array}$$

There is a canonical bijection between such geometric operations θ and "abstract" stable operations $\mathcal{O}$. Given $\mathcal{O}$, define $\theta(K) = \mathcal{O}[K]$, where $[K] \in H_*(K, \partial K)$ is the fundamental class. Conversely, given θ, let $f : (K, \partial K) \to (X, A)$ represent $\alpha \in H_*(X; A)$. Then $\mathcal{O}(\alpha) = f_* \theta(K)$.

REMARK. In fact, every generalized homology theory is a bordism theory with certain singularities [**8**]. The correspondence between geometric and abstract operations holds for any such theory.

According to Thom, mod 2 homology is representable by smooth manifolds. Therefore, Conjecture A is equivalent to the following.

CONJECTURE B. The geometric homology operation $\theta_i(Z^n) = \bar{\mathscr{S}}_{n-1}(Z^n)$ equals Thom's operation $\mathcal{O}_i$.

Of course, we let θ_0 be the identity, and θ_1 is the Bockstein homomorphism, so $\theta_0 = \mathcal{O}_0$ and $\theta_1 = \mathcal{O}_1$. It is not hard to show, from the definition of $\bar{S}$, that if Z is an n-circuit, $\theta_n(Z) = 0$. On the other hand, $\mathcal{O}_n(x) = 0$ if x has degree $n > 0$ since $\langle \mathcal{O}(x), 1 \rangle = \langle x, c(\mathrm{Sq})1 \rangle = 0$.

Since stable homology and cohomology operations correspond by Lefschetz duality, as above, Conjecture B is equivalent to the following assertion.

CONJECTURE C. The stable cohomology operation Lefschetz dual to θ_i is the Steenrod operation Sq^i.

An obvious method for proving this statement is to check that the axioms for Sq [**10**, p. 1] hold for the operation dual to θ. Axioms 1, 2, and 4 are clear. Axiom 3, which says that if x has degree n, then $\mathrm{Sq}^n(x) = x^2$, can also be proved from the definition of θ by a geometrical argument. However, it is not clear how to prove Axiom 5, the Cartan formula, which is equivalent to the product formula for θ,

$\theta(X \times Y) = \theta(X) \times \theta(Y)$, where X, Y are circuits. Or, in other words

$$\theta_k(X \times Y) = \sum_{i+j=k} \theta_i(X) \times \theta_j(Y), \qquad k \geqq 0.$$

ADDED IN PROOF. Axiom 5 follows from Axioms 1–4 plus the fact that Sq is stable.

Wu [13] has defined mod 2 homology operations on a space X by applying P. A. Smith theory to the involution $(a, b) \to (b, a)$ of $X \times X$. These *Smith operations* Sm_i have an elegant combinatorial definition when X is a complex. Wu showed that $\langle \mathrm{Sm}(x), y\rangle = \langle x, Q(y)\rangle$, where $Q = c(\mathrm{Sq})$, so $\mathrm{Sm} = \mathcal{O}$. Thus, another way to prove our conjectures would be to show directly that $\mathrm{Sm} = \theta$.

Closely related to the problem of proving the product formula for θ is the following intriguing question. Suppose that V is a purely n-dimensional Euler space. The fact that $\mathcal{S}_{n-k}(V)$ is represented by the singularity of a map $V \to R^{n-(k-1)}$, while $\bar{\mathcal{S}}_{n-k}(V)$ is represented by the singularity of a map $V \to R^{n+(k-1)}$, "jette une lumière curieuse sur ce qu'on a appelé 'la dualité de Whitney' " [11, p. 56]. Is there an analogue of Whitney duality for algebraic varieties with singularities? (Of course the "normal" classes cannot *determine* the "tangential" classes, since the former are natural with respect to maps of degree 1 mod 2.)

REFERENCES

1. E. Akin, *Stiefel-Whitney homology classes and bordism theories* (to appear).

2. T. Banchoff, *Stiefel-Whitney homology classes and singularities of projections for polyhedral manifolds*, these PROCEEDINGS.

3. S. S. Cairns, *A simple triangulation method for smooth manifolds*, Bull. Amer. Math. Soc. **67** (1961), 389–390. MR **26** #6978.

4. S. Halperin and D. Toledo, *Stiefel-Whitney homology classes*, Ann. of Math. (2) **96** (1972), 511–525.

5. J. Mather, *Stratifications and mappings*, Dynamical Systems (M. Peixoto, Ed.), Academic Press, New York, 1973, pp. 195–232.

6. J. Milnor and J. Stasheff, *Characteristic classes*, Ann. of Math. Studies, no. 76, Princeton Univ. Press, Princeton, N. J., 1974.

7. L. S. Pontrjagin, *Vector fields on manifolds*, Mat. Sb. **24** (**66**) (1949), 129–162; English transl., Amer. Math. Soc. Transl. (1) **7** (1962), 220–278. MR **10**, 727.

8. S. Buoncristiano, C. P. Rourke, and B. J. Sanderson, *A geometric approach to homology theory* (to appear).

9. N. E. Steenrod, *Cohomology operations*, Lectures by N. E. Steenrod and revised by D. B. A. Epstein, Ann. of Math. Studies, no. 50, Princeton Univ. Press, Princeton, N. J., 1962. MR **26** #3056.

10. D. Sullivan, *Combinatorial invariants of analytic spaces*, Proc. Liverpool Singularities Sympos. I (1969/70), Lecture Notes in Math., vol. 192, Springer-Verlag, Berlin and New York, 1971, pp. 165–168. MR **43** #4063.

11. R. Thom, *Les singularités des applications différentiables*, Ann. Inst. Fourier (Grenoble) **6** (1955/56), 43–87. MR **19**, 310.

12. ———, *Espaces fibrés en sphères et corrés de Steenrod*, Ann. Sci. École Norm. Sup. (3) **69** (1952), 109–182. MR **14**, 1004.

13. Wu Wen-Tsün, *A theory of imbedding, immersion, and isotopy of polytopes in a Euclidean space*, Science Press, Peking, 1965.

14. J. Cheeger, *A combinatorial formula for Stiefel-Whitney classes*, Topology of Manifolds, Markham, New York, 1970.

INSTITUTE FOR ADVANCED STUDY

Proceedings of Symposia in Pure Mathematics
Volume 27, 1975

ISOLATED CRITICAL POINTS OF COMPLEX FUNCTIONS

JOHN MILNOR

A complex valued complex analytic function $f(z_1, \cdots, z_n)$ is said to have a *critical point* wherever $\partial f/\partial z_1 = \cdots = \partial f/\partial z_n = 0$. Let V be the vector space consisting of all germs of analytic functions $f(z_1, \cdots, z_n)$ which vanish, and have a critical point, at the origin. Let G be the group consisting of all germs of complex analytic diffeomorphisms from a neighborhood of the origin to a neighborhood of the origin in C^n. Then G operates on V by composition, and an orbit $f \circ G \subset V$ is called *a right equivalence class* of germs. Such an orbit has finite codimension in V if and only if the origin is isolated as a critical point of f.

Following V. Arnol'd an orbit $f \circ G$ is called *simple* if it possesses a neighborhood intersecting only finitely many other orbits. Arnol'd shows that every simple orbit is represented by one and only one of the following polynomials:

$$\begin{array}{lll} A_\mu & z_1^{\mu+1} + z_2^2 + \cdots + z_n^2, & \mu \geqq 1, \\ D_\mu & z_1^{\mu-1} + z_1 z_2^2 + z_3^2 + \cdots + z_n^2, & \mu \geqq 4, \\ E_6 & z_1^4 + z_2^3 + z_3^2 + \cdots + z_n^2, & \\ E_7 & z_1^3 z_2 + z_2^3 + z_3^2 + \cdots + z_n^2, & \\ E_8 & z_1^5 + z_2^3 + z_3^2 + \cdots + z_n^2. & \end{array}$$

Here the subscript μ is equal to the codimension plus one. Those orbits with codimension $\mu - 1 \leqq 4$ (or rather their real analogues) play a central role in Thom's theory of elementary catastrophes. In Thom's terminology, A_2, A_3, A_4, and A_5 are called repectively the "fold", "cusp", "swallow's tail", and "butterfly";

AMS (MOS) subject classifications (1970). Primary 57D70, 14B05.

while D_4 is called an "elliptic or hyperbolic umbilic", and D_5 is called a "parabolic umbilic".

In the special case $n = 3$ these polynomials (or the related algebraic singularities) have been listed earlier by Hirzebruch, Artin, Tjurina and others. Tjurina's characterization is particularly elegant. Let $f(z_1, \cdots, z_n)$ have an isolated critical point at the origin; and suppose that n is odd, say $n = 2m + 1$. Let X_f be the compact $4m$-dimensional manifold-with-boundary obtained by intersecting a small disk $|z_1|^2 + \cdots + |z_n|^2 \leqq \varepsilon$ with the nonsingular hypersurface $f^{-1}(c)$, where $|c|$ is very small but not zero. Then the middle homology group $H_{2m}(X_f; \boldsymbol{Z})$ is free abelian of rank say μ, and is spanned by elements α whose self-intersection number $\alpha \cdot \alpha$ is either $+2$ (for m even) or -2 (for m odd). Tjurina shows that this intersection pairing is positive definite (for m even) or negative definite (for m odd) if and only if f is right equivalent to one of the polynomials listed above. The corresponding homology lattice $H_{2m}(X_f; Z)$ is isomorphic to the lattice generated by the corresponding classical root system A_μ, D_μ, E_6, E_7, or E_8.

In the case $n = 3$ these same simple orbits have been characterized by Brieskorn as the only ones for which the 3-dimensional manifold ∂X_f has finite fundamental group.

I want to thank Arnol'd for communicating his research to me before publication.

REFERENCES

V. I. Arnol'd, *Singularities of smooth mappings*, Uspehi Mat. Nauk **23** (1968), no. 1 (139), 3–44= Russian Math. Surveys **23** (1968), no. 1, 1–43. MR **37** #2243.

———, *Normal forms for functions near degenerate critical points, the Weyl groups of A_k, D_k, E_k and Lagrangian singularities*, Funkcional. Anal. i Priložen. **6** (1972), no. 4, 3–25 = Functional Anal. Appl. **6** (1972), 254–272.

———, *Lectures on bifurcations in versal families*, Uspehi Mat. Nauk **27** (1972), no. 5 (167), 119–183 = Russian Math. Surveys **27** (1972), no. 5, 54–123.

M. Artin, *On isolated rational singularities of surfaces*, Amer. J. Math. **88** (1966), 129–136. MR **33** # 7340.

N. Bourbaki, *Éléments de mathématique*. Fasc. XXXIV. *Groupes et algèbres de Lie*. Chaps. 4–6, Actualités Sci. Indust., no. 1337, Hermann, Paris, 1968. MR **39** # 1590.

E. Brieskorn, *Rationale Singularitäten komplexer Flächen*, Invent. Math. **4** (1967/68), 336–358. MR **36** # 5136.

F. Hirzebruch, *The topology of normal singularities of an algebraic surface (d'après Mumford)*, Séminaire Bourbaki: 1962/63, Exposé 250, Benjamin, New York, 1966. MR **33** # 5420j.

J. Milnor, *Singular points of complex hypersurfaces*, Ann. of Math. Studies, no. 61, Princeton Univ. Press, Princeton, N. J., 1968. MR **39** # 969.

R. Thom, *Topological models in biology*, Topology **8** (1969), 313–335. MR **39** # 6629.

G. N. Tjurina, *Topological properties of isolated singularities of complex spaces of codimension one*, Izv. Akad. Nauk SSSR Ser. Mat. **32** (1968), 605–620 = Math. USSR Izv. **2** (1968), 557–571. MR **37** # 3053.

P. Wagreich, *Singularities of complex surfaces with solvable local fundamental group*, Topology **11** (1972), 51–72.

INSTITUTE FOR ADVANCED STUDY

Proceedings of Symposia in Pure Mathematics
Volume 27, 1975

THE CHERN-WEIL CONSTRUCTION

HOWARD OSBORN

Abstract. If a smooth bundle is replaced by the endomorphism algebra $\bigoplus_r \operatorname{End} \bigwedge^r F$ it induces over the de Rham algebra, then a lemma of Nathan Jacobson assigns a curvature $K \in \operatorname{End} F$ to any derivation $\mathscr{D}$ such that $\operatorname{tr} \mathscr{D} = d \operatorname{tr}$. One then establishes that the resulting characteristic classes $[\operatorname{tr} K^p]$ are independent of $\mathscr{D}$ by defining transgression forms algebraically. The transgression forms can be computed by means of a new product in $\bigoplus_r \operatorname{End} \bigwedge^r F$ for which $\operatorname{tr}(A \times B) = (\operatorname{tr} A)(\operatorname{tr} B)$.

The classical Chern-Weil construction assigns differential forms to the curvature of a connection in a smooth vector bundle, and the Chern-Weil theorem asserts that the forms are closed and that the resulting cohomology classes depend only on the bundle itself; the standard proofs of the Chern-Weil theorem use an endomorphism algebra associated to the bundle, either explicitly or implicitly. The present exposition simplifies both the construction and the theorem by assigning the differential forms ab initio to the endomorphism algebra via a lemma of Nathan Jacobson; this procedure leads to algebraic expressions of the transgression forms relating the Chern-Weil forms which arise from different connections.

The classical construction is sketched in § 1 as background for later developments. Purely algebraic properties of certain endomorphism algebras are presented in §§ 2 and 3, and these properties are used in §§ 4 and 5 to construct characteristic classes and to prove the corresponding Chern-Weil theorems. Finally, some algebraic expressions for transgression forms are computed in § 6 by means of a surprising result proved in § 7: the endomorphism algebras possess a new product for which the trace is an *algebra* homomorphism.

AMS (MOS) subject classifications (1970). Primary 53C05, 57D20; Secondary 15A15.
Key words and phrases. Chern-Weil construction, connections, curvature, derivations, Euler class, Pontrjagin class, trace, transgression forms.

1. The classical construction. The following exposition of the classical Chern-Weil construction of Pontrjagin classes is based on suggestions appearing in **[1]**; it is presented here as a point of departure.

Let $\mathscr{F}$ be the $C^\infty(M)$-module of smooth sections of a smooth real n-plane bundle on a smooth manifold M; in particular, let $\mathscr{E}$ be the $C^\infty(M)$-module of smooth sections of the cotangent bundle of M, and let $\wedge\mathscr{E}$ be the de Rham algebra. One usually defines a connection in $\mathscr{F}$ to be a real linear map $\mathscr{F} \to \mathscr{E} \otimes \mathscr{F}$ satisfying certain conditions. Such a map instantly induces a real linear map of $\wedge\mathscr{E} \otimes \mathscr{F}$ into itself, however, so that for the left $\wedge\mathscr{E}$-module $F = \wedge\mathscr{E} \otimes \mathscr{F}$ we may define a *connection in* F to be any real linear map $D : F \to F$ of degree $+1$ such that $D(\theta \cdot s) = d\theta \cdot s + (-1)^p \theta \cdot Ds$ for any $\theta \in \wedge^p\mathscr{E}$ and any $s \in F$. The composition DD is the *curvature* $K : F \to F$ of D, which one easily verifies to be a $\wedge\mathscr{E}$-module homomorphism of degree $+2$, and one forms the determinant $\det(I - (1/2\pi)K) \in \wedge^{2*}\mathscr{E}$ by regarding K as a $\wedge^{2*}\mathscr{E}$-module homomorphism. The classical *Chern-Weil theorem* for Pontrjagin classes asserts (i) that $\det(I - (1/2\pi)K)$ is closed, (ii) that the inhomogeneous de Rham cohomology class $[\det(I - (1/2\pi)K)] \in H^{2*}(M; \boldsymbol{R})$ is independent of the choice of D, and (iii) that $[\det(I - (1/2\pi)K)] \in H^{4*}(M; \boldsymbol{R})$. In fact $[\det(I - (1/2\pi)K)]$ is the total real Pontrjagin class of the original bundle.

In what follows we ignore the normalizing factor $-1/2\pi$ and we recall that $\det(I + K) = \sum_p \operatorname{tr} \wedge^p K$ for the induced endomorphisms $\wedge^p K$ of $\wedge^p F$. With this understanding the Chern-Weil theorem for Pontrjagin classes asserts (i) that each $\operatorname{tr} \wedge^p K \in \wedge^{2p}\mathscr{E}$ is closed, (ii) that $[\operatorname{tr} \wedge^p K] \in H^{2p}(M; \boldsymbol{R})$ is independent of the choice of D, and (iii) that $[\operatorname{tr} \wedge^p K] = 0$ when p is odd.

One easily proves (iii) in the case $p = 1$ as follows:

LEMMA 1.1. *The 2-form* tr K *is exact for any connection* D *in* F.

PROOF. If $s_1, \cdots, s_n$ and $t_1, \cdots, t_n$ are bases for a common restriction of F, and if $Ds_i = \sum_j \theta_i^j s_j$ and $Dt_i = \sum_j \varphi_i^j t_j$, then $t_1 = As_1, \cdots, t_n = As_n$ for a unique $A \in \operatorname{End} \mathscr{F}$ which one uses to verify that $d \sum_j \theta_j^j = d \sum_j \varphi_j^j$. Hence tr D exists as an element of $\mathscr{E}$/cocycles, and since $d \sum_j \theta_j^j$ is the restriction of tr K it follows that $\operatorname{tr} K = d(\operatorname{tr} D)$.

Lemma 1.1 is one of two properties of connections in F which we use to motivate a new Chern-Weil construction. Here is the other one:

LEMMA 1.2. *Let* End F *denote the left* $\wedge\mathscr{E}$ *-module of right* $\wedge\mathscr{E}$ *-module endomorphisms of* F, *and for any* $A \in \operatorname{End} F$ *homogeneous of degree* $d(A)$ *set* $\mathscr{D}A = DA - (-1)^{d(A)}AD$. *Then* $\mathscr{D}A$ *is an element of* End F *homogeneous of degree* $d(A) + 1$, *and the real linear operator* $\mathscr{D}$: End $F \to$ End F *of degree* $+1$ *satisfies*

(i) $\mathscr{D}(AB) = (\mathscr{D}A)B + (-1)^{d(A)}A(\mathscr{D}B)$, *and*

(ii) $\mathscr{D}(\theta \cdot I) = (d\theta) \cdot I$ *for* $\theta \in \wedge\mathscr{E}$.

PROOF. Verification.

2. Jacobson's lemma. A *derivation of degree k* of a graded algebra $\mathscr{A}$ over $\boldsymbol{R}$ is any real linear map $\Delta : \mathscr{A} \to \mathscr{A}$ of degree k such that $\Delta(AB) = (\Delta A)B + (-1)^{kd(A)} A(\Delta B)$. For example, Lemma 1.2 provides a derivation $\mathscr{D}$ of degree $+1$ of End F, and one easily verifies that the composition $\mathscr{D}\mathscr{D}$ is a derivation of degree $+2$ such that $\mathscr{D}\mathscr{D}(\theta \cdot I) = 0$.

For any $K \in \mathscr{A}$ of degree k the *inner derivation* $\Delta_K : \mathscr{A} \to \mathscr{A}$ of degree k is given by $\Delta_K A = KA - (-1)^{kd(A)} AK$. For example, the derivation $\mathscr{D}\mathscr{D}$ of degree $+2$ arising from Lemma 1.2 is the inner derivation Δ_K for the curvature $K \in \text{End } F$ of the connection D in F, as one easily checks.

Conversely, the following lemma of Jacobson ([2] or [3]) will be used to capture the essential features of a curvature K directly from any $\mathscr{D} : \text{End } F \to \text{End } F$ satisfying the conclusions of Lemma 1.2, with no appeal to the original connection $D : F \to F$.

LEMMA 2.1. *Let* $\mathscr{A}$ *be the real algebra* End $\boldsymbol{R}^n$, *concentrated in degree* 0; *then any derivation* $\Delta : \mathscr{A} \to \mathscr{A}$ *is an inner derivation* Δ_K *for some* $K \in$ End $\boldsymbol{R}^n$.

PROOF. Since $\Delta(AB) = (\Delta A)B + A(\Delta B)$ one can represent $\mathscr{A}$ in the endomorphism ring of $\boldsymbol{R}^n \oplus \boldsymbol{R}^n$ by

$$A \mapsto \begin{pmatrix} A & \Delta A \\ 0 & A \end{pmatrix}.$$

If $\boldsymbol{R}^n \oplus \boldsymbol{R}^n$ is represented by column vectors then $\boldsymbol{R}^n \oplus 0$ is an invariant subspace, and since representations of $\mathscr{A}$ are completely reducible one can add elements of $\boldsymbol{R}^n \oplus 0$ to the canonical basis vectors of $0 \oplus \boldsymbol{R}^n$ to conclude that

$$\begin{pmatrix} I & B \\ 0 & C \end{pmatrix}^{-1} \begin{pmatrix} A & \Delta A \\ 0 & A \end{pmatrix} \begin{pmatrix} I & B \\ 0 & C \end{pmatrix} = \begin{pmatrix} A & 0 \\ 0 & A' \end{pmatrix}$$

for a similarity $A \mapsto A'$ and elements $B, C \in \mathscr{A}$ with C invertible. By applying C to the new basis vectors one can then replace $\left(\begin{smallmatrix} I & B \\ 0 & C \end{smallmatrix}\right)$ by some $\left(\begin{smallmatrix} I & K \\ 0 & I \end{smallmatrix}\right)$, and by solving for $\left(\begin{smallmatrix} A & \Delta A \\ 0 & A \end{smallmatrix}\right)$ one finds $A' = A$ and $\Delta A = KA - AK$ as desired.

LEMMA 2.2. *Let* $\mathscr{D} : \text{End } F \to \text{End } F$ *be any derivation of degree* $+1$ *such that* $\mathscr{D}(\theta \cdot I) = (d\theta) \cdot I$; *then* $\mathscr{D}\mathscr{D}$ *is an inner derivation* Δ_K *for some* $K \in$ End F *of degree* $+2$.

PROOF. $\mathscr{D}\mathscr{D}$ is uniquely defined by its action on the subring of endomorphisms of degree 0, and by composing $\mathscr{D}\mathscr{D}$ with evaluations on 2-vector fields one obtains derivations of the subring into itself. Since $dd = 0$ and $\mathscr{D}$ is of degree $+1$ it also follows that the preceding derivations are linear over $C^\infty(M)$, and one then applies Jacobson's lemma.

The element K is not uniquely defined, either in Lemma 2.1 or in Lemma 2.2; for the same results hold if one replaces K by K plus a multiple of the identity with the appropriate degree, 0 or 2. In the latter case Lemma 1.1 suggests a partial normalization, however, which we incorporate in the following definition:

DEFINITION 2.3. A *connection in* End F is any derivation $\mathscr{D}$ of degree $+1$ such that $\mathscr{D}(\theta \cdot I) = (d\theta) \cdot I$ for $\theta \in \wedge\mathscr{E}$. A *curvature* of $\mathscr{D}$ is any $K \in$ End F of degree $+2$ such that $\mathscr{D}\mathscr{D} = \Delta_K$ and such that $\operatorname{tr} K \in \wedge^2 \mathscr{E}$ is exact.

Lemma 1.2 guarantees the existence of connections in End F; alternatively, one can use partitions of unity to construct connections in End F in precisely the same fashion normally applied to construct connections in F itself. In any event Definition 2.3 contains all the data needed for the later Chern-Weil construction.

3. The endomorphism algebra. Composition is one product in the direct sum $\bigoplus_p \operatorname{End} \wedge^p \boldsymbol{R}^n$, and we now define a second product in the same direct sum. There is a corresponding second product for which $\bigoplus^p \operatorname{End} \wedge^p F$ becomes an algebra over $\wedge\mathscr{E}$. (Third products will be defined in § 6.)

Suppose that $A \in \operatorname{End} \wedge^p \boldsymbol{R}^n$ and that $B \in \operatorname{End} \wedge^q \boldsymbol{R}^n$. Then there is a unique $A\cdot B \in \operatorname{End} \wedge^{p+q} \mathbf{R}^n$ such that

$$A\cdot B(s_1 \wedge \cdots \wedge s_{p+q}) = \sum_{\pi} \varepsilon_\pi\, A(s_{\pi 1} \wedge \cdots \wedge s_{\pi p}) \wedge B(s_{\pi(p+1)} \wedge \cdots \wedge s_{\pi(p+q)}),$$

where the sum is computed over all permutations π of $(1, \cdots, p+q)$ such that $\pi 1 < \cdots < \pi p$ and $\pi(p+1) < \cdots < \pi(p+q)$, and where $\varepsilon_\pi = \pm 1$ according to the parity of π. For convenience we call $A\cdot B$ the *second product* of A and B, the *first product* being the composition AB in the special case $p = q$.

LEMMA 3.1. $\bigoplus_p \operatorname{End} \wedge^p \boldsymbol{R}^n$ *is an associative and commutative algebra over* $\boldsymbol{R}$ *with respect to the second product, commutativity meaning* $A\cdot B = B\cdot A$ *with no* $\pm$ *signs.*

PROOF. Verification.

If $A \in \operatorname{End} \boldsymbol{R}^n$ then the p-fold second product $A\cdot \cdots \cdot A \in \operatorname{End} \wedge^p \boldsymbol{R}^n$ will be denoted A^p; the identity in $\operatorname{End} \wedge^p \boldsymbol{R}^n$ will be denoted I_p. Trivially $I_1^p = p!I_p$, so that $I_p \cdot I_q = \binom{p+q}{p} I_{p+q}$. Furthermore $A^n = n!(\det A)I_n$ and more generally $A^p = p! \wedge^p A$ for any $A \in \operatorname{End} \boldsymbol{R}^n$ and any $p \leqq n$. Finally, if $B \in \operatorname{End} \wedge^p \boldsymbol{R}^n$ then $B\cdot I_{n-p} = (\operatorname{tr} B) I_n$, an identity which may be regarded as the definition of the trace.

LEMMA 3.2. $\bigoplus_p \operatorname{End} \wedge^p \boldsymbol{R}^n$ *is generated by* $\operatorname{End} \boldsymbol{R}^n$ *with respect to second products; that is, each vector space* $\operatorname{End} \wedge^p \boldsymbol{R}^n$ *is spanned by elements* $A_1 \cdot \cdots \cdot A_p$, *where* $A_1, \cdots, A_p \in \operatorname{End} \boldsymbol{R}^n$.

PROOF. Let $s_1, \cdots, s_n$ be any basis of $\boldsymbol{R}^n$ and define $E_{i,j} \in \operatorname{End} \boldsymbol{R}^n$ by $E_{i,j}(s_i) = s_j$ and $E_{i,j}(s_k) = 0$ for $k \neq i$. Then if $s_{i_1} \wedge \cdots \wedge s_{i_p} \neq 0$ the product $E_{i_1,j_1} \cdot \cdots \cdot E_{i_p,j_p}$ carries $s_{i_1} \wedge \cdots \wedge s_{i_p}$ into $s_{j_1} \wedge \cdots \wedge s_{j_p}$ and annihilates all other basis elements of $\operatorname{End} \wedge^p \boldsymbol{R}^n$.

One effect of Lemma 3.2 is that any derivation of $\operatorname{End} \boldsymbol{R}^n$ extends to a unique derivation Δ of $\bigoplus_p \operatorname{End} \wedge^p \boldsymbol{R}^n$, satisfying both $\Delta(AB) = (\Delta A)B + A(\Delta B)$ for the composition $AB \in \operatorname{End} \wedge^p \boldsymbol{R}^n$ of two elements of $\operatorname{End} \wedge^p \boldsymbol{R}^n$, as well as $\Delta(A\cdot B) = (\Delta A)\cdot B + A\cdot(\Delta B)$ for the second product $A\cdot B \in \operatorname{End} \wedge^{p+q} \boldsymbol{R}^n$ of $A \in \operatorname{End} \wedge^p \boldsymbol{R}^n$ and $B \in \operatorname{End} \wedge^q \boldsymbol{R}^n$.

We now replace $\boldsymbol{R}^n$ by the $\wedge\mathscr{E}$-module F in the preceding construction to obtain the *endomorphism algebra* $\bigoplus_p \operatorname{End} \wedge^p F$ over $\wedge\mathscr{E}$. The definition of the *second product* $A\cdot B \in \operatorname{End} \wedge^{p+q} F$ of $A \in \operatorname{End} \wedge^p F$ and $B \in \operatorname{End} \wedge^q F$ is the same as the definition of second prodcts in $\bigoplus_p \operatorname{End} \wedge^p \boldsymbol{R}^n$ for elements $s_1, \cdots, s_{p+q} \in F$ of the special form $1 \otimes \sigma_1, \cdots, 1 \otimes \sigma_{p+q}$, where $\sigma_1, \cdots, \sigma_{p+q} \in \mathscr{F}$. The additional $\pm$ signs resulting from (homogeneous) coefficients in $\wedge\mathscr{E}$ cause no difficulty in the general case; but the endomorphisms A and B themselves may have coefficients in $\wedge\mathscr{E}$, which we assume to be homogeneous of degrees $d(A)$ and $d(B)$, respectively. One verifies that the second product in $\bigoplus_p \operatorname{End} \wedge^p F$ is commutative in the expected sense that $A \cdot B = (-1)^{d(A)d(B)} B \cdot A$ for endomorphisms $A \in \operatorname{End} \wedge^p F$ and $B \in \operatorname{End} \wedge^q F$ of degrees $d(A)$ and $d(B)$, respectively. The degrees $d(A)$ and $d(B)$ may be unrelated to p and q.

Lemma 3.2 implies that $\bigoplus_p \operatorname{End} \wedge^p F$ is generated by End F with respect to second products, and this permits one to extend any connection in End F to a unique *connection* $\mathscr{D}$ *in* $\bigoplus_p \operatorname{End} \wedge^p F$, satisfying (i) $\mathscr{D}(AB) = (\mathscr{D}A)B + (-1)^{d(A)}A(\mathscr{D}B)$ for compositions $AB \in \operatorname{End} \wedge^p F$, (ii) $\mathscr{D}(A \cdot B) = (\mathscr{D}A) \cdot B + (-1)^{d(A)} A \cdot \mathscr{D}B$ for second products, and (iii) $\mathscr{D}(\theta I_p) = (d\theta) I_p$ for any $\theta \in \wedge\mathscr{E}$ and the identity $I_p \in \operatorname{End} \wedge^p F$, for each $p = 0, \cdots, n$.

LEMMA 3.3. *For any* $A \in \operatorname{End} \wedge^p F$ *and any connection* $\mathscr{D}$ *in* $\bigoplus_p \operatorname{End} \wedge^p F$ *it follows that* $\operatorname{tr} \mathscr{D}A = d \operatorname{tr} A$.

PROOF. A standard argument gives $\mathscr{D}I_{n-p} = 0$ for the identity $I_{n-p} \in \operatorname{End} \wedge^{n-p} F$, and condition (iii) then implies that

$$(\operatorname{tr} \mathscr{D}A)\, I_n = \mathscr{D}A \cdot I_{n-p} = \mathscr{D}\,(A \cdot I_{n-p}) = \mathscr{D}\,((\operatorname{tr} A)\, I_n) = d(\operatorname{tr} A)\, I_n.$$

Conversely, if a real linear map $\mathscr{D}$ satisfies the conclusion of Lemma 3.3 one trivially has $\mathscr{D}(\theta I_p) = (d\theta)I_p$. Thus *a connection in* $\bigoplus_p \operatorname{End} \wedge^p F$ *is any derivation* $\mathscr{D}$ (with respect to both first and second products) *such that* $\operatorname{tr} \mathscr{D} = d \operatorname{tr}$.

Lemma 3.3 also implies the following *Bianchi identity:*

LEMMA 3.4. *Let* K *be a curvature of a connection* $\mathscr{D}$ *in* End F; *then* $\mathscr{D}K = 0$.

PROOF. For any $A \in \operatorname{End} F$ we compute with first product (composition) to find

$$\begin{aligned} K(\mathscr{D}A) - (\mathscr{D}A)K &= \Delta_K(\mathscr{D}A) = \mathscr{D}\mathscr{D}\mathscr{D}A = \mathscr{D}(\Delta_K A) \\ &= (\mathscr{D}K)A + K(\mathscr{D}A) - (\mathscr{D}A)K - (-1)^{d(A)} A(\mathscr{D}K) \end{aligned}$$

so that $(\mathscr{D}K)A = (-1)^{d(A)} A(\mathscr{D}K)$. Hence $\mathscr{D}K$ lies in the center of End F, giving

$$\mathscr{D}K = (1/n)\,(\operatorname{tr} \mathscr{D}K)\, I_1 = (1/n)\, d(\operatorname{tr} K)\, I_1 = 0$$

since $\operatorname{tr} K$ is exact by hypothesis.

4. Pontrjagin classes. The first of the following lemmas is a special case of the standard lemma used to show for any Lie algebra of derivations that the inner derivations form an ideal:

LEMMA 4.1. *For any connection* $\mathscr{D}$ *in* End F *and any* $R \in \operatorname{End} F$ *of degree* $+1$ *the inner derivations* Δ_R *and* $\Delta_{\mathscr{D}R}$, *of degrees* $+1$ *and* $+2$, *respectively, satisfy* $\Delta_R \mathscr{D} + \mathscr{D}\Delta_R = \Delta_{\mathscr{D}R}$.

PROOF. Verification.

LEMMA 4.2. *For the preceding* $\mathscr{D}$ *and* R *the sum* $\mathscr{D} + \Delta_R$ *is a connection* $\mathscr{D}'$ *in* End F; *conversely, for any two connections* $\mathscr{D}$ *and* $\mathscr{D}'$ *there is an* $R \in$ End F *of degree* $+1$ *such that* $\mathscr{D}' - \mathscr{D} = \Delta_R$.

PROOF. The first statement is a trivial verification. Conversely, $(\mathscr{D}' - \mathscr{D})(\theta I) = 0$ for any $\theta \in \wedge\mathscr{E}$, so that $\mathscr{D}' - \mathscr{D}$ is a derivation over $\wedge\mathscr{E}$ of the algebra End F. For the subalgebra of elements of degree 0 one then composes $\mathscr{D}' - \mathscr{D}$ with evaluations on vector fields to obtain derivations of the subalgebra into itself to which Jacobson's lemma applies in an evident fashion.

Here is a more precise version of Lemma 4.2:

LEMMA 4.3. *For any connections* $\mathscr{D}_0$ *and* $\mathscr{D}_1$ *in* End F *with curvatures* K_0 *and* K_1, *respectively, and for any* $t \in [0, 1]$, *let* $\mathscr{D}_t$ *be the connection* $(1-t)\,\mathscr{D}_0 + t\,\mathscr{D}_1$ *with curvature* K_t *such that* $\operatorname{tr} K_t = \operatorname{tr}((1-t)K_0 + t\,K_1)$; *then* $\mathscr{D}_1 - \mathscr{D}_0 = \Delta_R$ *for an* $R \in$ End F *of degree* $+1$ *such that* $(d/dt)K_t = \mathscr{D}_t R$.

PROOF. Since $\operatorname{tr} K_0$ and $\operatorname{tr} K_1$ are exact by hypothesis it follows that $\operatorname{tr}(d/dt)K_t = \operatorname{tr}(K_1 - K_0) = d\theta$ for some $\theta \in \wedge^1\mathscr{E}$. According to Lemma 4.2 there is some $R \in$ End F such that $\mathscr{D}_1 - \mathscr{D}_0 = \Delta_R$, and since $\Delta_I = 0$ one can prescribe $\operatorname{tr} R = \theta$. Then Lemma 4.1 gives

$$\begin{aligned}\Delta_{(d/dt)K_t} &= (d/dt)(\mathscr{D}_t\,\mathscr{D}_t) = ((d/dt)\,\mathscr{D}_t)\,\mathscr{D}_t + \mathscr{D}_t((d/dt)\,\mathscr{D}_t)\\ &= \Delta_R\,\mathscr{D}_t + \mathscr{D}_t\,\Delta_R = \Delta_{\mathscr{D}_t R}\end{aligned}$$

in addition to $\operatorname{tr}\,(d/dt)\,K_t = d\theta = d\operatorname{tr} R = \operatorname{tr}\mathscr{D}_t R$ so that $(d/dt)\,K_t = \mathscr{D}_t R$ as claimed.

LEMMA 4.4. End F *admits an involution* τ ; *that is, there is a real linear map* τ : End $F \to$ End F *such that*

$$^\tau(AB) = (-1)^{d(A)d(B)}\,{}^\tau B\,{}^\tau A, \qquad \tau^2 = I, \qquad \textit{and} \qquad \operatorname{tr}{}^\tau A = (-1)^{d(A)} \operatorname{tr} A.$$

PROOF. Any metric in the original bundle induces a nondegenerate bilinear map $\langle\ ,\ \rangle : F \otimes F \to \wedge\mathscr{E}$, and for any $A \in$ End F one defines the transpose ${}^\tau A$ as usual.

LEMMA 4.5. *For the preceding involution there is a connection* $\mathscr{D}$ *in* End F *such that* $\tau\mathscr{D} = \mathscr{D}\tau$.

PROOF. Let D be a connection in F itself which is metric with respect to $\langle\ ,\ \rangle$ in the usual sense that $\langle Ds, t\rangle \pm \langle s, Dt\rangle = d\langle s, t\rangle$, and let $\mathscr{D}$ be the induced connection in End F as in Lemma 1.2.

LEMMA 4.6. *Let* K *be the curvature of the preceding connection* $\mathscr{D}$ *such that* $\operatorname{tr} K = 0$; *then* ${}^\tau K = -K$.

PROOF. For any $A \in$ End F it follows that $K\,{}^\tau A - {}^\tau A K = \mathscr{D}\mathscr{D}{}^\tau A = {}^\tau(\mathscr{D}\mathscr{D}A) = {}^\tau(KA - AK) = {}^\tau A\,{}^\tau K - {}^\tau K\,{}^\tau A$, hence that $(K + {}^\tau K)\,{}^\tau A = {}^\tau A(K + {}^\tau K)$. Thus $K + {}^\tau K$ lies in the center of End F and $\operatorname{tr}(K + {}^\tau K) = 2 \operatorname{tr} K = 0$, so that $K + {}^\tau K = 0$.

Here is the Chern-Weil construction of Pontrjagin classes, where we ignore the normalizing factor $-1/2\pi$ as explained in § 1.

PROPOSITION 4.7. *Let K be a curvature of a connection $\mathscr{D}$ in* End F, *and let K^p denote the p-fold second product* $K \cdot \cdots \cdot K \in \operatorname{End} \wedge^p F$; *then* (i) $d \operatorname{tr} K^p = 0$, (ii) *the de Rham cohomology class* $[\operatorname{tr} K^p] \in H^{2p}(M; \mathbf{R})$ *depends only on* End F, *and* (iii) $[\operatorname{tr} K^p] = 0$ *for odd p.*

PROOF. (i) $d \operatorname{tr} K^p = \operatorname{tr} \mathscr{D} K^p = \operatorname{tr} p\, K^{p-1} \cdot \mathscr{D} K = 0$ by the Bianchi identity. (ii) Define $\mathscr{D}_t = \mathscr{D} + t\,\Delta_R$ and K_t as in Lemma 4.3. According to Lemma 4.2 it suffices to show that $\operatorname{tr} K_1^p - \operatorname{tr} K_0^p$ is exact; equivalently, it suffices to show that $(d/dt) \operatorname{tr} K_t^p$ is exact for any t. But $(d/dt) \operatorname{tr} K_t^p = \operatorname{tr} p\, K_t^{p-1} \cdot \mathscr{D}_t R = \operatorname{tr} \mathscr{D}_t (p\, K_t^{p-1} \cdot R)$ $= d \operatorname{tr} (p\, K_t^{p-1} \cdot R)$ by Lemma 4.3, the Bianchi identity, and Lemma 3.3. (iii) For the curvature K appearing in Lemma 4.6 it follows that $\operatorname{tr} K^p = \operatorname{tr} {}^\tau(K^p) = \operatorname{tr} ({}^\tau K)^p$ $= (-1)^p \operatorname{tr} K^p$.

It follows from part (ii) of the preceding Chern-Weil theorem that $\operatorname{tr} K_1^p - \operatorname{tr} K_0^p$ $= d\Theta^p$ for a *transgression form* $\Theta^p = p \int_{t=0}^1 K_t^{p-1} \cdot R \; dt \in \wedge^{2p-1} \mathscr{E}$. In fact, in § 6 we shall show that $\operatorname{tr} K_t^p = \operatorname{tr} K_0^p + \sum_{q=1}^{2p-1} t^q\, d\, \Theta_q^p$ for any real number t and *transgression coefficients* $\Theta_q^p \in \wedge^{2p-1}\mathscr{E}$ which can be expressed algebraically in terms of K, R, and $\mathscr{D} R$.

5. Euler classes. If F is constructed from the smooth sections of a smooth n-plane bundle then End $\wedge^n F$ is free of rank 1 over $\wedge \mathscr{E}$, even when the original bundle is not oriented. In particular, if $n = 2m$ the second product End $\wedge^m F \otimes$ End $\wedge^m F \to$ End $\wedge^{2m} F$ provides a nondegenerate bilinear form, and hence a unique involution σ : End $\wedge^m F \to$ End $\wedge^m F$ such that $({}^\sigma A)B \cdot C = (-1)^{d(A)d(B)}$ $B \cdot AC$ for every $A, B, C \in \operatorname{End} \wedge^m F$, where $({}^\sigma A)B$ and AC are compositions.

LEMMA 5.1. $\sigma \mathscr{D} = \mathscr{D} \sigma$ *for any connection $\mathscr{D}$ in* End $\wedge^m F$.

PROOF. $\mathscr{D}$ extends to End $\wedge^{2m} F$ in the obvious way, and one expands the identity $\mathscr{D}({}^\sigma AB \cdot C) = (-1)^{d(A)d(B)}\, \mathscr{D}(B \cdot AC)$ to find

$$(\mathscr{D}^\sigma A)B \cdot C = (-1)^{(d(A)+1)d(B)}\; B \cdot (\mathscr{D} A)\, C$$

as required.

Now let τ : End $\wedge^m F \to$ End $\wedge^m F$ be the involution induced by a metric as in Lemma 4.4; then the composition $\sigma\tau$ is an automorphism of End $\wedge^m F$. A well known analog of Jacobson's lemma implies that every automorphism of End $\wedge^m F$ is an inner automorphism; hence there is an invertible $* \in \operatorname{End} \wedge^m F$ such that ${}^\sigma({}^\tau A) = *^{-1} A\, *$ for every $A \in \operatorname{End} \wedge^m F$. Finally, one normalizes $*$ up to a $\pm$ sign by requiring $\det * = 1$; this can be accomplished globally whenever the original $2m$-plane bundle is oriented.

DEFINITION 5.2. Given a metric in an oriented $2m$-plane bundle, the *Hodge operator* $* \in \operatorname{End} \wedge^m F$ is defined (up to ± 1) by the conditions ${}^\sigma({}^\tau A) = *^{-1} A\, *$ and $\det * = 1$.

One easily verifies that $\sigma\tau = \tau\sigma$, which implies $** A = A **$ for every $A \in$

End $\wedge^m F$. Thus $**$ is an element of the center of End $\wedge^m F$ with det $** = 1$; hence $** = \pm I_m$.

LEMMA 5.3. *If $\mathscr{D}$ is a metric connection in* End $\wedge^m F$, *satisfying $\tau\mathscr{D} = \mathscr{D}\tau$ as in Lemma* 4.5, *then $\mathscr{D} * = 0$.*

PROOF. Lemma 5.1 and the condition $\tau\mathscr{D} = \mathscr{D}\tau$ imply that $\mathscr{D}(*^{-1} A*) = *^{-1}(\mathscr{D} A) *$ so that

$$(*^{-1} A*)(*^{-1}\mathscr{D}*) - (*^{-1}\mathscr{D}*)(*^{-1} A*) = \mathscr{D}(*^{-1} A*) - *^{-1}(\mathscr{D}A)* = 0$$

It follows that $*^{-1}\mathscr{D}*$ lies in the center of End $\wedge^m F$ so that $\mathscr{D}* = \theta *$ for some $\theta \in \wedge^1 \mathscr{E}$; but $** = \pm I_m$, so that $2\theta I_m = \pm \mathscr{D}(**) = \mathscr{D} I_m = 0$, and hence $\theta = 0$.

PROPOSITION 5.4. *Let $n = 2m$, and let K be a curvature of any metric connection $\mathscr{D}$ in* End $\wedge^m F$; *then* (i) tr $(*K^m)$ *is closed in* $\wedge^{2m}\mathscr{E}$, *and* (ii) *the cohomology class* $[\mathrm{tr}\,(* K^m)] \in H^{2m}(M; \mathbf{R})$ *does not depend on the choice of $\mathscr{D}$.*

PROOF. (i) Since $\mathscr{D}* = 0$ one computes

$$d\,\mathrm{tr}\,(* K^m) = \mathrm{tr}\,\mathscr{D}(* K^m) = \mathrm{tr}\,(m*K^{m-1}\cdot \mathscr{D}K) = 0$$

by the Bianchi identity. (ii) As in the proof of Proposition 4.7 one has $(d/dt)\,\mathrm{tr}(* K_t^m) = d(m * (K_t^{m-1}\cdot R))$.

One can show (as in **[4]**) that tr $(* K^m)$ is itself independent of the metric, so that the cohomology class [tr $(*K^m)$] is thus determined by the bundle itself, up to a $\pm$ sign. The definition of $*$ is equivalent to the definition given in **[4]**, and up to a constant factor tr$(* K^m)$ is the Pfaffian of K; in fact,

$$(4\pi)^{-m}(-1)^{m(m-1)/2}(1/m!)\,[\mathrm{tr}(* K^m)]$$

is the usual Euler class in this case.

6. Transgression coefficients. We now re-examine the transgression form $\Theta^p \in \wedge^{2p-1}\mathscr{E}$ defined up to closed forms in § 4 by the relation tr $K_1^p -$ tr $K_0^p = d\Theta^p$.

PROPOSITION 6.1. *Let $\mathscr{D}_t$ be the connection $\mathscr{D} + t\Delta_R$ and let K_t be the curvature satisfying $(d/dt)K_t = \mathscr{D}_t R$ as in Lemma* 4.3; *then*

$$\mathrm{tr}\,K_t^p - \mathrm{tr}\,K_0^p = \sum_{q=1}^{2p-1} t^q\, d\,\Theta_q^p$$

for any $p \geqq 0$, where each transgression coefficient $\Theta_q^p \in \wedge^{2p-1}\mathscr{E}$ is of the form $\mathrm{tr}(\pi_q^p(K, \mathscr{D}R, RR)\cdot R)$ *for a polynomial π_q^p of degree $p-1$ (with respect to second products) in the degree 2 endomorphisms K_0, $\mathscr{D}R$, $RR \in$* End F.

PROOF. By integrating the relation $(d/dt)K_t = \mathscr{D}_t R$ one obtains

$$K_t - K_0 = \int_{\tau=0}^{t} (\mathscr{D} + \tau\Delta_R)R\,d\tau = t\mathscr{D}R + (t^2/2)\,\Delta_R R = t\mathscr{D}R + t^2 RR;$$

that is, $K_t = K_0 + t\mathscr{D}R + t^2RR$. Since (d/dt) tr $K_t^p = d\,\mathrm{tr}\,(p\,K_t^{p-1}\cdot R)$ as in Proposition 4.7 it then follows that

$$\operatorname{tr}(K_t^p - K_0^p) = d \operatorname{tr}\left(p \int_{\tau=0}^{t} (K_0 + \tau \mathscr{D}R + \tau^2 RR)^{p-1} \, d\tau \cdot R\right).$$

For example, $\Theta_1^p = \operatorname{tr}(p\, K^{p-1} \cdot R)$ for every $p \geqq 0$, where we now write K in place of K_0. The other transgression coefficients are more complicated, of course, since they result from powers of $K + t\mathscr{D}R + t^2\, RR$ rather than from powers of a polynomial of degree 1 in t. There may be some simplifications, however, and with that in mind we present an entirely different way of computing transgression coefficients, using a surprising result: there is a third product in $\bigoplus_r \operatorname{End} \bigwedge^r F$ for which the trace is an *algebra* homomorphism.

To define the third product we recall that the first product was defined as composition in each summand $\operatorname{End} \bigwedge^r F$. This extends to a product in the direct sum such that the image of $\operatorname{End} \bigwedge^p F \otimes \operatorname{End} \bigwedge^q F$ vanishes for $p \neq q$; that is, if $A = \sum_r A_r$ and $B = \sum_r B_r$ then $AB = \sum_r A_r B_r$. Trivially $\bigoplus_r \operatorname{End} \bigwedge^r F$ is an associative $\bigwedge\mathscr{E}$-algebra with respect to the (extended) first product, and the direct sum $\sum_r I_r$ of the identity endomorphisms $I_r \in \operatorname{End} \bigwedge^r F$ is the identity with respect to the (extended) first product. However, $\sum_r I_r$ is *not* the identity with respect to the second product, and one obtains a $\bigwedge\mathscr{E}$-module endomorphism α of $\bigoplus_r \operatorname{End} \bigwedge^r F$ by setting $\alpha A = A \cdot \sum_r I_r$. In fact α is an automorphism with inverse given by $\alpha^{-1} A = A \cdot \sum_r (-1)^r I_r$, as we shall verify in § 7.

DEFINITION 6.2. The *third product* in $\bigoplus_r \operatorname{End} \bigwedge^r F$ is given by $A \times B = \alpha^{-1}((\alpha A)(\alpha B))$, where $(\alpha A)(\alpha B)$ is a first product.

Although $\bigoplus_r \operatorname{End} \bigwedge^r F$ is graded over $\bigwedge\mathscr{E}$ with respect to third products as it was with respect to first and second products, the direct sum decomposition does not provide an additional grading with respect to third products, and for any elements A and B we write $A \times B = \sum_r (A \times B)_r \in \bigoplus_r \operatorname{End} \bigwedge^r F$; specifically if $A \in \operatorname{End} \bigwedge^p F$ and $B \in \operatorname{End} \bigwedge^q F$ then $(A \times B)_r$ does not necessarily vanish for $r \neq p + q$. The following result will be proved in § 7:

PROPOSITION 6.3. (i) $\bigoplus_r \operatorname{End} \bigwedge^r F$ *is an associative* $\bigwedge\mathscr{E}$*-algebra with respect to the third product.*

(ii) *The trace is an algebra homomorphism* $\operatorname{tr}\colon \bigoplus_r \operatorname{End} \bigwedge_r F \to \bigwedge\mathscr{E}$; *specifically, both* $\operatorname{tr}(A + B) = \operatorname{tr} A + \operatorname{tr} B$ *and* $\operatorname{tr}(A \times B) = (\operatorname{tr} A)(\operatorname{tr} B)$.

(iii) *If* $A \in \operatorname{End} \bigwedge^p F$ *and* $B \in \operatorname{End} \bigwedge^q F$ *then* $(A \times B)_{p+q} = A \cdot B$ *(second product), and* $(A \times B)_r = 0$ *except when* $\max(p, q) \leqq r \leqq p + q$.

For example, one easily verifies for any $A \in \operatorname{End} F$ and $B \in \operatorname{End} F$ that $A \times B = AB + A \cdot B$, and since $\bigoplus_r \operatorname{End} \bigwedge^r F$ is anticommutative with respect to second products one obtains a familiar result:

$$\begin{aligned} \operatorname{tr} AB - (-1)^{d(A)d(B)} \operatorname{tr} BA &= \operatorname{tr}(A \times B) - (-1)^{d(A)d(B)} \operatorname{tr}(B \times A) \\ &= (\operatorname{tr} A)(\operatorname{tr} B) - (-1)^{d(A)d(B)} (\operatorname{tr} B)(\operatorname{tr} A) = 0. \end{aligned}$$

In the following lemma we use both first and second products in $\bigoplus_r \operatorname{End} \bigwedge^r F$ to define polynomials in the degree 2 endomorphism $RR \in \operatorname{End} F$; for example, both $RRRR \in \operatorname{End} F$ and $RR \cdot RR \in \operatorname{End} \bigwedge^2 F$ are polynomials in RR which will be assigned degree 4.

LEMMA 6.4. *For any $R \in \text{End } F$ of degree $+1$ the trace annihilates every polynomial in RR.*

PROOF. First consider the composition $R^{(2q)}$ of q copies of RR; in this case $\text{tr } R^{(2q)} = \text{tr } RR^{(2q-1)} = -\text{tr } R^{(2q-1)} R = -\text{tr } R^{(2q)}$ so that $\text{tr } R^{(2q)} = 0$, by the identity $\text{tr } AB = (-1)^{d(A)d(B)} \text{tr } BA$. Then

$$\text{tr } RR \cdot RR = \text{tr } RR \times RR - \text{tr } RRRR = (\text{tr } RR)(\text{tr } RR) - \text{tr } RRRR = 0.$$

More generally, by induction on the degree, any polynomial of degree $2p$ can be expressed as a linear combination of the $2p$-fold composition $(RR)^{(p)}$ plus third products $A \times B$ where A and B are polynomials of degree $2q < 2p$; but $\text{tr}(A \times B) = (\text{tr } A)(\text{tr } B) = 0$ by the inductive hypothesis.

As an application of Lemma 6.4 observe that

$$\text{tr } K_t^p - \text{tr } K^p = \text{tr}(K + t\,\mathscr{D} R + t^2 RR)^p - \text{tr } K^p,$$

from which one would expect an additional transgression coefficient to account for the term in t^{2p}; but $\text{tr}(RR)^p = 0$ by Lemma 6.4.

As a further illustration of the use of third products we compute all transgression coefficients for $p = 2$ directly from the preceding expansion. Although this computation is more complicated than that arising from Proposition 6.1, it provides an alternative which may lead to simpler expressions for other transgression coefficients.

LEMMA 6.5. $\text{tr } RKR = \text{tr } K \cdot RR$.

PROOF. Since $d(R) = 1$ and $d(KR) = 3$ one has $\text{tr } R(KR) = -\text{tr } (KR)R$, hence

$$\begin{aligned} \text{tr } K \cdot RR - \text{tr } RKR &= \text{tr}(K \cdot RR + K(RR)) \\ &= \text{tr}(K \times RR) \\ &= (\text{tr } K)(\text{tr } RR) = 0 \end{aligned}$$

by Lemma 6.4.

LEMMA 6.6. $2 \ \text{tr } K \cdot RR = \text{tr}(KR - RK) \cdot R$.

PROOF. For any $A, B, C \in \text{End } F$ the three-fold third product $A \times B \times C$ is

$$ABC + [AB \cdot C + A \cdot BC + (-1)^{d(A)d(B)} B \cdot AC] + A \cdot B \cdot C,$$

and in particular

$$R \times K \times R = RKR + [RK \cdot R + R \cdot KR + K \cdot RR] + R \cdot K \cdot R.$$

Since $d(R) = 1$ one has $R \cdot R = 0$ and $(\text{tr } R)(\text{tr } R) = 0$, hence $R \cdot K \cdot R = 0$ and $\text{tr}(K \times K \times R) = (\text{tr } R)(\text{tr } K)(\text{tr } R) = 0$ so that

$$\text{tr } RKR + \text{tr } RK \cdot R + \text{tr } R \cdot KR + \text{tr } K \cdot RR = 0;$$

but $\text{tr } RKR = \text{tr } K \cdot RR$ by Lemma 6.5, and $RK \cdot R = -R \cdot RK$ since $d(R) = 1$ and $d(RK) = 3$.

LEMMA 6.7. $d \, \text{tr } RR \cdot R = 3 \ \text{tr } \mathscr{D}R \cdot RR$.

PROOF. Since

$$\operatorname{tr}(RR)R + \operatorname{tr} RR\cdot R = \operatorname{tr} RR \times R = (\operatorname{tr} RR)(\operatorname{tr} R) = 0$$

and

$$\operatorname{tr}(\mathscr{D} R)RR + \operatorname{tr} \mathscr{D}R\cdot RR = \operatorname{tr} \mathscr{D}R \times RR = (\operatorname{tr} \mathscr{D}R)(\operatorname{tr} RR) = 0$$

by Lemma 6.4, it suffices to show that $d \operatorname{tr} RRR = 3 \operatorname{tr}(\mathscr{D}R)RR$; but $\mathscr{D}(RRR) = (\mathscr{D}R)RR - R(\mathscr{D}R)R + RR(\mathscr{D}R)$, and since $\operatorname{tr} R((\mathscr{D}R)R) = -\operatorname{tr}((\mathscr{D}R)R)R$ and $\operatorname{tr}(RR)(\mathscr{D}R) = +\operatorname{tr}(\mathscr{D}R)RR$ one has $d \operatorname{tr} RRR = \operatorname{tr} \mathscr{D}(RRR) = 3 \operatorname{tr}(\mathscr{D}R)RR$ as desired.

We now verify Proposition 6.1 in the special case $p = 2$ directly from the expansion of $K_t^2 - K^2$, using the preceding lemmas as an illustration of the method of third products.

PROPOSITION 6.8.

$$\operatorname{tr} K_t^2 - \operatorname{tr} K^2 = td\,\Theta_1^2 + t^2 d\,\Theta_2^2 + t^3 d\Theta_3^2$$

for $\Theta_1^2 = \operatorname{tr}(2K\cdot R)$, $\Theta_2^2 = \operatorname{tr}(\mathscr{D}R\cdot R)$, *and* $\Theta_3^2 = \operatorname{tr}(\tfrac{2}{3} RR\cdot R)$.

PROOF.

$$\begin{aligned}\operatorname{tr} K_t^2 - \operatorname{tr} K^2 &= t\operatorname{tr}(2K\cdot\mathscr{D}R) + t^2 \operatorname{tr}(\mathscr{D}R\cdot\mathscr{D}R + 2K\cdot RR)\\ &\quad + t^3 \operatorname{tr}(2\,\mathscr{D}R\cdot RR) + t^4 \operatorname{tr}(RR\cdot RR),\end{aligned}$$

where $\operatorname{tr}(RR\cdot RR) = 0$ by Lemma 6.4 and $\operatorname{tr}(2K\cdot\mathscr{D}R) = d \operatorname{tr}(2K\cdot R) = d\Theta_1^2$ as in the general case. It remains to note that

$$\begin{aligned}&\operatorname{tr}(\mathscr{D}R\cdot\mathscr{D}R + 2K\cdot RR)\\ &\qquad = \operatorname{tr}(\mathscr{D}R\cdot\mathscr{D}R + (KR - RK)\cdot R) = \operatorname{tr}(\mathscr{D}R\cdot\mathscr{D} R + \mathscr{D}\mathscr{D}R\cdot R)\\ &\qquad = \operatorname{tr}\mathscr{D}(\mathscr{D}R\cdot R) = d \operatorname{tr}(\mathscr{D}R\cdot R) = d\Theta_2^2\end{aligned}$$

by Lemma 6.6, and that $\operatorname{tr}(2\mathscr{D}R\cdot RR) = d\operatorname{tr}(\tfrac{2}{3} RR\cdot R) = d\Theta_3^2$ by Lemma 6.7.

7. The trace is an algebra homomorphism. The first step in proving Proposition 6.3 is to verify that if $\alpha A = A\cdot \sum_r I_r$ then α is an automorphism as claimed.

LEMMA 7.1. $\sum_s I_s\cdot\sum_t(-1)^t I_t = I_0 + 0 \in \operatorname{End} \wedge^0 F \oplus \bigoplus_{r>0} \operatorname{End} \wedge^r F.$

PROOF. The 0th entry is trivially I_0, and for $r > 0$ one computes

$$\sum_{s+t=r} (-1)^t I_s\cdot I_t = \sum_{s=0}^{r} (-1)^{r-s}\binom{r}{s} I_r = (1 - 1)^r I_r = 0.$$

Since I_0 is the identity in $\bigoplus_r \operatorname{End} \wedge^r F$ with respect to second products it follows that α is an automorphism with $\alpha^{-1} A = A\cdot\sum_r(-1)^r I_r$, and we now prove the three parts of Proposition 6.3.

PROOF OF (i). Since $\bigoplus_r \operatorname{End} \wedge^r F$ is associative with respect to the first product, associativity with respect to the third product is the result of the trivial computation $\alpha((A \times B) \times C) = (\alpha A)(\alpha B)(\alpha C) = \alpha(A \times (B \times C))$, to which one applies α^{-1}.

PROOF OF (ii). By definition of the trace, $(\text{tr } A) I_n = A \cdot I_{n-p}$ for $A \in \text{End} \wedge^p F$, it follows that $(\text{tr } A) I_n$ is merely the nth component $(\alpha A)_n \in \text{End} \wedge^n F$ of αA, where $\text{End} \wedge^n F$ is free of rank one by hypothesis. But $\sum_r \alpha(A \times B)_r = \alpha(A \times B) = (\alpha A)(\alpha B) = \sum_r (\alpha A)_r (\alpha B)_r$, by definition of the first and third products, so that

$$\begin{aligned} \text{tr}(A \times B) I_n &= (\alpha(A \times B)) I_n = (\alpha A)_n (\alpha B)_n \\ &= (\text{tr } A)(\text{tr } B) I_n I_n = (\text{tr } A)(\text{tr } B) I_n. \end{aligned}$$

The proof of (iii) requires the following result:

LEMMA 7.2. *If $A \in \text{End } F$ then $A \times B = A \cdot B + (\alpha A) B$ for any $B \in \bigoplus_r \text{End} \wedge^r F$, where $A \cdot B$ is a second product and $(\alpha A) B$ is a first product.*

PROOF. Trivially αA acts like a derivation of $\bigoplus_r \text{End} \wedge^r F$, with respect to second products in the sense that $(\alpha A)(C \cdot B) = (\alpha A) C \cdot B + (-1)^{d(C)} C \cdot (\alpha A) B$ when C is homogeneous of degree $d(C)$. It follows for any $B \in \text{End} \wedge^r F$ (fixed r) and $C = I_{s-r} \in \text{End} \wedge^{s-r} F$ that $(\alpha A)(\alpha B)_s = (\alpha A)(I_{s-r} \cdot B) = (\alpha A) I_{s-r} \cdot B + I_{s-r} \cdot (\alpha A) B = (I_{s-r-1} \cdot A) \cdot B + I_{s-r} \cdot (\alpha A) B = I_{s-r-1} \cdot (A \cdot B) + I_{s-r} \cdot (\alpha A) B = \alpha(A \cdot B)_s + (\alpha A) B_s$, and by summing on s one obtains $(\alpha A)(\alpha B) = \alpha(A \cdot B) + \alpha((\alpha A) B)$, hence $A \times B = \alpha^{-1}((\alpha A)(\alpha B)) = A \cdot B + (\alpha A) B$.

PROOF OF (iii). According to Lemma 7.2, if $A \in \text{End } F$ and $B \in \text{End} \wedge^q F$ then all the summands $(A \times B)_r$ of $A \times B$ vanish except $(A \times B)_q = (\alpha A) B$ and $(A \times B)_{q+1} = A \cdot B$. By induction, if $A_1, \cdots, A_p$ belong to End F then all summands $(A_1 \times \cdots \times A_p \times B)_r$ of $(A_1 \times \cdots \times A_p) \times B$ vanish for $r > p + q$, and $(A_1 \times \cdots \times A_p \times B)_{p+q} = A_1, \cdots, A_p \cdot B$. It follows that all summands of $(A_1, \cdots, A_p) \times B$ above $p + q$ vanish, and that $((A_1, \cdots, A_p) \times B)_{p+q} = A_1, \cdots, A_p \cdot B$. Since every element of $\text{End} \wedge^p F$ is a sum of elements $A_1, \cdots, A_p$ this implies for any $A \in \text{End} \wedge^p F$ that $(A \times B)_r = 0$ for $r > p + q$ and that $(A \times B)_{p+q} = A \cdot B$. Finally, expanding $A \times B$ directly from the construction of α and α^{-1}, one finds an explicit formula $(A \times B)_r = \sum_s (-1)^s (A \cdot I_{r-s-p})(B \cdot I_{r-s-q}) \cdot I_s$ for $A \in \text{End} \wedge^p F$ and $B \in \text{End} \wedge^q F$, where $I_t = 0$ for $t < 0$; this yields $(A \times B)_r = 0$ for $r < \max(p, q)$.

As another application of the method of third products we prove a simple result which trivializes the computations needed for the details of Lemma 1.1, for example.

PROPOSITION 7.3. *If $A \in \text{End } F$ is nonsingular (of degree 0) then $\text{tr}(A^{-1} \mathscr{D} A) = d \ln |\det A|$ for any connection $\mathscr{D}$.*

PROOF. Let A^n be the n-fold second product $A \cdot \cdots \cdot A = n! (\det A) I_n$ as in § 3. Then since $\wedge^r F = 0$ for $r > n$ one has $(A^n \times A^{-1} \mathscr{D} A)_r = 0$ except for $r = n$, so that Lemma 7.2 implies $A^n \times A^{-1} \mathscr{D} A = A^n (I_{n-1} \cdot A^{-1} \mathscr{D} A) = n A^{n-1} \cdot \mathscr{D} A = \mathscr{D} A^n$. Hence $(\text{tr } A^n)(\text{tr } A^{-1} \mathscr{D} A) = \text{tr}(A^n \times A^{-1} \mathscr{D} A) = \text{tr}(\mathscr{D} A^n) = d(\text{tr } A^n)$, and one divides by $\text{tr } A^n$ $(= n! \det A)$ to complete the proof.

References

1. R. Bott and S.S. Chern, *Hermitian vector bundles and the equidistribution of the zeroes of their holomorphic sections*, Acta Math. **114** (1965), 71–112. MR **32** #3070.

2. N. Jacobson, *Abstract derivations and Lie algbras*, Trans. Amer. Math. Soc. **42** (1937), 206–224.

3. ———, *Theory of rings*, Math. Surveys, No. I, Amer. Math. Soc., Providence, R.I., 1943. MR **5**, 31.

4. H.Osborn, *Representations of Euler classes*, Proc. Amer. Math. Soc. **31** (1972), 340–346.

University of Illinois, Urbana-Champaign

Proceedings of Symposia in Pure Mathematics
Volume 27, 1975

CHARACTERISTIC CLASSES AND SINGULARITIES OF MAPPINGS

RICHARD PORTER*

The purpose of this note is to give a description of the characteristic classes of a differentiable manifold in terms of singularities of mappings of the manifold into Euclidean space. The formula applies to maps which are not necessarily generic but whose singularities satisfy a dimension condition which holds for generic maps. The result is an extension of some of the comments in chapter five of René Thom's paper [**4**]. Parallel results for Stiefel-Whitney classes in the P.L. category have recently been obtained by Thomas Banchoff and by Clint McCrory, see [**1**]. It appears that the problem of finding an analogous result for the Pontrjagin classes in the P.L. category has not yet been solved. This paper was motivated by some lectures of Robert MacPherson, and it is a pleasure to acknowledge the many helpful conversations with both Robert MacPherson and Thomas Banchoff.

Statement of the result for Stiefel-Whitney classes. Let $f: M^q \to R^{q-i+1}$ ($1 \leqq i \leqq q$) be a continuously differentiable map. Assume M^q is compact and let (X, Y) be a closed locally connected pair in M^q satisfying:

(a) $\{x \in M^q: \operatorname{rank}(df)_x \leqq q - i\} \subseteq X$.

(b) $\{x \in M^q: \operatorname{rank}(df)_x \leqq q - i - 1\} \subseteq Y$.

(c) $X - Y$ is a $(q - i)$-dimensional submanifold of M^q with a finite number of arc-components $A_1 \cdots A_l$.

(d) There is a neighborhood UY of Y in X such that the inclusions $Y \to UY$ and $X - UY \to X - Y$ are homotopy equivalences.

AMS (MOS) subject classifications (1970). Primary 57D20, 57D45; Secondary 55C05.

*Partially supported by grant NSF GP-34184.

It follows that $H_{q-i}(X, Y\colon Z_2)$ has rank l, with one generator $[A_k]$ for each arc-component of $X - Y$. There is a procedure, described below, which assigns to each arc-component A_k of $X - Y$ an integer mod 2, a_k. It will be shown that there is an element a in $H_{q-i}(X : Z_2)$ whose image in $H_{q-i}(X, Y : Z_2)$ is $\sum a_k[A_k]$ and whose image in $H_{q-i}(M^q : Z_2)$ is the Poincaré dual of the ith Stiefel-Whitney class of M^q. Thus if $H_{q-i}(Y : Z_2) = 0$, the element $\sum a_k[A_k]$ determines the ith Stiefel-Whitney class of M^q. If the map f is generic and the pair (X, Y) is defined by requiring the inclusions in conditions (a) and (b) to be equalities, then the pair (X, Y) satisfies conditions (c) and (d) and $H_{q-i}(Y : Z_2) = 0$.

DEFINITION OF a_k. Choose a point x_k in the arc-component A_k of $X - Y$ and choose a chart $R^{q-i} \times R^i \to M^q$ so that $R^{q-i} \times \{0\}$ is a chart for $X - Y$ at x_k. Denote by $\bar{f}$ the composition $R^{q-i} \times R^i \to M^q \to R^{q-i+1}$. From the assumptions it follows that the rank of $d\bar{f}$ is always $\geqq q - i$. Thus there is an element A in $\mathrm{GL}(q - i + 1)$ and a continuous map $R^{q-i} \times R^i \to \mathrm{GL}(q)$, $x \to B(x)$ so that

$$B(x)\begin{pmatrix} \partial\bar{f}_1/\partial x_1 \cdots \partial\bar{f}_{q-i+1}/\partial x_1 \\ \vdots \qquad\qquad \vdots \\ \partial\bar{f}_1/\partial x_q \cdots \partial\bar{f}_{q-i+1}/\partial x_q \end{pmatrix} A$$

has the form

$$\begin{array}{c} \\ q-i \end{array}\begin{pmatrix} \overset{q-i}{I} & \overset{1}{*} \\ 0 & c(x) \end{pmatrix}.$$

Since $R^{q-i} \times \{0\}$ contains the zeroes of $c(x)$, $c(x)$ restricted to $\{0\} \times R^i$ yields a map of pairs $(R^i, R^i - \{0\}) \to (R^i, R^i - \{0\})$. Set a_k equal to the degree of this map reduced mod 2.

EXAMPLES. The first Stiefel-Whitney class of the torus and real projective plane.

1. Identify the torus, $S^1 \times S^1$, with the locus

$$z^2 = (3 - (x^2 + y^2)^{1/2})\,((x^2 + y^2)^{1/2} - 1) \text{ in } R^3,$$

and define f to be projection onto the (y, z) plane. Set X equal to the union of the two circles $(y - 2)^2 + z^2 = 1$ and $(y + 2)^2 + z^2 = 1$ in the (y, z) plane with the two circles $x^2 + y^2 = 4$, $z = \pm 1$. Set Y equal to the four points $(0, 2, 1)$, $(0, 2, -1)$, $(0, -2, 1)$, $(0, -2, -1)$. Then $X - Y$ has eight arc-components and each a_k is nonzero.

2. Define $RP^2 \to R^2$ by

$$[\sin(\phi), \cos(\phi)\cos(\theta), \cos(\phi)\sin(\theta)] \to (\sin(2\phi)\cos(\theta), \sin(2\phi)\sin(\theta)).$$

Set X equal to the union of the two circles $\phi = \pi/4$ and $\phi = 0$. Set $Y = \varnothing$. Again each a_k is nonzero.

Statement of the result for Pontrjagin classes. Assume the manifold M^q is compact and orientable. Let $f : M^q \to R^{q-2k+2}$ be continuously differentiable and let (X, Y) be a closed locally connected pair in M^q satisfying

(A) $\{x \in M^q : \operatorname{rank}(df)_x \leqq q - 2k\} \subseteq X$.

(B) $\{x \in M^q : \operatorname{rank}(df)_x \leqq q - 2k - 1\} \subseteq Y$.

(C) $X - Y$ is an orientable $(q - 4k)$-dimensional submanifold of M^q with a finite number of arc-components, $A_1 \cdots A_l$.

(D) There is a neighborhood UY of Y in X such that the inclusions $Y \to UY$ and $X - UY \to X - Y$ are homotopy equivalences.

Orient $X - Y$, then $H_{q-4k}(X, Y : Z)$ is free and has rank l, with one generator $[A_i]$, for each arc-component of $X - Y$. To the oriented arc-component A_i. of $X - Y$ assign the integer a_i defined as follows. Pick a point x_i in the arc-component A_i and choose an orientation preserving chart $R^{q-4k} \times R^{4k} \to M^q$ so that $R^{q-4k} \times \{0\}$ is an orientation preserving chart for $X - Y$ at x_i. Denote by $\bar{f}$ the composition $R^{q-4k} \times R^{4k} \to M^q \to R^{q-2k+2}$. From the assumptions, it follows that the rank of $d\bar{f}$ is always $\geqq q - 2k$. Thus there is an element A in $\mathrm{GL}(q - 2k + 2)$ and a continuous map of $R^{q-4k} \times R^{4k} \to \mathrm{GL}(q)$, $x \to B(x)$ so that

$$B(x)\begin{pmatrix} \partial\bar{f}_1/\partial x_1 & \cdots & \partial\bar{f}_{q-2k+2}/\partial x_1 \\ \vdots & & \vdots \\ \partial\bar{f}_1/\partial x_q & \cdots & \partial\bar{f}_{q-2k+2}/\partial x_q \end{pmatrix} A$$

has the form

$$\begin{array}{c} \\ q-2k \\ 2k \end{array}\!\!\begin{array}{c} \begin{array}{cc} q-2k & 2 \end{array} \\ \left(\begin{array}{c|c} I & * \\ \hline 0 & c(x) \end{array}\right) \end{array}.$$

Note that $R^{q-4k} \times \{0\}$ contains the zeroes of $c(x)$. So that after identifying the space of $2k \times 2$ matrices with R^{4k}, the restriction of $c(x)$ to $\{0\} \times R^{4k}$ yields a map of pairs $(R^{4k}, R^{4k} - \{0\}) \xrightarrow{c(x)} (R^{4k}, R^{4k} - \{0\})$. Set a_i equal to the degree of this map. Then there is an element a in $H_{q-4k}(X : Z)$ whose image in $H_{q-4k}(X, Y: Z)$ is $\sum a_i[A_i]$ and whose image in $H_{q-4k}(M^q : Z)$ is the Poincaré dual of the kth Pontrjagin class of M^q. The proof is similar to the one for Stiefel-Whitney classes and is left to the reader.

EXAMPLE. The first Pontrjagin class of complex projective space. Define $f : CP^2 \to R^4$ by

$$[z_0, z_1, z_2] \to (\mathrm{Re}(z_0\,\bar{z}_1), \mathrm{Im}(z_0\,\bar{z}_1), \mathrm{Re}(z_0\,\bar{z}_2), \mathrm{Re}(z_1\,\bar{z}_2)).$$

Set $X = \{[0, 0, 1], [0, 1/\sqrt{2}, \pm 1/\sqrt{2}], [1/\sqrt{2}, 0, \pm 1/\sqrt{2}], [1/\sqrt{2}, \pm 1/\sqrt{2}, 0]\}$ and set $Y = \varnothing$. The procedure, described above, assigns the number -1 to the points $[0, 0, 1], [0, 1/\sqrt{2}, \pm 1/\sqrt{2}], [1/\sqrt{2}, 0, \pm 1/\sqrt{2}]$ and the number $+1$ to the points $[1/\sqrt{2}, \pm 1/\sqrt{2}, 0]$. Thus for one choice of orientation class, u, for CP^2, the first Pontrjagin class of CP^2 is the element $-3u$ in $H^4(CP^2 : Z)$.

Proof of the result for Stiefel-Whitney classes. Assume that M^q has been embedded in R^{q+N} so that the projection of M^q onto the first $q - i + 1$ coordinates of R^{q+N} is the map $f : M^q \to R^{q-i+1}$. Denote by F the corresponding classifying map of M^q into the Grassmannian, $G(q, N)$, of q planes in $q + N$ space. Set S_1 equal to the collection of planes, X^q, such that the dimension of the projection of X^q onto R^{q-i+1} is less than or equal to $q - i$, and set S_2 equal to $\{X^q \in G(q, N): \dim(\mathrm{Proj}\ X^q — R^{q-i+1}) \leq q - i - 1\}$. Next choose a neighborhood UY of Y satisfying condition (d) stated in the first section. The duality theorems in Chapter 6 of [3] yield the commutative diagram with Z_2 coefficients throughout.

$$\begin{array}{ccccc}
H_{q-i}(M^q) & \xrightarrow{\text{Poincaré Duality}} & H^i(M^q) & \xleftarrow{F^*} & H^i(G(q,N)) \\
\uparrow & & \uparrow & & \uparrow j^* \\
H_{q-i}(X) & \xrightarrow{\cong} & H^i(M^q, M^q - X) & \xleftarrow{F^*} & H^i(G(q,N), G(q,N) - S_1) \\
\downarrow & & \downarrow & & \downarrow \cong \\
H_{q-i}(X, Y) & \xrightarrow{\cong} & H^i(M^q - Y, M^q - X) & \xleftarrow{F^*} & H^i(G(q,N) - S_2, G(q,N) - S_1) \\
\downarrow \cong & & \downarrow \cong & & \\
H_{q-i}(X, UY) & & H^i(D(X - UY), (X - UY)^c) & & \\
\uparrow \cong & & \uparrow & & \\
H_{q-i}(X - Y, UY - X) & & & & \\
\downarrow \cong & & \cong \text{ Thom Isomorphism} & & \\
H_{q-i}(X - Y, X - Y - (X - UY)) & \longrightarrow & H^0(X - UY) & &
\end{array}$$

where $D(X - UY)$ denotes the normal bundle of the embedding of $X - Y$ in M^q restricted to $X - UY$, $(X - UY)^c$ denotes the complement of $(X - UY)$ in $D(X - UY)$.

From [2] it follows that

$$H^i(G(q,N), G(q,N) - S_1 : Z_2) \xrightarrow{\cong} H^i(G(q,N) - S_2, G(q,N) - S_1 : Z_2) \xrightarrow{\cong} Z_2$$

and that if 1 denotes the nonzero element, then $j^*(1)$ equals the ith universal Stiefel-Whitney class, w^i. Thus, it suffices to show that the element $F^*(1)$ in $H^i(M^q - Y, M^q - X)$ is the dual of the element $\sum a_k [A_k]$ in $H_{q-i}(X, Y)$.

Given the arc-component A_k of $X - Y$, choose a point x_k in A_k, and a chart $R^{q-i} \times R^i \to M^q$ so that $R^{q-i} \times \{0\}$ is a chart for $X - Y$ at x_k. Denote by $\bar{f}$ the composition $R^{q-i} \times R^i \to M^q \to R^{q+N}$. Assume that an element A in $\mathrm{GL}(q - i + 1)$ and a map $R^{q-i} \times R^i \to \mathrm{GL}(q)$, $x \to B(x)$ have been chosen so that

$$B(x) \begin{pmatrix} \dfrac{\partial \bar{f}_1}{\partial x_1} & \cdots & \dfrac{\partial \bar{f}_{q+N}}{\partial x_1} \\ \vdots & & \vdots \\ \dfrac{\partial \bar{f}_1}{\partial x_q} & \cdots & \dfrac{\partial \bar{f}_{q+N}}{\partial x_q} \end{pmatrix} \begin{pmatrix} A & 0 \\ 0 & I \end{pmatrix}$$

has the form

$$\begin{matrix} q-i \\ i \end{matrix} \left(\overset{q-i}{\begin{matrix} I \\ 0 \end{matrix}} \;\middle|\; \overset{1}{\begin{matrix} (*) \\ c(x) \end{matrix}} \;\; \overset{N+i-1}{\begin{pmatrix} * & * \\ * & * \end{pmatrix}} \right).$$

Next choose an element B in GL $(N + i - 1)$ such that

$$B(x)\begin{pmatrix} \dfrac{\partial \bar{f}_1}{\partial x_1} \cdots \dfrac{\partial \bar{f}_{q+N}}{\partial x_1} \\ \vdots \qquad \vdots \\ \dfrac{\partial \bar{f}_1}{\partial x_q} \cdots \dfrac{\partial \bar{f}_{q+N}}{\partial x_q} \end{pmatrix}\begin{pmatrix} A & 0 \\ 0 & B \end{pmatrix}$$

has the form

$$\begin{array}{r} q-i \\ i \end{array}\left(\overset{\begin{array}{cccc} q-i & 1 & i & N-1 \end{array}}{\begin{array}{c|c|c} I & (*) & E(x) \\ \hline 0 & c(x) & D(x) \end{array}\begin{pmatrix} * & * \\ * & * \end{pmatrix}}\right) \quad \text{with } D(0) = I.$$

Then we can assume $D(x)$ has rank i for all x. Next identify $R^{qN-i} \times R^i$ with the collection of $q \times (q + N)$ matrices of the form

$$\begin{array}{r} q-i \\ i \end{array}\left(\overset{\begin{array}{cccc} q-i & 1 & i & N-1 \end{array}}{\begin{array}{c|c|c} I & K_1 & 0 \\ \hline 0 & K_2 & I \end{array}\ (K_3)}\right)$$

so that the $i \times 1$ block K_2 is identified with R^i. Define $h : R^{q-i} \times R^i \to R^{qN-i}$ by

$$x \to \begin{pmatrix} I, & -\ E(x)D^{-1}(x) \\ 0, & D^{-1}(x) \end{pmatrix} B(x)\begin{pmatrix} \dfrac{\partial \bar{f}_1}{\partial x_1} \cdots \dfrac{\partial \bar{f}_{q+N}}{\partial x_1} \\ \vdots \qquad \vdots \\ \dfrac{\partial \bar{f}_1}{\partial x_q} \cdots \dfrac{\partial \bar{f}_{q+N}}{\partial x_q} \end{pmatrix}\begin{pmatrix} A & 0 \\ 0 & B \end{pmatrix}$$

$$= \left(\begin{array}{c|c|c} I & (*) & 0 \\ \hline 0 & D^{-1}(x)c(x) & I \end{array}\begin{pmatrix} * & * \\ * & * \end{pmatrix}\right).$$

This yields the commutative diagram

$$\begin{array}{ccc}
M^q & \xrightarrow{\quad F \quad} & G(q, N) \\
\uparrow & & \downarrow \text{Projection onto row space} \\
 & & q \times (q + N) \text{ matrices of rank } q \\
j_1 & j_2 & \downarrow \text{multiplication by } \begin{pmatrix} A^{-1} & 0 \\ 0 & B^{-1} \end{pmatrix} \\
 & & q \times (q + N) \text{ matrices of the form } \left(\begin{array}{c|c|c} I & K_1 & 0 \\ \hline 0 & K_2 & I \end{array}(K_3)\right) \\
 & & \updownarrow \\
R^{q-i} \times R^i & \xrightarrow{\quad h \quad} & R^{qN-i} \times R^i \\
 & \searrow (\text{proj}\circ h) & \downarrow \text{projection onto the second factor} \\
 & & R^i
\end{array}$$

Applying cohomology with Z_2 coefficients and restricting j_1 to $\{0\} \times R^i$ we get the commutative diagram

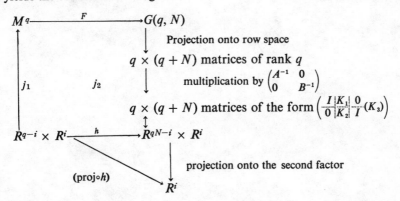

$$
\begin{array}{ccc}
H^i(M^q - Y, M^q - X) & \xleftarrow{\ F^*\ } & H^i(G(q, N) - S_2, G(q, N) - S_1) \\
\Big\downarrow j_1^* & & \Big\downarrow j_2^* \\
H^i(R^i, R^i - \{0\}) & \xleftarrow{\ h^*\ } & H^i(R^{qN-i} \times R^i, R^{qN-i} \times (R^i - \{0\})) \\
& \searrow^{(\mathrm{proj} \circ h)^*} & \cong \Big\downarrow (\mathrm{proj})^* \\
& & H^i(R^i, R^i - \{0\})
\end{array}
$$

Since $j_2 : R^{qN-i} \times R^i \to G(q, N)$ is a chart with $R^{qN-i} \times \{0\} \to G(q, N)$ a chart for $S_1 - S_2$, $j_2{}^*$ is an isomorphism; and since $j_1 : R^i \to M^q - Y$ is a normal to $X - Y$ at x_k, $j_1{}^*$ can be identified with the projection of $H^i(M^q - Y, M^q - X)$ onto the factor determined by the arc-component A_k. Thus it remains to show that the degree of the map

$$(R^i, R^i - \{0\}) \xrightarrow{(\mathrm{proj} \circ h) = D^{-1}(x)c(x)} (R^i, R^i - \{0\})$$

reduced mod 2 is a_k. But since $D(0) = I$, we can assume $\{D^{-1}(x) : x \in R^i\}$ is contained in a contractible neighborhood of I in $\mathrm{GL}(i)$, so that $c(x)$ and $D^{-1}(x)c(x)$ are homotopic as maps of pairs of $(R^i, R^i - \{0\})$ into itself.

References

1. T. Banchoff, *Stiefel-Whitney homology classes and singularities of projections for polyhedral manifolds*, these Proceedings.

2. S. Chern, *On the multiplication in the characteristic ring of a sphere bundle*, Ann. of Math. (2) **49** (1948), 362–372. MR **9**, 456.

3. E. H. Spanier, *Algebraic topology*, McGraw-Hill, New York, 1966. MR **35** #1007.

4. R. Thom, *Les singularitiés des applications différentiables*, Ann. Inst. Fourier (Grenoble) **6** (1955/56), 43–87. MR **19**, 310.

Brown University

MISCELLANEOUS

Proceedings of Symposia in Pure Mathematics
Volume 27, 1975

RIEMANNIAN SUBMERSIONS FROM SPHERES

RICHARD H. ESCOBALES, JR.

Let π be a Riemannian submersion from the unit n sphere S^n onto B. A classification of all such submersions is given under the assumption that the fibers are connected, totally geodesic and $1 \leqq \dim \text{ fiber} \leqq n - 1$.

Details will appear in the following article: R. Escobales, Jr., *Riemannian submersions with totally geodesic fibers,* J. Differential Geometry.

AMS (MOS) subject classifications (1970). Primary 53C20; Secondary 55F05.

Proceedings of Symposia in Pure Mathematics
Volume 27, 1975

SOME OPEN PROBLEMS IN DIFFERENTIAL GEOMETRY

COMPILED BY LEON GREEN

These are some problems submitted by participants in the Symposium. No attempt has been made to cull problems from the papers in this volume; it should be realized that they contain many questions and conjectures at least as interesting as the following.

Since the date of the Symposium, progress has been made on several of these problem. The most obvious example that has come to my mind is Thurston's solution of Problem 1 in Section 5, Foliations. Moreover, Koschorke has announced solutions to Problem 2 in that same section [Bull. Amer. Math. Soc. **80** (1974), 760–765], at least for $n>2q-2$.

This all emphasizes the fact that the researcher consulting this problem list must be aware of the time that has elapsed since it was proposed.

1. Riemannian metrics.

1. Give examples of compact Riemannian manifolds with nonnegative Ricci curvature that do not admit a metric with nonnegative sectional curvature. (Gromoll)

2. Do all vector bundles over the n-sphere admit a complete metric with nonnegative sectional curvature? (Gromoll)

3. Let M be a compact oriented simply connected Riemannian n-manifold with boundary B of dimension $n-1$. If M is δ-pinched with a certain δ satisfying $\frac{1}{4}<\delta\leqq 1$ with respect to a Riemannian metric g, find the most general g and boundary conditions on B such that M is diffeomorphic to a compact subset C of an n-sphere with vanishing homology groups $H_i(C)$, $i=1,\cdots,n-1$. (Hsiung)

Problems 4–8 were proposed by Kazdan and Warner. References are to the bibliography of their paper in these PROCEEDINGS.

AMS (MOS) subject classifications (1970). Primary 53C10, 53C15, 53C20, 53C35, 53C40, 57D30.

4. What are necessary and sufficient conditions for a function to be the Gaussian curvature of a *complete* metric on an open (noncompact) 2-manifold?

REMARK. This is known only if $M = R^2$ **[15]**.The Cohn-Vossen "Gauss-Bonnet Inequality" gives a necessary condition.

5. (a) If a compact manifold M (dim $M \geqq 3$) has a metric with scalar curvature zero, does it also admit a metric with positive scalar curvature?

(b) In particular, does the 3-torus, T^3, admit a positive scalar curvature metric?

REMARK. See § 5 of our paper in these PROCEEDINGS.

6. On R^3, any function is the scalar curvature of some metric **[16]**. Which functions are scalar curvatures of complete metrics? Is there an analogue of the Bonnet-Meyers theorem for scalar curvature?

7. Let (M, g) be a compact Riemannian manifold (dim $M \geqq 3$) of positive scalar curvature.

(a) Does M admit a metric of positive constant scalar curvature?

(b) Even stronger, does M admit a metric conformal to g, that has positive constant scalar curvature? This is the only unresolved case of the Yamabe problem.

8. Consider S^2 with the standard metric g_0 (whose Laplacian is Δ). For which $f \in C^\infty(S^2)$ can one solve $\Delta u = 1 - fe^{2u}$? Geometrically, this asks which functions f are curvatures of metrics pointwise conformal to g_0. (See **[14]** for some background.)

9. On a Riemannian manifold M, each eigenfunction φ_i of the Laplacian Δ determines certain nodal hypersurfaces $\varphi_i^{-1}(0)$ of codimension one. Cutting these hypersurfaces, one breaks M into a number of pieces. Is it always possible, given M, to choose some particular eigenfunction so that all the associated pieces are cells? (Knutson)

2. Immersions and embeddings.

1. Does there exist a complete hypersurface M^m isometrically immersed in $\boldsymbol{R}^{m+1}$ such that all the eigenvalues of the Ricci tensor are $\leqq -c < 0$ for some constant c? A negative answer would yield a striking generalization of the Hilbert-Efimov theorem. Chern has furnished a proof in the special case where M is the graph $x_{m+1} = f(x_1, \cdots, x_m)$ of a globally defined C_3 function f. (Reilly)

2. Consider a set of $n + 1$ points $\{P_0, \ldots, P_n\}$ on a sphere of radius r in Euclidean n-space. Let a_{ij} denote the geodesic distance between P_i and P_j on that sphere. It is well known that the $(n + 1) \times (n + 1)$ symmetric matrix (x_{ij}), $x_{ij} = \cos(a_{ij}/r)$ $(x_{ii} = 1)$, is necessarily degenerate positive semidefinite. The points are not on a great-sphere ($(n-2)$-dimensional) if and only if the rank of (x_{ij}) is exactly n; in this case the matrix of cofactors is of rank 1 and its entries are all positive if and only if the center of the sphere is interior to the convex hull of $\{P_0, \cdots, P_n\}$. In this case we say that the matrix of distances (a_{ij}) is *taut*, meaning that there is no small displacement of the points $\{P_0, \cdots, P_n\}$ along the sphere that can simultaneously either increase or decrease all geodesic distances.

Let Σ be an ovaloid in $\boldsymbol{R}^n$ (i.e., a smooth, closed, convex hypersurface) and let the Gaussian curvature K of Σ ($K =$ product of the principal curvatures) satisfy the inequalities $0 < r_0^{-n+1} \leqq K \leqq r_1^{-n+1}$. We say that a set of $n+1$ points $\{P_0, \cdots, P_n\}$ on Σ is *tight* if the pairwise geodesic distances (a_{ij}) between P_i and P_j cannot be decreased by small displacements of the points in Σ; similarly, we say that the set is *compressed* if these distances cannot be simultaneously increased.

CONJECTURE. For any set of points $\{P_0, \cdots, P_n\}$ on Σ there exists a unique number $r > 0$ such that there exists a set of points $\{P_1', \cdots, P_n'\}$ on a sphere of radius r, with the same pairwise geodesic distances as there are between the corresponding points on Σ; furthermore, at least if the given set of points is either tight or compressed, one has the estimate $r_1 \leqq r \leqq r_0$. (Calabi)

3. Geodesics.

1. Let $\mathscr{C} \subseteq C^\infty(S^2)$ be the set of functions φ for which the metric conformally related to the standard metric by φ has all its geodesics closed with length 2π. Is $\mathscr{C}$ a manifold near zero? If f is a sufficiently small odd function, can one find an even-function g so that $f + g \in \mathscr{C}$? If f is invariant under a 1-parameter group of rotations, the answer is yes (Funk). (Weinstein)

2. Let (M, g) be a Riemannian manifold such that the first conjugate point along each geodesic occurs at a fixed distance. Is (M, g) necessarily a symmetric space of rank one? The answer is yes if dim $M = 2$ (Green) or if $M = p^n(\boldsymbol{C})$ and g is a Kähler metric near the standard one and compatible with the standard complex structure (Berger). (Weinstein)

3. Let M be a Riemannian manifold and let $A: M \to M$ be an isometry on M. A geodesic $\gamma: \boldsymbol{R} \to M$ is said to be *A-invariant* if there is a $\theta \geqq 0$ such that $A(\gamma(t)) = \gamma(t + \theta)$ for all $t \in \boldsymbol{R}$.

(i) Does any isometry A on a compact 1-connected Riemannian M have (non-constant) invariant geodesics? (If not, then A has exactly one fixed point and $1 - A_* : \pi_q(M) \to \pi_q(M)$ is an isomorphism for all $q \in \boldsymbol{N}$— compare [**2**] and [**3**].)

(ii) Let $\beta_k = \dim H_k(C^0_{G(A)}(M); \boldsymbol{R})$ be the kth Betti number of $C^0_{G(A)}(M) = \{f : [0, 1] \to M ; f \text{ continuous and } f(1) = A(f(0))\}$. Does "$\{\beta_k\}$ unbounded" imply the existence of infinitely many (nonconstant) A-invariant geodesics? (Compare [**2**], [**3**] and the corresponding theorem for closed geodesics ($A = 1_M$) by Gromoll and Meyer [**1**].)

REFERENCES

[1] D. Gromoll and W. Meyer, *Periodic geodesics on compact riemannian manifolds*, J. Differential Geometry **3** (1969), 493–510. MR **41** #9143.

[2] K. Grove, *Condition* (C) *for the energy-integral on certain path-spaces and applications to the theory of geodesics*, J. Differential Geometry **8** (1973), 207–223.

[3] ——, *Isometry-invariant geodesics* (to appear).

(Grove)

4. Let M be a complete, simply-connected Riemannian manifold without conjugate points. Do geodesic rays with the same initial point diverge? If the dimension is two and the curvature bounded away from $-\infty$, this is true (Trans. Amer. Math. Soc. **76** (1954)), but P. Eberlein (and, independently, T. Otsuki) have pointed out that the "proof" of the similar assertion in Proc. Amer. Math. Soc. **7** (1956), 438–448, is grievously in error. (L.W. Green)

4. Other structures.

1. Is there a simply-connected symplectic manifold which admits no Kähler structure? (Thurston has observed that a certain T^2 bundle over T^2 has a symplectic structure although its first Betti number is three.) (Weinstein)

2. Consider $\boldsymbol{R}^{2n}$ with the symplectic structure $\Omega = \sum_{i=1}^{n} dq_i \wedge dp_i$. Which compact manifolds M admit a Lagrangian embedding in $\boldsymbol{R}^{2n}$; i.e., an embedding $f: M \to \boldsymbol{R}^{2n}$ such that $f^* \Omega = 0$? For the existence of a Lagrangian immersion, a necessary and sufficient condition is that the complexified tangent bundle of M be trivial (Gromov). (Weinstein)

3. Let $P_{n+p}(\boldsymbol{C})$ be an $(n+p)$-dimensional complex projective space with the Fubini-Study metric of constant holomorphic sectional curvature 1 and let M be an n-dimensional complete Kähler submanifold immersed in $P_{n+p}(\boldsymbol{C})$. Is each of the following sufficient for M to be totally geodesic?

(1) Every holomorphic sectional curvature of $M > 1/2$.

(2) Every sectional curvature of M is positive and $p < n(n+1)/2$.

(3) The scalar curvature of $M > n^2$. (Ogiue)

4. Let (M, ∇, s) be an AR-manifold with regular s-structure $\{s_x\}$. (See the article in these Proceedings by G. Tsagas.) Is there another s-structure on (M, ∇) which has finite order? (Tsagas)

5. If (M, g) is a symmetric space, then $\pi_1(M)$ is Abelian. Is this true for any Riemannian s-manifold? (Tsagas)

6. It is known that any two-point homogeneous Riemannian manifold is either Euclidean space or a symmetric space of rank one, but in the compact case the only known proof is by classification arguments. Give an intrinsic proof.

REFERENCES

(i) H. Freudenthal, *Zweifache Homogenität und Symmetrie*, Nederl. Akad. Wetensch Proc. Ser. A **70** = Indag. Math. **29** (1967), 18–22. MR **35** #6090.

(ii) S. Helgason, *The Radon transform on Euclidean spaces, compact two-point homogeneous spaces and Grassmann manifolds*, Acta Math. **113** (1965), 153–180. MR **30** #2530.

(iii) J. Tits, *Sur certaines classes d'espaces homogènes de groupes de Lie*, Acad. Roy. Belg. Cl. Sci. Mém. Coll. in 8°**29** (1955), no. 3, 268 pp. MR **17**, 874.

(iv) S. Varma, Nederl. Akad. Wetensch. Proc. Ser A **68** (1965), 746–753.

(v) H.-C. Wang, *Two-point homogeneous spaces*, Ann. of Math. (2) **55** (1952), 177–191. MR **13**, 863.

(vi) J. Wolf, *Spaces of constant curvature*, 2nd ed., Chap. 8. (Chavel)

5. Foliations.

1. Does the existence of a foliation of codimension one on a compact manifold without boundary imply any condition on the signature? If the manifold has a boundary, it does not. In all the known examples on closed manifolds, the signature is zero. P. Schweitzer has proved recently that for the transversally orientable case, the existence of one foliation implies the existence of a foliation whose tangent plane field belongs to any prescribed homotopy class. Some references are:

B. L. Reinhart, Topology **6** (1967), 467–471. (The lemma on p. 469 is incorrect, which invalidates the example for closed manifolds, but not for manifolds with boundary.)

M. H. Freedman, *The signature of manifolds admitting codimension* 1 *foliations* (to appear).

(Reinhart)

2. Consider, on the one hand, bordism groups of manifolds with a foliation of codimension q and, on the other hand, the bordism groups of Haefliger's classifying space $B\Gamma_q$. When does the forgetful homomorphism from one into the other have finite kernel and co-kernel? (This is always the case, e.g., if $q = 2$.) (Koschorke)

3. Find computable invariants (other than the characteristic numbers derived from the normal line field) which can detect torsion in the bordism groups of manifolds with a foliation of codimension one. (Koschorke)

6. Minimal submanifolds.

1. Let $M^{n-1} \subset S^n$ be a compact minimal hypersurface of the unit n-sphere ($n \geqq 7$) such that the cone $C(M^{n-1}) = \{tx \in \boldsymbol{R}^{n+1} : t \geqq 0 \text{ and } x \in M^{n-1}\}$ is mass minimizing in $\boldsymbol{R}^{n+1}$. Is $C(M^{n-1})$ a homogeneous algebraic variety? Determine as far as possible the allowable topological types of these manifolds M^{n-1}. (Lawson)

2. Does every compact smooth manifold admit a minimal immersion into some euclidean sphere? (Lawson)

3. Is the conjecture stated at the end of §5 of the lectures on minimal varieties true? In particular, does it hold for flat tori? (Lawson)

4. Let D be a domain on an m-dimensional minimal submanifold of $\boldsymbol{R}^n$, V its volume, A the (m-1)-dimensional measure of its boundary, ω_m the volume of the unit ball in $\boldsymbol{R}^m$. Does one have the same isoperimetric inequality $(m^{-1}A)^m \geqq \omega_m V^{m-1}$ that holds for domains D in $\boldsymbol{R}^m$? (Osserman)

5. Let D be a domain in an m-dimensional minimal submanifold of $\boldsymbol{R}^n$, V its volume, d the distance of D to the origin. If the boundary of D lies on the sphere of radius r about the origin, can one show that $V \geqq \omega_m (r^2 - d)^{m/2}$? (Osserman)

6. Let C be a Jordan curve lying on the surface of a sphere in $\boldsymbol{R}^3$. Let D be the unit disk in $\boldsymbol{R}^2$ and $f : D \to \boldsymbol{R}^3$ a Douglas solution to the Plateau problem for C. Is f an embedding? (Osserman)

7. Suppose that the graph of $f : \boldsymbol{R}^n \to \boldsymbol{R}$, $n > 7$, is a minimal hypersurface in $\boldsymbol{R}^{n+1}$. If f is a polynomial, must it be linear? (Poritz)

8. Can one estimate the number of distinct solutions of Plateau's problem for a given Jordan curve Γ in terms of the geometric properties of Γ alone? It has been proved that a regular analytic Jordan curve Γ whose total curvature does not exceed the value 4π bounds precisely one solution surface of Plateau's problem. This surface is analytic up to its boundary and everywhere free of branch points. The bound 4π is sharp. (J.C.C. Nitsche, *A new uniqueness theorem for minimal surfaces*, Arch. Rational Mech. Anal. **52** (1973), 319–329.) (Nitsche)

9. Let Γ be an analytic Jordan curve in E^3 of total curvature $\leqq 4\pi$. Γ is the boundary of a unique, immersed minimal surface S of the topological type of the disk (see problem 8). Must S be an embedded surface? (Gulliver)

UNIVERSITY OF MINNESOTA

Proceedings of Symposia in Pure Mathematics
Volume 27, 1975

SOME GEOMETRICAL ASPECTS OF GEODESY

NATHANIEL GROSSMAN*

We investigate the relation of geometry and physics in geodesy, especially the question of whether the current geodetic geometric hypotheses and techniques are consonant with current instrumental techniques of geodetic measurement. We conclude they are not, and introduce a new geometric formulation of geodesy, based on general linear connnections. These connections come from parallelisms induced by the physical, physiological, and psychological aspects of geodetic measurement. The physical information resides in the torsion of the connections, which is responsible for the misclosure of measurements in surveying around closed loops. We show, if the measuring process obeys infinitesimally a certain consistency axiom, that all geodetic measurables can be obtained locally from a single function, which can be identified with the gravitational potential. This clarifies the geometrical meaning of Marussi's calculation of geodetic measurables from the potential, and illuminates the relation of local to global properties of the potential. We obtain a lower bound for the total misclosure uncertainty in mapping the earth.

UNIVERSITY OF CALIFORNIA, LOS ANGELES

AMS (MOS) subject classifications (1970). Primary 86A30, 53B05.

*The material on which this abstract is based is part of a paper *Holonomic measurables in geodesy*, published in the J. Geophys. Research **79** (1974), 689–694.

Proceedings of Symposia in Pure Mathematics
Volume 27, 1975

THE INDEX THEOREM FOR CLOSED GEODESICS

W. KLINGENBERG

Let c be a closed geodesic of length $\omega > 0$ on a riemannian manifold M. Then the index of c is defined as the index of c, considered as a critical point for the energy integral in the Hilbert manifold of H^1- maps of the circle into M.

As in the Morse index for a geodesic segment one may ask whether there is another description of index c due to the fact that c determines an orbit $\dot{c}$ of the geodesic flow in the tangent bundle.

Such a description does indeed exist, although it is more complicated than in the case of a segment: A crucial role is played by the (linear) Poincaré map P associated to c: If P has an invariant lagrangian (= maximal isotropic) subspace then things are relatively easy, as for instance in the case that P is hyperbolic. In general one has to make a detailed study of P and then the index c can be defined as the intersection number of a certain pair of lagrangian bundles plus the index of a quadratic form determined by P alone.

As a first application of this theory one obtains a formula for the index of the iterated closed geodesics c^m, $m = 2,3,\cdots$, of c which makes precise a formula of Bott who gave index c^m only modulo a constant of the form mi. It turns out that the integer i is essentially the intersection number mentioned above.

An announcement of these results was given in W. Klingenberg, *Le théorème de l'indice pour les géodésiques fermées*, C.R. Acad. Sci. Paris **276** (1973), 1005-1008. A complete proof for the case c nondegenerate is given in: W. Klingenberg, *The index theorem for closed geodesics*, Tôhoku Math. J. (to appear). A complete exposition of the general case together with the application to iterated closed geodesics has appeared under the title *Der Indexsatz für geschlossene Geodatische* in Math. Z. **137** (1974).

UNIVERSITY OF BONN

AMS (MOS) subject classifications (1970). Primary 53C20, 58B20, 79F15, 53C15.

Proceedings of Symposia in Pure Mathematics
Volume 27, 1975

THE TOPOLOGY OF THE SOLUTIONS OF A LINEAR HOMOGENEOUS DIFFERENTIAL EQUATION ON $\boldsymbol{R}^n$

NICOLAAS H. KUIPER

The set of solutions (orbits) of

$$\frac{dx}{dt} = \sigma x, \qquad x \in V = \boldsymbol{R}^n,\ \sigma \in GL(\boldsymbol{R}, n),\ (x, t) \to e^{t\sigma}x = \sum_0^\infty t^k\sigma^k x/k!,$$

is called the *orbit system* of σ. The orbit systems of σ and σ' are called *linearly equivalent resp. topologically equivalent* in case some linear automorphism resp. homeomorphism of V exists that carries one orbit system onto the other. In the linear case then and only then, $\varphi \in GL(\boldsymbol{R}, n)$ and $c > 0$ exist such that $\sigma' = c\varphi\sigma\varphi^{-1}$.

Concerning the classification under topological equivalence we have the following:

THEOREM. *Call the eigenvalues of σ, $\lambda = \alpha + i\beta$ and let $V = V_+ \oplus V_- \oplus V_0$ be a direct sum decomposition corresponding to eigenvalues with real part $\alpha > 0$, < 0 and $= 0$ respectively. Then the pair of dimensions* ($\dim V_+$, $\dim V_-$) *and the linear equivalence class of $\sigma_0 = \sigma | V_0$ is a complete set of topological invariants for σ. In particular the matrices*

$$\sigma = \begin{bmatrix} 0 & 1 & 0 & 0 \\ 0 & 0 & 1 & 0 \\ 0 & 0 & 0 & 0 \\ 0 & 0 & 0 & 0 \end{bmatrix} \quad \textit{and} \quad \sigma' = \begin{bmatrix} 0 & 1 & 0 & 0 \\ 0 & 0 & 0 & 0 \\ 0 & 0 & 0 & 1 \\ 0 & 0 & 0 & 0 \end{bmatrix}$$

AMS (MOS) subject classifications (1970). Primary 58F99.

give rise to nonhomeomorphic orbit systems, although the fixed point sets are homeomorphic.

The main part of the proof consists in the description of the linear invariants of σ_0 in terms of topological properties of the orbit system of σ.

The work is related to the paper of Kuiper and Robbin, *Topological classification of linear endomorphisms*, Invent. Math. **19** (1972), 83–100. A complete exposition will appear in the Proceedings of the International Colloquium on Manifolds in Tokyo, 1973.

INSTITUT DES HAUTES ÉTUDES SCIENTIFIQUES

Proceedings of Symposia in Pure Mathematics
Volume 27, 1975

UNIQUE STRUCTURE OF SOLUTIONS TO A CLASS OF NONELLIPTIC VARIATIONAL PROBLEMS

JEAN E. TAYLOR

In a previous paper **[TJ]**, it was proved that given any continuous function F mapping the Grassmannian $\boldsymbol{G}(n, n-1)$ of unoriented $(n-1)$-plane directions in $\boldsymbol{R}^n$ into the positive real numbers, there exists a set W in $\boldsymbol{R}^n$ with $\mathscr{L}^n(W) = 1$ which minimizes the integral of F over its boundary, compared to any other set in $\boldsymbol{R}^n$ of $\mathscr{L}^n$ volume 1 with piecewise C^1 boundary; that is,

$$\int_{x\in\partial W} F(\mathrm{Tan}(\partial W, x))\, d\mathscr{H}^{n-1}x \leqq \int_{x\in\partial S} F(\mathrm{Tan}(\partial S, x))\, d\mathscr{H}^{n-1}x$$

whenever $S \subset \boldsymbol{R}^n$ with $\mathscr{L}^n(S) = 1$ and ∂S piecewise C^1. ($\mathscr{H}^{n-1}$ is Hausdorff $(n-1)$-dimensional measure; it agrees with any other reasonable definition of $(n-1)$-area on ∂S.) Moreover, it was shown that W is unique up to translation and is easily constructed by the *Wulff construction* in $\boldsymbol{R}^n$ (which is restated in the Appendix) followed by a homothety; the construction shows directly that W must be convex.

This problem of finding a set of given volume whose boundary minimizes a given integral is known as the *prescribed volume problem* for the integrand F. If F is interpreted as being a function giving the energy per unit of surface lying in a given tangent plane direction, then the problem can be considered as that of finding the existence and shape of the solid of least total surface energy having a given mass.

One often wishes to use the *direct method* to solve a problem in the calculus of

---

AMS (MOS) subject classifications (1970). Primary 49F22.

variations. This method, applied to this prescribed volume problem, is as follows:

(1) In the class of all "nice" sets satisfying the prescribed conditions (here bounded piecewise C^1 sets which are the boundaries of regions in $\boldsymbol{R}^n$ of $\mathscr{L}^n$ measure 1), one finds the infimum of the integral of F.

(2) One takes a sequence $\{\partial S_\nu\}_{\nu=1,2,\cdots}$ of "nice" sets from (1) such that their F integrals converge to the infimum.

(3) One associates a measure $|\partial S_\nu|$ on $\boldsymbol{R}^n \times \boldsymbol{G}(n, n-1)$ (i.e. a varifold) to each set in the sequence, by setting

$$|\partial S_\nu|(A) = \mathscr{H}^{n-1}\{x : x \in \partial S_\nu \text{ and } (x, \operatorname{Tan}(\partial S_\nu, x)) \in A\}$$

for any set $A \subset \boldsymbol{R}^n \times \boldsymbol{G}(n, n-1)$. Note that

$$\int_{(x,T)\in \boldsymbol{R}^n \times \boldsymbol{G}(n,n-1)} F(T)\, d|\partial S_\nu|(x, T) = \int_{x\in\partial S_\nu} F(\operatorname{Tan}(\partial S_\nu, x))\, d\mathscr{H}^{n-1}x.$$

(4) One extracts a convergent subsequence of the measures $|\partial S_\nu|$ and finds the limit measure μ.

(5) One attempts to show that there is set a Q in $\boldsymbol{R}^n$ which has tangent $(n-1)$-planes $\mathscr{H}^{n-1}$ almost everywhere, which is the support of μ in $\boldsymbol{R}^n$ and the boundary of a set S in $\boldsymbol{R}^n$ of volume 1; one desires of course that this set S be W, the set obtained by the Wulff construction and that $\mu = |\partial W|$ (where $|\partial W|$ is the varifold associated to ∂W (as in (3))).

This procedure works easily for the prescribed volume problem for any continuous integrand F as far as finding a set Q which is the support of μ and which has $(n-1)$-dimensional tangent planes at almost all points. The type of uniqueness of W proved in [**TJ**] is not sufficient, however, to prove that the set S is W and the set Q must be (up to a set of $\mathscr{H}^{n-1}$ measure 0) the boundary of W. (A full discussion of the kind of problems that might conceivably arise with regard to Q is in [**TJ**].)

In this paper, it is shown that in fact S must equal W (up to a set of measure 0).

These results, the first of their kind for variational problems without the hypothesis of ellipticity, encourage the hope that other general variational problems may also be amenable to geometric solutions.

In the following theorem, the sets in the subsequence are taken to be polyhedral; that this can be done is a consequence of the strong approximation theorem [**F**, 4.2.20]. They are also assumed to be contained in a bounded region in $\boldsymbol{R}^n$; it can easily be shown as in [**AF**] that for any minimizing sequence, the volume of the set outside some large ball goes uniformly to zero. Therefore this is in fact no restriction. Finally, it is assumed that $\mathscr{L}^n(W) = 1$ which is a condition on F; however, any function F can be made to satisfy this condition simply by multiplying it by the right constant.

THEOREM. *Let* $F\colon \boldsymbol{G}(n, n-1) \to \boldsymbol{R} \cap \{t : 0 < t < \infty\}$ *be continuous, let* W *be the set resulting from the Wulff construction (see Appendix) for* F, *and assume* $\mathscr{L}^n(W) = 1$. *Let* $\{S_\nu\}_{\nu=1,2\cdots}$ *be a sequence of polyhedral sets contained in a bounded region* *of $\boldsymbol{R}^n$, each with center of gravity at the orgin and $\mathscr{L}^n$ volume* 1, *and satisfying*

$$\lim_\nu \boldsymbol{F}(\partial S_\nu) = \boldsymbol{F}(\partial W),$$

where $\boldsymbol{F}(A) = \int_{x\in A} F(\mathrm{Tan}\,(A, x))\, d\mathscr{H}^{n-1}x$ for A either ∂W or ∂S_ν.
Then the sequence $\{S_\nu\}$ converges to the set W in the sense that

$$\lim_\nu \int_{\boldsymbol{R}^n} |\chi_{S_\nu} - \chi_W|\, d\mathscr{L}^n = 0$$

where for any set A in $\boldsymbol{R}^n$, χ_A is its characteristic function.

PROOF. By [**F**, 4.2.17] and [**F**, 4.1.12], there exist an $\mathscr{L}^n$ measurable set S in $\boldsymbol{R}^n$ and a subsequence of the sequence $\{S_\nu\}$ such that

$$\lim_\nu \int |\chi_{S_\nu} - \chi_S|\, d\mathscr{L}^n = 0.$$

Either the theorem is true, in which case $\int |\chi_S - \chi_W|^2\, d\mathscr{L}^n = 0$ or, by the existence of inverse Radon transforms and the analog of the Plancherel theorem [**GG**], the functions χ_S and χ_W do not have the same Radon transform—that is, there exist an $\varepsilon_1 > 0$, a rotation of the original coordinate system to a system $(x_1, x_2, \cdots, x_n)$, and a Borel set V_1 in $\boldsymbol{R}$ of $\mathscr{L}^1$ measure greater than ε_1 such that

$$\mathscr{L}^{n-1}(S \cap \{x\colon x_1 = t\}) - \mathscr{L}^{n-1}(W \cap \{x\colon x_1 = t\}) > \varepsilon_1$$

for every t in V_1. We assume in the following that the theorem is *not* true, and hence that there exist such an ε_1 and V_1, and eventually arrive at a contradiction. The proof is in eight steps.

STEP 1. (1) *Let $\varphi\colon [0, \frac{1}{2}] \to \boldsymbol{R}^+$, $1 + \varphi(t) = t^{(n-1)/n} + (1 - t)^{(n-1)/n}$. Then $\varphi(0) = 0$ and φ is a strictly increasing continuous function with $\varphi'(t) > 0$ for $0 < t \leqq \frac{1}{2}$.*

(2) *Let A, B be subsets of $\boldsymbol{R}^n$ with piecewise smooth boundaries such that*

$$0 < t = \mathscr{L}^n(A) \leqq \tfrac{1}{2} \leqq \mathscr{L}^n(B) = 1 - t < 1.$$

Then

$$\boldsymbol{F}(\partial A) + \boldsymbol{F}(\partial B) \geqq [1 + \varphi(t)]\, \boldsymbol{F}(\partial W).$$

(3) *Let A, B, t be as in* (2) *and assume that*

$$\mathscr{H}^{n-1}(\partial A \cap \partial B) \leqq \left(\tfrac{1}{4} \Big/ \sup_{G(n,n-1)} F\right) \varphi(t) \boldsymbol{F}(\partial W).$$

Then $\boldsymbol{F}[\partial(A \cup B)] \geqq (1 + \varphi(t)/2)\, \boldsymbol{F}(\partial W)$.

(4) *Let $0 < t \leqq \frac{1}{2}$ and $s \in \boldsymbol{R}$ such that $\mathscr{L}^n(S \cap \{x\colon x_1 < s\}) = t$. Then*

$$\liminf_{\Delta s \downarrow 0}\, (\Delta s)^{-1} \mathscr{L}^n(S \cap \{x\colon s < x < s + \Delta s\}) \geqq (\tfrac{1}{4}/\sup F)\, \varphi(t)\, \boldsymbol{F}(\partial W).$$

PROOF. (1) Compute φ'.

(2) Without loss of generality one can assume clearly that $A = \mu(t^{1/n})W$, $B = \mu((1-t)^{1/n})W$ (where $\mu(r)(x) = rx$ for $r > 0$ and x in $\boldsymbol{R}^n$) so that

$$\boldsymbol{F}(\partial A) = t^{(n-1)/n}\,\boldsymbol{F}(\partial W), \qquad \boldsymbol{F}(\partial B) = (1-t)^{(n-1)/n}\,\boldsymbol{F}(\partial W).$$

Conclusion (2) now follows from conclusion (1).

(3) One notes

$$2\boldsymbol{F}(\partial A \cap \partial B) \leqq 2(\sup F)\,\mathscr{H}^{n-1}(\partial A \cap \partial B) \leqq \tfrac{1}{2}\varphi(t)\boldsymbol{F}(\partial W)$$

and estimates

$$\begin{aligned}\boldsymbol{F}(\partial(A \cap B)) &= \boldsymbol{F}(\partial A) + \boldsymbol{F}(\partial B) - 2\boldsymbol{F}(\partial A \cap \partial B)\\ &\geqq (1+\varphi(t))\,\boldsymbol{F}(\partial W) - \tfrac{1}{2}\,\varphi(t)\boldsymbol{F}(\partial W)\\ &= (1+\varphi(t)/2)\boldsymbol{F}(\partial W).\end{aligned}$$

(4) The negation of conclusion (4) implies

$$\liminf_{\Delta s \downarrow 0}\,(\Delta s)^{-1}\,\mathscr{L}^n(S \cap \{x\colon s < x_1 < s + \Delta s\}) < (\tfrac{1}{4}/\sup F)\,\varphi(t)\,\boldsymbol{F}(\partial W).$$

For large ν one sets

$$A = S_\nu \cap \{x\colon x_1 \leqq s_\nu\}, \qquad B = S_\nu \cap \{x\colon x_1 > s_\nu\}$$

for suitably chosen $s_\nu > s$ near s such that $\mathscr{L}^n(A)$ is nearly equal to t and $\mathscr{H}^{n-1}(S_\nu \cap \{x\colon x = s_\nu\})$ is less than $(\tfrac{1}{4}/\sup F)\,\varphi(t)\,\boldsymbol{F}(\partial W)$ (this is possible by [**F**, 3.2.22]). Conclusion (3) now contradicts the near minimality of S_ν.

STEP 2. *There exists* $\varepsilon_2 > 0$ *such that*

$$|\mathscr{L}^{n-1}(S \cap \{x\colon x_1 = t\}) - \mathscr{L}^{n-1}(W \cap \{x\colon x_1 = [\varphi_1(W)^{-1}\circ\varphi_1(S)](t)\})| > \varepsilon_2$$

for all t in some Borel set V_2 *of* $\mathscr{L}^1$ *measure greater than* ε_2, *where for any* $\mathscr{L}^n$ *measurable set A, we define*

$$\varphi_1(A)\colon \boldsymbol{R} \to \boldsymbol{R}^+, \qquad \varphi_1(A)(t) = \mathscr{L}^n(A \cap \{x\colon x_1 < t\}).$$

PROOF. By conclusion (4) of the previous proposition, the x_1 axis can be reparametrized by a locally Lipschitz map σ to be the volume of S swept out by a plane perpendicular to the x_1 axis. Since W is convex, the corresponding map ω for W is bilipschitzian and hence the composition $\omega^{-1}\circ\sigma$ is locally Lipschitz. Thus the original inequality can be recomputed in terms of the reparametrizations.

Step 3. *There exists* $0 < \varepsilon < \tfrac{1}{2}$ (min $\{1, (\text{diameter } S)^{-1}\}$) *such that, renumbering the coordinate axes if necessary,*

$$|\mathscr{L}^1\{y\colon (t_1,\cdots,t_{n-1},y) \in S\} - \mathscr{L}^1\{y\colon (\tau_1,\cdots,\tau_{n-1},y) \in W\}| > \varepsilon$$

for all $(t_1,\cdots,t_{n-1})$ *in some Borel set* V_3 *of measure greater than* ε. *Here, the numbers* $\tau_1,\cdots,\tau_{n-1}$ *are functions of* $t_1,\cdots,t_{n-1}$ *defined inductively by* $\tau_1 = [\varphi_1(W)^{-1}\circ\varphi_1(S)](t_1)$ *and for* $i = 2,\cdots,n-1$,

$$\tau_i \equiv \tau_i(t_1,\cdots,t_i) = [\varphi_{i-1}(W,\tau_1,\cdots,\tau_{i-1})^{-1}\circ\varphi_{i-1}(S,t_1,\cdots,t_{i-1})](t_i)$$

where

$$\varphi_{i-1}(A, a_1,\cdots, a_{i-1})(t) = \mathscr{L}^{n-i+1}(A \cap \{x: x_1 = a_1,\cdots, x_{i-1} = a_{i-1}, x_i < t\})$$

for $A = W$ *or* S.

PROOF. The contradiction of this lemma would contradict the previous lemma.

STEP 4. *There exists a polyhedral set* T *(which may be taken to be* S_ν *for some large integer* ν*) such that*

(1) $\boldsymbol{F}(\partial T) - \boldsymbol{F}(\partial W) < (\varepsilon/10)^4$.

(2) $|f(T)(t_1,\cdots, t_{n-1}) - f(W)(\tau_1,\cdots, \tau_{n-1})| > \varepsilon/2$ *on a Borel set* V_3' *of measure greater than* $\varepsilon/2$, *where* $\tau_1,\cdots, \tau_{n-1}$ *are as in Step* 3 *and* $f(A)(a_1, \cdots, a_{n-1}) = \mathscr{L}^1\{y: (a_1,\cdots, a_{n-1},y) \in A\}$ *for any set* $A \subset \boldsymbol{R}^n$ *and point* $(a_1,\cdots, a_{n-1}) \in \boldsymbol{R}^{n-1}$.

(3) *On an open set* V_4 *of measure greater than* $\varepsilon/4$, *the quantity inside the absolute value signs in* (2) *above has contant sign and* $V_4 \subset V_3'$.

PROOF. There exists a number $N_1 < \infty$ such that (1) is true with $T = S_\nu$ whenever $\nu > N_1$. There exists a number $N_2 < \infty$ such that, if $\nu > N_2$, $\int|\chi_{S_\nu} - \chi_S|\, d\mathscr{L}^n < (\varepsilon/10)^4$. By Fubini's theorem,

$$\int|\chi_{S_\nu} - \chi_S|\, d\mathscr{L}^n = \int\cdots\int|f(S_\nu)(x_1,\cdots, x_{n-1}) - f(S)(x_1,\cdots, x_{n-1})|dx_{n-1} \cdots dx_1.$$

Hence, if $\nu > N_2$, $|f(S_\nu) - f(S)|$ can be greater than ε^2 only on a Borel set of measure less than ε^2. Therefore, $|f(S_\nu) - f(W)| > \varepsilon - \varepsilon^2$ on a Borel set of measure at least $\varepsilon - \varepsilon^2$. Since $\varepsilon < \frac{1}{2}$, we have both (1) and (2) if T is S_ν for any $\nu > \max(N_1, N_2)$.

Since such a T is polyhedral, its boundary contains at most a finite number of hyperplane segments parallel to the x_n axis. Therefore for each $i = 1,\cdots, n-1$, $f(T)(x_1,\cdots, x_{n-1})$ is continuous, as a function of x_i, except perhaps on a finite set. Since W is convex, $f(W)$ is continuous at all points in the interior of the projection of W on the $(x_1,\cdots, x_{n-1})$ plane. Conclusion (3) now follows.

STEP 5. *Let* P: $\boldsymbol{R}^{n-1} \times \boldsymbol{R} \to \boldsymbol{R}^{n-1}$ *be the orthogonal projection. There exist partitions of* T *and* W, *which are the images under* P^{-1} *of partitions of* $P(T)$ *and* $P(W)$ *by* $(n-1)$*-dimensional cubes and which are indexed by a multi-index* $\alpha = (\alpha_1,\cdots,\alpha_{n-1})$, *which satisfy the following properties:*

(1) *There is a one-to-one correspondence between elements of the partitions of* T *and* W *and corresponding elements have equal volumes.*

(2) *Let* $\Gamma = \{\alpha: P(T(\alpha)) \subset V_4\}$. *Then* $\mathscr{L}^n(\bigcup \{T(\alpha): \alpha \in \Gamma\}) > \varepsilon/5$.

(3) *Let* $\Delta = \{\alpha: T(\alpha)$ *is the finite union of "trapezoidal" sets*$\}$. *(By a "trapezoidal" set we mean a set which has as boundary* $2n$ *planar segments, of which* 2 *are perpendicular to the ith coordinate axis for each* $i = 1,\cdots, n-1$.) *Then*

$$\mathscr{L}^n (\bigcup \{T(\alpha): \alpha \neq \Delta\}) < (\varepsilon/10)^4.$$

(4) *The set sum* $T(\alpha) + W(\alpha) = \{x + y: x \in T(\alpha), y \in W(\alpha)\}$ *of any two corresponding elements of the partitions of* T *and* W *is disjoint from the sum of any other two corresponding elements.*

(5) *If we define* $f(W)$ *and* $f(T)$ *as in Lemma* 4, *then for each* α,

$$\sup\{f(W)(t_1,\cdots,t_{n-1}):(t_1,\cdots,t_{n-1})\in P(W(\alpha))\}$$
$$-\inf\{f(W)(t_1,\cdots,t_{n-1}):(t_1,\cdots,t_{n-1})\in P(W(\alpha))\}\leqq n^{-1/2}(\varepsilon/10)^4[\mathscr{H}^{n-1}(\partial T)]^{-1};$$

the same holds true with W *replaced by* T *provided* $\alpha\in\Delta$.

(6) *For any* α, *let* $p_1(\alpha),\cdots,p_{n-1}(\alpha)$ *be the dimensions of* $P(T(\alpha))$ *and* $r_1(\alpha),\cdots,r_{n-1}(\alpha)$ *be the dimensions of* $P(W(\alpha))$. *Let*

$$p_n(x)=\mathscr{L}^n(T(\alpha))\,(p_1(\alpha)\cdots p_{n-1}(\alpha)),\qquad r_n(\alpha)=\mathscr{L}^n(W(\alpha))/(r_1(\alpha)\cdots r_{n-1}(\alpha)).$$

Then if $\alpha\in\Gamma$, $|p_n(\alpha)-r_n(\alpha)|>\varepsilon/3$.

PROOF. We obtain this partition by the following construction:

Let k be any positive integer. Define for any bounded $\mathscr{L}^n$-measurable set A and each $i=1,\cdots,n-1$ the functions $\Psi_i(A)\colon \boldsymbol{R}\to\boldsymbol{R}\cap\{v\colon v>0\}$ by

$$\Psi_i(A)(\sigma)=\mathscr{L}^n(A\cap\{x\in\boldsymbol{R}^n\colon x_i<\sigma\})/\mathscr{L}^n(A)\qquad\text{for any }\sigma\in\boldsymbol{R}$$

and the numbers

$$\sigma_i(A)(j)=\sup\{\sigma\colon\Psi_i(A)(\sigma)<j/k\}\quad\text{for }j=1,\cdots,k,$$
$$\sigma_i(A)(0)=\inf\{\sigma\colon\Psi_i(A)(\sigma)>0\}.$$

Additionally, for any multi-index $\alpha=(\alpha_1,\cdots,\alpha_{n-1})$, where each α_i varies from 1 to k, we define the intervals $\boldsymbol{I}_i(A,\alpha)$ and the sets $A_i(\alpha)$, $i=1,\cdots,n-1$, inductively by

$$\boldsymbol{I}_1(A,\alpha)=\{t\colon\sigma_1(A)(\alpha_1-1)<t<\sigma_1(A)(\alpha_1)\},$$
$$A_1(\alpha)=A\cap(\boldsymbol{I}_1(A,\alpha)\times\boldsymbol{R}^{n-1}),$$
$$\boldsymbol{I}_i(A,\alpha)=\{t\colon\sigma_i(A_{i-1}(\alpha))(\alpha_i-1)<t<\sigma_i(A_{i-1}(\alpha))(\alpha_i)\},$$
$$A_i(\alpha)=A\cap(\boldsymbol{I}_1(A,\alpha)\times\cdots\times\boldsymbol{I}_i(A,\alpha)\times\boldsymbol{R}^{n-i}).$$

Finally, abbreviate $A_{n-1}(\alpha)$ by $A(\alpha)$.

By construction, for any choice of k the resulting sets $T(\alpha)$ and $W(\alpha)$ satisfy (1) and (4) (in fact, *all* the elements of the partitions have the same volume $k^{-(n-1)}$). By choosing k large enough, they satisfy (2) (since the sets are bounded), (3) (since T is polyhedral), and (5) (since T is polyhedral and W is convex). Conclusion (6) follows from conclusion (5) above and conclusion (2) of Lemma 4.

STEP 6. *We let* W' *be the polyhedral approximation to* W *obtained by approximating each element* $W(\alpha)$ *of the partition of* W *by the rectilinear solid*

$$\boldsymbol{I}_1(W,\alpha)\times\cdots\times\boldsymbol{I}_{n-1}(W,\alpha)\times J,$$

where J *is an interval of length* $r_n(\alpha)$ *and is centered so that*

$$\mathscr{L}^n((W'(\alpha)\sim W(\alpha))\cup(W(\alpha)\sim W'(\alpha)))$$

is a minimum. Also, we define the map

$$\hat{w}\colon\boldsymbol{G}(n,n-1)\to\boldsymbol{R}\cap\{v\colon v>0\}$$

by $\hat{w}(T) = \sup \{x \cdot z\colon z \in \partial W\}$ *for any* $T \in \boldsymbol{G}(n, n-1)$ *and* x *a vector of length* 1 *perpendicular to* T, *and the map* $\hat{w}'$ *by replacing* W *by* W' *in the above formula. Then*

$$\hat{w}(T) \geqq \hat{w}'(T) - 2\, n^{1/2}\, (\varepsilon/10)^4\, [\mathscr{H}^{n-1}(\partial T)]^{-1}.$$

The proof follows from Step 5.

STEP 7. *For any* $h > 0$, *define* $W'_h = \{x\colon h^{-1} x \in W'\}$. *Then if* $\alpha \in \varDelta$,

(a) $$\mathscr{L}^n(T(\alpha) + W'_h(\alpha)) \geqq \prod_{1=i}^{n} (p_i(\alpha) + h r_i(\alpha)),$$

(b) $$\mathscr{L}^n(T(\alpha) + W'_h(\alpha)) \geqq \mathscr{L}^n(T(\alpha))\left(1 + nh + h \sum_{i=1}^{n} r_i(\alpha)/p_i(\alpha) - n\right) + o(h).$$

PROOF. (a) This can be shown by direct computation, since $W'_h(\alpha)$ is rectilinear and $T(\alpha)$ is trapeziodal.

(b) Observing that $W'_h(\alpha) = (W'(\alpha))_h$, we see that the numbers $q_i(\alpha) = h\, r_i(\alpha)$ for $i = 1,2,\cdots, n$ give the dimensions of $W'_h(\alpha)$. Fixing α in Δ, we make the following computation. (All sums and products are over $i = 1,\cdots, n$ and the index α is to be understood.)

$$\begin{aligned}
1 &- [\mathscr{L}^n(T(\alpha))^{1/n} + \mathscr{L}^n(W'_h(\alpha))^{1/n}][\mathscr{L}^n(T(\alpha) + W'_h(\alpha))]^{-1/n} \\
&\geqq n^{-1} \textstyle\sum (p_i/(p_i + q_i) + q_i/(p_i + q_i)) \\
&\qquad - [(\Pi p_i)^{1/n} + (\Pi q_i)^{1/n}]\, [\Pi(p_i + q_i)]^{-1/n} \\
&= [\Pi(p_i + q_i)]^{-1} \Big[n^{-1} \textstyle\sum \Big(p_i \prod_{j \neq i} (p_j + q_j) + q_i \prod_{j \neq i} (p_j + q_j) \Big) \\
&\qquad - (\Pi p_j^{1/n} + \Pi q_j^{1/n})\, \Pi(p_j + q_j)^{(n-1)/n} \Big] \\
&= \Big[\Pi p_j - h \textstyle\sum (r_i/p_i) \prod_j p_j \Big] \\
&\quad \cdot \Big[\Pi p_j + (n^{-1}(n-1) + n^{-1}) \textstyle\sum h\, r_i \prod_{j \neq i} p_j - (\Pi p_j^{1/n} + h\, \Pi r_j^{1/n}) \\
&\qquad \cdot (\Pi p_j^{(n-1)/n}) (1 + n^{-1}(n-1) \textstyle\sum h\, r_i/p_i) \Big] + o(h) \cdot \\
&= (1 - h \textstyle\sum r_i/p_i) \Big[1 + h \textstyle\sum r_i/p_i - (1 + h)(1 + n^{-1}(n-1)\, h \textstyle\sum r_i/p_i) \Big] + o(h) \\
&\qquad (\text{note that } \ \Pi r_i/p_i = 1 \text{ since } \ \Pi r_i = \mathscr{L}^n(W'(\alpha)) = \mathscr{L}^n(T(\alpha)) = \Pi p_i) \\
&= h[n^{-1} \textstyle\sum r_i/p_i - 1] + o(h) \cdot \text{Hence}
\end{aligned}$$

$$\begin{aligned}
&\mathscr{L}^n(T(\alpha) + W'_h(\alpha)) \\
&\qquad \geqq [h(n^{-1} \textstyle\sum r_i/p_i - 1)(\Pi p_i)^{1/n} + \Pi p_i^{1/n} + h\, \Pi p_i^{1/n}]^n + o(h) \\
&\qquad \geqq (\Pi p_i)\, (1 + n\, h + h\, (\textstyle\sum r_i/p_i)) + o(h),
\end{aligned}$$

which is the inequality we desired.

STEP 8. *By the Wulff construction (see Appendix),* $F(Y) \geqq \hat{w}(Y)$ *for any* Y *in* $\boldsymbol{G}(n, n-1)$, *whereas* $F(\mathrm{Tan}(\partial W, x)) = \hat{w}(\mathrm{Tan}(\partial W, x))$ *for every* x in ∂W. *Therefore*

$$\begin{aligned} F(\partial T) - F(\partial W) &\geqq \int_{\partial T} \hat{w}(\mathrm{Tan}(\partial T, \cdot))\, d\mathscr{H}^{n-1} - \int_{\partial W} \hat{w}(\mathrm{Tan}(\partial W, \cdot))\, d\mathscr{H}^{n-1} \\ &\geqq \int_{\partial T} \hat{w}'(\mathrm{Tan}(\partial T, \cdot))\, d\mathscr{H}^{n-1} \\ &\quad - \int_{\partial W} \hat{w}(\mathrm{Tan}(\partial W, \cdot))\, d\mathscr{H}^{n-1} - 2(\varepsilon/10)^4 \\ &\geqq \lim_{h\to 0} h^{-1}(\mathscr{L}^n(T + W_h') - \mathscr{L}^n(T)) - n - 2(\varepsilon/10)^4 \end{aligned}$$

(the last inequality holds since T is piecewise C^1 and since the volumes of both W' and T are one).

Now

$$\mathscr{L}^n(T + W_h') \geqq \textstyle\sum_\alpha \mathscr{L}^n(T(\alpha) + W_h'(\alpha))$$

since $T + W_h' \supseteq \bigcup_\alpha (T(\alpha) + W_h'(\alpha))$ and the sets $\{T(\alpha) + W_h'(\alpha)\}$ are pairwise disjoint (5(4)). So by 7 and 5(3), the above difference dominates

$$\sum_{\alpha\in\Delta} n\Big(\prod_{i=1}^n p_i(\alpha)\Big)\Big[n^{-1}\sum_{i=1}^n r_i(\alpha)/p_i(\alpha) - 1\Big] - 3(\varepsilon/10)^4.$$

The heart of the proof is the recognition that the quantity in square brackets above is the difference between the arithmetic mean and the geometric mean of the quantities $r_1(\alpha)/p_1(\alpha), \cdots, r_n(\alpha)/p_n(\alpha)$ (again, since their product is the ratio of the volumes of $W'(\alpha)$ and $T(\alpha)$, which is one). We calculate below a positive lower bound to this difference for a fixed α such that $|r_n(\alpha) - p_n(\alpha)| > \varepsilon/3$.

Let $a = (n-1)/n$, $b = 1/n$, $u = \prod_{i=1}^{n-1} [r_i(\alpha)/p_i(\alpha)]^{1/(n-1)}$, and $v = r_n(\alpha)/p_n(\alpha)$. Then

$$\begin{aligned} &\log(au + bv) - a\log u - b\log v \\ &\qquad \geqq (au + bv - 1) - \tfrac{1}{2}(au + bv - 1)^2 - a[(u-1) - \tfrac{1}{2}(u-1)^2] \\ &\qquad\quad - b[(v-1) - \tfrac{1}{2}(v-1)^2] \\ &\qquad = \tfrac{1}{2}[u^2a(1-a) + v^2b(1-b) - 2\,abuv] \\ &\qquad = \tfrac{1}{2}\,ab(u-v)^2, \end{aligned}$$

so

$$(au + bv)(u^av^b)^{-1} \geqq \exp(\tfrac{1}{2}\,ab(u-v)^2) \geqq 1 + \tfrac{1}{2}\,ab(u-v)^2.$$

Let $[r_n(\alpha) - p_n(\alpha)]/p_n(\alpha) = \kappa$. Now

$$(u-v)^2 = (1 + \kappa - (1+\kappa)^{-1/(n-1)})^2 \geqq (\tfrac{1}{2}\,n/(n-1)\,\kappa)^2$$

and u^av^b is one. So

$$\begin{aligned} n^{-1}\sum_{i=1}^n r_i(\alpha)/p_i(\alpha) &\geqq n^{-1}(n-1)\Big(\prod_{i=1}^{n-1}(r_i(\alpha)/p_i(\alpha))^{n/(n-1)}\Big) + n^{-1}\,r_n(\alpha)/p_n(\alpha) \\ &\geqq 1 + \tfrac{1}{2}n^{-2}(n-1)\,(n(n-1)^{-1}\tfrac{1}{2}\,\kappa)^2 \\ &\geqq 1 + (1/8)(1/(n-1))\,\kappa^2 \\ &\geqq 1 + (1/8)(1/(n-1))\,(\varepsilon/3)^2\,(p_n(\alpha))^{-2}. \end{aligned}$$

Therefore,

$$\begin{aligned} \boldsymbol{F}(\partial T) - \boldsymbol{F}(\partial W) &\geqq \sum_{\alpha \in \Delta \cap \Gamma} n\left(\prod_{i=1}^{n-1} p_i(\alpha)\right) p_n(\alpha) \\ &\quad \cdot \left[n^{-1} \sum_{i=1}^{n} r_i(\alpha)/p_i(\alpha) - 1\right] - 3(\varepsilon/10)^4 \\ &\geqq (\varepsilon/6)\,(1/8)\,(1/9)\,\varepsilon^2/\sup_\alpha p_n(\alpha) - 3(\varepsilon/10)^4. \end{aligned}$$

But $\boldsymbol{F}(\partial T) - \boldsymbol{F}(\partial W) < (\varepsilon/10)^4$; contradiction. Therefore in fact χ_S and χ_W cannot have different Radon transforms, and hence $\int |\chi_S - \chi_W| \, d\mathscr{L}^n = 0$.

APPENDIX

Wulff's construction in dimension n. Let F: $\boldsymbol{G}(n, n-1) \to \boldsymbol{R}^+$ be a continuous function. Since an $(n-1)$-plane π is uniquely specified by its unoriented unit normal (denoted $^*\pi$), and *vice versa*, a continuous positive function F on $\boldsymbol{G}(n, n-1)$ is equivalent to a continuous function

$$F^*: \boldsymbol{B}^n(0, 1) \to \boldsymbol{R} \cap \{t: t > 0\}$$

satisfying $F^*(x) = F^*(-x)$ for any $x \in \partial \boldsymbol{B}^n(0, 1)$, the boundary of the unit ball in $\boldsymbol{R}^n$. (F^* is given by $F^*(x) = F(\pi)$ where $^*\pi = \pm x$.) The Wulff construction is as follows:

Plot F^* radially as a function of the direction $x \in \partial \boldsymbol{B}^n(0, 1)$. At each point of this "polar plot," draw the hyperplane through that point perpendicular to the radius vector x to that point (*not* usually tangent to the "polar plot"). The Wulff shape W is the compact set bounded by these planes (its boundary is the inner envelope of these planes) [WG].

BIBLIOGRAPHY

[AF] F. J. Almgren, Jr., *Existence and regularity almost everywhere of solutions to elliptic variational problems among surfaces of varying topological type and singularity structure*, Ann. of Math. (2) **87** (1968), 321–391. MR **37** #837.

[F] H. Federer, *Geometric measure theory*, Die Grundlehren der math. Wissenschaften, Band 153, Springer-Verlag, New York, 1969. MR **41** #1976.

[GG] I. M. Gel'fand, M. I. Graev and N. Ja. Vilenkin, *Generalized functions*. Vol. 5: *Integral geometry and representation theory*, Fizmatgiz, Moscow, 1962; English transl., Academic Press, New York, 1966. MR **34** #7726.

[TJ] Jean E. Taylor, *Existence and structure of solutions to a class of nonelliptic variational problems*, Symposia Math. (to appear).

[WG] G. Wulff, *Zur Frage der Geschwindigkeit des Wachsthums und der Auflosung der Krystallflachen*, Z. Krist. **34** (1901), 44–530.

DOUGLASS COLLEGE (RUTGERS UNIVERSITY)

Proceedings of Symposia in Pure Mathematics
Volume 27, 1975

TWIST INVARIANTS AND THE PONTRYAGIN NUMBERS OF IMMERSED MANIFOLDS

JAMES H. WHITE*

In this paper we give two applications of the geometry of secant spaces ([4], [5]). The first section, entitled "Twist invariants" studies the total twist (or torsion) of an n-dimensional differentiable manifold imbedded in Euclidean $(2n + 1)$-space and computes critical points of this total twist by developing a new invariant normal vector field for such imbeddings and investigating when this field vanishes everywhere.

The second section, entitled "The Pontryagin numbers of immersed manifolds," relates the algebraic number of triple points of an immersed $4k$-manifold in Euclidean $6k$-space to certain polynomials in the Pontryagin numbers of this manifold.

1. Twist invariants.

1. *The total twist.* Let M be a compact orientable n-dimensional differentiable manifold and let $S(M)$ denote the space of secants of M, i.e.

$$S(M) = M \times M - \Delta \cup T(M),$$

where Δ is the diagonal in $M \times M$, and where $T(M)$ is the space of oriented tangent directions of M. $S(M)$ is a differentiable manifold with boundary $T(M)$ ([4], [5]).

Let $f: M \to E^{2n+1}$ be a smooth imbedding into Euclidean $(2n + 1)$-space. We define a map $e : S(M) \to S^{2n}$, the unit $2n$-sphere in E^{2n+1}, as follows:

$$e(x, y) = \frac{f(y) - f(x)}{\|f(y) - f(x)\|} \quad \text{for } (x, y) \in M \times M - \Delta,$$

AMS (MOS) subject classifications (1970). Primary 53C65, 57DXX.

*This research was supported in part by NSF grant GP-27576.

and

$$e(t) = \frac{f_*(t)}{\|f_*(t)\|} \quad \text{for } t \in T(M)$$

(cf. [4]).

Let $e^*(dS^{2n})$ denote the pull-back of the volume element of S^{2n} under e; let O_k denote the volume of the k-sphere. Finally, let ν be a normal vector field of unit length on the imbedded manifold M. Then one may prove [5]

$$\text{(1)} \qquad \frac{1}{O_{2n}} \int_{S(M)} e^*(dS^{2n}) + \frac{1}{O_n} \int_M \tau_\nu \, dV = L(f, f_{\varepsilon_\nu}),$$

where $\tau_\nu \, dV$ is the "torsion (or twist) form of the imbedded manifold with respect to the vector field ν," and where $L(f, f_{\varepsilon_\nu})$ is the linking number of the imbedded manifold with the same manifold deformed a small distance ε along the vector field ν [5].

PROPOSITION. *$\tau_\nu \, dV$ is a Chern-Simons type invariant.*

PROOF. First, the exterior derivative of $\tau_\nu \, dV$ is, in fact, the Euler class form of the normal bundle of M in E^{2n+1} and hence does not depend on the choice of ν. Second, mod Z, $(1/O_n) \int_M \tau_\nu \, dV$ does not depend on the choice of the vector field ν. For suppose μ is another normal vector field of unit length. Then

$$\text{(2)} \qquad \frac{1}{O_{2n}} \int_{S(M)} e^*(dS^{2n}) + \frac{1}{O_n} \int_M \tau_\mu \, dV = L(f, f_{\varepsilon_\mu}).$$

Subtracting (2) from (1) we obtain that $(1/O_n) \int_M \tau_\nu \, dV - (1/O_n) \int_M \tau_\mu \, dV$ is an integer. Q.E.D.

The integral $(1/O_n) \int_M \tau_\nu \, dV$ will called the total twist of the imbedding with respect to the vector field ν.

2. *The first variation of the total twist.* It is interesting to study the first variation of the total twist for, although this twist depends on the choice of the normal vector field ν, the first variation does not. Indeed, the computation leads to a new invariant normal vector field for n-manifolds in Euclidean $(2n + 1)$-space. To begin this study, we express $\tau_\nu \, dV$ in terms of differential forms as in [5]. Let $f_1, \cdots, f_{2n+1}$ be local orthonormal frame fields on M such that $f_1, \cdots, f_n$ are tangent, f_{n+1} is along ν, and such that $f_{n+2}, \cdots, f_{2n+1}$ span the remaining normal space. Set $\pi_{ij} = df_i \cdot f_j$ and $\Lambda_{\alpha\beta} = \sum_{i=1}^n \pi_{\alpha i} \wedge \pi_{i\beta}$. Let

$$\Phi_K = \varepsilon_{\alpha_1 \cdots \alpha_n} \pi_{n+1\alpha_{n-2k}} \wedge \Lambda_{\alpha_{n-2K+1}\alpha_{n-2K+2}} \wedge \cdots \wedge \Lambda_{\alpha_{n-1}\alpha_n},$$

where

$$\begin{aligned} \varepsilon_{\alpha_1 \cdots \alpha_n} &= +1 \quad \text{if } \alpha_1 \cdots \alpha_n \text{ is an even permutation of } n+2, \cdots, 2n+1, \\ &= -1 \quad \text{if } \alpha_1 \cdots \alpha_n \text{ is an odd permutation of } n+2, \cdots, 2n+1, \\ &= \;\;\, 0 \quad \text{otherwise.} \end{aligned}$$

Define

$$(1)\qquad \Lambda = \frac{1}{\pi^{(n-1)/2}} \sum_{K=0}^{(n+1)/2} (-1)^K \frac{1}{1\cdot 3\cdots(n-2K)\, 2^{(n+1)/2+K}K!}\Phi_K.$$

Then we may write $\tau_\nu\, dV = O_n\Lambda$, where O_n is the volume of the unit n-sphere. To compute the first variation of $\int_M \tau_\nu\, dV$ would seem to be a formidable task, but using the techniques of Chern [1] the calculation is computationally difficult but straightforward. Following Chern [1], we apply a variation of M as follows: let I be the interval $-\frac{1}{2} < t < \frac{1}{2}$. Let $F\colon M \times I \to E^{2n+1}$ be a differentiable mapping such that its restriction to $M \times \{t\}$, $t \in I$, is an immersion and such that $F(m,0) = f(m)$, $m \in M$. We consider a frame field $e_1(m,t),\cdots, e_{2n+1}(m,t)$ over $M \times I$ such that $e_1(m,t),\cdots, e_n(m,t)$ are tangent vectors to $F(M \times \{t\})$ at (m,t) and such that $e_{n+1}(m,t),\cdots, e_{2n+1}(m,t)$ are normal vectors, and such that e_A at $(m,0)$ is f_A, $A = 1,\cdots, 2n+1$. Then the Cartan forms may be written

$$(2)\qquad \omega_i = \pi_i + a_i dt,\quad \omega_\alpha = a_\alpha dt,\quad \omega_{AB} = \pi_{AB} + a_{AB}dt,$$

where i has range 1 to n, α has range $n+1$ to $2n+1$, and A, B range 1 to $2n+1$. (In what follows we will assume latin letters, greek letters and capital letters will have the ranges indicated above.) The π_i and π_{AB} are linear differential forms in M with coefficients which may depend on t. For $t = 0$, they reduce to the forms with the same notation on M. The vector $\mathfrak{A} = \sum_A a_A e_A$ at $t = 0$ is called the deformation vector. Since normal variations give sufficient information, in what follows we will assume $a_1 = \cdots = a_n = 0$ without loss of generality, so that $\mathfrak{A} = \sum_\alpha a_\alpha e_\alpha$ at $t = 0$.

Now the operator d on $M \times I$ may be written $d = d_M + dt(\partial/\partial t)$. We first compute $d\Pi$ where Π is the same form as Λ in equation (1) except that ω_{AB} replaces π_{AB}. The form Λ (or Π) has the interesting property that its exterior derivative is essentially the differential form associated with the Euler class of the bundle whose fibre is spanned by $e_{n+1},\cdots, e_{2n+1}$, i.e. the "normal" bundle, that is

$$(3)\qquad d\Pi = c_0\, \varepsilon_{\alpha_{n+1}\cdots\alpha_{2n+1}}\, \Omega_{\alpha_{n+1}\alpha_{n+2}} \wedge \cdots \wedge \Omega_{\alpha_{2n}\alpha_{2n+1}}$$

where c_0 is a universal constant, where $\Omega_{\alpha\beta} = \sum_{i=1}^n \omega_{\alpha i} \wedge \omega_{i\beta}$, and where

$$\begin{aligned}\varepsilon_{\alpha_{n+1}\cdots\alpha_{2n+1}} &= +1 \quad \text{if } \alpha_{n+1}\cdots\alpha_{2n+1} \text{ is an even permutation of } n+1,\cdots,2n+1,\\ &= -1 \quad \text{if } \alpha_{n+1}\cdots\alpha_{2n+1} \text{ is an odd permutation of } n+1,\cdots,2n+1,\\ &= 0 \quad\ \ \text{otherwise.}\end{aligned}$$

Now

$$(4)\qquad d\Pi = d_M\Pi + dt \wedge \partial\Pi/\partial t.$$

We substitute into (4) the equations of (2) and equate the terms in dt. The right-hand side term in dt is

$$\partial\Pi/\partial t + (\text{terms arising from } d_M\Pi),$$

the important part being the $\partial\Pi/\partial t$. The left-hand side term is

$$(5)\quad c_0\varepsilon_{\alpha_{n+1}\cdots\alpha_{2n+1}} \sum_{K=(n+1)/2}^{n} \Lambda_{\alpha_{n+1}\alpha_{n+2}} \wedge \cdots \wedge \sum_{j=1}^{n}(a_{\alpha_{2k}j}\pi_{j\alpha_{2k+1}} - a_{j\alpha_{2k+1}}\pi_{\alpha_{2k}j}) \wedge \cdots \wedge \Lambda_{\alpha_{2n}\alpha_{2n+1}}.$$

If we integrate both sides of (4) over M, we find that the term arising from $d_M\Pi$ integrates to zero and we obtain

$$\text{(6)} \qquad \frac{\partial}{\partial t}\int_M \Pi = \int_M (\text{expression in (5)}).$$

The next step is to simplify the right-hand side of equation (6). To this end, we observe that the form

$$\pi_{j\alpha_{n+2}} \wedge \Lambda_{\alpha_{n+3}\alpha_{n+4}} \wedge \cdots \wedge \Lambda_{\alpha_{2n}\alpha_{2n+1}}$$

is an n-form on M and is thus a multiple of the volume element dV of M. We write

$$\pi_{j\alpha_{n+2}} \wedge \Lambda_{\alpha_{n+3}\alpha_{n+4}} \wedge \cdots \wedge \Lambda_{\alpha_{2n}\alpha_{2n+1}} = \varphi_{j\alpha_{n+2}\ldots\alpha_{2n+1}}\, dV.$$

An extremely lengthy computation yields the following result:

$$\int_M (\text{expression in (5)}) = \int_M c_0 \mathfrak{A} \cdot K\, dV$$

where

$$K\, dV = (n+1)\varepsilon_{\alpha_{n+1}\ldots\alpha_{2n+1}} d\Big\{e_{\alpha_{n+1}} \sum_{j=1}^{n} (-1)^j \varphi_{j\alpha_{n+2}\ldots\alpha_{2n+1}} \pi_1 \wedge \cdots \wedge \hat{\pi}_j \wedge \cdots \wedge \pi_n\Big\},$$

where $\mathfrak{A}$ is the variation vector, and where $\hat{\pi}_j$ means to omit π_j. Thus, the first variation becomes

$$\text{(7)} \qquad \frac{\partial}{\partial t}\int_M \Pi \Big|_{t=0} = c_0 \int_M \mathfrak{A} \cdot K\, dV$$

where $\mathfrak{A}$, K, dV are evaluated at $t = 0$. The normal component $K^\perp$ of the vector field K on M is the globally defined normal vector field on M mentioned at the outset of this section. We wish to study next when this global field $K^\perp$ vanishes.

3. *The critical points of the total twist.* We investigate now the imbeddings f for which the left-hand side of (7) is zero for all variations. Such imbeddings will be called critical points of the total twist. For such imbeddings we must have $K^\perp \equiv 0$.

For the case of curves, when $n = 1$, $K\, ds = 2\varepsilon_{\alpha_2\alpha_3} d\{e_{\alpha_2}\varphi_{1\alpha_3}\}$, where s is arc-length along the curve. Hence,

$$K\, ds = 2d\{e_2\varphi_{13} - e_3\varphi_{12}\}$$

where $\varphi_{13}\, ds = \pi_{13}$ and $\varphi_{12}\, ds = \pi_{12}$. Now, de_1 on the curve satisfies $de_1 = \pi_{12}e_2 + \pi_{13}e_3$. Hence $K\, ds = 2d\{de_1 \times e_1\}$. The curve is a critical point, therefore, when $K = K^\perp = 0$ or when $de_1/ds \times e_1 = N_0$, where N_0 is a constant vector, and e_1 is the unit tangent to the curve. This implies that $e_1 \cdot N_0 = 0$ or that the curve lies in a plane; further, it also implies $\|de_1/ds \times e_1\|$ is constant or that the curve has constant curvature. Thus, the result is:

THEOREM 1. *The critical points of the total twist of a closed space curve are circles.*

The higher dimensional cases are much more difficult to handle but there are a few partial results as follows:

THEOREM 2. *Among the critical points of the total twist of a compact n-manifold in* E^{2n+1}, $n \geqq 3$, *n odd, are*

(a) *any compact n-manifold imbedded in* $E^{2n-1} \subset E^{2n+1}$,

(b) *any totally geodesic compact n-dimensional submanifold in an immersed* $(n+2)$*-manifold in* E^{2n+1},

(c) *any compact n-manifold imbedded on an* $S^{2n-1} \subset E^{2n+1}$, S^{2n-1} *being the unit* $(2n-1)$*-sphere in* E^{2n}.

REMARK. It is an open question as to whether any imbedding of an n-manifold in an $E^{2n} \subset E^{2n+1}$ is a critical point for $n \geqq 3$. Note that in the case of curves in E^3, not every planar curve is a critical point but only circles.

PROOF OF THEOREM 2. To prove the theorem we must show that the normal component of K, which we call $K^{\perp}$, vanishes identically in each case. Note that $K^{\perp}$ does not depend on the choice of vector field, so in each case we will choose special frames and show $K^{\perp}$ vanishes. For case (a) we choose e_{2n} and e_{2n+1} constant vector fields orthogonal to E^{2n+1} in E^{2n+1}. Then every $\varphi_{j\alpha_{n+2}\cdots\alpha_{2n+1}}$ vanishes since it must contain one of the indices $2n$ or $2n+1$ and its definition above implies it must vanish. Hence, $K = 0 \Rightarrow K^{\perp} = 0$.

In case (b) we choose e_{n+1} and e_{n+2} orthogonal to M but tangent to the $(n+2)$-manifold. Then since M is totally geodesic π_{in+1} and π_{in+2} are all zero, $i = 1,\cdots,n$. Hence, $\Lambda_{\alpha n+1} = \Lambda_{\alpha n+2} = 0$ for all α. Hence, once again every $\varphi_{j\alpha_{n+2}\cdots\alpha_{2n+1}}$ must vanish since it must contain one of the indices $n+1$ or $n+2$. Thus $K = 0 \Rightarrow K^{\perp} = 0$.

Finally case (c) needs a little more analysis. We choose e_{2n+1} constant and orthogonal to E^{2n} and choose e_{2n} orthogonal to the S^{2n-1} in E^{2n}. Thus any $\varphi_{j\alpha_{n+2}\cdots\alpha_{2n+1}}$ which contains an index $2n+1$ must vanish since e_{2n+1} is constant and $de_{2n+1} = 0$. $K\,dV$ then becomes

$$K\,dV = (n+1)\varepsilon_{\alpha_{n+2}\cdots\alpha_{2n+1}}\, e_{2n+1}\, d\left\{\sum_j (-1)^j \varphi_{j\alpha_{n+2}\cdots\alpha_{2n+1}}\, \pi_1 \wedge \cdots \wedge \hat{\pi}_j \wedge \cdots \wedge \pi_n\right\}$$

where

$$\begin{aligned} \varepsilon_{\alpha_{n+2}\cdots\alpha_{2n+1}} &= +1 \quad \text{if } \alpha_{n+2}\cdots\alpha_{2n+1} \text{ is an even permutation of } n+1,\cdots,2n,\\ &= -1 \quad \text{if } \alpha_{n+2}\cdots\alpha_{2n+1} \text{ is an odd permutation of } n+1,\cdots,2n,\\ &= 0 \quad\ \ \text{otherwise.} \end{aligned}$$

Further, since S^{2n-1} is a unit sphere, $\pi_{A2n} = \pi_A$ on M, and hence since $\pi_\alpha = 0$, $\pi_{\alpha 2n} = 0$ for $\alpha = n+1,\cdots,2n$. Therefore, any expression of the form $\Lambda_{\alpha 2n} = 0$. In fact,

$$\Lambda_{\alpha 2n} = \sum_{i=1}^{n} \pi_{\alpha i} \wedge \pi_{in} = \sum_{i=1}^{n} \pi_{\alpha i} \wedge \pi_i = \sum_{A=1}^{2n+1} \pi_{\alpha A} \wedge \pi_A = d\pi_\alpha = 0.$$

Therefore, the only $\varphi_{j\alpha_{n+2}\cdots\alpha_{2n+1}}$ that are not zero are those for which $\alpha_{n+2} = 2n$. Hence

$$K\,dV = (n+1)\varepsilon_{\alpha_{n+3}\cdots\alpha_{2n+1}}\, e_{2n+1}\, d\left\{\sum_j (-1)^j \varphi_{j_{2n}\alpha_{n+3}\cdots\alpha_{2n+1}}\, \pi_1 \wedge \cdots \wedge \hat{\pi}_j \wedge \cdots \wedge \pi_n\right\},$$

where

$$\begin{aligned}\varepsilon_{\alpha_{n+3}\cdots\alpha_{2n+1}} &= +1 \quad \text{if } \alpha_{n+3}\ldots\alpha_{2n+1} \text{ is an even permutation of } n+1, \cdots, 2n-1,\\ &= -1 \quad \text{if } \alpha_{n+3}\ldots\alpha_{2n+1} \text{ is an odd permutation of } n+1, \cdots, 2n-1,\\ &= 0 \quad \text{otherwise.}\end{aligned}$$

Consider the term

$$\Gamma = \varepsilon_{\alpha_{n+3}\cdots\alpha_{2n+1}} \sum_j (-1)^j \varphi_{j_{2n}\alpha_{n+3}\cdots\alpha_{2n+1}} \pi_1 \wedge \cdots \wedge \hat{\pi}_j \wedge \cdots \wedge \pi_n.$$

Then, for all $k = 1, \cdots, n$,

$$\begin{aligned}\Gamma \wedge \pi_k &= \varepsilon_{\alpha_{n+3}\cdots\alpha_{2n+1}} \varphi_{k_{2n}\alpha_{n+3}\cdots\alpha_{2n+1}}\, dV\\ &= \varepsilon_{\alpha_{n+3}\cdots\alpha_{2n+1}} \pi_{k2n} \wedge \Lambda_{\alpha_{n+3}\alpha_{n+4}} \wedge \cdots \wedge \Lambda_{\alpha_{2n}\alpha_{2n+1}}\\ &= \varepsilon_{\alpha_{n+3}\cdots\alpha_{2n+1}} \pi_k \wedge \Lambda_{\alpha_{n+3}\alpha_{n+4}} \wedge \cdots \wedge \Lambda_{\alpha_{2n}\alpha_{2n+1}}.\end{aligned}$$

Hence, on M,

$$\Gamma = \varepsilon_{\alpha_{n+3}\cdots\alpha_{2n+1}} \Lambda_{\alpha_{n+3}\alpha_{n+4}} \wedge \cdots \wedge \Lambda_{\alpha_{2n}\alpha_{2n+1}}.$$

By definition, $\Lambda_{\alpha\beta} = \sum_{i=1}^{n} \pi_{\alpha i} \wedge \pi_{i\beta}$. But since M lies on S^{2n-1}, $\pi_{\alpha 2n} = \pi_{\alpha 2n+1} = 0$ for all α and

$$\Lambda_{\alpha\beta} = \sum_{i=1}^{n} \pi_{\alpha i} \wedge \pi_{i\beta} + \pi_{\alpha 2n} \wedge \pi_{2n\beta} + \pi_{\alpha 2n+1} \wedge \pi_{2n+1\beta}.$$

Given this observation Γ can be recognized as the Euler class form of the bundle over M whose fibre is spanned by e_{n+1}, e_{2n-1}, i.e. whose fibre is the normal space to M which is tangent to S^{2n-1}. Hence, Γ is closed and $d\Gamma = 0$. Therefore, $K\, dV = (n+1)\, e_{2n+1}\, d\Gamma = 0$ and so $K^{\perp} = 0$. This concludes part (c) and hence Theorem 2.

2. The Pontryagin numbers of immersed manifolds. Let $f : M^{4k} \to E^{6k}$ be an immersion of a compact orientable differentiable manifold $M^{4k} \equiv M$ into Euclidean $6k$-space. Let

$$\Sigma = \{(x, y) \in M \times M - \Delta \mid f(x) = f(y)\},$$

where $\Delta = \{(x, y) \in M \times M \mid x = y\}$. Let

$$D_1 = \{((x, y), z) \in \Sigma \times M \mid z = x\},$$

and

$$D_2 = \{((x, y), z) \in \Sigma \times M \mid z = y\}.$$

We define $S(\Sigma, M)$ to be the directed dilation of $\Sigma \times M$ along D_1 and D_2 **[4]**, that is,

$$S(\Sigma, M) = \Sigma \times M - D_1 - D_2 \cup N(D_1) \cup N(D_2),$$

where $N(D_i)$ is the space of oriented normal directions of D_i in $\Sigma \times M$, $i = 1, 2$.

Let $\pi_1, \pi_2 : M \times M \to M$ be the usual projection mappings onto the first and second factors respectively and denote by the same π_1 and π_2 their restriction to Σ. Thus,

$$\pi_1 \times \mathrm{id}: \Sigma \times M \to \pi_1(\Sigma) \times M$$

is defined by $\pi_1 \times \mathrm{id}\,((x,y), z) = (x, z)$, and

$$\pi_2 \times \mathrm{id}: \Sigma \times M \to \pi_2(\Sigma) \times M$$

is defined by $\pi_2 \times \mathrm{id}\,((x, y), z) = (y, z)$. Finally, for $i = 1,2$, let

$$\Delta_i = \{(x, y) \in \pi_i(\Sigma) \times M \mid x = y\}.$$

Clearly, $\pi_i \times \mathrm{id}$ maps D_i onto Δ_i but may not be one-one, and, in fact, the lack of one-one-ness of the $\pi_i \times \mathrm{id}$ measures in some sense the amount of triple points of f.

In [5], the directed dilation of $\pi_i(\Sigma) \times M$ along Δ_i was seen to be

$$S(\pi_i(\Sigma), M) = \pi_i(\Sigma) \times M - \Delta_i \cup N(\Delta_i),$$

and $N(\Delta_i)$ may be identified with $TM_{\pi_i(\Sigma)}$, where $N(\Delta_i)$ is the space of oriented normal directions of Δ_i in $\pi_i(\Sigma) \times M$, and where $TM_{\pi_i(\Sigma)}$ is the space of oriented tangent directions of M restricted to $\pi_i(\Sigma)$.

From now on we assume that the immersion f is self-transversal so that the triple points are isolated and hence finite in number. Let I be the set of triple points; more precisely, let

$$I = \{((x, y), z) \in \Sigma \times M \mid f(x) = f(y) = f(z) \text{ and } x \neq y,\ y \neq z,\ x \neq z\}.$$

Note that in some sense each triple point of f is essentially counted $3! = 6$ times in I.

We define a mapping $e : S(\Sigma, M) - I \to S^{6k-1}$ into the unit $(6k - 1)$-sphere as follows: for $((x, y), z) \in \Sigma \times M - D_1 - D_2 - I$,

$$e((x, y), z) = (f(z) - f(x))/|f(z) - f(x)|$$

(recall that $f(x) = f(y)$). We must still define e on $N(D_1)$ and $N(D_2)$. We first remark that $(\pi_i \times \mathrm{id})_*$ maps $N(D_i)$ onto $N(\Delta_i)$, $i = 1,2$. Let $n_{((x,y),x)}$ be an oriented unit normal vector in $N(D_1)$, and let $n_{(x,x)}$ represent the normal direction determined by $(\pi_1 \times \mathrm{id})_*(n_{((x,y),x)})$. Finally, let t_x be the unit tangent vector in TM_x identified with $n_{(x,x)}$ under the identification made between $N(\Delta_1)$ and $TM_{\pi_1(\Sigma)}$. Then, for $n_{((x,y),x)}$ in $N(D_1)$, we define

$$e(n_{((x,y),x)}) = f_*(t_x)/|f_*(t_x)|.$$

We proceed similarly for $N(D_2)$. Let $n_{((x,y),y)}$ be an oriented unit normal vector in $N(D_2)$, let $n_{(y,y)}$ represent the normal direction determined by $(\pi_2 \times \mathrm{id})_* \cdot (n_{((x,y),y)})$, and let t_y be the unit tangent vector in TM_y identified with $n_{(y,y)}$ under the identification made between $N(\Delta_2)$ and $TM_{\pi_2(\Sigma)}$. Then, for $n_{((x,y),y)}$ in $N(D_2)$, we define

$$e(n_{((x,y),y)}) = f_*(t_y)/|f_*(t_y)|.$$

LEMMA. *The mapping* $e : S(\Sigma, M) - I$ *is a differentiable map.*

This lemma is proved using techniques similar to those in [4] and [5].

We wish now to relate the algebraic number of triple points to certain integrals over $N(D_1)$ and $N(D_2)$. To this end we observe again that since f is self-transversal the number of triple points is finite, and hence the number of points in I is finite. Let us suppose there are r points in I. Surround each of these points by small disjoint balls B_a in $\Sigma \times M$. Now $S(\Sigma, M)$ is an oriented manifold with boundary consisting of $N(D_1) \cup N(D_2)$. This orientation induces an orientation on B_a and hence on ∂B_a, $a = 1, \cdots, r$. We are now ready to apply Stokes' theorem as in [4], [5]. Let dO_{6k-1} be the pull-back under e^* of the volume element of S^{6k-1}, and let O_{6k-1} be the volume of the $(6k-1)$-sphere. Then Stokes' theorem yields

$$0 = \frac{1}{O_{6k-1}} \int_{N(D_1)} dO_{6k-1} + \frac{1}{O_{6k-1}} \int_{N(D_2)} dO_{6k-1} - \sum_{a=1}^{r} \frac{1}{O_{6k-1}} \int_{\partial B_a} dO_{6k-1}.$$

The sum assigns to each point $\{((x, y), z)\}$ in I an integer ± 1 which is the intersection number of $f(\pi_1(\Sigma))$ and $f(M)$ at the point $f(x) = f(y) = f(z)$. Each triple point of f is clearly counted $3! = 6$ times in I; further, because of the dimension of the intersecting manifold, the integer ± 1 assigned to this triple point each of the six times is the same. Hence, we may write

$$\sum_{a=1}^{r} \frac{1}{O_{6k-1}} \int_{\partial B_a} dO_{6k-1} = 6I(\Sigma, M),$$

where $I(\Sigma, M)$ represents the algebraic number of triple points of f.

To evaluate the two integrals over $N(D_i)$, $i = 1,2$, we note the connection between $N(D_i)$ and $N(\Delta_i)$ and that between $N(\Delta_i)$ and $TM_{\pi_i(\Sigma)}$. Using the methods of [5], we obtain

$$\frac{1}{O_{6k-1}} \int_{N(D_i)} dO_{6k-1} = \chi(N(M))[\pi_i(\Sigma)],$$

where $\chi(N(M))[\pi_i(\Sigma)]$ is the Euler class of the normal bundle of M evaluated on the fundamental class of $\pi_i(\Sigma)$. Collecting our results, we have

$$6I(\Sigma, M) = \chi(N(M))\,[\pi_1(\Sigma)] + \chi(N(M))\,[\pi_2(\Sigma)].$$

Since the map $t : M \times M \to M \times M$ defined by $t(x, y) = (y, x)$ is orientation preserving, $\chi(N(M))\,[\pi_1(\Sigma)] = \chi(N(M))\,[\pi_2(\Sigma)]$. Hence,

$$(*) \qquad 3I(\Sigma, M) = \chi(N(M))\,[\pi_1(\Sigma)].$$

We recall that $\pi_1(\Sigma)$ measures in some sense the self-intersection locus of f and we use the fact that the Poincaré dual of the class of $\pi_1(\Sigma)$ is $\chi(N(M))$ ([2], [5]). Combining this result with equation $(*)$, we have

$$3I(\Sigma, M) = \chi(N(M)) \cup \chi(N(M))[M],$$

where the right-hand side is the Euler class of the normal bundle of M cupped with itself once evaluated on the fundamental class of M. We abbreviate this formula and write $3I(\Sigma, M) = \bar{\chi} \cup \bar{\chi}$, where $\bar{\chi}$ represents the normal bundle Euler class. It is well known [3] that $\bar{\chi} \cup \bar{\chi} = \bar{P}_k$, the kth normal Pontryagin class of M.

THEOREM. *Let $f : M^{4k} \to E^{6k}$ be a smooth self-transversal immersion of an oriented compact 4k-dimensional differentiable manifold into Euclidean 6k-space. Then, three times the algebraic number of triple points of f is equal to the normal kth Pontryagin class evaluated on the fundamental class of M.*

REMARK. This theorem corrects a formula of Lashof-Smale [3]. A full correction of that paper will appear in a forthcoming paper of R. Herbert, University of Minnesota.

Using the duality between the normal Pontryagin classes and the regular Pontryagin classes of the manifold, we obtain a relationship between the algebraic number of triple points and certain polynomials in the Pontryagin numbers of the immersed manifolds. For example, we have

COROLLARY 1. *Let $f : M^4 \to E^6$ be a smooth self-transversal immersion of an oriented compact 4-dimensional differentiable manifold into Euclidean 6-space. Then, three times the algebraic number of triple points of f is equal to minus the Pontryagin number of M^4.*

REMARK. This gives some geometric insight into why, at least for our special case, the Pontryagin number of a 4-manifold is divisible by three.

Finally, we remark by stating that all the ideas in this section may be generalized to arbitrary self-intersections.

BIBLIOGRAPHY

1. S. S. Chern, *Minimal submanifolds in a Riemannian manifold*, Technical Report 19, University of Kansas, Lawrence, Kan., 1968.

2. R. K. Lashof and S. Smale, *On the immersion of manifolds in euclidean space*, Ann. of Math. (2) **68** (1958) 562–583. MR **21** #2246.

3. ———, *Self-intersections of immersed manifolds*, J. Math. Mech. **8** (1959), 143–157. MR **21** #332.

4. W. F. Pohl, *Some integral formulas for space curves and their generalization*, Amer. J. Math. **90** (1968), 1321–1345. MR **38** #6523.

5. J. H. White, *Self-linking and the Gauss integral in higher dimensions*, Amer. J. Math. **91** (1969), 693–728. MR **40** #6479.

UNIVERSITY OF CALIFORNIA, LOS ANGELES

INDEXES

AUTHOR INDEX

Italic numbers refer to pages on which a complete reference to a work by the author is given.

Roman numbers refer to pages on which a reference is made to a work of the author. For example, under Oppenheim would be the page on which a statement like the following occurs: "This result had been conjectured by Oppenheim **[13]** in 1929, ..."

Boldface numbers indicate the first page of the articles in this volume.

SUBJECT INDEX